POLYMER SCIENCE AND TECHNOLOGY

CONDUCTING POLYMERS

SYNTHESIS, PROPERTIES AND APPLICATIONS

POLYMER SCIENCE AND TECHNOLOGY

Additional books in this series can be found on Nova's website
under the Series tab.

Additional e-books in this series can be found on Nova's website
under the e-book tab.

MATERIALS SCIENCE AND TECHNOLOGIES

Additional books in this series can be found on Nova's website
under the Series tab.

Additional e-books in this series can be found on Nova's website
under the e-book tab.

POLYMER SCIENCE AND TECHNOLOGY

CONDUCTING POLYMERS

SYNTHESIS, PROPERTIES AND APPLICATIONS

LUIZ CARLOS PIMENTEL ALMEIDA
EDITOR

New York

For permission to use material from this book please contact us:
Telephone 631-231-7269; Fax 631-231-8175
Web Site: http://www.novapublishers.com

Library of Congress Cataloging-in-Publication Data

ISBN: 978-1-62618-119-9

Published by Nova Science Publishers, Inc. † New York

CONTENTS

PREFACE

Conducting polymer field date back to the late 1970s when Alan G. MacDiarmid, Hideki Shirakawa and Alan J. Heeger discovered the high electrical conductivity in doped polyacetylene. Achieving electrical conductivity in polyacetylene as high as that of copper metal inspired the development of new conducting polymers such as polyacetilene, polythiophene, polypyrrole, poly(p-phenilene), poly(p-phenilene sulphide) and polyaniline and a large number of their derivatives and copolymers. It has consequently spawned new interdisciplinary research activities which collectively contributed to the success of conducting polymers as molecular semiconductor materials enabling the development of new technologies in electronic and optoelectronic field.

This book, written by leading experts of the international scientific community, is divided in 10 chapters and gives a comprehensive coverage of important aspects of the conducting polymers. Synthetic methodologies of these polymers and their nanocomposites along with their electrical and electrochemical properties are highlighted herein. Application of the conducting polymers for sensors, solar cells and lithium batteries are also presented.

The subjects quoted above are important topics in the field of the conducting polymers and make this book a very useful scientific support to a large audience of readers, from students to senior researchers in the academic community and from engineer to business people in different industrial sectors.

I cannot finish this book without expressing my gratitude to our distinguished authors who have dedicated their valuable time and effort to write high-quality chapters.

Finally, my special thanks go to Nova Science for the opportunity of organizing this book.

Luiz Carlos Pimentel Almeida
Serra-ES, Brazil

In: Conducting Polymers
Editor: Luiz Carlos Pimentel Almeida
ISBN: 978-1-62618-119-9

Chapter 1

CURRENT SYNTHETIC METHODOLOGIES FOR SEMICONDUCTING POLYMERS

Paulo C. M. L. Miranda*[*1] *Lindomar A. dos Reis*[2] *and Jason G. Taylor*[3]

[1]Instituto de Química, UNICAMP, Campinas, SP, Brazil
[2]INMETRO, Xerém, RJ, Brazil
[3]Departamento de Química, UFOP, Ouro Preto, MG, Brazil

ABSTRACT

Since the publication of pioneering work in the 1970's, which described and evaluated conducting polymers, the preparation of precisely substituted polymers for the purpose of obtaining materials with tuned electrical and mechanical properties has seized great attention from the scientific community. The arsenal of synthetic methodologies available to access archetypal or tailored polymer structures can be either wide or limited depending upon the nature of the desired monomer. These type of materials include polyacetylene itself or its related poly(hetero)arylenevinylenes analogues and the major synthetic approaches towards polymer synthesis comprise Ullmann, Heck or Wittig type reactions, olefin metathesis, aldol condensation and, of course, electrochemical polymerizations. The regio- and chemoselectivity of these methodologies will be discussed here with special emphasis covering synthetic approaches to homopolymer preparation, although some of these reactions can be effectively used to prepare different types of copolymers.

INTRODUCTION

Semiconducting polymers typically have an extended conjugated double bond framework and we will therefore focus on just some of the numerous synthetic methodologies that are employed to construct these bonds with high selectively and yields. Commonly used

* E-mail address: miranda@iqm.unicamp.br

approaches usually involve three different strategies (Scheme 1): i) carbon-heteroatom single bond construction followed by oxidation, such as in the case of Ullmann-Goldberg type reaction or electrochemical reactions; ii) carbon-carbon single bond formation, such as in the case of cross-coupling reactions; or iii) carbon-carbon double bonds construction, such as in the case of addition-elimination or metathesis reactions. Given the fact that Ullmann-Goldberg type reactions just create a carbon-heteroatom single bond, this step of course must be followed by a mild oxidation step to provide the semiconducting π-conjugated system.

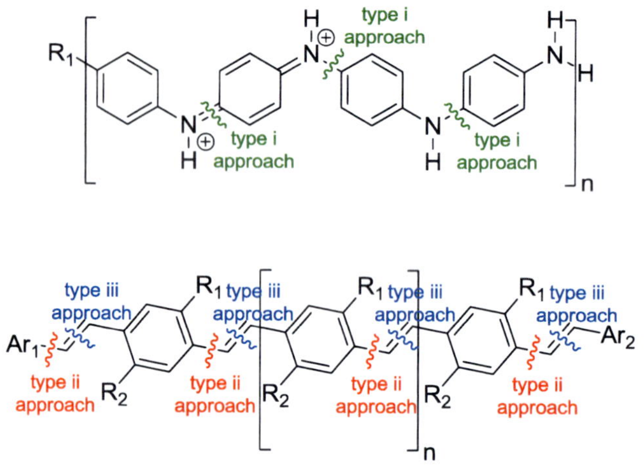

Scheme 1. Different synthetic approaches to polymeric semiconductor frameworks.

ULLMANN-GOLDBERG TYPE REACTIONS

Ullmann-Goldberg type reactions are amongst the most commonly used methodologies for nucleophilic substitution of aryl halides. Aryl halides are relatively inert to the halogen atom substitution unless it is activated by the presence of electron withdrawing groups (e.g. a nitro, sulfonyl or carbonyl group and aryl pyridines) [1]. However, metals and metallic complexes may facilitate substitution of deactivated aryl halides by several mechanistic pathways [2]. In the case of Ullmann-Goldberg type reactions, the suggested mechanism for nucleophilic substitution facilitated by the transition metal involves oxidative addition/reductive elimination [3], single electron transfer, and halogen atom transfer [4].

At the beginning of last century, the pioneering work of Ullmann and Goldberg on copper catalyzed coupling reactions led to new ways to form C-N and C-O bonds [5]. These studies are considered to be the basis for all subsequent work on Cu-promoted nucleophilic substitution of aryl halides, and the importance of their results is demonstrated by the wide application of these methodologies in various fields of Chemistry, particularly, in the conjugated polymers synthesis. Since the discovery of Shirakawa, MacDiarmid, and Heeger in the 1970s that electrically insulating polyacetylene can be made highly conducting by chemical or electrochemical doping [6,7], efforts have been devoted to developing conjugated materials. In this context, there are many examples in the literature that describe polycondensation methods based on the catalytic action of copper and its salts in coupling reactions.

Goto and coworkers reported the synthesis of polyanilines by the polycoupling reaction of dibromobenzene with diaminobenzene or polyaminopyridines and by coupling of dibromopyridine with diaminopyridine (Scheme 2). Copper(I) iodide was used as catalyst and nitrobenzene as solvent [8]. In this case, the regioregulation of the polymer structures was not evaluated and the possibility of crosslinking reactions occurring during the polymerization of monomers could not be ruled out or discarded.

Despite the several applications of these Cu-promoted nucleophilic coupling reactions, up until 2000, the full potential of these methodologies has far from been completely exploited. As shown in the above examples, these methods usually require harsh reaction conditions (strong bases, high temperatures, and stoichiometric amounts of copper or copper salts, long reaction times) to effect these transformations, which restrict the scope of suitable substrates [9].

However, in the 1990's, improvements to those aforementioned procedures were developed for the Ullmann-Goldberg coupling reactions based on the utilization of some bidentate ligands such as 1,10-phenanthroline [10, 11]. These modified methods made it possible for a catalytic amount of the copper complex to be used at lower reaction temperatures, thus broadening the applicability of the reaction to a wider variety of functional groups. In recent years, other combinations with different copper sources and ligands were used in the synthesis of many classes of molecules [5, 9, 12].

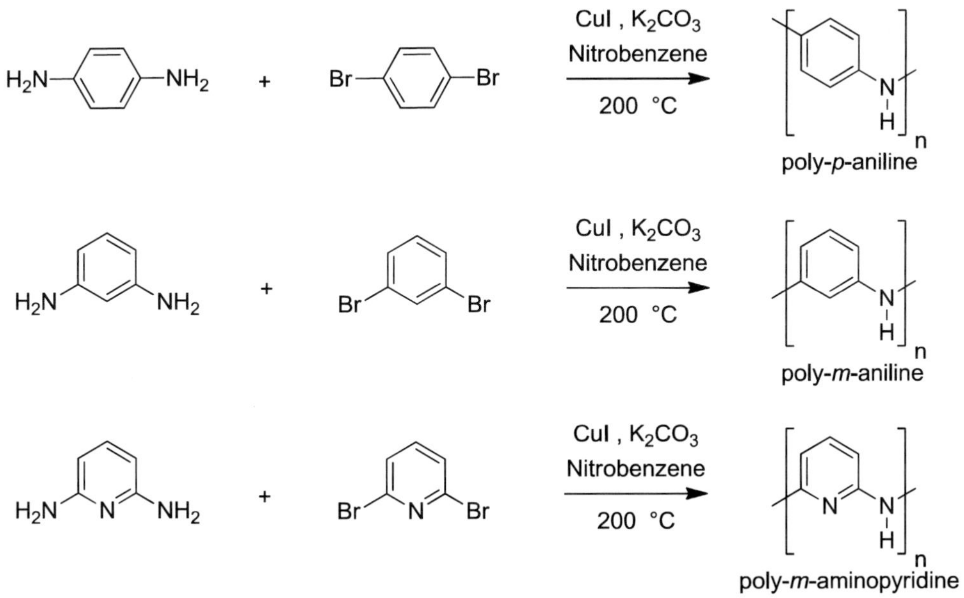

H_2N–C$_6$H$_4$–NH_2 + Br–C$_6$H$_4$–Br → (CuI, K_2CO_3, Nitrobenzene, 200 °C) poly-*p*-aniline

H_2N, NH_2 + Br, Br → (CuI, K_2CO_3, Nitrobenzene, 200 °C) poly-*m*-aniline

H_2N, N, NH_2 + Br, N, Br → (CuI, K_2CO_3, Nitrobenzene, 200 °C) poly-*m*-aminopyridine

Scheme 2. Goto route to polyanilines and polyaminopyridines.

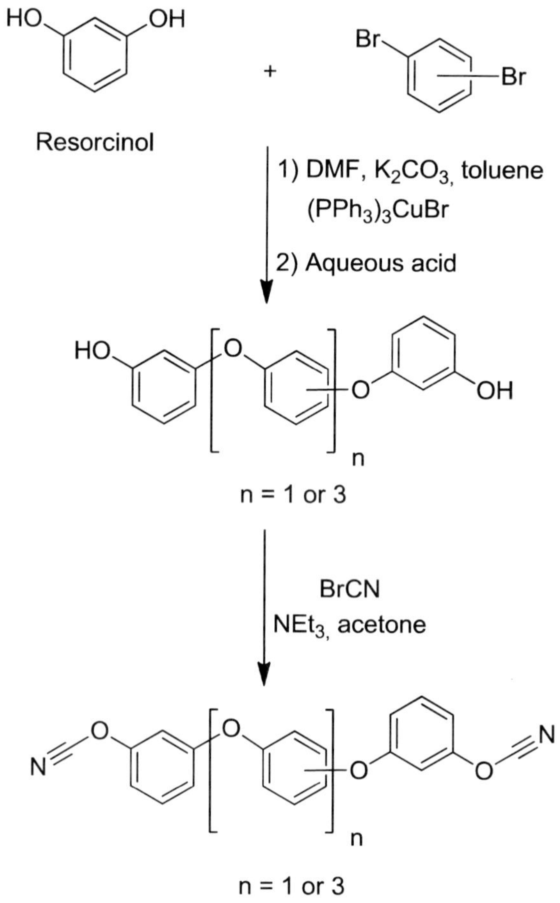

HO OH
Resorcinol
+
Br Br
1) DMF, K_2CO_3, toluene
$(PPh_3)_3CuBr$
2) Aqueous acid
HO O O OH
n
n = 1 or 3
BrCN
NEt_3, acetone
N O O O O N
n
n = 1 or 3

Scheme 3. Laskoski route to cyano ester resin systems.

Some examples on the use of catalyzed coupling reactions by Cu/ligand systems in conjugated polymers synthesis have already reported in the literature. Laskoski and coworkers prepared bifunctional aryl ether oligomers by a reaction between resorcinol and 1,3- or 1,4-dibromobenzene in the presence of K_2CO_3 and catalytic amount of $(PPh_3)_3CuBr$, Scheme 3. These aryl ether oligomers were converted into novel cyano ester resins that have innumerous potential applications, such as printed circuit boards, and radomes [13].

In 2008, Wang and coworkers reported the synthesis of a new class of poly-(phthalazinone ether)s from modified Ullmann-Goldberg coupling between a potassium salt of 1,2-dihydro-4-(4-hydroxyphenyl)-1-(2H)-phthalazinone and aromatic dibromide compounds, Scheme 4. CuCl was used as the copper source and quinoline as ligand making it possible to obtain materials with molecular weights up to 20,000 [14]. The inherent properties of these materials potentiate its use in optical systems [14].

Metals other than copper can be used in Ullmann-Goldberg type aryl aminations such as palladium. In 2002, Chou and coworkers disclosed the synthesis of aminopyridine oligomers in moderate yields using $Pd_2(dba)_3$ as catalyst [15]. Such oligomers showed a highly complex hydrogen bond network and thus reducing their solubility in common organic solvents (Scheme 5).

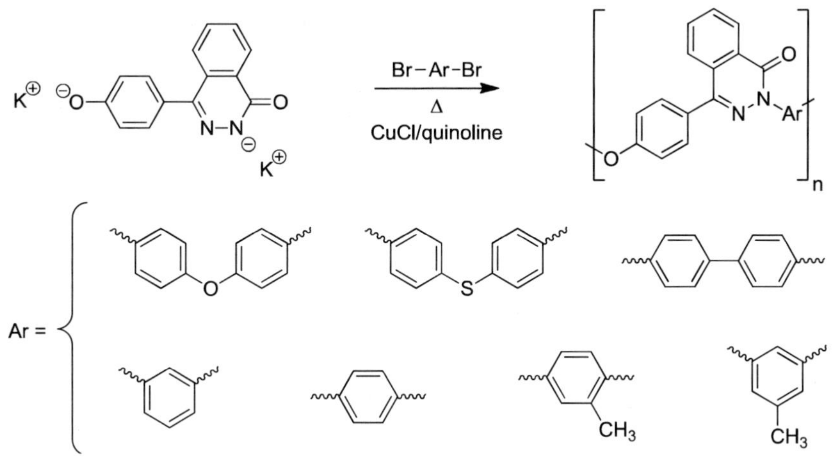

$K^{\oplus}$ $^{\ominus}O$ O N–N $\ominus$ $K^{\oplus}$

Br–Ar–Br
Δ
CuCl/quinoline

O N N Ar O n

Ar = O S CH_3 CH_3

Scheme 4. Wang route to poly(phthalazinone) ethers.

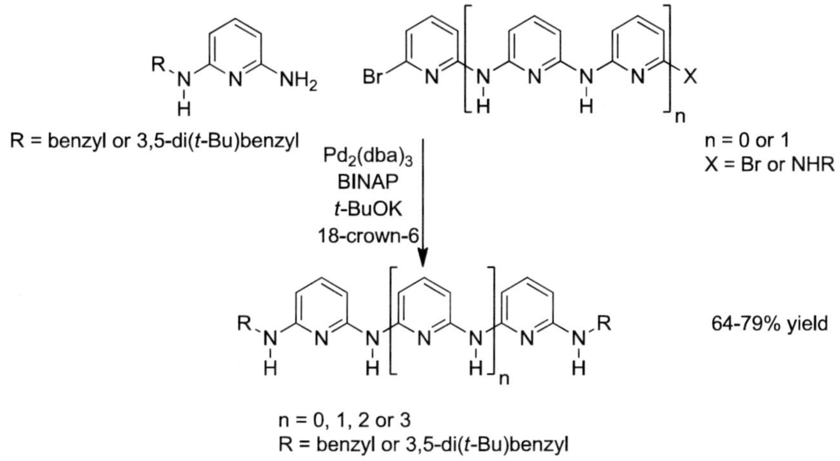

R N H N NH_2

Br N N H N N H N X n

R = benzyl or 3,5-di(*t*-Bu)benzyl

n = 0 or 1
X = Br or NHR

$Pd_2(dba)_3$
BINAP
t-BuOK
18-crown-6

R N H N N H N N H N N H R n

64-79% yield

n = 0, 1, 2 or 3
R = benzyl or 3,5-di(*t*-Bu)benzyl

Scheme 5. Chou route to oligo(aminopyridines).

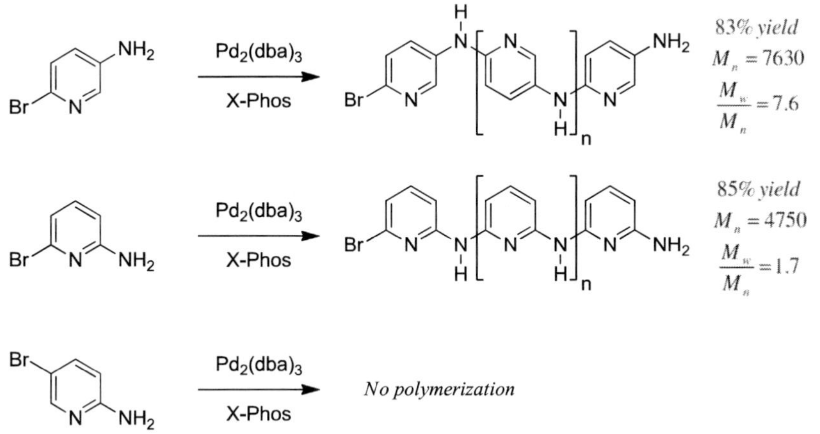

Br N NH_2 → $Pd_2(dba)_3$ / X-Phos → Br N H N N N H N NH_2 n

83% yield
$M_n = 7630$
$\frac{M_w}{M_n} = 7.6$

Br N NH_2 → $Pd_2(dba)_3$ / X-Phos → Br N N H N N H N NH_2 n

85% yield
$M_n = 4750$
$\frac{M_w}{M_n} = 1.7$

Br N NH_2 → $Pd_2(dba)_3$ / X-Phos → *No polymerization*

Scheme 6. Kanbara route to polyaminopyridines.

This work was one of the first examples of hetetoatom-arylation cross coupling reactions to produce π-conjugated extended systems. A few years later, in 2009, Kanbara and coworkers in a similar approach showed that the polymerization of two bromoaminopyridine isomers could be accomplished when using $Pd_2(dba)_3$ as catalyst (Scheme 6). The yield, molecular mass and weight dispersion of polymers depend strongly on the isomer structure [16].

Cross-Coupling Reactions

The classical reductive coupling called "Ullmann Reaction" was first reported in 1901 [17] and employs metallic copper or salts of this metal to make $C_{sp^2}-C_{sp^2}$ single bond between two aromatic nuclei. Normally, two equivalents of aryl halide are reacted with stoichiometric amounts of copper at high temperature (above 200 °C) to form a biaryl and a copper halide [18]. For these catalyzed reactions, the ease of halogen substitution from the aromatic ring follows the order I > Br >Cl>> F [2].

Among the numerous families of conjugated polymers, poly(thiophene)s (PTs) have been one of the most studied. PT have been considered as a model for the study of charge transport in conjugated heteroaromatic polymers with a nondegenerate ground state, or in other words, polymers with quinoid and aromatic structures that are not energetically equivalent [19].

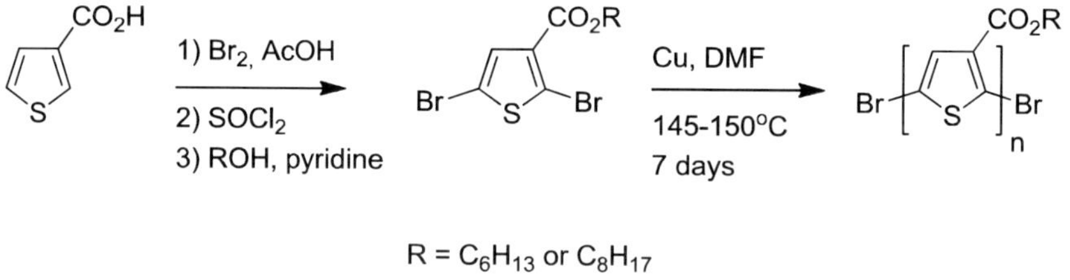

Scheme 7. Pomerantz route to 3-substituted polythiophenes.

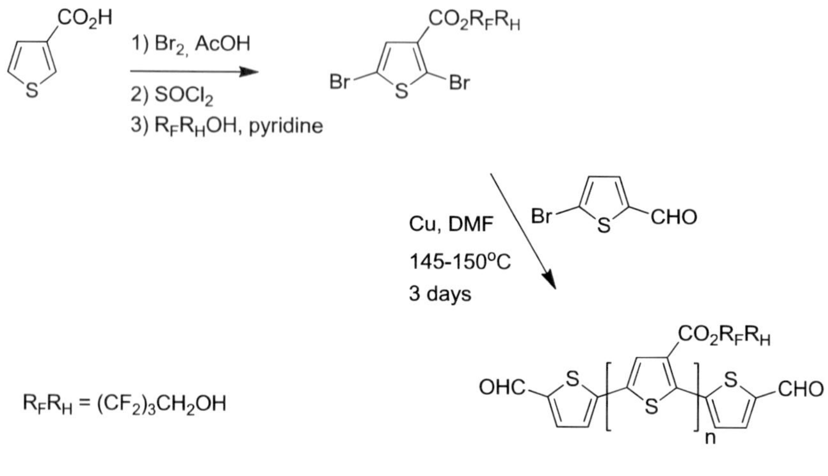

Scheme 8. Thomas route to n-type polythiophenes.

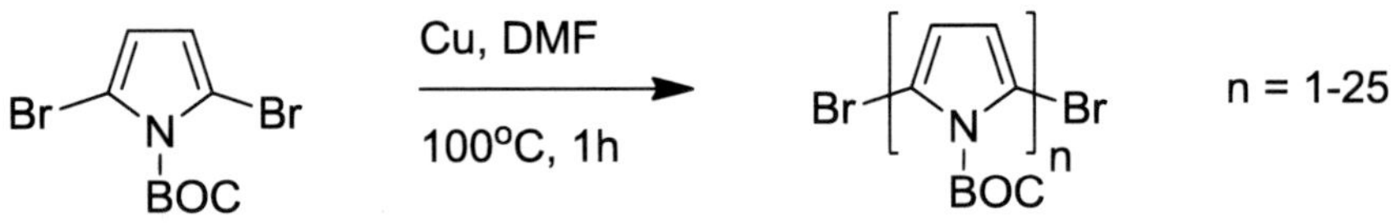

Br N Br BOC — Cu, DMF / 100°C, 1h → Br [N BOC]n Br n = 1-25

Scheme 9. Groenendaal route to oligo(pyrrole-2,5-diyl).

In addition, the high environmental stability of both its doped and undoped states together with its structural versatility, have allowed PT to be applied in electrochromic and electric devices [20]. The preparation of these polymers by metal catalyzed polycondensation is the method of choice for obtaining materials with high structural quality. Pomerantz and coworkers were the first to report poly(thiophene) synthesis by Cu-catalyzed polycondensation of substituted thiophene rings at the 3-position (Scheme 7) [21]. The authors employed metallic copper in DMF and they obtained polymers with low polydispersity and a low degree of polymerization (DP): in between 14-17. Theses PTs were soluble in common organic solvents such as $CHCl_3$, THF, DMF, CH_2Cl_2, benzene, toluene and xylene.

More recently, employing a similar methodology to that described by Pomerantz, Thomas and collaborators prepared a new terminal-functionalized and side-chain-fluorinated acceptor-type (n-Type) PT with a molecular weight up to 15,000 $g{\cdot}mol^{-1}$ (Scheme 8) [22]. The presence of fluorine groups in the polymer backbone may improve some material characteristics such as thermal stability, chemical resistance, and a high tendency for phase separation with hydrocarbon segments. Moreover, electron-withdrawing groups such as fluorine, provide an acceptor character to the polymer, which highlights the potential use of this type PT in light-harvesting applications [23].

Another important family of conjugated polymers that can be prepared by Cu-catalyzed polycondensations are the poly(pyrrole)s. These polymers are characterized by high conductivity and good stability of their doped forms. Groenendaal and coworkers prepared well-defined oligo(pyrrole-2,5-diyl) by coupling of *N-t*-BOC-2,5-dibromopyrrol using Cu-bronze in DMF (Scheme 9). In their protocol, oligomers with up to 25 pyrrole repeating units were obtained [24]. The isolated oligomers were used to study their optical and electrical properties as a function of chain length. The results showed a linear relationship between the band gap energy and the degree of polymerization, confirming that the electronic and optical properties of conjugated polymers are largely dependent on the extent of conjugation.

Tour and coworkers described the synthesis of low-band gap (1.1 eV) zwitterionic pyrrole-derived polymer by Ullmann coupling (Scheme 10), however, the use of common solvents in this reaction (DMF, quinoline, pyridine) did not allow for polymerization to occur. The effective coupling was only achieved through the use of 1,2-dimethoxyethane as solvent and copper-bronze as catalyst after 18 h at high temperature [25]. The obtained polymer presented molecular weights of approximately 4,000 Da and polydispersity between 1.15-1.25. This compound can be reversibly converted to a linear or a planar conjugated polymer and its absorption spectral ranges from the UV to the near-IR spectral region. It also possesses an enormous pH and solvent dependences which expands the potential for optical and electrical material applications. During the first 70 years after the discovery of the Ullmann reaction, copper was almost the only metal employed to catalyze C-C bond formation among two C_{sp^2} fragments.

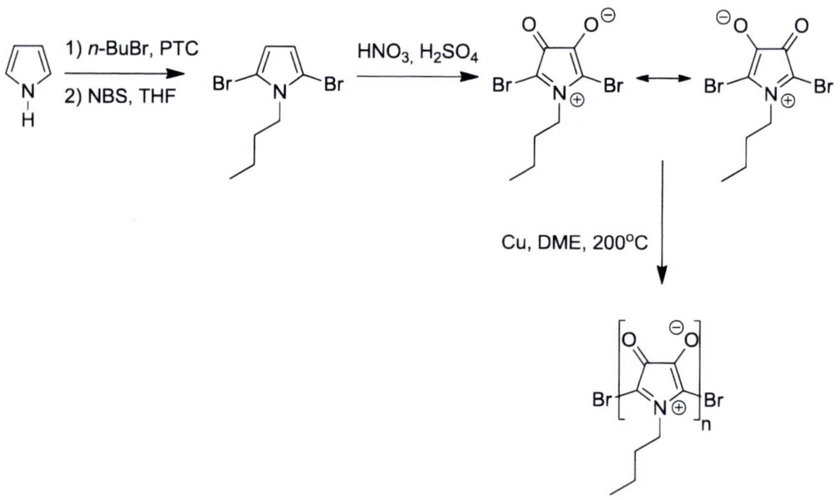

Scheme 10. Tour route to zwitterionic conjugated pyrrole-derived polymers.

In this period, numerous modifications in relation to the classical Ullmann and Goldberg methods resulted in the increasing of the scope of Cu-catalyzed reactions, however, the major advance was achieved in the replacement of copper by nickel as the catalyst in coupling reactions. Despite the use of terms such as "Ullmann reaction" and "Ullmann condensation" which relate to the use of copper and its derivatives in coupling reactions, from a broader perspective of "Ullmann Chemistry", the similarity of the initial procedures which utilize Ni(0) with those employing copper, and the fact that in both cases the product of reaction is a result of reductive elimination step and thus these nickel methods fall within the class of an "Ullmann type reaction".

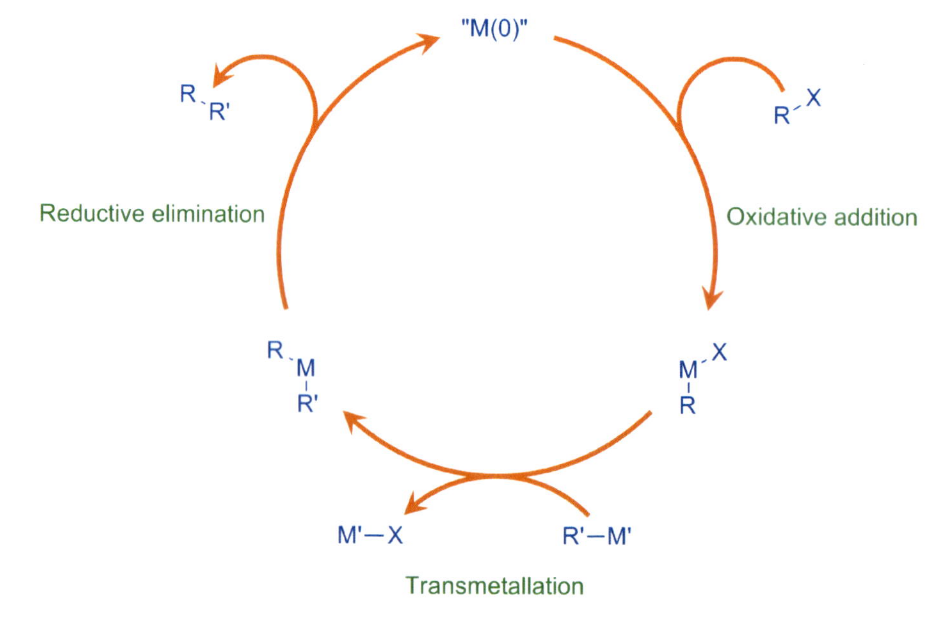

Figure 1. Simplified metal(0)-catalyzed coupling catalytic cycle.

In 1972, Kumada [26] and Corriu [27] separately reported the development of an efficient method for C-C bond formation by Ni-phosphine complex catalyzed coupling of Grignard reagents with organic halides. At the time of its development, the milder conditions of the Kumada coupling methodology, when compared with copper methods available at that time, rendered this procedure a very practical option for C-C bond formation. In addition, the possibility of using catalytic amounts of Ni(II) complex in the presence of a chemical reducing agent (mostly Zn) increased the interest in the use of nickel [28]. In this case, the active species, Ni(0), is generated *in situ*. Figure 1 shows a general catalytic cycle for metal catalyzed cross coupling reactions, which can be used to understand the behavior of different metals used in these reactions. Initially, the metal undergoes oxidative addition into a carbon-halide bond of the electrophile. After that, an organic scaffold is transferred to the metal center with the participation of a secondary species complex in the "transmetallation step". A reductive elimination restores the reduced metal back to its initial oxidation state.

Nowadays, due to the popularity of Ni coupling reactions, a large number of examples towards the synthesis of conjugated polymer by Ni-catalyzed polycondensation reactions have been reported. The regioregular poly(3-alkylthiphene)s (rP3ATs) polymer synthesis is an important example that demonstrates the great influence of reaction conditions on the final structure of the conjugated polymer system and consequently, the optical and electrical properties presented by the material.

Head-to-tail coupling | Head-to-head coupling | Tail-to-tail coupling

Scheme 11. Possible couplings patterns in P3AT backbones.

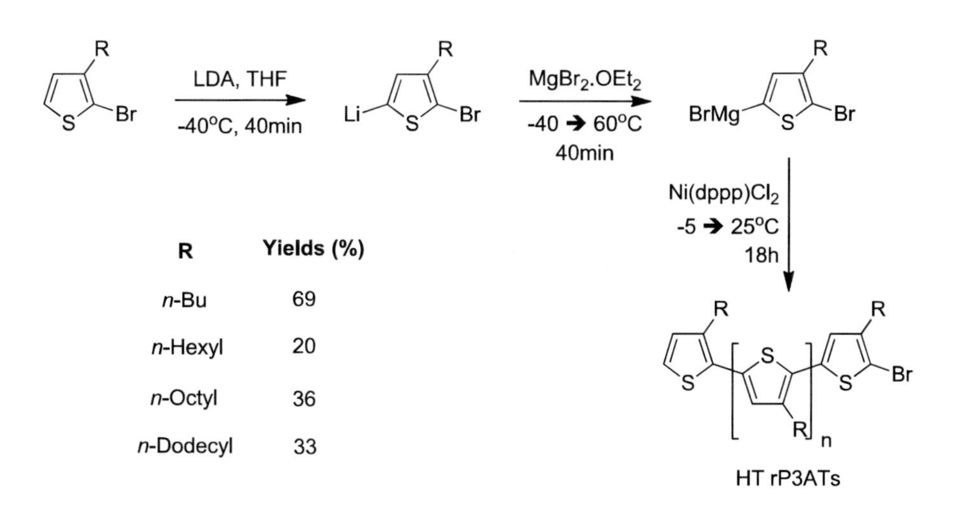

R	Yields (%)
n-Bu	69
n-Hexyl	20
n-Octyl	36
n-Dodecyl	33

Scheme 12. McCullough route to HT regioregular poly(3-alkylthiophene)s.

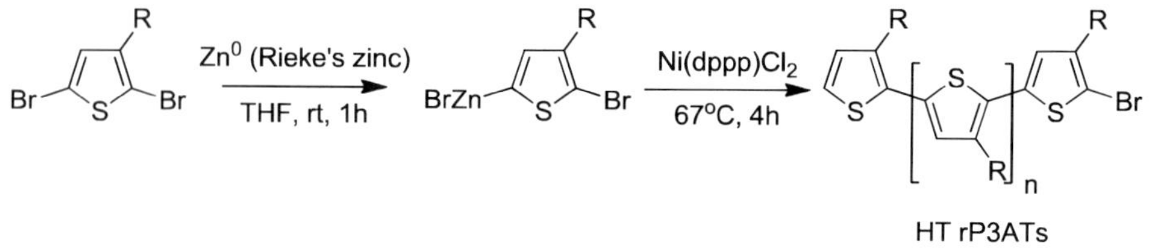

Scheme 13. Rieke route to HT regioregular poly(3-alkylthiophene)s.

As mentioned previously, PTs are one of the main conjugated polymer families, however, until to 1980's, electrochemical and chemical methods created P3ATs at only about 50 % Head-to-Tail (HT) couplings. Due to steric effects, the presence of Head-to-Head (HH) and Tail-to-Tail (TT) couplings in the polymer backbone caused a decrease in overlap between the π atomic orbitals on adjacent rings, which leads to a loss of π–conjugation (Scheme 11) [29]. McCullough and Lowe were the first to report the HT rP3ATs synthesis. The polymerization step was performed using Kumada cross-coupling methods and a catalytic amount of Ni(dppp)Cl_2, where dppp is 1,3-diphenylphosphinopropane (Scheme 12). NMR analysis showed that this method led predominantly to the formation of rP3ATs with HT-HT regioregularity of 91% in 20 to 69% yields. The molecular weights were around 20,000 g·mol^{-1} with polydispersities of about 1.4 [30]. Oxidation with I_2 produced polymers with conductivities up to 1000 S·cm^{-1}: 50- to 60-fold higher than P3ATs synthesized by the usual chemical or electrochemical methods [31].

In the same year of McCullough's publication, Rieke and coworkers reported the synthesis of HT rP3AT using regiocontrolled organozinc reagents to perform Negishi couplings with Ni(dppp)Cl_2 as catalyst (Scheme 13). The polymers presented 98.5% HT-HT regioregularity and molecular weights were around 15,000 g·mol^{-1} [32].

McCullough's group also described another interesting method for preparing HT rP3AT by a method known as the Grignard metathesis (GRIM). This inexpensive protocol is highly efficient and allowed the obtention of polymers with HT-HT regioregularity higher than 99%, molecular weights up to 35,000 g·mol^{-1}, and polydispersities between 1.0-1.47. In GRIM method, 2,5-dibromo-3-alkylthiophene can be treated with any Grignard reagent, then the intermediate formed is polymerized using Ni(dppp)Cl_2 (Scheme 14) [33]. This reaction can be carried out on very large scales, making this method potentially applicable to industry.

The Ni catalyzed coupling reactions are also applied to the synthesis of many other conjugated polymer families such as poly(pyrrole)s, poly(phenylene)s, poly(pyridine)s, and poly(aniline)s. Recent reviews concerning the synthetic approaches to conjugated polymers by Ni-catalyzed polycouplings have been reported [34, 35].

Shortly after Kumada and Corriu reported C-C bond formation by a Ni-phosphine catalyzed coupling reaction between Grignard reagents and organohalides, Pd-catalyzed versions of this reaction began to be appear. The great success achieved by other related methodologies such as Suzuki, Negishi, Heck, and Stille couplings, lead to a direct association of the term "coupling reactions" with palladium catalysis today. In recognition of their pioneerism, Richard F. Heck, Akira Suzuki and Ei-ichi Negishi were awarded with the 2010 Nobel Prize in Chemistry for their contributions on palladium-catalyzed cross-coupling reactions in organic synthesis.

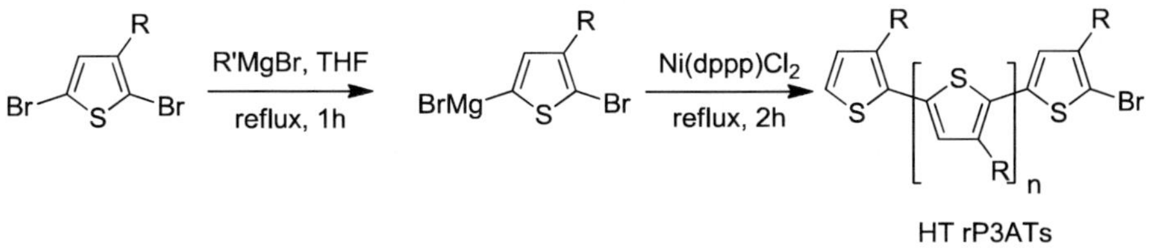

Scheme 14. GRIM method to HT regioregular poly(3-alkylthiophene)s.

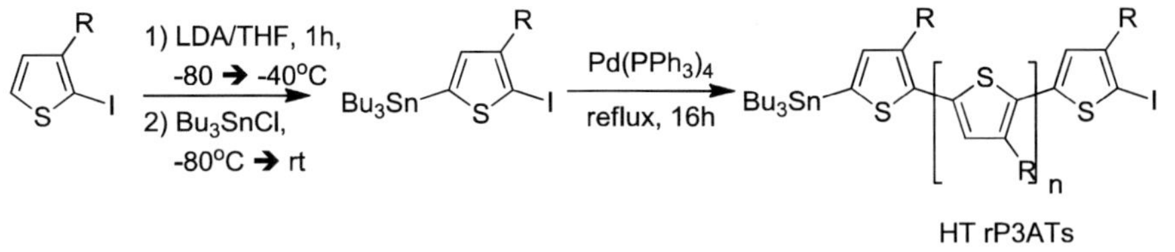

Scheme 15. Iraqi route to HT regioregular poly(3-alkylthiophene)s.

The catalytic cycle of Pd catalyzed coupling reactions is similar to that shown in Figure 1, and different species can undergo the required "transmetallation step" for the coupling reaction, among them zinc (Negishi coupling), tin (Stille coupling), and boranes or boronic acids (Suzuki couplings).

In a large number of cases, methodologies employing Pd complexes as catalysts were far more effective than analogous methods using Cu or Ni, allowing the use of milder reaction conditions. Due to great interest in this area, several of these methods with palladium were employed towards the preparation of conjugated polymers.

The Stille coupling reaction [36] is one of the most versatile palladium-mediated methods for a wide variety of conjugated polyaromatic systems, particularly polythiophene families. The advantages of this reaction are its tolerance to different functional groups and the requirement of mild reaction conditions. Furthermore, Stille couplings are stereospecific, regioselective, and give good yields [37]. Iraqi and Barker applied the Stille reaction to prepare HT rP3ATs using 2-iodo-3-alkyl-5-tri-*n*-butylstannyl-thiophene derivatives in a variety of solvents (Scheme 15). The polymers presented up to 96% HT-HT regioregularity, polydispersities between 1.22-1.44, and molecular weights were 14,000 to 20,000 $g{\cdot}mol^{-1}$ [38]. The poly(phenylenevinylene)s (PPVs) and their derivatives are another important class of conjugated polymers that can be obtained by Stille couplings. PPVs were the first class of conjugated polymers to demonstrate electroluminescence properties [39], and they are the most studied class of electroluminescent polymers. These polymers were formerly prepared by carbon-carbon double bonds construction strategy, as in the case of addition-elimination reactions, but nowadays cross coupling reactions are the most widespread approach since they have the advantage of avoiding the thermal elimination step which could jeopardize formation of the final polymer [40]. For example, Babudri and coworkers reported the synthesis of a soluble PPV by Stille coupling between 2,5-dialkoxy-1,4-diiodobenzene and *E*-1,2-bis(tributylstannyl)-ethene in the presence of Pd complexes (Scheme 16). The molecular weights of the polymers were around 3,500 $g{\cdot}mol^{-1}$ [41]. Recent reviews discussing a wide range of examples concerning the synthetic approaches to conjugated polymers using Stille polycondensation have been published [42].

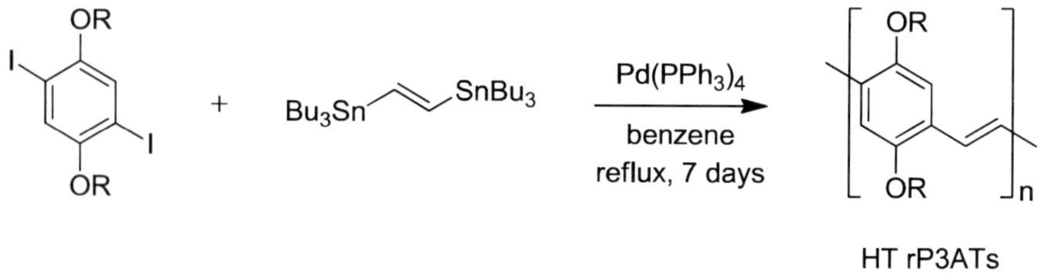

OR, I, I, OR + Bu_3Sn, $SnBu_3$ → $Pd(PPh_3)_4$, benzene, reflux, 7 days → OR, OR, n
HT rP3ATs

Scheme 16. Babudri route to soluble poly(phenylenevinylene)s.

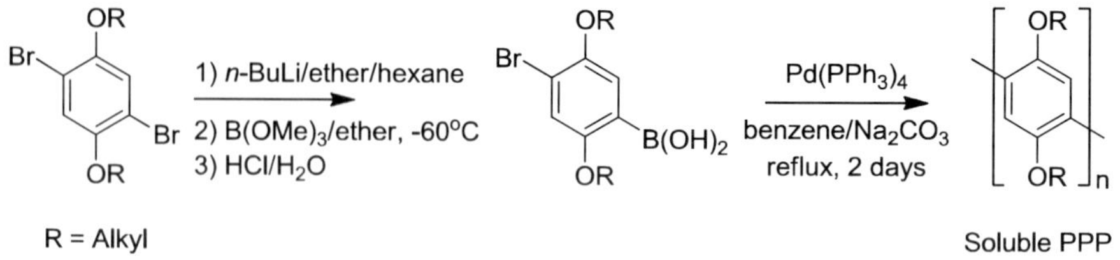

Br, OR, Br, OR → 1) *n*-BuLi/ether/hexane 2) $B(OMe)_3$/ether, -60°C 3) HCl/H_2O → Br, OR, $B(OH)_2$, OR → $Pd(PPh_3)_4$, benzene/Na_2CO_3, reflux, 2 days → OR, OR, n

R = Alkyl

Soluble PPP

Scheme 17. Schluter route to soluble poly(*p*-phenylene)s.

Probably, Suzuki couplings are the major improvement in the Pd-catalyzed coupling reactions. The use of boronic acid derivatives has several advantages over all the other organometallic derivatives such as their higher tolerance for a wide variety of functional groups and lower toxicity of those reagents [43]. Moreover, the Suzuki polycondensation (SPC) permits the preparation of regiospecific conjugated polymers with high molecular weights.

Poly(*p*-phenylene)s (PPPs) are among the main conducting polymer families which are prepared by SPC. The optical and electrical properties of these polymers allow their application in light emitting diodes (LEDs). Nevertheless, conjugated and conformationally rigid backbones reduce the solubility of unsubstituted PPPs in common organic solvents. In order to overcome this drawback, the introduction of side chains in PPPs is necessary. Schluter and coworkers were the first to report the synthesis of soluble poly(*p*-2,5-dialkylphenylene)s by SPC of 4-bromo-2,5-di-alkylbenzeneboronic acid, (Scheme 17). This method led to materials bearing a degree of polymerization of 30 [44]. Several examples of PPPs and others conjugated polymers synthesis by Suzuki policondensation are given in reference 45.

Virtually any type π-conjugated polymer can be prepared by Pd-catalyzed coupling reactions, and there are a large number of articles concerning this subject in the literature [42, 45].

Ullmann-Type and cross-coupling reactions have proven to be two of the most powerful tools for obtaining conjugated polymers, allowing the refinement of the architecture of the molecular structure of these materials in order to improve their optoelectronic properties. As presented here, numerous methods of cross-coupling reactions catalyzed by Cu, Ni and Pd provide several synthetic routes to preparation of new and already established π-conjugated systems. In addition, methodological advances such as the use of ligands in the Cu-catalyzed reactions and also the employment of boronic derivatives have increased the efficiency of both of these catalytic systems. Consequently, there is an increase in the scope of substrates susceptible to polycondensation. Apparently, despite the strong growth of cross-coupling reactions, it seems that based on the volume of work published in this area of science, Ullmann-Goldberg type reactions will continue to play an important role in the synthesis of conjugated polymers for a long time.

CARBON-CARBON DOUBLE BONDS CONSTRUCTION

The joining of aromatic carbo or heterocyclic monomers through a bridging carbon-carbon double bond is one of the most highly exploited strategies for the preparation of novel polymeric materials. The Wittig reaction, and it's variants such as the Wittig-Schlosser, Horner–Wadsworth–Emmons and Still-Gennari reactions, constitute some of the most synthetically convenient and selective methodologies in organic synthesis for the construction of simple alkenes or *α,β*-unsaturated carbonyl compounds (Scheme 18).

Yields are usually high and reactions often only requiring hours to go to completion. In general, the stereoselectivity is very high when stabilized phosphorous ylides are employed. In particular, the Horner-Wadsworth-Emmons (HWE) reaction between aldehydes or ketones with stabilised phosphonate carbanions is very popular amongst polymer scientist given that it leads to olefins with excellent *E*-selectivity. Moreover, the reaction has several advantages over the use of other phosphorus ylides traditionally employed in Wittig reactions.

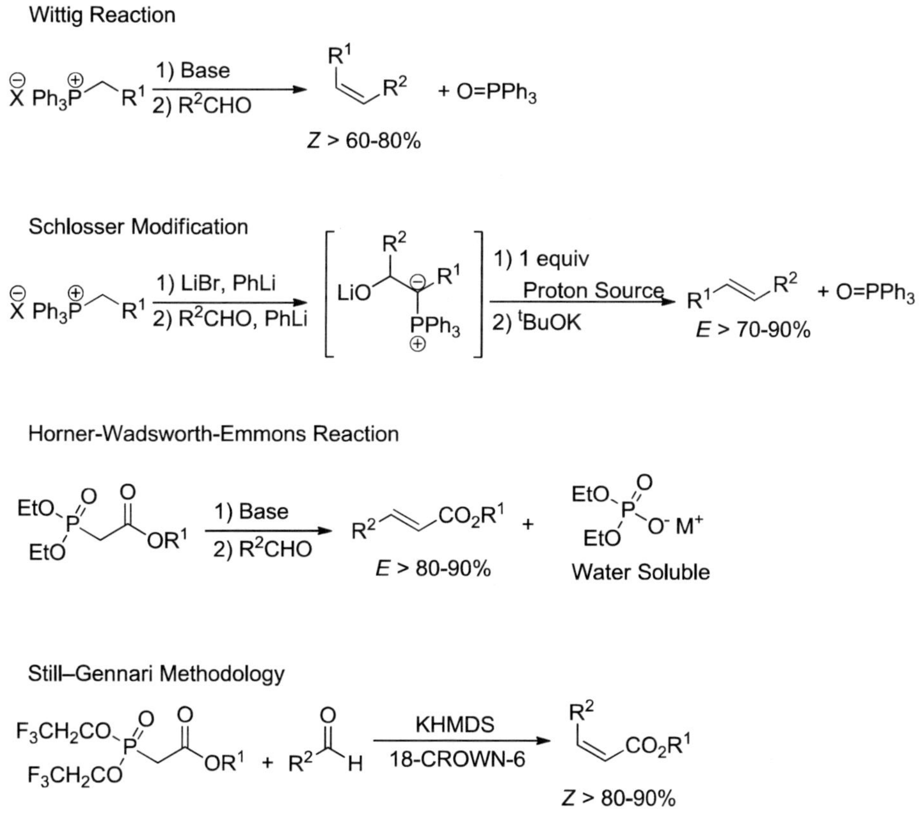

Wittig Reaction

1) Base; 2) R^2CHO — + $O{=}PPh_3$ — $Z > 60\text{-}80\%$

Schlosser Modification

1) LiBr, PhLi; 2) R^2CHO, PhLi — 1) 1 equiv Proton Source; 2) ^tBuOK — + $O{=}PPh_3$ — $E > 70\text{-}90\%$

Horner-Wadsworth-Emmons Reaction

1) Base; 2) R^2CHO — $R^2CH{=}CHCO_2R^1$ + $(EtO)_2P(O)O^- M^+$ (Water Soluble) — $E > 80\text{-}90\%$

Still–Gennari Methodology

KHMDS, 18-CROWN-6 — $Z > 80\text{-}90\%$

Scheme 18. The Wittig reaction and its variants.

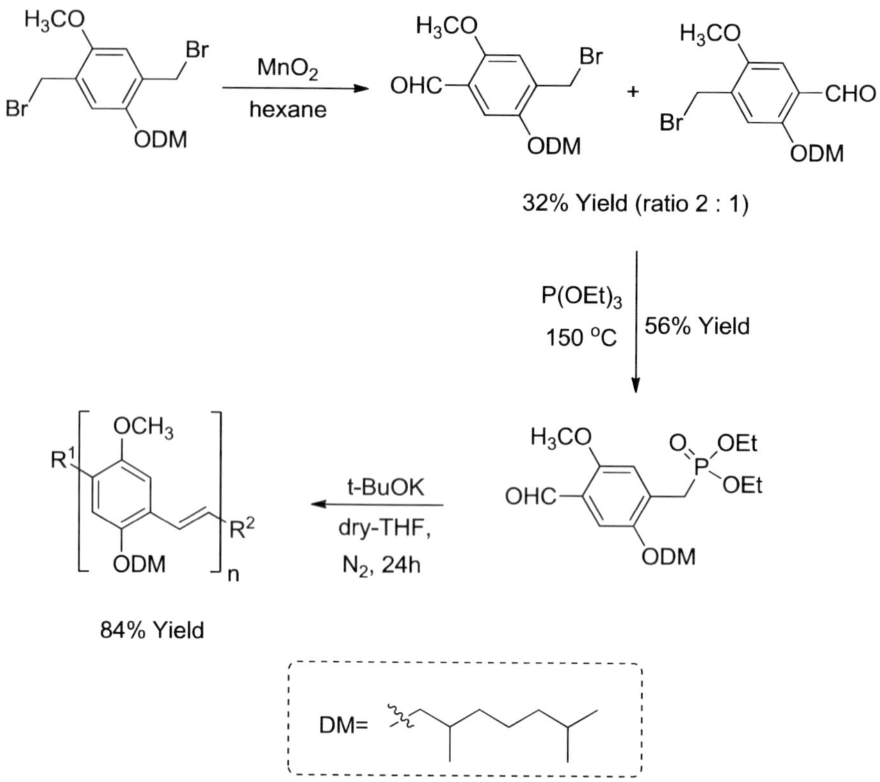

Scheme 19. Synthesis of PPV by HWE reaction.

The phosphorus byproduct is a water-soluble phosphonate ester, which makes it easier to separate from the olefin product than the rather intractable phosphine oxide. A straight forward application of this approach is exemplified in the preparation of a "regioregular" poly-phenyl-vinyl (PPV) type polymer developed for its interesting photovoltaic properties (Scheme 19) [46, 47].

Synthesis of the monomer was achieved by MnO_2 oxidation of a bis(bromomethyl)benzene compound and subsequent introduction of a phosphonate ester group was realized by the Arbuzov reaction. Synthesis of the monomer was plagued by the formation of mixtures, but pleasingly, these isomers could be conveniently separated by column chromatography and carried through to be polymerized. Polymerization of the dialdehydebisphosphonium ylide monomer under basic conditions at room temperature affords the desired regioregular polymer as a red fibrous solid in milligram quantities. The HWE reactions is often carried out in analytical grade solvents and open to air when applied to the construction of low molecular weight compounds. However, when employed for polymer synthesis, anhydrous solvents and an inert atmosphere are standard operating conditions in this case.

The Wittig reaction can also be a key step in the synthesis of arene vinyl monomers that are polymerized via Pd-catalysed Heck reaction to provide PPV copolymers (Scheme 20) [48]. This strategy was executed for the preparation of thin polymer films displaying a wide absorption range for the visible spectrum with optical band gaps of 2.19–2.35 eV.

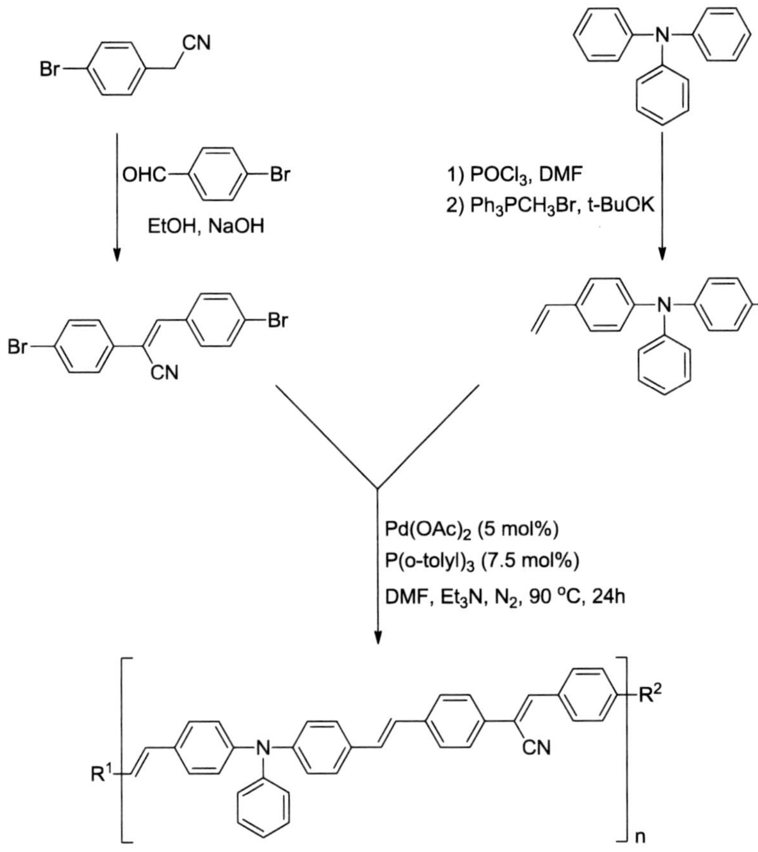

Scheme 20. Synthesis of vinyl arenes for polymerization by Heck reactions.

The aryl bromide is obtained *via* a Knoevenagel condensation reaction under standard basic conditions between 4-bromobenzaldehyde and a benzylnitrile. The other requisite monomer is prepared by Vilsmeyer formylation of triphenylamine to yield a dialdehyde which is then reacted with a simple phosphonium salt to afford a vinyl arene.

Finally, Heck coupling was realized at 90 °C in 24h and thus effectively providing milligram quantities of a highly thermally stable orange coloured polymer with low photovoltaic efficiency. The choice for monodentate ligand P(*o*-tolyl)$_3$ is not discussed by the authors but it is presumable that an extensive ligand screen was carried out and an optimal yield of 63% was achieved.

Continuing the theme of the application of transition metal crossed coupling reactions of carbon-carbon multiple bonds for the development of novel polymer materials, an interesting utilization of modified cellulose was reported in which poly(fluorene), poly(fluorenevinylene), and a poly (fluorene-ethynylene) were all successfully grafted onto cellulose (Scheme 3) [49]. Non-specific esterification of cellulose's free hydroxyl groups permits the introduction of either an arylbromide tether or an acetylene group for Suzuki, Heck, and Sonogashira-type polymerizations (Scheme 21).

A possible downfall of an approach that involves transition metal catalysed polymerization is the potential for the polymer product to retain the metal which can be on occasion, very difficult to wash out depending on the strong binding ability of the polymer itself. In this respect, it was found that although a $Ni(COD)_2$ catalysed coupling was very robust, the necessity for stoichiometric amounts of metal rendered the procedure unpractical and removal of metal salts an extremely labour intensive process. On the other hand, a Suzuki polymerization of 9,9-dihexyl-2,7-dibromofluorene and 9,9-dihexylfluorene-2,7-diboronic acid bis(1,3-propanediol) ester using $Pd(PPh_3)_4$ (3mol%) and K_2CO_3 in toluene/water with a catalytic amount of a phase transfer catalyst afforded the desired polymer with relatively ease of purification. Interestingly, performing the synthesis under microwave heating and suspending the paper material above the aqueous layer in the toluene phase avoided the problem of disintegration of the polymer compound that occurred in hot high pH solutions.

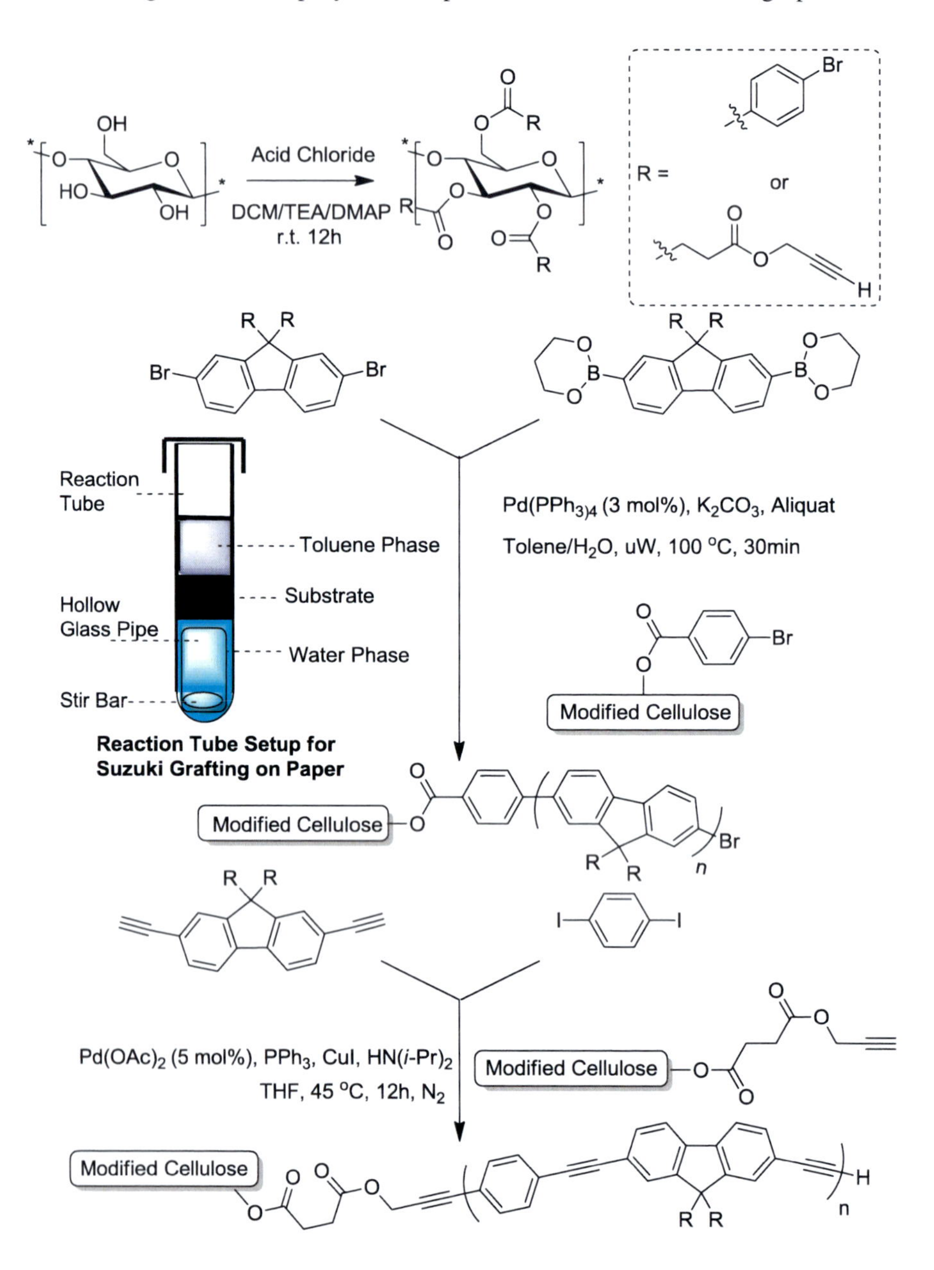

Scheme 21. Grafting of polymers on to cellulose by Suzuki and Sonogashira reactions.

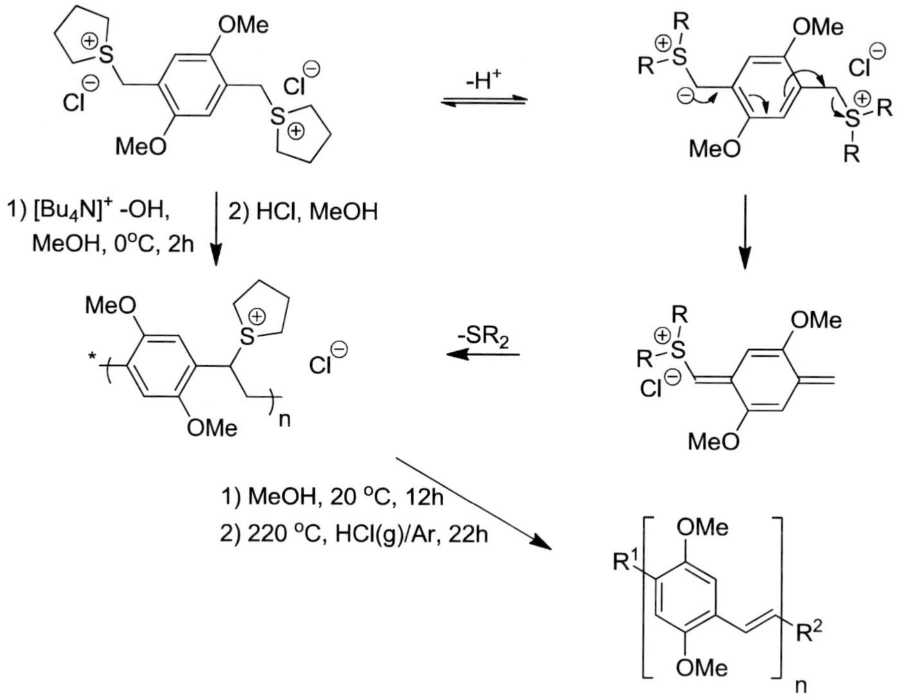

Scheme 22. Synthesis of mixed conjugated-non conjugated PPV *via* the Wessling-Zimmerman route.

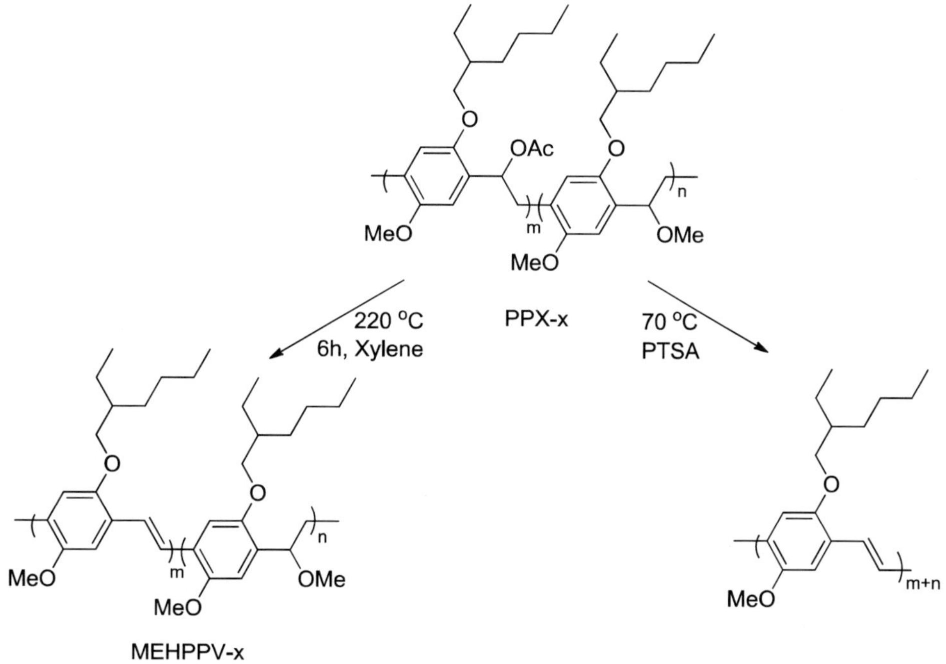

Scheme 23. Padmanaban synthesis of soluble MEHPPV-x.

The Wessling and Zimmerman approach to PPVs involves the elimination of sulfonium salts to form polymeric materials under strongly basic conditions or by simple thermal treatment. Copolymers related to PPV and MEH-PPV were prepared with intermitted sequences of conjugated and non conjugated segments by elimination of certain functional groups from a precursor polymer synthesised from a dimethoxybis(sulfonium) salt monomer (Scheme 22) [50]. Elimination of the sulfonium group is easily achieved by heating, furnishing polymers bearing electron donating alkoxy substituents that exhibit high efficiency in electroluminescent devices.

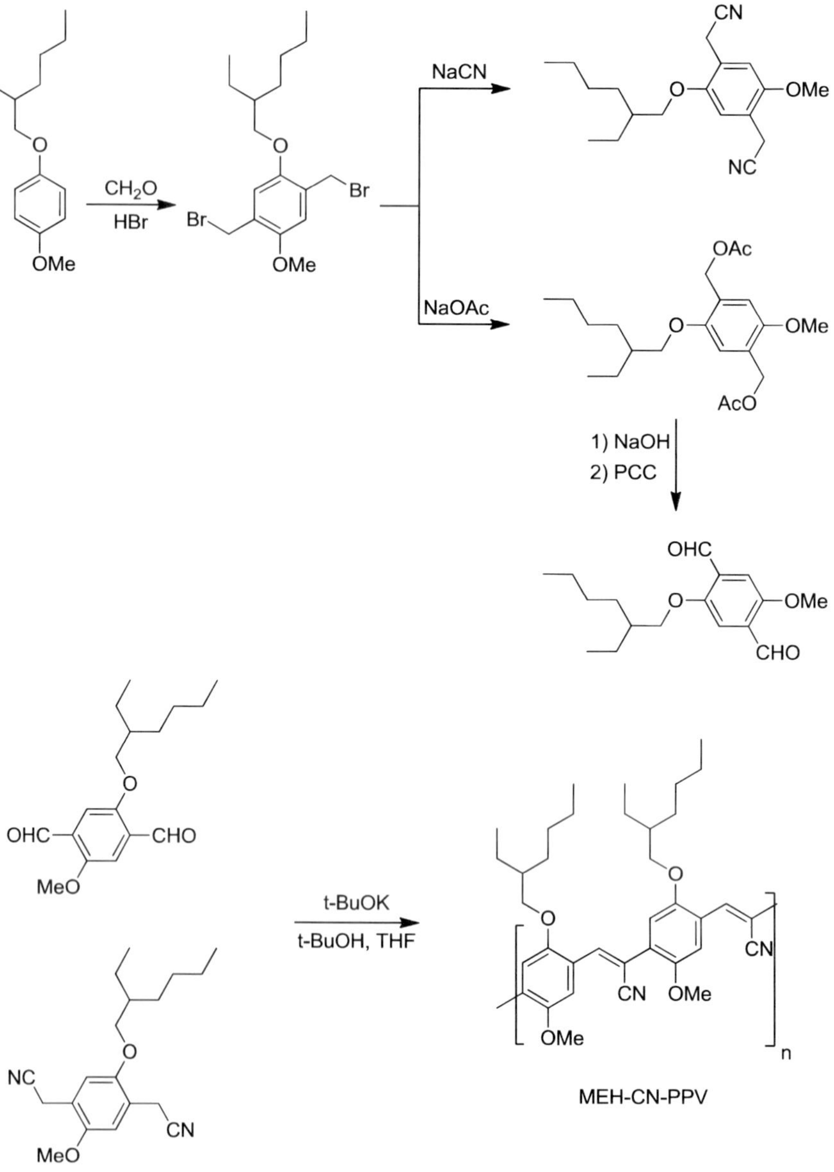

Scheme 24. Synthesis of MEH-CN-PPV by Knoevenagel condensation.

The synthesis of soluble MEHPPV samples with varying extents of conjugation was reported in 2000 by Padmanaban and Ramakrishnan and once again takes advantage of the Wessling-Zimmerman approach. In this example, acetate groups are eliminated preferentially to methoxy groups by thermal elimination (Scheme 23) to afford a MEHPPV-x polymer [51].

Thermal elimination apparently requires rigorous conditions as evidenced by the fact that synthesis of the MEHPPV-x polymer is carried out in anhydrous xylene which is used to prepare a solution of the monomer to be initially thoroughly degassed by freeze-thaw techniques and finally polymerization is effected in a heated vacuum sealed tube.

Polymer materials were developed with usable quantum efficiency values, low-temperature processing properties and high illumination levels which are all of interest for the development of organic photovoltaic devices. Here, we highlight a cyano derivative of poly(*p*-phenylenevinylene), MEH-CN-PPV used as an electron acceptor in a laminated diode [52]. The synthesis as described by Holmes and co-workers begins with bis-bromomethylation of the appropriate alkoxydisubstituted compound which can then either undergo a nucleophilic substitution by cyanide anion to afford the first monomer, or substitution by an acetate anion followed by hydrolysis and oxidation to provide the dialdehyde (Scheme 24) [53]. Apparently, the conditions for Knoevenagel condensation polymerization reaction require special care in order to avoid possible side-reactions such as Michael addition by various nucleophiles to the newly formed vinylene linkage or direct attack on the nitrile. This is mainly resolved by avoiding excessive reaction times, elevated temperatures and controlling precisely the stoichiometry of the reaction.

METATHESIS

Ring opening metathesis polymerization also known as ROMP, takes advantage of strained alkene rings by opening them up through a ruthenium-catalyzed cross metathesis reaction in which obviously the driving force is the relief of ring strain (Scheme 25). In other circumstances, a strained mono alkene can be polymerized with a second alkene to afford a copolymer (Scheme 25). Acyclic diene metathesis polymerization (ADMET) generally prefers terminal dienes as monomers as they entropically and sterically favour the polymerization reaction. In the case of ADMET, special consideration must be given to the choice of catalyst given the fact that certain neighbouring group effects can alter the reaction mechanism pathway and so therefore not favouring formation of a linear polymer but instead, affording the cyclic dimer. This is illustrated in the case of 1,5-hexadiene which gives different reaction outcomes depending on the use of either a molybdenum or ruthenium catalyst (Scheme 26) [54].

The ROMP and ADMET mechanism is characterized by a series of 2+2 cycloadditions, the formation of a metallacyclobutane ring intermediate and its subsequent collapse placing the electron deficient metal centre at the end of the monomer and propagating the polymerization cycle.

Thus, for ROMP and ADMET type reactions, a quenching agent is needed to limit the molecular weight or end cap the polymerchains and this is typically achieved by addition of monofunctional ethyl vinyl ether to replace the metal alkylidene active chain (Scheme 26).

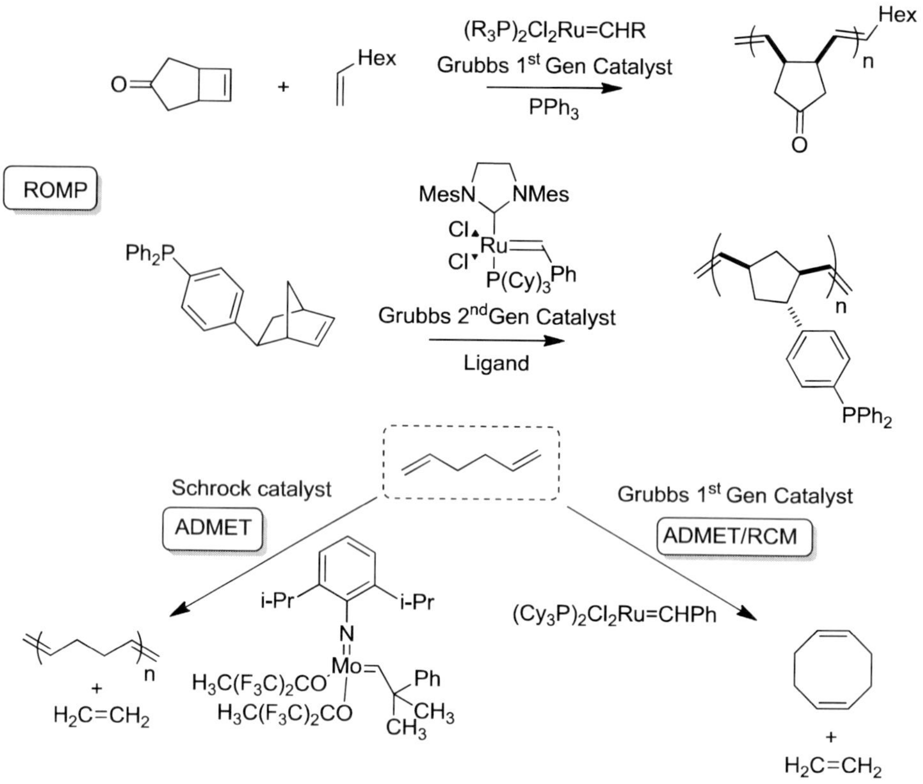

Scheme 25. Synthesis of MEH-CN-PPV by Knoevenagel condensation.

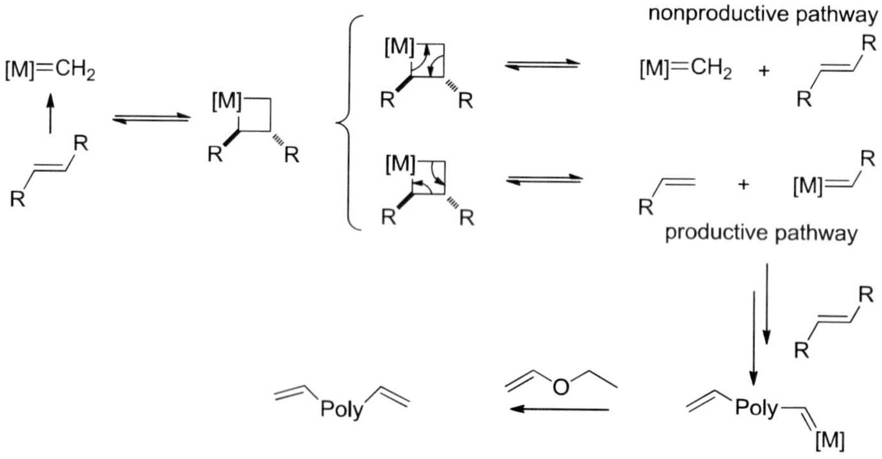

Scheme 26. General mechanism for polymerization by olefin metathesis polymerization.

Polymerization of a substituted [2.2]paracyclophanedienes using Grubbs 2nd generation metathesis catalysts to give a monodispersed, soluble phenylenevinylene was effective in affording a well-defined molecular weight homopolymer with an alternating *cis-trans* microstructure (Scheme 27)[55].

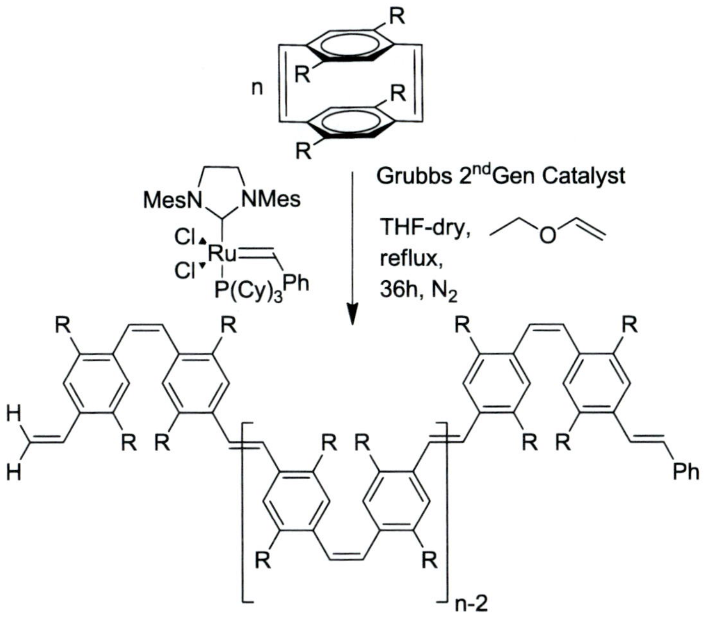

Scheme 27. ROMP of cyclophanediene to give monodisperse, soluble PPVs ($R=OC_8H_{17}$).

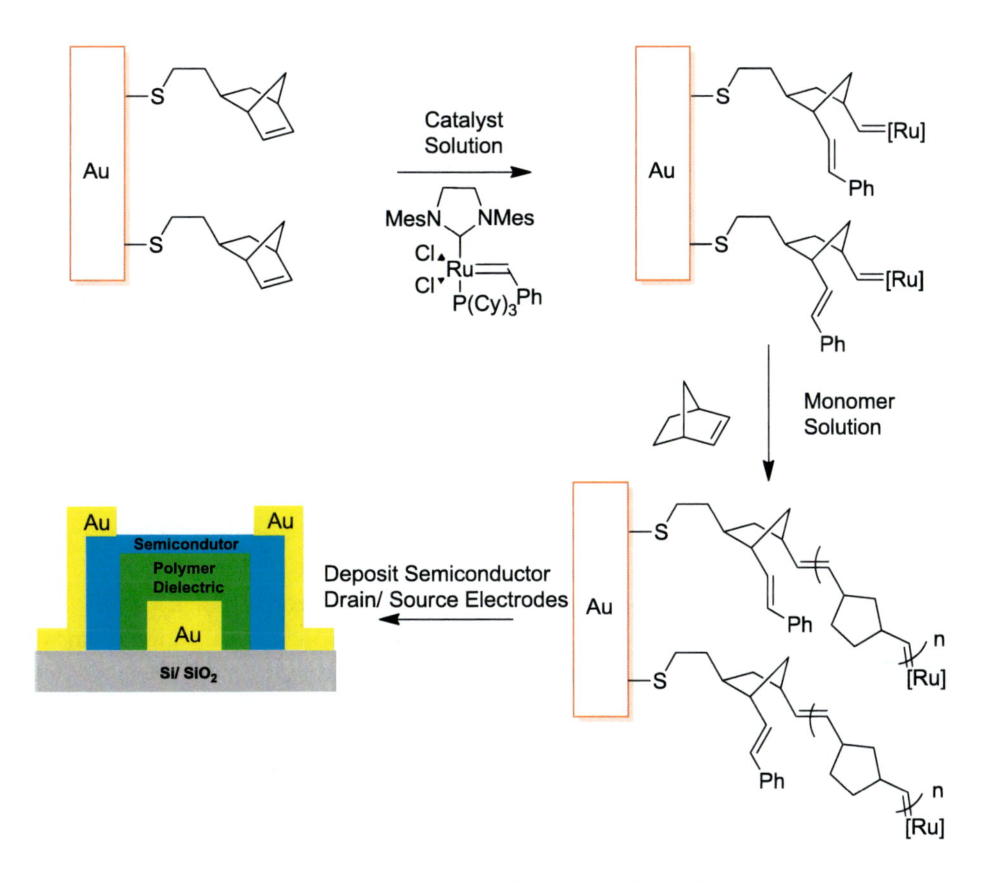

Scheme 28. Construction of an FET using an SI-ROMP polymer dielectric layer.

The authors state that a range of polymers with varying molecular weights could be conveniently obtained by simply altering the monomer-to-catalyst ratio and in general, a 36 hour reaction time was necessary to achieve complete conversion of monomer thus indicating a slow rate of polymerization. Standard Schlenk techniques using dry solvents and reagents in an inert atmosphere are once again standard operating procedures for polymerization as noted before for other reaction classes discussed earlier.

Construction of a field effect transistor (FETs) was accomplished by using a novel surface-initiated ring opening metathesis polymerization (SI-ROMP) technique under mild reaction conditions, short reaction times, and by a simple solution processing method [56]. These polymer dielectric layers covalently attached to Au or Si/SiO_2 surfaces are obtained by a ROMP surface-tethered metathesis in which the carbene bonded ruthenium catalyst was found to provide the greatest film thicknesses (up to 300 nm) (Scheme 28).

The SI-ROMP polymer dielectric layer was grown on an Au strip gate electrode (1000 Å thick, 1 mm wide) using a thiolnorbornene linker and a 3 M norbornene solution giving a film with a capacitance of 3 $nF{\cdot}cm^{-2}$ measured at 20 Hz.

Electro Polymerizations

Several examples of conducting polymers were originally prepared by electrochemical synthesis [57, 58]. In this technique the active species (anions, cations or radicals) are formed on the electrode surface where polymerization also occurs. Monomers can be oxidized electrochemically to obtain the respective polymer by an anodic route. The cathodic electropolymerization can also be employed in a few cases [59, 60], but its use is quite limited when compared to the anodic process.

The anodic electropolymerization is one of the most reported methodologies for conducting polymer synthesis, and it has assumed an important position amongst known methods capable of being used to prepare polymers of electron rich aromatic compounds such as thiophene, pyrrole, and aniline [61]. The electrochemical route has several advantages such as absence of catalyst, direct grafting of the doped form of the polymer onto the electrode surface, and easy control of the film thickness by the deposition charge [20]. However, chemical methods, such as organometallic couplings, lead to more regular and better-defined polymers than electrochemical synthesis.

The anodic electropolymerization mechanism [62] of aromatic compounds involves both electrochemical and chemical steps as shown in Scheme 29. The oxidation of the monomer requires 2 $Faradays{\cdot}mol^{-1}$, and the excess of charge is spent in order to achieve the reversible oxidation of the polymer leading to its doping, which occurs simultaneously with the synthesis. The mechanism suggests that an initial formation of the radical cation by monomer oxidation. After that, the coupling of two radical cations followed by subsequent loss of two protons leads to the system rearomatizing. The reaction proceeds with new couplings between the radical cations of the monomers or oligomers to produce more extended species until precipitation takes place at the electrode-electrolyte interface. It is important to emphasize that the electrical and physical-chemical properties of the electrosynthesized polymers will directly depend on the synthetic conditions such as solvent, monomer concentration, temperature and electrical conditions [63].

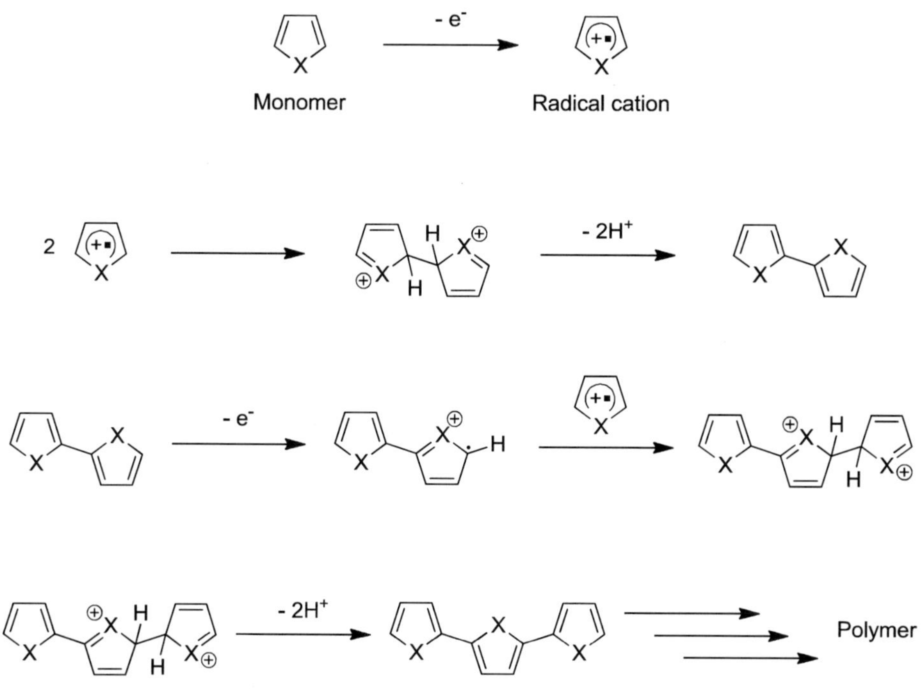

Scheme 29. Anodic electropolymerization mechanism for aromatic compounds.

Due to the better structure quality of polymers prepared from chemical methods, the electrochemical route has become less frequently utilized for conjugated polymer synthesis. However, recent advances such as the use of ionic liquids as electrolytes gave new life to this area. An ionic liquid is a salt in the liquid state. This type of salt has the advantage of being a better electrical conductor, but only a little more viscous, than usual electrolytes at room temperature [II]. Lu and Mattes were the first to describe the use of ionic liquid in electrochemical synthesis of conjugated polymers [64, 65]. The authors prepared several examples of conducting polymers using principally *N*-butyl-*N*-methylimidazoliumtetra-fluoroborate ([Bmim][BF_4]), which is the most common ionic liquid. Usually, films prepared into ionic liquids showed a larger doping level than those prepared by conventional electrosynthesis conditions [66, 67]. Experimental advances such as that mentioned above, still make the electrochemical synthesis one of most useful ways in the preparation of conjugated polymers.

REFERENCES

[1] J. F. Bunnett and R. E. Zahler, *Chem. Rev*. 49, 273 (1951).
[2] J. Lindley, *Tetrahedron* 40, 1433 (1984).
[3] T. Cohen, J. Wood and A. G. Dietz, *Tetrahedron Lett.* 15, 3555 (1974).
[4] G. O. Jones, P. Liu, K. N. Houk and S. L. Buchwald, *J. Am. Chem. Soc*. 132, 6205 (2010).
[5] K. Kunz, U. Scholz and D. Ganzer, *Synlett* 15, 2428 (2003).

[6] H. Shirakawa, E. J. Louis, A. G. MacDiarmid, C. K. Chiang and A. J. Heeger, *J. Chem. Soc., Chem. Commun.* 578 (1977).
[7] C. K. Chiang, C. R. Fincher, Y. W. Park, A. J. Heeger, H. Shirakawa, E. J. Louis, S. Gau and A. G. MacDiarmid, *Phys. Rev. Lett.* 39, 1098 (1977).
[8] H. Goto, K. Iino, K. Akagi and H. Shirakawa, *Synth. Met.* 85, 1683 (1997).
[9] F. Monnier and M. Taillefer, *Angew. Chem. Int. Ed.* 48, 6954 (2009).
[10] J. F. Marcoux, S. Doye and S. L. Buchwald, *J. Am. Chem. Soc.* 119, 10539 (1997).
[11] A. Kiyomori, J. F. Marcoux and S. L. Buchwald, *Tetrahedron Lett.* 40, 2657 (1999).
[12] F. Monnier and M. Taillefer, *Angew. Chem. Int. Ed.* 47, 3096 (2008).
[13] M. Laskoski, D. D. Dominguez and T. M. Keller, *J. Polym. Sci. Part A: Polym. Chem.* 44, 4559 (2006).
[14] J. Wang, Y. Gao, A. R. Hlil and A. S. Hay, *Macromolecules* 41, 298 (2008).
[15] M. K. Leung, A. B. Mandal, C. C. Wang, G. H. Lee, S. M. Peng, H. L. Cheng, G. R. Her, I. Chao, H. F. Lu, Y. C. Sun, M. Y. Shiao and P. T. Chou, *J. Am. Chem. Soc.* 124, 4287 (2002).
[16] J. Kuwabara, H. Mori, T. Teratani, M. Akita and T. Kanbara, *Macromol. Rapid Commun.* 30, 997 (2009).
[17] F. Ullmann, *Chem. Ber.* 34, 2174 (1901).
[18] J. Hassan, M. Sevignon, C. Gozzi, E. Schulz and M. Lemaire, *Chem. Rev.* 102, 1359 (2002).
[19] J. L. Bredas, *J. Chem. Phys.* 82, 3808 (1985).
[20] J. Roncali, *Chem. Rev.* 92, 711 (1992).
[21] M. Pomerantz, H. Yang and Y. Cheng, *Macromolecules* 28, 5706 (1995).
[22] S. Thomas, C. Zhang and S. S. Sun, *J. Polym. Sci. Part A: Polym. Chem.* 43, 4280 (2005).
[23] S. Percec, R. Getty. W. Marshall, G. Skidd and R. French, *J. Polym. Sci. Part A: Polym. Chem.* 42, 541 (2004).
[24] L. Groenendaal, H. W. I. Peerlings, J. L. J. van Dongen, E. E. Havinga, J. A. J. M. Vekemans and E. W. Meijer, *Macromolecules* 28, 116 (1995).
[25] T. W. Brockmann and J. M. Tour, *J. Am. Chem. Soc.* 116, 7435 (1994).
[26] K. Tamao, K. Sumitani and M. Kumada, *J. Am. Chem. Soc.* 94, 4374 (1972).
[27] R. J. P. Corriu and J. P. Masse, *J. Chem. Soc., Chem. Commun.* 144 (1972).
[28] M. Zembayashi, K. Tamao, J. Yoshida and M. Kumada, *Tetrahedron Lett.* 18, 4089 (1977).
[29] I. Osaka and R. D. McCullough, *Acc. Chem. Res.* 41, 1202 (2008).
[30] R. D. McCullough and R. D. Lowe, *J. Chem. Soc., Chem. Commun.* 70 (1992).
[31] R. D. McCullough, S. Tristam-Nagle, S. P. Williams and R. D. Lowe, *J. Am. Chem. Soc.* 115, 4910 (1993).
[32] T. A. Chen and R. D. Rieke, *J. Am. Chem. Soc.* 114, 10087 (1992).
[33] R. S. Loewe, S. M. Khersonsky and R. D. McCullough, *Adv. Mater.* 11, 250 (1999).
[34] T. Yamamoto and T. A. Koizumi, *Polymer* 48, 5449 (2007).
[35] A. Kiriy, V. Senkovskyy and M. Sommer, *Macromol. Rapid Commun.* 32, 1503 (2011).
[36] K. Stille, *Angew. Chem. Int. Ed.* 25, 508 (1986).
[37] Z. Bao, W. Chan and L. Yu, *Chem. Mater.* 5, 2 (1993).
[38] A. Iraqui and G. W. Barker, *J. Mater. Chem.* 8, 25 (1998).

[39] H. Burroughes, D. D. C. Bradley, A. R. Brown, R. N. Marks, K. Macay, R. H. Friend, P. L. Burn and A. B. Holmes, *Nature* 347, 539 (1990).
[40] C. Grimsdale, K. L. Chan, R. E. Martin, G. Pawel, P. G. Jokisz and A. B. Holmes, *Chem. Rev.* 109, 897 (2009).
[41] F. Babudri, S. R. Cicco, G. M. Farinola, F. Naso, A. Bolognesi, and W. Porzio, *Macromol. Rapid Commun.* 17, 905 (1996).
[42] B. Carsten, F. He, H. J. Son, T. Xu and L. Yu, *Chem. Rev.* 111, 1493 (2011).
[43] A. Suzuki, *Angew. Chem. Int. Ed.* 50, 6723 (2011).
[44] M. Rehahn, A. D. Schluter, G. Wegner and M. J. Feast, *Polymer* 30, 1060 (1989).
[45] J. Sakamoto, M. Rehahn, G. Wegner and A. D. Schluter, *Macromol. Rapid Commun.* 30, 653 (2009).
[46] Y. Suzuki, K. Kashimoto and K. Tajima, *Macromolecules* 40, 6521 (2007).
[47] K. Tajima, Y. Suzuki and K. Kashimoto, *J. Phys. Chem. C.* 112, 8507 (2008).
[48] J. A. Mikroyannidisa, Q. Dong, B. Xub and W. Tian, *Synth. Met.* 159, 1546 (2009).
[49] J. J. Peterson, M. Willgert, S. Hansson, E. Malmström and K. R. Carter, *J. Polym. Sci. Part A: Polym. Chem.* 49, 3004 (2011).
[50] P. L. Burn, A. Kraft, D. R. Baigent, D. C. Bradley, A. R. Brown, R. H. Friend, R. W. Gymmer, A. B. Holmes and R. W. Jackson, *J. Am. Chem. Soc.* 115, 10117 (1993).
[51] G. S. Padmanaban and S. Ramakrishnan, *J. Am. Chem. Soc.* 122, 2244 (2000).
[52] M. Granström, K. Petritsch, A. C. Arias, A. Lux, M. R. Andersson and R. H. Friend, *Nature* 395, 257 (1998).
[53] S. C. Moratti, R. Cervini, A. B. Holmes, D. R. Baigent, R. H. Friend, N. C. Greenham, J. Grüner and P. J. Hamer, *Synth. Met.* 71, 2117 (1995).
[54] K. B. Wagner, J. T. Patton and J. M. Boncella, *Macromolecules* 25, 3862 (1992).
[55] Y. Yu and M. L. Turner, *Angew. Chem. Int. Ed.* 45, 7797 (2006).
[56] M. Rutenberg, O. A. Scherman, R. H. Grubbs, W. R. Jiang, E. Garfunkel and Z. Bao, *J. Am. Chem. Soc.* 126, 4062 (2004).
[57] G. Tourillonand F. Garnier, *J. Electroanal. Chem.* 135, 173 (1982).
[58] F. Diaz and J. A. Logan, *J. Electroanal. Chem.* 111, 111 (1980).
[59] G. Zotti and G. Schiavon, *J. Electroanal. Chem.* 163, 385 (1984).
[60] G. Schiavon, G. Zotti and G. Bontempelli, *J. Electroanal. Chem.* 161, 323 (1984).
[61] Y. A. Udum, K. Pekmez and A. Yildiz, *Synth. Met.* 142, 7 (2004).
[62] P. Andrieux, P. Audebert, P. Hapiot and J. M. Saveant, *J. Phys. Chem.* 95, 10158 (1991).
[63] T. A. Skotheim and J. R. Reynolds, *Handbook of conducting polymers - Conjugated polymers: theory, synthesis, properties, and characterization.* 3rd Ed., CRC Press, Boca Raton, FL (2007).
[64] R. Sheldon, *Chem. Commun.* 2399 (2001).
[65] W. Lu, A. G. Fadeev, B. H. Qi, E. Smela, B. R. Mattes, J. Ding, G. M. Spinks, J. Mazurkiewicz, D. Z. Zhou, G. G. Wallace, D. R. MacFarlane, S. A. Forsyth and M. Forsyth, *Science* 297, 983 (2002).
[66] W. Lu, A. G. Fadeev, B. H. Qi and B. R. Mattes, *Synth. Met.* 135, 139 (2003).
[67] M. Pringle, J. Efthimiadis, P. C. Howlett, D. R. MacFarlane, A. B. Chaplin, S. B. Hall, D. L. Officer, G. G. Wallace and M. Forsyth, *Polymer* 45, 1447 (2004).

In: Conducting Polymers
Editor: Luiz Carlos Pimentel Almeida

ISBN: 978-1-62618-119-9

Chapter 2

BINARY CONDUCTING COPOLYMERS BASED ON BENZENE, BIPHENYL, THIOPHENE AND PYRROLE

Johannis Simitzis**, *Despina Triantou*† and Spyridon Soulis
National Technical University of Athens, School of Chemical Engineering, Athens, Greece

ABSTRACT

Conducting polymers (CPs) were first produced in the mid-1970s as a novel generation of organic materials with electrical properties similar to those of metals or inorganic semiconductors. They simultaneously exhibit the attractive properties of conventional polymers, such as ease of synthesis and flexibility in processing. CPs can be synthesized by chemical polymerization or by electropolymerization, with both methods having advantages and disadvantages. Chemical polymerization can be scaled up to produce large quantities of a product but generally needs catalysts which are not easily separated. The purification of the polymer consists of a time consuming process. On the other hand, electropolymerization offers several advantageous features, such as absence of catalyst, direct formation of the doped polymer film onto the electrode surface and easy control of the film thickness. Furthermore, it gives the possibility to perform in situ characterization of the deposited film by electrochemical and other techniques. However, the electropolymerization cannot be easily scaled up for mass production.

Some of the most important categories of CPs are the polypyrroles, the polythiophenes and the polyphenylenes. Polythiophenes and polypyrroles, as well as their derivatives based on heterocyclic monomers are stable in air and they don't absorb moisture, both in undoped and doped states. Polyphenylenes have been the subject of extensive research, particularly as active materials for use in light-emitting diodes (LEDs) and polymer lasers. Apart from the conducting homopolymers, copolymers based on different types of monomers have gained great scientific interest, because new electrically active materials can be produced, combining characteristic properties of both of the homopolymers. By electrochemical copolymerization, a variety of conducting materials with different electrical and morphological properties have been produced. By

* Professor of NTUA, e-mail:simj@chemeng.ntua.gr
† Dr. Chemical Engineer, e-mail address: dstrian@gmail.com

copolymerization of biphenyl with 3-octyl-thiophene, soluble copolymers were synthesized. However, compared to thiophene, 3-octyl thiophene is a quite expensive reagent. By electrochemical copolymerization of biphenyl with either thiophene or pyrrole, a series of copolymers with very interesting properties were synthesized. These novel materials have different structure, morphology, thermal stability and electrical conductivity compared to that of the corresponding homopolymers. In below, the synthesis, the structure, the properties and the applications of such copolymers are described.

1. Introduction

Since the discovery of conducting polyacetylene in late 1970s, the focus on the research for similar materials based on π-conjugated structures led to the development of the electrically conducting polymers (CP). The latter, due to their low cost, low density, mechanical flexibility and easy processabilty, are promising candidates for electronic applications, e.g. electromagnetic interference shielding, flexible "plastic" transistors, for electrooptical applications, e.g. electroluminescent polymer displays, polymer light-emitting diodes, photovoltaic solar cells and for electrochemical applications, e.g. rechargeable batteries, corrosion inhibitors, chemical and biochemical sensors [1-6].

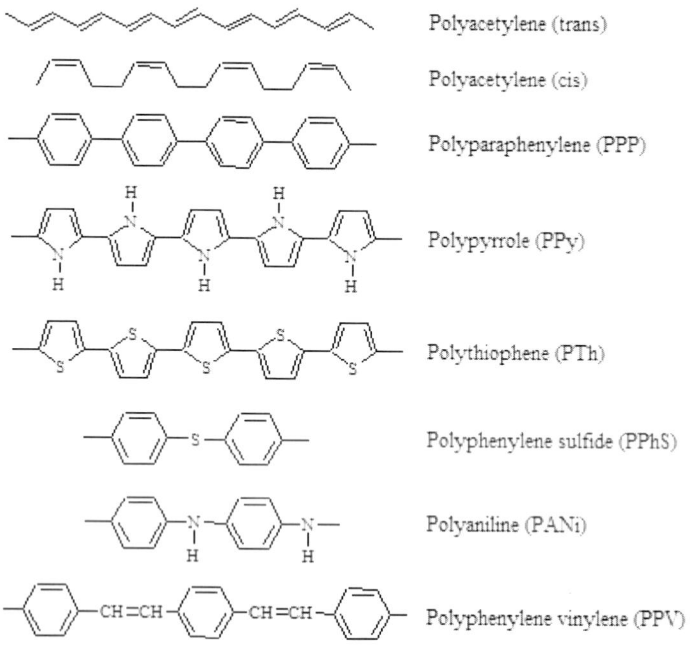

Figure 1. Most known conducting (homo)polymers [4].

The origin of the appealing properties of π-conjugated polymers (Figure 1) that enable their application is related to the extended nature of the electronic wave functions that is created by the alternating single and double bonds of their molecular structure.

This provides the basis for charge transport and gives rise to a range of linear and nonlinear optical properties. Semiconducting polymers offer the promise of achieving a new generation of materials, exhibiting the electrical and optical properties of metals or semiconductors and retaining the attractive mechanical properties and processing advantages of polymers [7].

The use of organic materials may enable electronic devices to be manufactured at a lower cost and to have a larger area of application. The ease of synthesis and purification is another factor that must be considered in regard to the utility of organic semiconductors [1,3]. CPs can be synthesized by chemical polymerization or by electropolymerization. A principal advantage of chemical polymerization compared to the electrochemical one is the possibility of mass production at a reasonable cost. On the other hand, electropolymerization offers several advantageous features, such as absence of catalyst, direct formation of the doped polymer film on the electrode surface, easy control of the film thickness and the possibility of performing in situ characterization of the deposited film by electrochemical and other techniques [6,8].

Polyphenylenes are one of the most important classes of conjugated polymers and have been the subject of extensive research, particularly as active materials for use in light-emitting diodes (LEDs) and polymer lasers. Polyphenylenes can be prepared using benzene or other aromatic compounds, such as biphenyl, leading to poly(p-phenylene)s, PPP, or to isomeric polyphenylenes, PP, respectively. PPP contains only para- couplings and is an infusible and insoluble polymer, however PP contains ortho-, meta- and para- couplings and is more processable than PPP [6,8,9].

Polythiophene (PTh) and its derivatives have been widely studied during the past years, owing to their large application possibilities, including sensors, photovoltaic cells, transistor, electroluminescent diodes, etc. PThs have shown considerable promise for materials applications, due to exceptional electrical properties and their environmental stability. However, the high oxidation potential of thiophene monomer compared to the polymer causes some degradation of the polymeric film, which has been the subject of a wide variety of studies [2,4,10-13].

Polypyrrole (PPy) and its derivatives are counted among the most stable CPs and are of particular interest since their films can be formed from aqueous and non-aqueous solutions at relatively low oxidation potentials of monomers [11,14]. However, PPy's are brittle, insoluble and infusible due to the delocalized π- bonds along the backbone which leads to poor processibility characteristics. As a result, intensive research is performed to improve processibility, such as modification of the monomer, utilization of a soluble precursor and formation of a blend or composite. Both chemical and electrochemical methods are mostly used to prepare conducting composite films [15].

By its very nature, copolymerization offers the unique capability or opportunity for chemists to design and construct molecules with special electronic properties using established techniques. Copolymerization enables the molecular architect or tailor to incorporate various molecules, e.g. biological components, into electroactive PTh and PPy polymers. The number of potential applications for these unique materials is unlimited [16]. One important aspect of conducting polymer technology is the modification and control of their electrical, optical and mechanical properties for specific technological applications [11], as well as to enhance the processability (e.g. increase the solubility) [16] and the stability in air [13].

Table 1. Monomers combined as binary systems to synthesize corresponding copolymers

	Benzene	Thiophene	Pyrrole
Biphenyl	+		
Biphenyl		+	
Biphenyl			+
Benzene		-	
Benzene			-
Thiophene			+

+ or - : described or not in the literature, respectively.

In the field of material design, copolymerization is a powerful approach for modification of the backbone structure and properties of electrically conducting polymers [17]. Copolymers based on different types of monomers have gained great scientific interest, because tailored conducting materials can be produced with the desired physical, chemical, and conduction properties. By copolymerizing, a variety of conducting materials with different electrical and morphological properties can be produced [8,14,16,18-20]. Table 1 presents the monomers biphenyl, benzene, thiophene and pyrrole combined as binary systems to synthesize corresponding copolymers.

In the literature, there are no reports about the production of copolymers by combination of benzene with thiophene or pyrrole. This is not surprising, since benzene is considered as very stable. As a matter of fact, in organic chemistry benzene is mostly known as a non-reactive solvent and less as a reagent. For example, owing to high oxidation potential of benzene, its electropolymerization in solution would cause either solvent breakdown or reactions of the very energetic cations with the solvent, so it is avoided as monomer. Therefore, in order to import aryl rings in a structure, biphenyl is preferred as monomer for the copolymerization reactions [21].

2. Synthesis of Copolymers

Two different methods can be used to prepare copolymers: (A) direct copolymerization of mixtures of specific monomers and (B) copolymerization through the use of polycyclic starting materials [4,17]. The first method can be applied either through electropolymerization or by chemical methods, opposite to the second that is mostly applied through electropolymerization. In this chapter, the first approach will be mainly analyzed.

2.1. Direct Copolymerization of Mixture of Two Different Monomers

Chemical polymerization methods are catalytic oxidation reactions, such as direct oxidation of monomers with a suitable catalyst-oxidant system, or metal-catalyzed coupling reactions (Grignard, Ullmann, etc) [21,22]. Figure 2 presents the oxidative cationic polymerization of benzene to produce polyparaphenylene (PPP) [21] and Figure 3 the corresponding polymerization of biphenyl that leads to isomeric polyphenylenes PP with ortho-, meta- and para-linkages [22, 23]. The polymers prepared by chemical synthesis are semiconductors (tending to insulator) or insulators, o they need a second processing step, i.e. doping, in order to become semiconductors or conductors. The doping of the conjugated polymers leads to high conductivity, mainly by increasing the concentration of the charge carries. The chemical reaction is an oxidation or reduction by using electron acceptors (e.g. Lewis acids) or electron donors (e.g. alkalimetals), respectively. The main difference between the addition of the doping agent (dopant) in a typical semiconductor, like silicone, and in a conjugated (organic) polymer refers the substitution of some silicone atoms into the lattice by the atoms of the dopant (having higher or lower valence electrons than four, the electron valence of silicone atom), whereas the dopant in a conjugated polymer does not participate in the structure of the macromolecule, but it acts as oxidative or reductive agent withdrawing or offering electrons in the macrolmolecule, respectively.

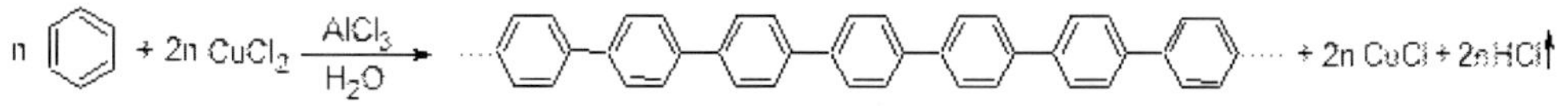

Figure 2. Oxidative cationic polymerization of benzene to produce polyparaphenylene PPP [21].

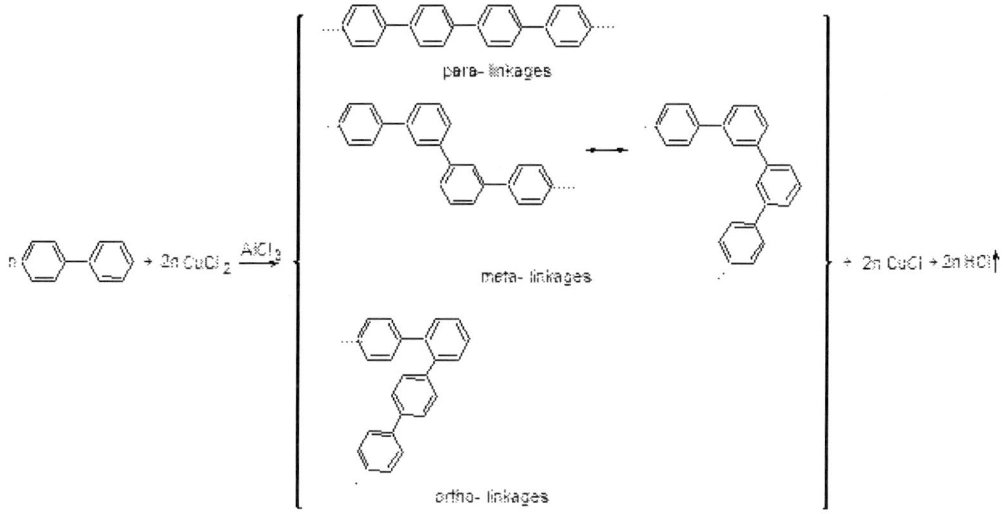

Figure 3. Oxidative cationic polymerization of biphenyl to produce isomeric polyphenylenes PP [22,23].

Figure 4. Formation of a structural defect, a radical cation, in a macromolecule of a polyparaphenylene by doping with an electron acceptor [6,21].

The oxidation of a macromolecule corresponds with p-type doping and its reduction with n-type doping. For example, a conjugated polymer is oxidized by withdrawing an electron (after breaking of a double bond, i.e. dissociation of the two electrons of a π-bond) forming a radical cation (a cation represents a hole) into its macromolecule and this electron is accepted from the acceptor dopant. By rearragment of the remaing double bonds in the macromolecule a quinoid bond is also created. The radical cation is named polaron by the solid state terminology and presents an energetic level in the gap between LUMO and HOMO (see below at 3. cyclic voltammetry) of the doped polymer (energetics : $E_{aromatic} < E_{quinoid}$). Figure 4 presents the formation of a structural defect, a radical cation, in a macromolecule of a polyparaphenylene by doping them with an electron acceptor [6,21,24,25].

Electrochemical polymerization is a useful and widely applied technique for synthesizing conducting polymer (CP) films. The CP films prepared are usually formed in a doped state and used without further treatment. The doping levels of CP films and their properties depend strongly on the experimental conditions such as applied potential, electrolyte, supporting salt, and monomer structure [26]. The electrochemical polymerization is performed using a three-electrode configuration (working, counter, and reference electrodes) in a solution of the monomer, appropriate solvent and electrolyte (dopant). Current is passed through the solution and electrodeposition occurs at the working electrode (or anode, oxidative electropolymerization). Figure 5 presents the basic reactions taking place during electropolymerization of thiophene [8]. This mechanism is similar to that of the other aromatic monomers (e.g. biphenyl, pyrrole). Monomers at the working electrode surface undergo oxidation to form radical cations (Figure 5a) that react with other radical cations (Figure 5b), forming a dihydrodimer dication.

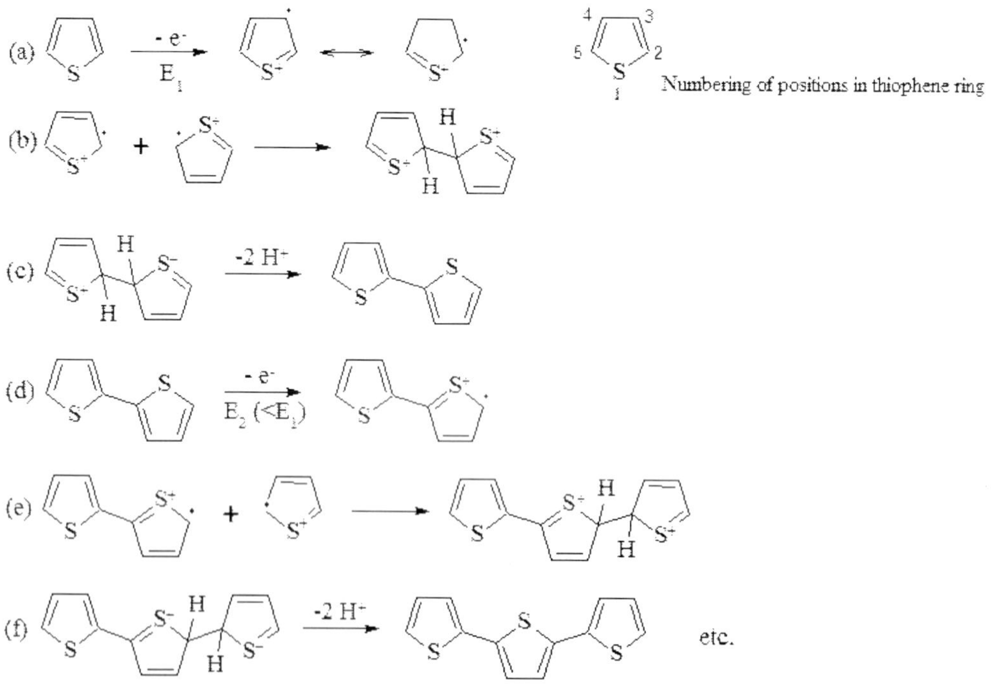

Figure 5. Mechanism of electropolymerization of thiophene [8].

The latter, after losing two protons and re-aromatizing, leads to the dimer (Figure 5c). Due to the potential applied the dimer, which is more easily oxidized than the monomer, reacts to form its radical-cation (Figure 5d) and the latter is coupled with the radical-cation of the monomer (Figure 5e). Then, it loses two protons and re-aromatizes, this way forming the trimer (Figure 5f) [8]. With the same reactions, the polymer chain keeps on growing. Using halogenated monomers the film is deposited on the negatively charged counter electrode (or cathode, reductive electropolymerization) [22,27]. The properties of the copolymer depends on the molar ratio of the monomers, and can be altered by other experimental conditions such as potential applied, scan rate, pH, etc., since generally the electrooxidation of one of the comonomers is much faster than that of the other one [4].

2.1.1. Copolymers of Biphenyl with Benzene

The copolymerization of aryl derivatives through oxidative cationic polymerization was investigated in the early 1970s from Bilow and co-workers. The copolymers produced were studied under the aspect of applying polyphenylenes as ablative polymers suitable for use in erosive hypethermal environments, although their potential application as conductive materials has not been investigated thereafter [6,28,29]. Recently, the copolymerization of benzene with biphenyl has been reported using oxidative cationic polymerization based on Kovacic method. Given that, the polymerization procedure of benzene alone differs significantly from that of biphenyl (polymerization temperature, etc), a series of preliminary experiments were carried out in order to define the copolymerization conditions. Copper chloride, $CuCl_2$, was used as the oxidizer and aluminum chloride, $AlCl_3$, as catalyst [6]. The copolymers were purified from oligophenylenes or other mixtures by extraction with n-hexane. Then, the copolymers were separated in the insoluble and the soluble in chlorobenzene polyphenylenes by extraction with chlorobenzene. The copolymers were doped using a solution of anhydrous ferric chloride, $FeCl_3$, in acetonitrile at room temperature [6].

2.1.2. Copolymer of Biphenyl with Thiophene

The copolymerization of biphenyl with thiophene [8] or 3-octyl-thiophene [19,20,30,31] has been reported. The monomers can be electropolymerized under potentiostatic [8] or galvanostatic [19,20,30] conditions or by cyclic voltammetry [20,31]. The electropolymerization solution consisted of the monomers, tetrabutylammonium tetrafluoroborate ($TBABF_4$), or lithium hexafluoroarsenate ($LiAsF_6$) or lithium tetrafluoroborate ($LiBF_4$) or lithium perchlorate ($LiClO_4$) as supporting electrolyte and acetonitrile or propylene carbonate as solvent. All electropolymerizations took place at room temperature, in one-compartment electrochemical cell. The working electrode was platinum, the counter electrode was platinum or glassy carbon and the reference electrode was calomel/SCE or Ag/AgCl. The reference electrode was placed in a Luggin capillary [8,19,20,30]. The surface of the working electrode should be polished mechanically with alumina powder or by heating over a flame in order to remove residues before each polymerization. The electrolyte was dried at 100 ^{o}C before use and the solvent was stored over molecular sieves. The electropolymerization solutions were deaerated with nitrogen, prior to the electropolymerization, for 10 min [8,19,20,30]. In one case, a spectroelectrochemical cell was used for in situ resonance Raman spectroscopy analysis of the biphenyl and 3-octyl-thiophene copolymers [31]. After polymerization, the films synthesized were immersed in acetonitrile or propylene carbonate to remove electrolyte residues and the soluble oligomers and then were vacuum dried up to constant weight. The thickness of the films synthesized was estimated from the amount of charge during the electropolymerization [8,20,30].

The chemical synthesis of functional biphenyl derivatives containing thiophene groups have been reported using a two-step procedure [32]. The palladium-catalysed coupling reactions it is well known for the preparation of new compounds, given that the aryl C–C bond construction can be induced either by palladium catalysts or nickel catalysts. The first step is the synthesis of the compound (Z)-2-(4-bromophenyl)-3-(thiophen-2-yl) acrylonitrile using CH_3ONa. In the second step, various substituted arylboronic acids were reacted with the afore mentioned compound by a Suzuki reaction in the presence of the catalyst $Pd(PPh_3)_4$ and a series of novel biphenyl derivatives containing thiophene rings were produced [32].

2.1.3. Copolymer of Biphenyl with Pyrrole

The copolymerization of biphenyl with pyrrole has been reported under potentiostatic polymerization [14]. The electropolymerization solution consisted of the monomers, $TBABF_4$ and acetonitrile. The electropolymerizations took place at room temperature, in one-compartment electrochemical cell. The latter includes the working and the counter electrodes (both of platinum) and the reference electrode (calomel/SCE). The thickness of the films synthesized was estimated from the amount of charge during the electropolymerization [14].

2.1.4. Copolymer of Thiophene with Pyrrole

The copolymerization of pyrrole with thiophene [16,18,33,34] or with other thiophene derivatives [11,16,35,36] has been reported. The reactions were carried out by electropolymerization [11,16,34-36] or by double discharge technique (i.e. plasma polymerization) [18]. Moreover, the template electropolymerization by potentiodynamic conditions has also been used using anodic aluminum oxide (AAO) membrane, in order to synthesize a variety of nanofibrils [33].

2.1.5. Other Thiophene Based Copolymers

In addition to functionalization of the thiophene ring, another concept extensively applied to modify polythiophene properties is the copolymerization process with different alkyl substituted thiophene. The formation of a copolymer has been a common approach for materials with tailored properties [10]. The copolymerization of thiophene with 3-methyl thiophene [16,37] or 3-hexyl thiophene [16,38] have been reported, as well as the copolymerization between various 3-alkyl substituted thiophene derivatives [10,39,40]. These copolymers can be produced by electropolymerization or by chemical polymerization, mainly through Grignard metathesis (GRIM) polymerization which uses dibromosubstituted monomers [39]. Furhermore, the electropolymerization of ethylene dioxythiophene (EDOT) with 1-(phenyl)-2,5-di(2-thienyl)-1H-pyrrole has been reported [41]. Moreover, copolymers containing thiophenebenzothiadiazole-thiophene units, such as poly[1-(2,6-diisopropylphenyl)-2,5-bis(2-thienyl) pyrrole-alt-4,7-bis(3-octyl-2-thienyl)benzothiadiazole] and poly[1-(p-octylphenyl)-2,5-bis(2-thienyl)pyrrole-alt-4,7-bis(3-octyl-2-thienyl) benzothia-diazole] were prepared via Suzuki polycondensation using appropriate thiophene based monomers [42].

2.2. Synthesis of Copolymers Using a Monomer Containing Units from Different Monomers

The monomers have complicated structures, comprising of various aromatic rings (phenyl, thiophene, pyrrole, carbazole, etc) and substitutes. These are produced through reactions of organic synthesis and then they are electrochemically polymerized. This way, a range of phenyl and thiophene containing monomers, such as 2-biphenyl- 3-octylthiophene (BOT) [17,43], 1,4-di(2-thienyl)benzene [16,44-46] has been polymerized and the corresponding polymers were extensively studied [16,47,48]. Simirarly, polymers containing thiophene and pyrrole units [16,49-52] or thiophene and carbazole units [53,54] have been also produced. In Table 2, the monomers and the polymerization conditions used for the copolymer synthesis are summarized.

Table 2. Monomers and the polymerization conditions used for the copolymer synthesis

A. Direct copolymerization of mixture of two different monomers			
Monomers		Polymerization	References
1	2		
Benzene	Biphenyl	Oxidative cationic Oxidant : $CuCl_2$ Catalyst : $AlCl_3$ 55 °C	6
Biphenyl	Thiophene	Potensiostatic electropolymerization Solvent : Acetonitrile Supporting electrolyte : $TBABF_4$ Room temperature	8
Biphenyl	3-octyl thiophene	Electropolymerization by cyclic voltammetry or galvanostatic electropolymerization Solvent : Acetonitrile or propylene carbonate Supporting electrolyte : $TBABF_4$, $LiAsF_6$, $LiBF_4$, $LiClO_4$ Room temperature	19,20,30,31
2,(4-bromo) biphenyl	2-methoxy thiophene	Chemical polymerization by Suzuki reaction catalyst $Pd(PPh_3)_4$	32
Biphenyl	Pyrrole	Potensiostatic electropolymerization Solvent : Acetonitrile Supporting electrolyte : $TBABF_4$ Room temperature	14
Thiophene	Pyrrole	Electropolymerization Solvent : Acetonitrile Supporting electrolyte : $LiClO_4$ Room temperature	16,34
Thiophene	Pyrrole	Template electropolymerization by potentiodynamic conditions Template : Anodic aluminum oxide (AAO) membrane Solvent : Acetonitrile Supporting electrolyte : Et_4NClO_4 Room temperature	33
Thiophene	Pyrrole	Plasma polymerization Double discharge technique	18

Table 2. (Continued)

A. Direct copolymerization of mixture of two different monomers			
Monomers		Polymerization	References
Bithiophene	Pyrrole	Potentiostatic electropolymerization or by cyclic voltammetry Solvent : Acetonitrile, propylene carbonate Supporting electrolyte : $LiClO_4$, Et_4NBF_4	11,16,36,
Terthiophene	Pyrrole	Potensiostatic electropolymerization Solvent : Acetonitrile Supporting electrolyte : Et_4NBF_4 or Et_4NClO_4 Room temperature	16,35
Thiophene	3-Methyl thiophene	Chemical polymerization	16, 37
Thiophene	3-Hexyl thiophene		16, 38
3-hexyl thiophene	3-ethylhexyl-thiophene	Grignard metathesis (GRIM) polymerization	39
3-methyl thiophene	Methylthiophene -3-acetate	Electropolymerization by cyclic voltammetry	10
3-butyl thiophene	Methylthiophene -3-acetate	Electropolymerization by cyclic voltammetry	10
3-hexyl thiophene	3-phenoxymethylthiophene	Grignard metathesis (GRIM) polymerization	39
3-hexyl thiophene	3-triphenylamine-thiophenes	Grignard metathesis (GRIM) polymerization	39
3-hexyl-thiophene	ethylene dioxythiophene (EDOT)	Grignard metathesis (GRIM) polymerization	39
Thiophene	3-(5'-octylthienylenevinyl) thiophene]	Chemical polymerization Catalyst : $Pd(PPh_3)_4$ Solvent : Toluene	40
ethylene dioxythiophene (EDOT)	1-(phenyl)-2,5-di(2-thienyl)-1H-pyrrole	Electropolymerization by cyclic voltammetry Solvent : Acetonitrile Supporting electrolyte : $NaClO_4/LiClO_4$ Room temperature	41
poly[1-(2,6-diisopropylphenyl)-2,5-bis(2-thienyl) pyrrole-alt-4,7-bis(3-octyl-2-thienyl)benzothiadiazole] and poly[1-(p-octylphenyl)-2,5- bis(2-thienyl)pyrrole-alt-4,7-bis(3-octyl-2-thienyl)benzothiadiazole]		Suzuki coupling	42
B. Synthesis of copolymers using a monomer containing units from different monomers			
2-biphenyl-3-octylthiophene		Electropolymerization by cyclic voltammetry Solvent : Acetonitrile Supporting electrolyte:$LiAsF_6$ Room temperature	17, 43
1,4-di-(2-thienyl)benzene		Electropolymerization	16, 44-46
3-octyl-2,2′-bithiophene		Electropolymerization	16,47,48
3′-octyl-2,2′;5′,2′′terthiophene		Electropolymerization	16,47,48
N-methyl-2,5-di(2-thienyl)-pyrrole		Electropolymerization	16, 51,52
2-(2 thienyl)-1H pyrrole		Electropolymerization by cyclic voltammetry Solvent : Acetonitrile Supporting electrolyte:$LiClO_4$ Room temperature	50

B. Synthesis of copolymers using a monomer containing units from different monomers		
Monomers	Polymerization	References
N,N′-bis(2-pyrrolylmethylene)-3,4-dicyano-2,5-diaminothiophene Py2ThAz	Electropolymerization by cyclic voltammetry Solvent : Acetonitrile Supporting electrolyte:$TBABF_4$, $TBAClO_4$,$LiClO_4$,$NaClO_4$ Room temperature	49
3,6 bis(3,4-ethylenedioxythiophenyl)-9- ethylcarbazole	Potentiostatic electropolymerization Solvent : Propylene carbonate Supporting electrolyte:$LiClO_4$ Room temperature	54
triazole–thiophenes (various monomers)	Electropolymerization Solvent : Acetonitrile or dichloromethane Supporting electrolyte:$LiClO_4$	53

$CuCl_2$: Copper chloride, $AlCl_3$: Aluminium chloride, $TBABF_4$ (or Et_4NBF_4): Tetrabutylammonium tetrafluoroborate, $LiAsF_6$: Lithium hexaflouroarsenate, $LiBF_4$: Lithium tetrafluoroborate, $LiClO_4$: Lithium perchlorate, $Pd(PPh_3)_4$: Palladium-tetrakis(triphenylphosphine), Et_4NClO_4 : Tetraethylammonium perchlorate, $NaClO_4$: Sodium perchlorate.

3. Structure of Copolymers

The characterization of copolymers is important not only from the point of view of their properties but also it is important to clarify that they are real copolymers and not a mixture of two homopolymers [19]. Conducting polymers have been studied using the whole arsenal of methods available to chemists and physicists. Apart from the classical analytical methods, such as FTIR, UV-Vis, etc, electrochemical techniques (e.g. cyclic voltammetry CV), are used to follow the formation and deposition of polymers, as well as the kinetics of their charge transport processes. The application of combinations of electrochemical methods with non-electrochemical techniques, especially spectroelectrochemistry, has enhanced our understanding of the nature of charge transport and charge transfer processes, structure–property relationships, and the mechanisms of chemical transformations that occur during charging/ discharging processes [4].

3.1. Fourier Transform Infrared Spectroscopy (FTIR)

The FTIR spectra can be obtained for copolymers synthesized in the shape of powder (mixed with KBr and formed as discs) or film (as received or after being powdered, mixed with KBr and formed as disks). The various bands of the copolymers are attributed to chemical bonds according to the literature for the corresponding homopolymers. The results of the FTIR spectra of polyphenylene (PP) [6,8,14,19,20,30,31], polythiophene (PTh) [8,11,14,19, 20,30,31,55] and polypyrrole (PPy) [11,14,33,55] are summarized in Table 3.

The broad absorption band at 1600 cm^{-1} is particularly strong if a further conjugation with aromatic rings takes place. The intensity of this band characterizes the degree of condensed aromatic rings in the case of PP [6]. Based on the literature this band at 1600-1570 cm^{-1} is attributed to quinoid structures and the band at 1480 cm^{-1} to benzenoid structures [8,14].

Table 3. Description of FTIR bands assigned to PP, PTh and PPy

Wavenumber (cm^{-1})	Characteristic bonds		
	Polyphenylenes (PP)	Polythiophene (PTh)	Polypyrrole (PPy)
3400			N-H bending vibrations
3030	Aromatic stretching vibrations (C-H)		Aromatic stretching vibrations (C-H)
2925,2850	CH_2 :aliphatic (cyclic and linear) parts		
1740			C-O substitution and specifically to vibration of C=O of pyrrolidinone
1600-1570	Stretching vibrations C=C of aromatic ring		
1514		Stretching modes from the thiophene rings	
1480	C-C bending vibrations in aromatic ring		C-C bending vibrations in aromatic ring/ C-N vibrations
1400			C-C bending vibrations
1440	CH_2: aliphatic (cyclic and linear parts	Stretching modes from the thiophene rings	
1260		conjugated backbone of PTh	Characteristic band of stretching vibration of C-N
1220, 1140			
1110–1150	C-H "in plane" bending vibrations characteristic of para- substitution		C–H stretching of N-substituted pyrroles
Or			
1000–1094		C-H bending vibrations	C–H stretching of N-substituted pyrroles /N-H bending vibrations
876	C-H "out of plane"-bending vibrations characteristic of separated H in aromatic ring		C-H "out of plane"-bending vibrations
820		C-H out-of-plane band from three-substituted thiophene rings	
806	C-H "out of plane"-bending vibrations of two neighboring H in aromatic ring	two-substituted (terminal) thiophene rings	C-H "out of plane"-bending vibrations
765	C-H bending vibrations of 4 or 5 neighboring H in aromatic ring	conjugated backbone of PTh	
740	indicates ortho- substitution		C–H stretching of N-substituted pyrroles
722	CH_2: aliphatic (cyclic or linear) parts		C-H "out of plane"-bending vibrations
695	"ring puckering" in aromatic ring; observed in mono, 1,3-,1,3,5-and 1,2,3-substituted phenylene		
680		C–S–C ring deformations	
610		C-S bending vibrations	

The type of substitution in PP can be obtained from the intense bands below 900 cm^{-1}. The band at ~ 806 cm^{-1} is characteristic of the para-substitution and the bands at 765 cm^{-1} and

695 cm^{-1} are characteristic of the aromatic rings which are at the end of the macromolecules, i.e. (mono) end-groups. In the case of the presence of the bands at 880-870 cm^{-1}, the stronger bands at 765 cm^{-1} and 695 cm^{-1} can be attributed to meta-substitution. The presence of a small band at 740 cm^{-1} indicates ortho-substitution [6,8]. Copolymers made of monomer units containing both benzene/biphenyl and thiophene, poly(4,4'- di(2-thienyl)biphenyl) and poly(1,4-di(2-thienyl)benzene), show bands at 795 cm^{-1} from the 2,5-disubstituted thiophene ring and at 820 cm^{-1} from 1,4-disubstituted benzene. In addition, the spectrum of this copolymer also contains the same characteristic bands that can be found in spectra of pure PPP and POT films [19].

Usually, the FTIR spectra of copolymers compared to those of the homopolymers have a differnet shape, mainly due to shifting of the absorption bands and more rarely due to the appearance of new bands.

3.2. Nuclear Magnetic Resonance (NMR)

Copolymers of EDOT and 3HT were synthesized by ''quasiliving''GRIM polymerization. The analysis of 1H NMR spectra allows to evaluate the ratio of the monomers inside the copolymers and the degree of regioregularity. 1H NMR spectra of P3HT-co-EDOT show one multiplet at 4.6 ppm, similar to that of the PEDOT homopolymer, and another singlet at 2.79 ppm corresponding to the a-methylene protons of regioregular head-to-tail (H-T) linked poly(3-hexyl-thiophene). 1H NMR highlights the presence of highly regioregular H–T linkages of the thiophene units in P3HT-co-EDOT. On the basis of the relative areas of the peaks at 4.6 and 2.79 ppm, the molar ratio of 3HT to EDOT units in P3HT-co-EDOT was estimated to 2.5 units of 3HT for each unit of EDOT (2.5:1). This ratio is very close to the monomer feed ratio (2:1) used for the copolymerization. The high regioregularity indicates that 3HT forms blocks of several units and EDOT units (or blocks) are interdispersed among them; this finding, together with the evidence that the final composition is very close to the feed ratio of the comonomers, are compatible with the hypothesis that comonomers form blocks within the polymer chains. Another result is that 3HT and EDOT have similar reactivities (as they are present in the polymer with a ratio close to the feed ratio) [39]. This is in contrast with previous report [56], where the presence of electron-rich monomers caused the deactivation of the chain growth. On the other side, regioregular copolymers of 3-alkoxythiophene with a 3-alkyl-thiophene, were prepared by GRIM polymerization, but the content (1–2.5 of 3-alkylthiophene) of the electron-rich monomer (3-alkoxythiophene) is lower than the feed ratio (1:1) [39]. The ^{13}C NMR spectra of 4,4′-bis(3-octylthienyl)biphenyl and 4,4′-bis(thienyl)biphenyl have been measured in solution or in solid state (CP/MAS technique) and the aromatic carbon atoms (which were observed in the region of 120 up to 140 ppm) did not show any shift between the two methods [57].

3.3. X-Ray Diffraction (XRD)

The XRD can be obtained for conducting copolymers synthesized in the shape of powder or film. The physical properties of conducting polymers are dependent on the crystalline order of the material. High conductivity upon doping is observed for organized and crystalline materials [17]. The copolymers based on benzene and biphenyl (both insoluble and soluble) show crystallinity and they have three main reflections, i.e. in the regions 19.1- 20.4^o, 21.0-22.8^o and 26.3 - 28.0^o [6]. These reflections are indexed in order to define the appropriate crystal system and the unit cell dimensions. The crystal system of the insoluble copolymers is the orthorhombic, whereas that of the soluble ones is the monoclinic. The crystal system of doped copolymers is the same as that of the corresponding undoped copolymers; however their unit cell dimensions are slightly increased. It is worth noticing the significant crystallinity of the doped (with $FeCl_3$) copolymers, contrary to the lack of crystallinity of the doped polyphenylene homopolymers [6].

The crystallinity of the copolymer films based on biphenyl and thiophene, (PP-PTh) [8] or biphenyl and pyrrole, (PP-PPy) [14] synthesized by potentiostatic electropolymerization depends on the applied potential used. Generally, the increase of the applied potential of the electropolymerization, for (PP-PTh) : from 1.80 V up to 1.84 V vs SCE and for (PP-PPy) : from 0.80 V up to 0.86 V vs SCE, leads to the production of less crystalline or amorphous copolymers [8,14]. Moreover, the soluble fraction of the copolymers follows the reverse order than that of crystallinity (i.e. the insoluble fraction is in accordance with the order of the degree of crystallinity) following the general relation between crystallinity and solubility [8,14].

3.4. Scanning Electron Microscopy (SEM)

Macroscopically the structure of conducting copolymers depends on its morphology and shape (e.g. film, powder). The SEM can be obtained for polymers synthesized in the shape of powder or film. Generally, the copolymers have very different morphology compared to the corresponding homopolymers [6,8,11,14,41,50].

The insoluble copolymers based on benzene and biphenyl shows a highly ordered structure with crystalline regions [6]. The copolymers (PP-PPy) have morphology globular with medium size or fibrillar morphology with many round shaped pores, depending on the applied potential used for their synthesis. It can be deduced that the electropolymerization potential strongly affects the morphology of the copolymer films. Moreover, the type of monomers used for the copolymers, influence the morphology of the films. For example, by replacing the pyrrole with the thiophene, the copolymers (PP-PTh) have a totally different morphology, with round shaped cracks or with large size aggregates of cauliflower shape, opposite to (PP-PPy) that have globular or fibrillar morphology [8,14]. The copolymer based on ethylene dioxythiophene (EDOT) with 1-(phenyl)-2,5-di(2-thienyl)-1H-pyrrole exhibits homogenous and compact structure [41].

3.5. Cyclic Voltammetry (CV)

Cyclic voltammetry provides basic information on the oxidation potential of the monomers, on film growth, on the redox behavior of the polymer, and on the surface concentration (charge consumed by the polymer formed). Conclusions can also be drawn from the cyclic voltammograms regarding the rate of charge transfer, charge transport processes, and the interactions that occur within the polymer segments, at specific sites and between the polymer and the ions and solvent molecules [4]. The redox behavior of the aromatic based copolymers by CV has been extensively studied in the literature [19,20,30]. Moreover, the energy levels of the Highest Occupied Molecular Orbital (E_{HOMO}) and that of Lowest Unoccupied Molecular Orbital (E_{LUMO}), as well as the energy gap (E_g), can be also determined [8,11,14,58]. The copolymer film (as deposited film on the electrode) was placed in the solution containg electrolyte and solvent and underwent cyclic potential sweep, firstly from 0 to –2 V and then from 0 to +2 V (scan rate : 100 mV/s) [8,14]. Figure 6 shows the cyclic voltammogram of the copolymer film $(PP\text{-}PTh)_{1.82}$ synthesized synthesized by applying 1.82 V vs SCE for 30 min, The onset potentials of oxidation and reduction for the p-doping (E_{onset}^{ox}) and n-doping (E_{onset}^{red}), respectively, were determined graphically. Then, from the equations de Leeuw et al [59], E_{HOMO}, E_{LUMO} and E_g were calculated.

$$E_{HOMO} = -e(E_{onset}^{ox} + 4.4)$$

$$E_{LUMO} = -e(E_{onset}^{red} + 4.4)$$

$$E_g = e(E_{onset}^{ox} - E_{onset}^{red})$$

where

E_{onset}^{ox}: onset potential for p-doping,vs SCE

E_{onset}^{red} : onset potential for n-doping,vs SCE

E_{HOMO} : energy level of highest occupied molecular orbital (HOMO), i.e. of valence band

E_{LUMO} : energy level of lowest unoccupied molecular orbital (LUMO), i.e. of conduction band

E_g : energy gap

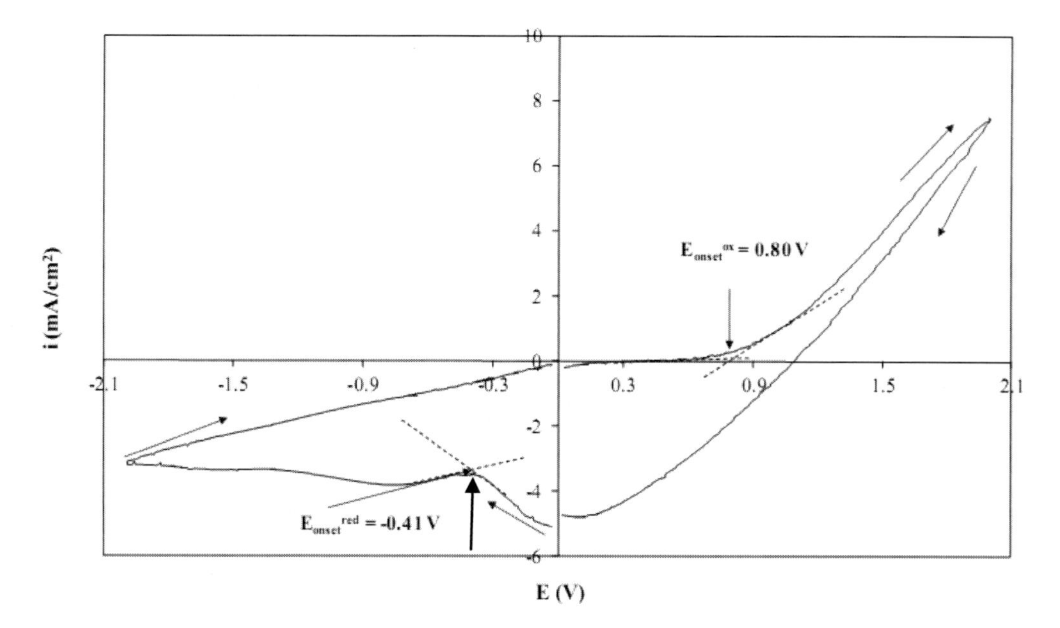

Figure 6. Cyclic voltammogram of the copolymer film based on biphenyl with thiophene (PP-PTh), synthesized by applying 1.82 V vs SCE for 30 min, in a solution of 0.1 M TBABF4 in ACN (scan rate=100 mV/s).

The E_g of the copolymers (PP-PTh) and (PP-PPy) varies from 1.19 up to 1.40 eV, whereas that of the copolymers based on pyrrole and 2,2-bithiophene are approx. 1.94 eV [11]. i.e. the values are in the range of semiconductors. E_g is reversely proportional to the electrical conductivity, thus the polymers with the higher conductivity had also the lower E_g [8,14]. Comparing the copolymers (PP-PPy) with the copolymers based on biphenyl with thiophene (PP-PTh), they have different HOMO and LUMO energy levels and, more specifically, the (PP-PPy) has lower energy levels (HOMO and LUMO) than the (PP-PTh). This difference means that the copolymers can be used in different electrooptical applications. For example, the efficiency of an organic solar cell is defined by the function of the p-n diode, which depends on the energy levels of p- and n-type semiconductor materials. Thus, both kinds of copolymers (which are p-type semiconductors) can be combined with corresponding proper n- type semiconductors [14].

3.6. Spectroelectrochemistry

Spectroelectrochemical experiments were performed in order to check the applicability of these polymers in future electrochromic devices. Several interband electronic transitions induced by the potential doping were observed for all the polymers, corresponding to the presence of polaronic species absorbing in the visible to near-infrared range. Electrochemically induced absorption bands that appear during both p- and n-doping of copolymers make them material attractive for the design of high-contrast dual dopable electrochromic devices [53]. Spectroelectrochemistry experiments were done to examine the change in optical properties which result from the redox switching of conjugated polymers [41,50,53,55].

In situ spectroelectrochemical studies of conducting polymers are important when trying to understand the basic electronic structure of the material. Spectroscopy is one of the most powerful methods to characterize the electronic processes taking place in a conducting polymer film. Changes in the spectra upon doping are very significant in elucidating the mechanism of structural changes taking place in the polymer chain. The most widely used spectroelectrochemical technique which yields information on electrochromic properties of conductive polymers is in situ UV–vis–NIR spectroelectrochemistry [19].

The copolymer based on biphenyl and 3-octyl thiophene showed a red shift in the absorbance maxima. The absorbance maxima of the bilayer films of pure homopolymers are at a lower wavelength than the poly(thienyl biphenyl) film indicating that the structure of the copolymer is not a mixture of pure PPP and POT [19]. In its neutral state the copolymer based on biphenyl and 3-octyl thiophene shows an absorption maximum at 465 nm. Electrochemical doping results in bleaching of the π–π^* transition with a simultaneous growth of two peaks at about 800 and > 1100 nm. Upon bleaching of the π–π^* transition an absorbance maximum at 340 nm becomes visible. The absorbance at 800 nm first grows quickly at the beginning of the doping process and reaches its maximum at 1034 mV. After this potential the absorbance starts to decrease while the absorbance at >1100 nm starts to increase continuously with increased applied potential [19]. With this method the band gap (E_g) and the intergap states that appear upon doping were determined [41].

Table 4. Methods for the characterization of the copolymers structure

A. Direct copolymerization of mixture of two different monomers		Characterizations						
Monomers		FTIR	NMR	XRD	SEM	CV	Spectroelectro-chemistry	References
1	2							
Benzene	Biphenyl	+		+	+	+		6
Biphenyl	Thiophene	+		+	+	+		8
Biphenyl	3-octyl thiophene	+			+	+		20
		+				+	+	19
		+				+		30
		+				+	+	31
Biphenyl	Pyrrole	+		+	+	+		14
Thiophene	Pyrrole	+						33
Bithiophene	Pyrrole	+				+		16,36
		+				+		11
3-hexyl-thiophene	ethylene dioxythiophene (EDOT)		+					39
Ethylene dioxythiophene (EDOT)	1-(phenyl)-2,5-di(2-thienyl)-1H-pyrrole				+		+	41

B. Synthesis of copolymers using a monomer contain units from different monomers	Characterizations						
Monomer	FTIR	NMR	XRD	SEM	CV	Spectroelectro-chemistry	References
2-biphenyl-3-octylthiophene		+	+		+		17
	+					+	43
	+	+					57
1,4-di-(2-thienyl)benzene	+	+					16, 44-46
	+	+					57
triazole–thiophenes (various monomers)						+	53

The copolymer based on ethylene dioxythiophene (EDOT) with 1-(phenyl)-2,5-di(2-thienyl)-1H-pyrrole exhibits homogenous and compact structure. The π–π* transition in the neutral state (-0.5V) is claret red, displaying a maximum at 480 nm. The electronic band gap, calculated from the onset of the π–π* transition is 1.9eV. Upon oxidation, the π–π* transition is depleted at the expense of a peak at about 600 nm , corresponding to the charge carrier (radical cations) and producing a blue color at 1.2 vs Ag wire [41].

In Table 4 the methods used for the characterization of the copolymers structure are summarized.

4. Properties and Applications of Copolymers

4.1. Properties

4.1.1 Electrical Conductivity

Copolymerization is an effective way to increase the conductivity of the electrically active polymers. For example, copolymerization with ethylene dioxythiophene, EDOT, increases the conductivity of poly(1-(phenyl)-2,5-di(2-thienyl)-1H-pyrrole), P(PTP), since the presence of EDOT moieties enhances the conjugation in P(PTP) chains [41], and copolymers of pyrrole with thiophene have higher conductivity than PPy or PTh prepared under the same conditions. This is attributed to the importance of the inter-chain (three-dimensional) hopping process over intra-chain hopping as the main conduction mechanism [55].

4.1.1.1. Copolymers of Benzene with Biphenyl

The electrical conductivity in constant current of the undoped polyphenylene copolymers produced by modified Kovacic method is in the order of 10^{-12} S/cm, i.e. they are insulators. The doped copolymers have conductivity in the order of 10^{-4} to 10^{-5} S/cm, i.e. they are closer to that of doped polyperaphenylene and they remain semiconductors even after 14 months. Generally, the copolymers are more stable than the homopolymers, since their electrical conductivity decreases with slower rate [6].

4.1.1.2. Copolymers of Biphenyl with Thiophene

The copolymer films produced by electropolymerization have conductivity one order of magnitude higher than homopolymers. The highest conductivity for the copolymers is exhibited in the film synthesized in the E_{ox} of the system of the monomers (which is 1.80 V). The electrical conductivity of the copolymers can be correlated to the morphology of the films and to their ratio of biphenyl to thiophene structural units (calculated from EDAX), with the higher conductivity appearing in the more compact films. This can be explained considering that the more compact films have a more extensive and continuous network of active paths, which facilitate the charge transfer. Considering that polythiophenes generally have lower conductivity than polyphenylenes, it can be expected that the higher content of thiophene units, the lower the conductivity of the copolymers would be. Thus, in copolymers, the higher conductivity appears in polymers having the lowest ratio of thiophene units. The electrical conductivity can be correlated with the structure of the polymers. The ratio R of the peak area of the quinoid (A_{1570}) to benzenoid (A_{1480}) correleates strongly with the electrical conductivity increases by increasing R, i.e. the electrical conductivity increases by increasing the quinoid structures in the macromolecules [8].

4.1.1.3. Copolymers of Biphenyl and Pyrrole

Contrarily to the homopolymers, the highest conductivity for the copolymer films produced by electropolymerization is exhibited in the film synthesized in the E_{ox} of the monomers system (which is 0.80 V). By comparing these copolymers with those of biphenyl with thiophene (PP-PTh), their electrical conductivity is in the same order of magnitude. As in the case of the biphenyl-thiophene copolymers, the electrical conductivity of the copolymers can be correlated to the morphology of the films and to their ratio of biphenyl to pyrrole structural units (calculated from EDAX), with the higher conductivity appearing in polymers having the higher ratio of pyrrole units. The ratio R of the quinoid to benzenoid rings is directry related with the conductivity. However, in contrast to the copolymers based on biphenyl and thiophene, in the case of (PP-PPy) copolymers it follows a ratio type relationship, indicating a different conductivity behavior in the two cases [14].

4.1.2. Thermal Stability

4.1.2.1. Copolymers of Benzene with Biphenyl

The undoped copolymers exhibit an abrupt loss (up to 80 % of their initial weight) in the region of 400-600 °C. Doped polyphenylenes show a similar behavior. Opposite to the weight loss behavior of the undoped and the doped homopolymers, the weight loss of the doped copolymers is lower than that of the corresponding undoped. Even though the total weight loss of the copolymers (both doped and undoped) is quite large, in the temperature region up to 400 °C they exhibit lower weight loss compared to the homopolymers, indicating that they are more stable in this region [6].

4.1.2.2. Copolymers of Biphenyl with Thiophene

The copolymer has a weight loss behavior similar to that of polythiophene produced under similar conditions. From the TGA results, the structure of the copolymers was established. If the copolymer was structured containing separate large segments of polyphenylene and polythiophene (resembling a block copolymer), then it would have been stable up to 200°C and would have had an abrupt weight loss at approximately 300°C, due to the decomposition of the large segments of polyphenylene. Thus, the copolymer macromolecules contain distributed monomeric units of biphenyl and thiophene. Therefore, there are not meta- linkages between the monomeric units of biphenyl [8].

4.1.2.3. Copolymers of Biphenyl with Pyrrole

The copolymers have lower total weight loss than the homopolymers and they are more stable up to 1000 °C, which is attribited to the stabilizing effect that the incorporated biphenyl units (consisting of two aryl rings) impart on the segments of pyrrole units [14].

4.1.3. Electrochemical Stability of the Synthesized Films

The stability of the films under repetitive potential cycling is very important for practical applications. Cyclic voltammetry was used as a method to evaluate the stability of the copolymer films synthesized. The film (as deposited film on the electrode) is placed in a solvent – electrolyte blank solution and undergoes repetitive cyclic potential sweeps and in every cycle the anodic and cathodic peaks (i.e. their corresponding potential and current), as well as the total charge of the cycle are determined [8,14].

4.1.3.1. Copolymers of Biphenyl with Thiophene (PP-PTh)

These copolymers are very stable, considerably better than the correspoding homopolymers (polyphenylenes and polythiophenes) produced under similar conditions. Specifically, in the broad potential region (from -2 V to +2 V) the copolymers are stable up to 60 cycles. The corresponding charge exhibited only a small decrease (0.17% per cycle with reference to the charge of the first cycle). In the narrow potential region (from 0 to +2 V), copolymers are even more stable, i.e. up to 250 cycles. The corresponding charge exhibited smaller decrease (0.024% per cycle with reference to the charge of the first cycle). In comparison, the correspoding homopolymers are stable only up to 15 cycles in the broad potential region and up to 20 cycles in the narrow one [8].

4.1.3.2. Copolymers of Biphenyl with Pyrrole (PP-PPy)

Similarly to the above mentioned system, these copolymers are too very stable, again considerably better than the correspoding homopolymers (polyphenylenes and polypyrroles) produced under similar conditions. Specifically, in the broad potential region (from -2 V to +2 V) they are stable up to 150 cycles and the corresponding charge exhibited only a small decrease (less than 0.15 % per cycle with reference to the charge of the first cycle). In the narrow potential region (from 0 to +2 V), copolymers are even more stable, i.e. up to 400 cycles and the corresponding charge exhibited smaller decrease (less than 0.015% per cycle with reference to the charge of the first cycle). In comparison, the homopolymers are stable only up to 15 cycles in the broad potential region and up to 35 cycles in the narrow one. The PP-PPy copolymers are profoundly more stable (about twice higher) than the PP-PTh copolymers. Given that both (PP-PTh) and (PP-PPy) copolymers have high conductivity, for an application where electrochemical stability is required, the latter are more suitable than the former [14].

4.1.3.3. Pyrrole–Thiophene Co-Polymer from an Oligomer Precursor

The monomer used was N, N'-bis(2-pyrrolylmethylene)-3,4-dicyano-2,5-diamino-thiophene (Py_2ThAz), which was electropolymerized under potentiostatic and potentiodynamic conditions. The effects of different electrolytes on the redox behaviour and the stability of p-doped polymer films were investigated by repetitive scanning (from -0.5 to 1.0 V) for up to 90 cycles. The most stable performance occurred when $LiClO_4$ was used as electrolyte, with charge loss of 14 % on the oxidation sweep and 20 % on the reduction sweep [49].

4.2. Applications

During recent years, there has been growing interest in electrically conducting polymers because of their potential applications for electrochromic displays and smart windows, light emitting diodes (LEDs), sensors, corrosion inhibitors, field effect transistors (FETs), electromagnetic interference (EMI) shielding etc [2,4,13,60]. This is not surprising, since many conducting polymers possess unsual combinations of properties, including high transparency and good solubility, which make them the best (and sometimes the unique) choice for many applications [4]. Specifically, PPy are used in electrochromic displays, batteries, corrosion inhibitors sensors, PTh in electrolumincense devices, electrochemical capacitors, batteries, corrosion inhibitors, and PP in electroluminence devices, photoconductors, laser materials, photovoltaic cells, etc [13]. The use of copolymers could increase the number of applications and allow better control over different properties of the material such as conductivity, filmability, processability, stability and morphology [19]. In below, some of the most important applications are described in detail.

4.2.1. Organic Light Emitting Diodes (OLED)

Conjugated polymers have been studied extensively, not only to explore their fundamental optical and electrical properties, but also to identify their potential applications in various devices, such as organic light emitting diodes (OLEDs), photovoltaic cells etc [4,13,39,40,57,61-66]. Ever since the first discovery of electroluminescence (EL) in semiconducting conjugated polymers, interest has grown rapidly and many polymers have been successfully used in OLEDs. The polymer based LEDs are especially attractive for use in display technology [13]. The great interest in such polymer based devices is understandable in terms of significant advantages that these systems have in possessing better mechanical properties and geometry possibilities as compared to conventional semiconductors. Another favourable aspect of the polymer LED is that it is possible to cover the spectral range from blue to near infrared, even within a single family of conductive polymers such as polythiophene. The recent demonstration of voltage-controlled electroluminescence colours from polymer blends in LEDs as well as the possibility of obtaining polarised light from oriented polymers in LED devices extend the possibilities of fabricating `exotic' polymer devices [13]. The oligo and polythiophene, as well as polyphenylene and their derivatives have been investigated as advanced and suitable materials for OLED. The thienylene and phenylene based polymers have been functionalized with different electron-withdrawing or donating groups and focused on luminescent materials. For instance, the emission color and lifetime besides emission efficiency and driving voltage are very important for devising light emitting diodes (LEDs) [57].

4.2.2. Organic Photovoltaic Solar Cells (OPV)

Research and development endeavours are focusing on continuous roll-to-roll printing of polymeric or organic compounds from solution—like newspapers—to produce flexible and lightweight devices at low cost. In particular, polymeric semiconductor-based solar cells are currently in investigation as potential low-cost devices for sustainable solar energy conversion. Because they are large-area electronic devices, readily processed polymeric semiconductors from solution have an enormous cost advantage over inorganic semiconductors. Further benefits are the low weight and flexibility of the resulting thin-film devices [64,67]. Organic photovoltaics (OPVs) are a promising cost-effective alternative to silicon-based solar cells, and possess low-cost, light-weight, and flexibility advantages. Since the discovery of photoinduced charge transfer in composites of conjugated polymers and fullerenes, OPVs have attracted considerable attention [40,42,63,64,68]. The synthesis of new conjugated polymers with small bandgaps, strong and broad absorption, appropriate energy levels, and high carrier mobilities will be the key to allowing plastic solar cells to become a viable energy source for this century [68].

Up till now, OPV based on poly(3-hexylthiophene) (P3HT) as the donor and methanofullerene (6,6)-phenyl-C61-butyric acid methyl ester (PC61BM) as the acceptor have been extensively stydied [13,40,42,62-64] and they have reached an efficiency of 8-10 % [66]. However, the narrow absorption spectrum of P3HT hampered the further improvement in the efficiency. To overcome this problem, some narrow band polymers as the donor have been synthesized and applied to photovoltaic devices. For example, side-chain conjugated polythiophene derivatives exhibited a broad response range to the solar irradiation spectrum in comparison with P3HT and 38% improvement of OPV was obtained. The polythiophene copolymer, poly[3-(5'-octylthienylenevinyl) thiophene]-thiophene (POTVTh-Th), was used as electron donor in OPV with very promising results [40].

4.2.3. Electrochromic Desplays (ECD)

Electrochromism is the reversible change in optical properties that can occur when a material is electrochemically oxidized or reduced. Conjugated polymers are used as electroactive layers in electrochromic devices due to their ease of color tuning properties, fast switching times and high contrast ratios [69]. An ECD is an electrochemical cell in which the electrochromic (EC) electrode is separated by an electrolyte from the charge-balancing counter electrode. The EC electrode constitutes of a conductive, transparent glass (like indium tin oxide, ITO) coated with the EC material [5,13,55]. Commercial applications of EC materials in devices include smart windows, anti-glare car rear-view mirrors, strips as battery state-of-charge indicators and sunglasses [70,71]. A variety of CPs have colours both in the oxidized and reduced states, since their band gap is in the visible region. After oxidation, the intensity of the π-π^* transition decreases, and two low energy transitions emerge to produce a second colour [70].

This, combined with the fact that CPs can repeatedly undergo electrochemical doping/undoping processes, makes them the most promising class of materials to be used in ECDs [72]. For example, PPy has been extensively utilised as an EC material and thin films of neat PPy are yellow in the undoped insulating state and black in the doped conductive state [50]. Some of the most promising conducting polymers for use in ECDs are based on poly(3,4 ethylenedioxythiophene) (PEDOT) [5,69] and copolymers such as 2,5-di(thiophen-2-yl)-1-p-tolyl-1Hpyrrole and that of 2-(2-thienyl)-1H-pyrrole [50,69]. Moreover, the ECD based on the copolymer of 1-(phenyl)-2,5-di(2-thienyl)- 1H-pyrrole with EDOT shows a maximum absorption at 545 nm due to π to π^* transition and it exhibits good stability and optical memory [41].

4.3.4. Rechargeable Batteries

One of the most important energy storage applications of conducting polymer film is their use as cathode material for rechargeable battery in view reversible doping. Certainly, the concept of polymer batteries is very attractive in terms of the various interesting applications that such a electrochemical power source could offer [13]. The first prototypes of commercial batteries with conductive polymers used Li/polypyrrole [73] (Varta-BASF) or Li/polyaniline [74,75].

It was demonstrated that high charge densities can only be achieved in Li|polyaniline (PANI)|propylene carbonate-based batteries [4]. Currently, development is focused on new cathode materials for lithium batteries. Systems such as the fullerene functionalized poly(terthiophenes) have been proposed as cathode materials for Li batteries [76]. Good results were obtained with substituted polythiophenes and poly(1,2-di(2-thienyl)ethylene). A flexible fiber battery has been constructed consisting of a polyphenylene (PP)/PF_6 (PF_6 : Hexafluorophosphate) cathode and a PP/PSS (PSS : poly(styrene sulfonate)), anode [77]. Unresolved problems include the insufficient cycle stability of the system compared with inorganic systems and its high discharge rate [4].

REFERENCES

[1] O. A. Guskova, P. G. Khalatur, A. R. Khokhlov, *Macromolecular Theory and Simulations*, 18, 219–246 (2009).

[2] F. Liu, Y. Chen, Y. Wei, L. Li, S. Shang, *Journal of Applied Polymer Science*, 123, 2582–2587 (2012).

[3] H. K. Choi, S. H. Jin, J. W. Park, S. Y. Kim, Y. S. Gal, *Journal of Industrial and Engineering Chemistry*, 18, 814–817 (2012).

[4] G. Inzelt, *Conducting Polymers, In Monographs in Electrochemistry*, Springer, Berlin, 1-122, 255-264 (2008).

[5] M. Ak, A. Cirpan, F. Yılmaz, Y. Yagci, L. Toppare, *European Polymer Journal*, 41, 967–973 (2005).

[6] J. Simitzis, D. Triantou, S. Soulis, *Journal of Applied Polymer Science*, 110, 356-367 (2008).

[7] W. Beek, R. Janssen, *Hybrid Polymer-Inorganic Photovoltaic Cells, In Hybrid Nanocomposites for Nanotechnology*, ed. by L. Merhari, Springer, Berlin, 321-385 (2009).

[8] J. Simitzis, D. Triantou, S. Soulis, *Journal of Applied Polymer Science*, 118, 1494–1506 (2010).

[9] P. Kovacic, M. B. Jones, *Chemical Reviews*, 87, 357-379 (1987).

[10] M. Ates, A. S. Sarac, *Polymer-Plastics Technology and Engineering*, 50, 1130–1148 (2011).

[11] X. D. Dang, C. M. Intelmann, U. Rammelt, W. Plieth, *Journal of Solid State Electrochemistry*, 8, 727–734 (2004).

[12] E. Salatelli, L. Angiolini, A. Brazzi, M. Lanzi, E. Scavetta, D. Tonelli, *Synthetic Metals*, 160, 2681–2686 (2010).

[13] K. Gurunathan, A. Vadivel Murugan, R. Marimuthu, U. P. Mulik, D. P. Amalnerkar, *Materials Chemistry and Physics*, 61, 173-191 (1999).

[14] J. Simitzis, S. Soulis, D. Triantou, *Journal of Applied Polymer Science*, DOI: 10.1002/app.36301 (2012).

[15] T. Balkan, A. S. Sarac, *Fibers and Polymers*, 12, 565-571 (2011).

[16] R. W. Gumbs, *Polythiophene and Polypyrrole Copolymers, In Handbook of Organic Conductive Molecules and Polymers*, ed. by H.S. Nalwa, Wiley, Chichester, 469-504 (1997).

[17] R. M. Latonen, J. E. Lonnqvist, L. Jalander, K. Froeberg, C. Kvarnstroem, A. Ivaska. *Synthetic Metals*, 156, 878–884 (2006).

[18] H. Goktas, T. Gunes, B. Atalay, A. Oguz Er, I. Kaya, *IEEE transactions on plasma science*, 39, 2578-2579 (2011).

[19] R. M. Latonen, C. Kvarnstroem, A. Ivaska, *Journal of Electroanalytical Chemistry*, 512, 36–48 (2001).

[20] R. S. Latonen, C. Kvarnstrom, A. Ivaska, *Electrochimica Acta*, 44, 1933-1943 (1999).

[21] J. Simitzis, S. Soulis, D. Triantou, *Benzene as a precursor for conducting* polymers, In *Benzene: Structure, Uses and Health Effects*, ed. by G. Tranfo, Nova Science Publishers, New York, 1-21 (2012).

[22] P. C Lacaze, S. Aeiyach, J. C Lacroix, *Poly(p-phenylenes): Preparation Techniques and Properties, In Handbook of Organic Conductive Molecules and Polymers*, ed. by H.S. Nalwa, Wiley, Chichester, 205-270 (1997).
[23] G. K. Noren, J. K. Stille, *Journal of Polymer Science, Part D: Macromolecular Reviews*, 5, 385-430 (1971).
[24] E. Riande, R. Díaz-Calleja, *Electrical properties of polymers*, Marcel Dekker, New York, 575-599 (2005).
[25] A. Moliton, *Ion implantation doping of electroactive polymers and device fabrication, In Handbook of Conducting Polymers*, 2nd ed., ed. by T. A. Skotheim, R. l. Elsenbaumer, J. R. Reynolds, Marcel Dekker, New York, 589- 638 (1998).
[26] G. Shi, J. Xu, M. Fu, *Journal of Physical Chemistry B*, 106, 288-292 (2002).
[27] S. Sarac, *Electropolymerization, In Encyclopedia of Polymer Science and Technology*, Wiley, New York (2006).
[28] N. Bilow, L. Miller, *Journal of Macromolecular Science (Chemistry)*, A1, 183-197 (1967).
[29] N. Bilow, L. J. Miller, *U.S. Pat.* 3, 578-611 (1971).
[30] R. S. Latonen, C. E. Kvarnstrom, M. Grzeszczuk, A. Ivaska, *Synthetic Metals*, 130, 257–269 (2002).
[31] R. S. Latonen, C. Kvarnstrom, A. Ivaska, *Synthetic Metal*s, 129, 135–145 (2002).
[32] B. Li, B. Liu, Q. Li, H. Fang, M. Yu, *Journal of Chemical Research*, 653–655 (2010).
[33] X. Li, M. Lu, H. Li, *Journal of Applied Polymer Science*, 86, 2403–2407 (2002).
[34] S. Kuwabata, H. Honeyama, *Journal of Electroanalytical Society*, 135, 1691-1695 (1988).
[35] O. Inganas, B. Liedberg, W. Chang-Ru, *Synthetic Metals*, 11, 239-249 (1985).
[36] E. M. Peters, J. D. Van Dyke, *Journal of Polymer Science, Part A: Polymer Chemistry*, 29, 1379-1386 (1991).
[37] S. R. Vadera, N. Kumar *Frontiers in Polymer Res.* ed. by P. N. Pasad, J. K. Nigam, Vol. 1, Plenum, New York, 443-447 (1991).
[38] S. Hotta, *Synthetic Metals*, 2, 103-113 (1987).
[39] L. Miozzo, N. Battaglini, D. Braga, L. Kergoat, C. Suspene, A. Yassar, *Journal of Polymer Science, Part A: Polymer Chemist*ry, 50, 534–541 (2012).
[40] B. Qu, Z. Jiang, Z. Chen, L. Xiao, D. Tian, C. Gao, W. Wei, Q. Gong, *Journal of Applied Polymer Science*, 124, 1186–1192 (2012).
[41] S. Tarkuc, E. Sahmetlioglu, C. Tanyeli, I. M. Akhmedov, L. Toppare, *Optical Materials*, 30, 1489–1494 (2008).
[42] V. Tamilavan, M. Song, S. H. Jin, M. H. Hyun, *Polymer*, 52, 2384-2390 (2011).
[43] R. S. Latonen, J. E. Lonnqvist, L. Jalander, C. Kvarnstrom, A. Ivaska, *Electrochimica Acta*, 51, 1244–1254 (2006).
[44] R. Danieli, P. Ostoja, M. Tiecco, R. Zamboni, C. Tailiani, *Journal of Chemical Society, Chemical Communications*, 1473-1474 (1986).
[45] S. Tanaka, M. Sato, K. Kaeriyama, *Journal of Macromolecular Science, Chemistry Edition*, A24, 749-761 (1987).
[46] J. R. Reynolds, J. P. Ruiz, A. D. Child, K. Nayak, D. S. Marynick, *Macromolecules*, 24, 678 (1991).

[47] Q. Pei, O. Inganas, G. Gustafsson, M. Granstrom, M. Andersson, T. Hjertberg, O. Wennerstrom, J. E. Osterholm, J. Lakso, H. Jarvinen, *Synthetic Metals*, 55, 1221-1226 (1993).
[48] M. R. Anderson, Q. Pei, T. Hjertberg, O. Inganas, O. Wennerstrom, *Synthetic Metals*, 55, 1227-1231 (1993).
[49] F. Al-Yusufy, S. Bruckenstein, W. Schlindwein, *Journal of Solid State Electrochemistry*, 11, 1263–1268 (2007).
[50] C. Pozo-Gonzalo, J. A. Pomposo, J. A. Alduncin, M. Salsamendi, A. I. Mikhaleva, L. B. Krivdin, B. A. Trofimov, *Electrochimica Acta*, 52, 4784–4791 (2007).
[51] J. P. Ferraris, G. E. Skiles, *Polymer*, 28, 179-182 (1987).
[52] J. P. Ferraris, T. R. Hanlon, *Polymer*, 30, 1319-1327 (1989).
[53] F. Montilla, F. Huerta, D. Salinas-Torres, E. Morallon, C. Cebrian, P. Prieto, A. Dvaz-Ortiz, A. de la Hoz, J. R. Carrillo, C. Romer, *Electrochimica Acta*, 58, 215– 222 (2011).
[54] E. Sezer, A. S. Sarac, A. E. Parlak, *Journal of Applied Electrochemistry*, 33, 1233–1237 (2003).
[55] M. Ak, L. Toppare, *Materials Chemistry and Physics*, 114, 789–794 (2009).
[56] S. Wu, L. Bu, L. Huang, X. Yu, Y. Han, Y. Geng, F. Wang, *Polymer*, 50, 6245–6251 (2009).
[57] F. Martinez, G. Neculqueo, S. O. Vasquez, R. Letelier, M. T. Garland, A. Ibanez, J. C. Bernede, *Journal of Molecular Structure*, 973, 56–61 (2010).
[58] D. Udayakumar, A. Vasudene Adhikari, *Synthetic Metals*, 156, 1168-1173 (2006).
[59] D. M. Leeuw, M. M. J. Simenon, A. R. Brown, R. E. F. Einerhand, *Synthetic Metals*, 87, 53-59 (1997).
[60] M. Verswyvel, P. Verstappen, L. De Cremer, T. Verbiest, G. Koeckelberghs, *Journal of Polymer Science, Part A: Polymer Chemistry*, 49, 5339–5349 (2011).
[61] Y. Y. Noh, D. Y. Kim, M. Misaki, K. Yase, *Thin Solid Films*, 516, 7505–7510 (2008).
[62] S. Gunes, H. Neugebauer, N. S. Sariciftci, *Chemical Reviews*, 107, 1324-1338 (2007).
[63] H. Hoppe, N. S. Sariciftci, *Advances in Polymer Science*, 214, 1–86 (2008).
[64] P. A. Troshin, R. N. Lyubovskaya, V. F. Razumov, *Nanotechnologies in Russia*, 3, 242–271 (2008).
[65] I. Kymissis, *Organic Field Effect Transistors Theory, Fabrication and Characterization, In Series on Integrated Circuits and Systems, Springer,* New York, 1-149 (2009).
[66] E. Von Hauff, *The Role of Molecular Structure and Conformation in Polymer Electronics, In Semiconductors and Semimetals*, Elsevier, Vol. 85, 231-260 (2011).
[67] J. Loos, A. Alexeev, *Scanning Probe Microscopy on Polymer Solar Cells, Applied Scanning Probe Methode X: Biomimetic and Industrial Applications*, 183-215 (2008).
[68] E. Sahmetlioglu, R. Varol, L. Toppare, H. Yuruk, *Journal of Macromolecular Science, Part A: Pure and Applied Chemistry*, 46, 584–590 (2009).
[69] B. Yigitsoy, S. Varis, C. Tanyeli, I. M. Akhmedov, L. Toppare, *Thin Solid Films*, 515, 3898–3904 (2007).
[70] M. Ak, M. Sulak Ak, G. Kurtay, M. Güllü, L. Toppare, *Solid State Sciences*, 12, 1199-1204 (2010).
[71] T. Ikeda, J. F. Stoddart, *Science and Technology of Advanced Materials*, 9, 014104, 7pp (2008).

[72] E. da CostaRios, A. V. Rosario, A. F. Nogueira, L. Micaroni, *Solar Energy Materials and Solar Cells*, 94, 1338–1345 (2010).
[73] G. Mengoli, M. M. Musiani, M. Fleischmann, D. Pletcher, *Journal of Applied Electrochemistry*, 14, 285-291 (1983).
[74] D. Naegele, R. Bittihn, *Solid State Ionics*, 28–30,983-989 (1988).
[75] W. Qiu, R. Zhou, L. Yang, Q. Liu, *Solid State Ionics*, 86-88, 903-906 (1996).
[76] J. Chen, G. Tsekouras, D. L. Officer, P. Wagner, C. Y. Wang, C. O. Too, G. G. Wallace, *Journal of Electroanalytical Chemistry*, 599, 79-84 (2007).
[77] J. Wang, C. O. Too, G. G. Wallace, *Journal of Power Sources*, 150, 223-228 (2005).

In: Conducting Polymers
Editor: Luiz Carlos Pimentel Almeida
ISBN: 978-1-62618-119-9

Chapter 3

BASICS AND NEW INSIGHTS IN THE ELECTROCHEMISTRY OF CONDUCTING POLYMERS

F. Miomandre and P. Audebert
PPSM – CNRS – ENS Cachan, Cachan, France

ABSTRACT

This chapter deals with conventional and innovative electrochemical properties of conducting polymers. In the first part, fundamentals will be reminded including electropolymerization mechanisms, techniques and the main features of the resulting material according to the technique used. Then the doping issue will be envisaged with highlight on the nature of charge carriers, mixed electronic and ionic conductivities and its consequences, typical features of the charge-discharge process. Relaxation effects typical of conducting polymers will also be shortly discussed. The major part of the review will be devoted to the new trends in the field of the electrosynthesis of conducting polymers, with the use of new monomers, new electrolytic media or hard and soft templates to generate special morphologies (nanoparticles, nanowires ...). Finally the main applications involving the electrochemical features of conducting polymers will be browsed: energy storage devices, actuators, biosensors, anti-corrosion coatings or drug delivery. In that final part, only recent developments will be highlighted since an extensive review of these applications would be out of the scope of this chapter.

INTRODUCTION

Conducting polymers, that should now be rather considered as conjugated polymers, have known a great success in fundamental research as well as – albeit not widespread - in some applications since their discovery in 1977 by Shirakawa, MacDiarmid and Heeger[1]. As oxidation – or more scarcely reduction – reactions are involved in their first historical syntheses and even if organometallic coupling have challenged this approach since the early 1990s, electrochemistry is naturally associated to their development, not only as a synthetic tool, but also as an insight method into their fundamental properties and as a wide application field through energy storage devices, sensors and actuators. This chapter intends to give the reader an overview of the way electrochemistry is involved in these various aspects, and complementary information can be found in a previously published review[2]. The outline will start with the fundamental knowledge to gradually browse more recent results in a non-exhaustive way, mainly targeted towards the applications that will be described in the final part. As most of other organic and inorganic materials, conjugated polymers have known their nano-size era, and the use of electrochemical methods to generate nanostructures, possibly associated with other materials in naocomposites, will not be forgotten. We hope to offer to the reader, beginner or already familiar with the field, both the basic knowledge and the present and future trends in the electrochemical properties of conjugated polymers.

1. FUNDAMENTALS

1.1. Electropolymerization: Mechanism, Techniques, Control

Electropolymerization of conjugated molecules to lead to conducting polymers answers to a now well established reaction scheme. We will consider the general case of polymers made by oxidation, which is by far more common than those made by reduction, but the following equations can be easily transposed. The general equation summarizing the reaction outcome can be written :

$$n(HMH) + n\delta X^- \rightarrow \left(M^{\delta+}, \delta X^-\right)_n + (2+\delta)ne^- + 2nH^+ \quad (1)$$

where M features the molecular repeating unit and X^- the counter-ion coming from the electrolyte salt.

Equation (1) shows that the electropolymerization process requires (2+δ) electrons to be removed per repeating unit, among which two electrons are required for the polymerization reaction, and an additional δ to create positive charges on the macromolecular structure. This δ number is called the doping level and corresponds to the average number of charge per repeating unit in the polymer. This equation also features that this charging process is usually accompanied by an anion insertion to maintain the polymer electroneutrality[1]. X^- is thus often called the dopant ion.

Applying Faraday's law to reaction (1) results in a relation between the surfacic coulombic charge (charge per electrode area unit) Q_s injected during the synthesis and the weight m of polymer produced:

$$m = M_{(M+\delta X)} \frac{AQ_s}{(2+\delta)F} \quad (2)$$

[1] In less common cases, cation removal instead of anion insertion can occur to balance the charge (see §1.2 for discussion of this point.)

A is the electrode area, $M_{(M+\delta X)}$ is the molar mass of the $(M^{\delta+}, \delta X^{-})$ moiety and F is the Faraday constant (96 500 C mol^{-1}).

Equation (2) can be easily converted into the following relation involving the theoretical polymer thickness l:

$$l = M_{(M+\delta X)} \frac{Q_S}{(2+\delta)\rho F} \quad (3)$$

where ρ is the polymer density.

Equation (3) is valid if one assumes that the polymer is dense, and the electropolymerization yield is 100%. In that case, the polymer thickness is proportional to the surfacic charge used for the synthesis. On the other hand, the doping process requires a surfacic charge Q_r that can also be drawn from Faraday's law:

$$Q_r = \delta F \frac{m}{AM_{(M+\delta X)}} = \frac{\delta}{(2+\delta)} Q_S \quad (4)$$

The doping level can be extracted from equation (4):

$$\delta = \frac{2Q_r}{Q_S - Q_r} \quad (5)$$

Finally, by injecting (4)-(5) into (3), the final expression of the polymer thickness is given by:

$$l = M_{(M+\delta X)} \frac{Q_S - Q_r}{2\rho F} \cong M_{(M+\delta X)} \frac{Q_S}{2\rho F} \quad (6)$$

For polypyrrole (PPy) doped by perchlorate, assuming $M_{(M+\delta X)} \approx 100\, gmol^{-1}$ and $\rho \approx 2gcm^{-3}$, equation (6) leads to $Q_s \approx 0.4\, C.cm^{-2}$ for l = 1 µm. This gives an interesting order of magnitude for the relation between synthetic charge and thickness, which can be generalized since molar masses and density do not vary on a large scale, when changing the molecular structure.

Thus, it is easy to estimate the film thickness from the coulombic charge passed during synthesis, which can be measured by integrating the current with time.

The electropolymerization mechanism is now well established since the early 1990s and the works of Savéant's group on oligopyrroles[3-4] and Garnier's group on oligothiophene[5]. The initial steps of the mechanism are represented as follows:

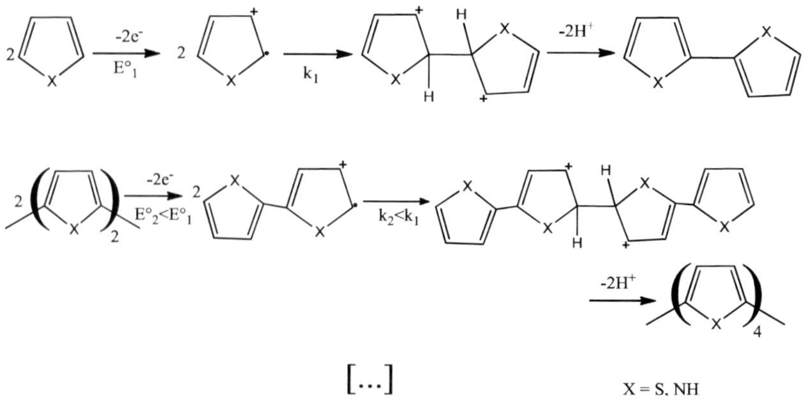

[...]

X = S, NH

The first step is the oxidation of the initial monomer into a cation radical followed by fast coupling of the two cation radicals and deprotonation to recover aromaticity in the neutral dimer. After these three steps, the dimer undergoes the same reaction path to give longer oligomers. There are several key points in this mechanism:

- The first straightforward one is the decreasing standard potential values with increasing chain lengths ($E°_n<E°_m$ for n>m), due to longer conjugation.
- The second point is the decreasing kinetic constant for the cation radical coupling with increasing chain lengths, due to higher stabilization through enhanced conjugation.
- The third point concerns the deprotonation step, which is usually the rate determining step once the cation radical has been formed, despite the rearomatization driving force[6]. The associated kinetics depends strongly on the reaction medium (solvent basicity, counter ion) and it is also decreasing with increasing chain lengths. Thus the mechanism is likely to stop at the oligomer level, which has been already evidenced for a long time: PPy does not form nice polymer films in very dry acetonitrile conversely to "wet acetonitrile" (i.e. acetonitrile with 1% water). A way to favor this deprotonation step for long oligomer chains is to increase their charge, which can be performed by increasing the applied potential.

It is now acknowledged that electropolymerization starts with oligomerization reactions in solution according to the scheme above, i.e. mainly involving oligomers with an even number of repetition units (dimers, tetramers, octamers…). As the solubility decreases with the increasing chain length, a deposition process follows rapidly the formation of these intermediate species. Thus, the following steps leading to polymer state involves solid state reactions, generally decomposed into nucleation and growth by analogy with metal electrodeposition. These three parts in the electropolymerization mechanism (oligomerization in solution, nucleation and growth) can be visualized by recording current vs. time variations (chronoamperograms) when applying a constant electrode potential (potentiostatic conditions). The classical shape of $i(t)$ is given in figure 1.

Figure 1 shows three successive regimes : a) fast current decrease associated with diffusion limited electron transfer at the electrode|electrolyte interface ; b) then fast increase of the current corresponding to instantaneous nucleation on the electrode ; c) finally slow increase (or stagnation) of the current associated with 3-dimensional growth mechanism.

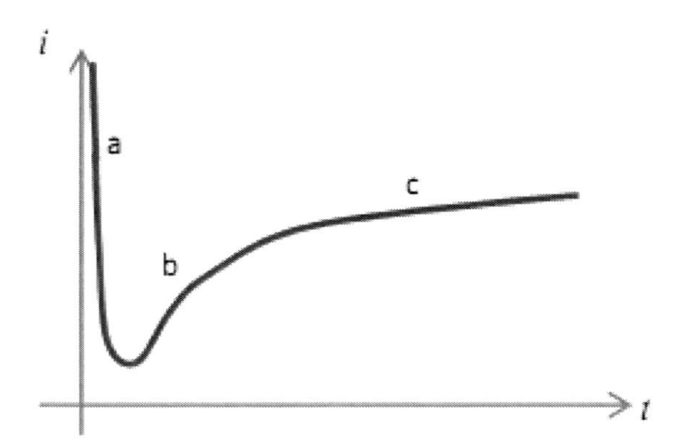

Figure 1. Classical chronoamperogram for electropolymerization at constant potential.

Nevertheless, one must not believe that coupling reactions only occur for short oligomers in solution and that further steps only involves precipitation phenomena. It was recently demonstrated by Heinze using oligophenylenes and oligothiophenes that dimerization reactions also occur at the solid state, after the first deposition steps[7]. One can imagine that besides oligomerization at the solid state, cross-linking reactions also contribute to increase the polymerization degree after electrodeposition. The final average chain length as commonly defined in polymer chemistry is usually difficult to measure since conducting polymers are scarcely soluble apart from specific conditions. However, it is acknowledged that this value is low (below 100) and dependent on the experimental conditions, especially the applied potential (the trend is to increase chain lengths when applying more positive potentials).

Another usual technique for electropolymerization is the potentiodynamic one, which consists in cycling the potential between the open circuit and monomer oxidation values at a moderate sweep rate (100 $mV.s^{-1}$) and recording the current as a function of the potential (cyclic voltammetry). The classical shape for the few initial cycles is given in figure 2 (example of polydiiodopyrrole). After the first forward scan, a new signal appears at lower potentials than the monomer oxidation, which can be ascribed to oligomer formation and electrodeposition. The polymer growth can be followed by the regular increase of the corresponding redox signal on cycling, the conductivity of the electrogenerated film impeding passivation phenomena that could block the growth (case of redox polymers). The growth of the polymer film on the electrode surface can normally proceed up to micrometer size thicknesses, but restrictions can occur according to the electrode surface cleanness, rugosity, and adherence of the film.

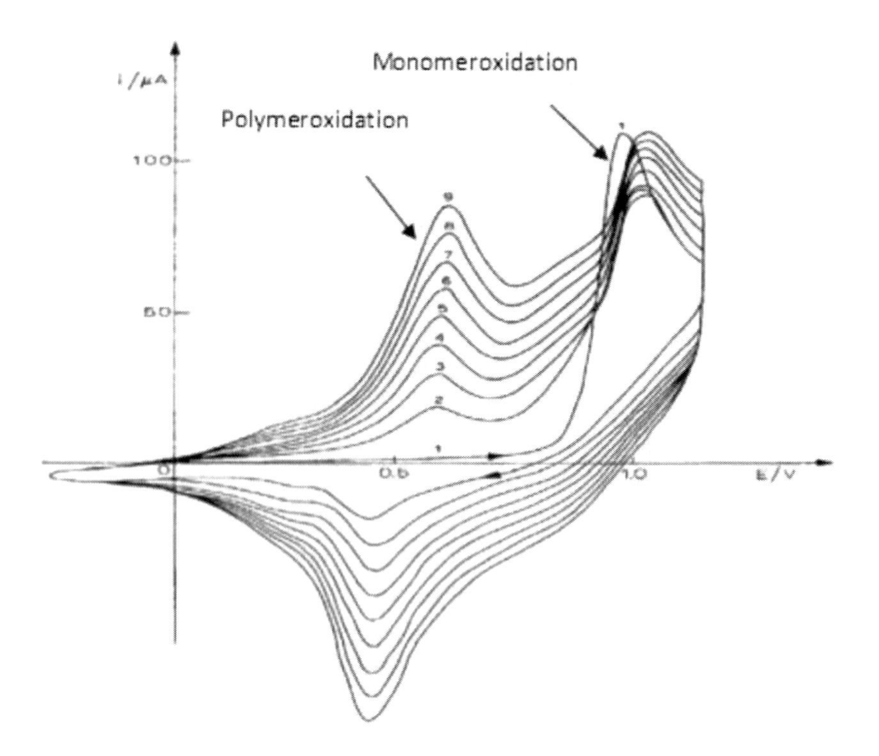

Figure 2. CVs for successive cycles upon potentiodynamic electropolymerization (adapted from ref. [8]). The polymer growth can be visually followed.

There is a third way to control the electropolymerization process, albeit less used than the two others, which is the galvanostatic mode. In that configuration, a constant current is applied and the variation of the electrode potential with the time (chronopotentiometry) is recorded. The general shape of the E(t) curve is the following:

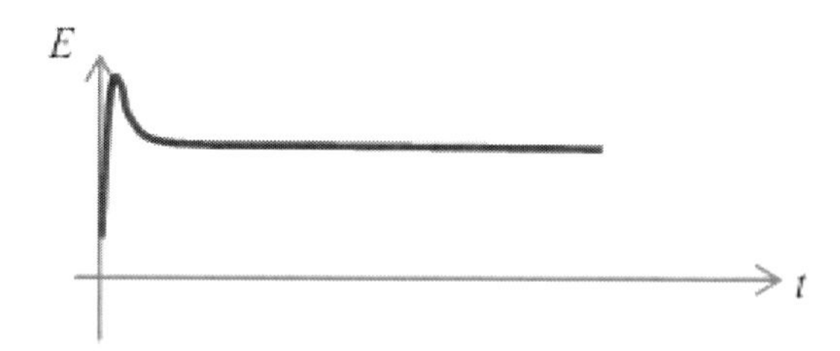

Figure 3. Classical chronopotentiogram for electropolymerization at constant current (galvanostatic).

A rapid increase of the potential up to the monomer oxidation is observed followed by a peak and a plateau. In that final part, the potential is fixed by the redox couple of the conducting polymer coated onto the electrode surface. This method has the advantage to directly control the injected charge through the elapsed time, since the current is constant.

However it requires choosing the applied current in the proper range to induce the polymerization while avoiding overoxidation, which is not always straightforward.

The electrogenerated conjugated polymers usually display various morphologies and properties according to the electropolymerization technique.

However it is not so easy to draw general conclusions as it depends on the polymer concerned and on other experimental conditions, among which the solvent. Although it was demonstrated for polybithiophene derivatives that the mechanism is similar with the same oligomers formed with both techniques [9], poly(3-methylthiophene) however exhibits better properties when synthesized under potentiostatic conditions [10].

A similar comparative study in the case of polyaniline (PANI) was performed using impedance spectroscopy as a way to estimate charge transfer kinetics and showed that potentiostatic and galvanostatic methods give identical results while potentiodynamic conditions lead to a conjugated polymer whose charge transfer resistance is strongly dependent on the scan rate [11].

1.2. Electrochemical Doping: Charge Carriers, Redox vs. Capacitive Behavior, Related Properties

After the electropolymerization process, electrochemical techniques can be used to characterize the electrogenerated polymer film and check its electroactivity. Cyclic voltammetry (CV) is the method of choice, as it allows recording several cycles to verify the polymer stability. Beyond this first glance analysis, it is interesting to pay attention to the shape of the voltammogram, which can vary significantly with the polymer thickness and morphology.

Typical CVs can be classified in three categories (Figure 4) : for very thin layers a symmetrical peak can be seen (Figure 4a) ; for thicker layers, the CV is close to a square without any peak (Figure 4b) ; finally an intermediate case is often observed with both a peak and a plateau (Figure 4c).

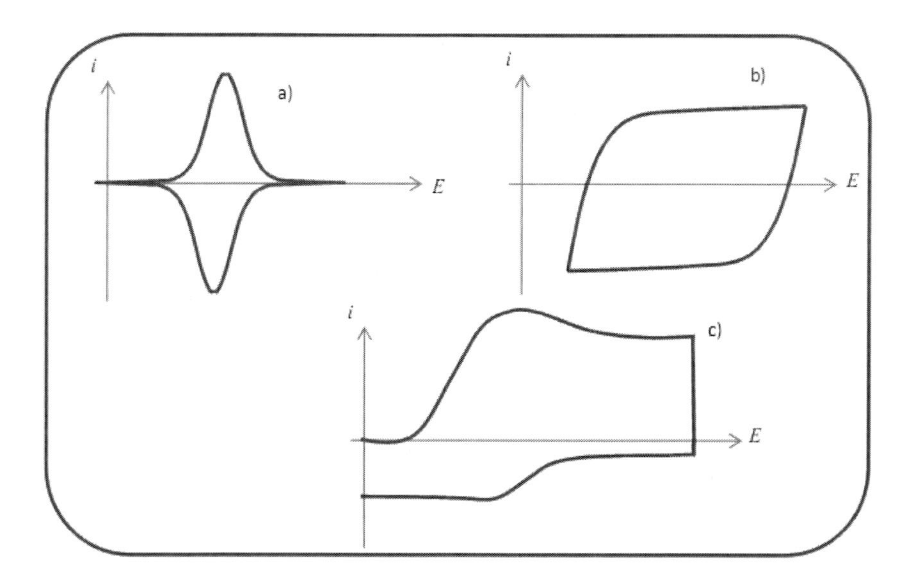

Figure 4. Various CV shapes for conducting polymers : a) thin layer behavior ; b) full capacitive behavior ; c) mixed redox and capacitive behavior.

For a long time there was a debate to discriminate between faradaic and capacitive components in the electrochemical response of conjugated polymers[12]. Indeed, these systems are able to display both behaviors:

- The Faradaic contribution is ascribed to the redox reaction associated with the polymer doping. In that case, the polymer reacts like a macromolecule exchanging electrons with the electrode according to a classical redox process; according to the layer thickness, limitation by mass transfer (usually ionic diffusion due to the counter-ion insertion, possibly coupled to migration) can modify the peak shape and the peak-to-peak separation. Discrimination between pure charge transfer kinetics (thin layers) and diffusion limited ones (thicker films) is evidenced by the dependence of the current with the scan rate. Proportionality with the scan rate is observed in the former case, while the current scales with the square root of the scan rate in the latter case.
- The capacitive contribution is due to the double layer charging with mainly two contributions : one from the polymer film and the other from the polymer|electrolyte interface. In that case, the conducting polymer behaves no more as a macromolecule but rather as a dielectric material, which can store charges as a capacitor. The capacitive contribution is very dependent on the doping level of the conjugated level for several reasons : one is that the transition from insulating to conducting state drastically increases the dielectric constant of the material, but additionally this transition is accompanied by swelling phenomena that strongly increases the specific area of the polymer|electrolyte interface, and thus the capacitance. At high doping level, the common picture is that the polymer behaves as a conductor with a ohmic contact with the electrode and highly capacitive interface with the electrolyte[13].

Thus the common interpretation of the CV of figure 4c consists in a mixed redox and capacitive charge transfer mechanism, the Faradaic component being predominant below the peak region and the capacitive one in the "tail"[14]. A way to discriminate between Faradaic and capacitive components relies on their different kinetics and uses Electrochemical Impedance Spectroscopy (EIS), which is much more adequate than CV for such an issue. Early works by Tanguy et al. demonstrated fast capacitive vs. slow faradaic processes in poly(3-methylthiophene)[15-16]. A recent contribution in that field from Abruña's group used EIS to quantify the relative amounts of capacitive vs. faradaic contributions according to the potential [17]. The capacitive component is usually predominant at high potential in PPy and polythiophene (PTh), as already known[18] but in polyethylenedioxythiophene (PEDOT) they are much more concomitant.

However, it must be mentioned that this vision is still controversial. This interpretation has been recently challenged by Heinze according to several experiments performed on oligomers. The conclusion is that the same behavior can be observed with only Faradaic processes, but considering overlapping of various standard potential redox reactions due to the various oligomeric chain lengths[19]. Additionally it must be mentioned that the classical 'tail' usually associated with the capacitive plateau can also be observed for pure redox processes provided that very high doping levels (close to 1) can be reached. Experimental evidence of this was given in the case of poly(4,4'-dimethoxybithiophene) which has the peculiarity to be able to stabilize highly charged redox states[20]. Thus it appears that the capacitive plateau may actually come from Faradaic processes, as already suggested by Dunsch for PANI using ESR-UV-Vis spectroelectrochemical measurements [21].

Charge carriers in conjugated polymers have puzzled the scientists since the discovery of electrical conductivity in these materials. Pioneering works of Brédas[22] introduced the concept of polarons and bipolarons, as the main charge carriers involved in the doping process of the macromolecular structure (figure 5). Actually polarons are lattice coupled charge carriers bearing a charge and an unpaired electron. Bipolarons are spinless double charged species obtained from polarons by one-electron removing. In each case the lattice deformation facilitates the ionization process and the resulting charge is therefore partially delocalized. This assumption relies on the band energy model, which requires a periodicity in the macromolecular structure and thus is especially valid for low defect materials with high conjugation lengths. The existence of polaronic and bipolaronic species was demonstrated by spectroelectrochemical (UV-vis and ESR[2]) techniques.

From the point of view of electrochemistry, one must keep in mind the necessity of balancing these charge carriers by counter ions to observe significant changes in the doping level (contrariwise to charge injection in LED or transistors where the ECP remains neutral or very slightly doped).

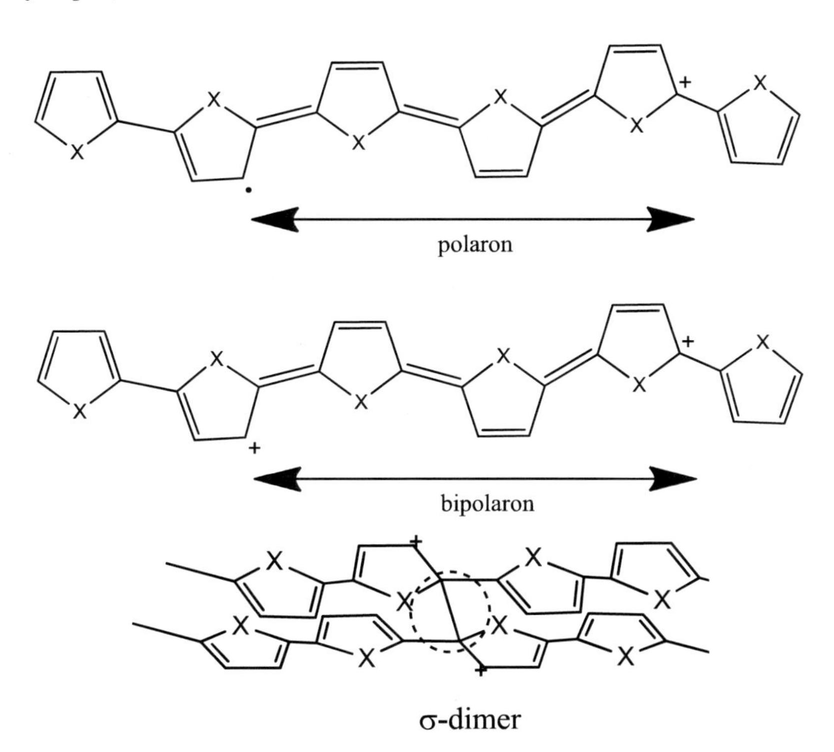

Figure 5. Polaron, bipolaron and interchain σ-dimer in conjugated polymers.

The most straightforward way to envisage the positive charge compensation is to insert anions in the polymer film upon oxidation (and cation insertion upon negative doping). However cations and solvent also play a role in this process [23]. Coupled techniques to

[2] Electron Spin Resonance

electrochemistry like EQCM[3] or PBD[4] have helped to better understand this question, but once again there are several behaviors according to the anion nature and doping levels[24-25] : the cation plays a significant role at the early stages of doping and for bulky or tightly bound anions. Another interesting question is the relation between the doping level and the observed properties like conductivity. As far as positive charge carriers are involved (oxidative doping), the general behavior is the existence of a percolation threshold, which leads to a sigmoidal shape for the conductivity vs. charge curve [26]. The high conductivity level is associated with the formation of bipolarons due to polaron recombination. But polarons are also likely to couple between each other through inter-chain interactions to form either σ- or π-dimers (figure 5), generally resulting in a conductivity drop. The dopant ion also plays an important role in the charge delocalization ability since some anions are known to have a pinning effect [26]. Reductive doping can also be obtained by injecting negative charge carriers, like for example in polythiophenes substituted by electrowithdrawing groups, but charge mobility is usually lower than for p-doped polymers and the resulting conductivity lower.

1.3. Relaxation Effects

A very specific trend in the electrochemistry of conjugated polymers is related to relaxation effects. This was demonstrated experimentally by observing hysteresis phenomena ('memory effects') when cycling a conducting polymer film, with a marked influence of the negative switching potential on the position and height of the peak current in CV [27]. After the first experimental results by Nechtschein et al. [28-29], a model attempt was proposed by Otero based on electro-mechanical considerations [30-31]. General dependence of the logarithm of the waiting time with the waiting potential was proposed to fit the experimental data [29], further corroborated by the ESCR[5] model [32]:

$$\tau = \tau_0 e^{\frac{\Delta H * + z_c(E_s - E_c) - z_r(E - E_0)}{RT}}$$

where E_s is the characteristic potential for closure of the polymer matrix and z_c is the charge required to compact one mole of polymeric segments. E_o and z_r are the similar quantities for the reverse relaxation process and ΔH* is the conformational energy consumed by one mole of polymeric segments in absence of applied electric field. This model quantitatively describes the volume changes observed upon oxidation-reduction steps in conjugated polymers accompanying the insertion-desinsertion of ions and solvent molecules. The shrinkage observed upon undoping (reducing) the polymer impedes the ionic exchanges, so that long durations in the fully reduced state requires anodic overvoltage at the following scan to obtain the same peak current. However this model is controversial due to similar observations in non-polymeric systems [33] and very different behaviors in other solvents than acetonitrile (like ionic liquids) [34]. This question is still not fully elucidated and

---

[3] Electrochemical Quartz Cristal Microbalance.

[4] Probe Beam Deflection (also known as Mirage effect)

[5] Electrochemically Stimulated Conformational Relaxation

remains the purpose of investigations by Otero's group [35-36] among others, as this phenomenon is a key point for the development of actuators or artificial muscles (see §3.2).

2. NEW TRENDS IN ELECTROSYNTHESIS OF CONDUCTING POLYMERS

This paragraph is devoted to the more recent developments concerning the electrosynthesis of new conducting polymers or electropolymerization in new electrolytic media. We will also describe the main pathways leading to nano-sized materials made of conducting polymers alone or associated in hybrids.

2.1. New Monomers

a) Classical Monomers and Derived Polymers

At the beginning of the conducting polymers era, the only electropolymerizable monomers were derived from pyrrole, thiophene or aniline. However, research on modified aniline monomers quickly paced off since almost all polymers from substituted anilines had uninteresting properties [37]; in addition, aniline oligomers are difficult to synthesize and weakly stable upon storage. In the field of heterocycles, on the other hand, the situation was exactly the opposite. A remarkable variety of new monomers and subsequent polymers were prepared in the late 80's and 90's by several groups [38-46]. However, maybe at some point because of the limitation of the synthetic possibilities, the synthetic activity has also taken a quieter pace. One should however acknowledge works of the groups of Reynolds [47-49], Zotti [50-52], Bidan [53], Roncali [54-57] and Audebert [38,58-61], who steadily continued to provide new functional monomers until quite recently, mainly based on polymerizable pyrroles, thiophenes, ethylenedioxythiophenes (EDOTs) and their extended analogues. Most of their works aimed at producing low band gap materials, and the alkyldioxypyrroles [48] were a noticeable success in that direction. The synthesis by Roncali's group of several rigid backbone polymers and monomers was a key step in the production of low bandgap polymers by electropolymerization. The last step has been recently brought on by Sessler who synthesized and polymerized the most recent low bandgap polypyrrole, namely poly(naphthobipyrrole) [62]. This polymer has a very rigid planar backbone, which results in a nice multicolor electrochromism (Figure 6).

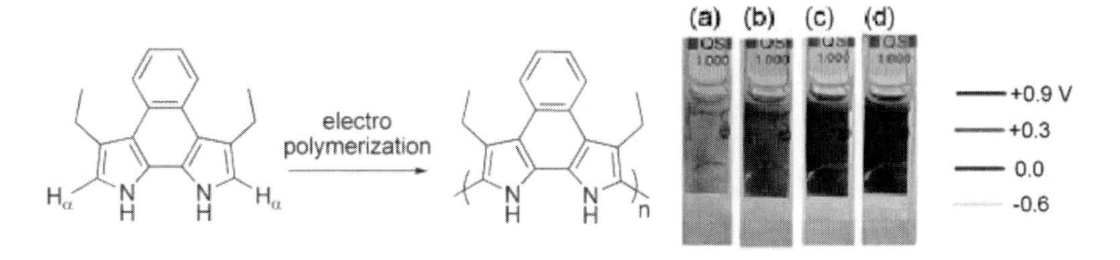

Figure 6. Left, electropolymerization of a β-protected naphthobipyrrole. Right, picture showing the multiple electrochromism of the film (adapted from ref. [62]).

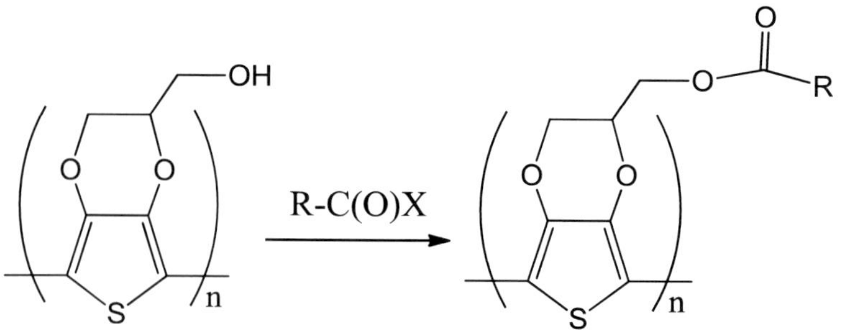

Figure 7. Post-functionalization of poly(hydroxymethylEDOT) with activated esters or acid chlorides.

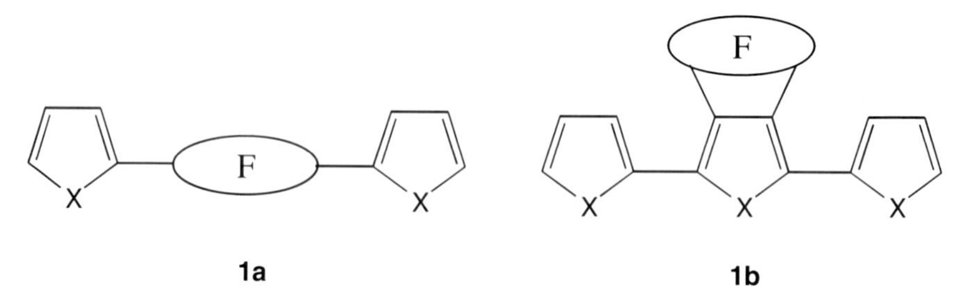

Figure 8. The two main strategies to add a new functionality to a conducting/conjugated polymer : inside the main chain (1a) or as a side group (1b).

Electrochromism [63] was actually a major goal within this quest for new monomers. An interesting case is poly(hydroxymethyl)EDOT (Figure 7), since this polymer contains an hydroxymethyl function which is useful for post-functionalization [64].

However, although started a long time ago, the addition of a property to a conducting electrodeposited polymer is still an interesting topic. This activity has never stopped, although it is not increasing any longer; this goal is also at the root of the synthetic effort towards several families of new monomers. Two strategies are possible for this aim (Figure 8): either the active moiety is inserted inside the chain (1a), or it is attached to the backbone through a conjugated or a non-conjugated link (1b).

Actually at the stage reached today, and somewhat in contradiction to occasional previous statements, it is not very clear if one strategy clearly primes over the other, and now things must clearly be thought in term of the expected properties. While a higher cooperativity of the co-monomers is expected if both are in the main chain [65], sometimes a clearer response is obtained with side groups. This has been well documented recently in a book chapter [66].

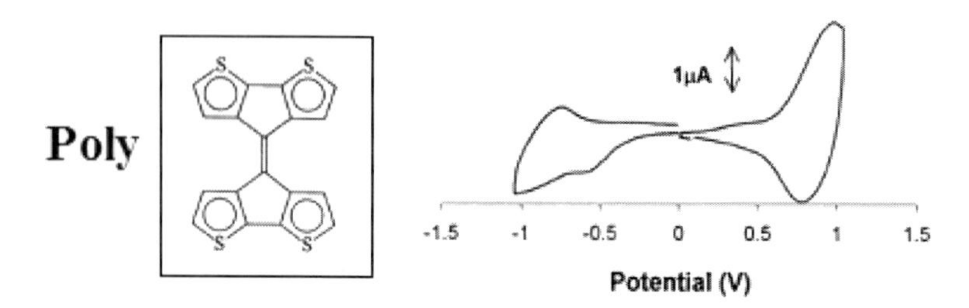

Figure 9. A typical low band-gap polymer with its CV feature (adapted from ref. [69]).

As previously signaled, a major aim at the origin of several works is again, and naturally, the quest for low band-gap polymers [67]. First Wudl [68], then Pickup [69] prepared very interesting monomers that they managed after to electropolymerize into low bandgap polymers, as shown for example in Figure 9.

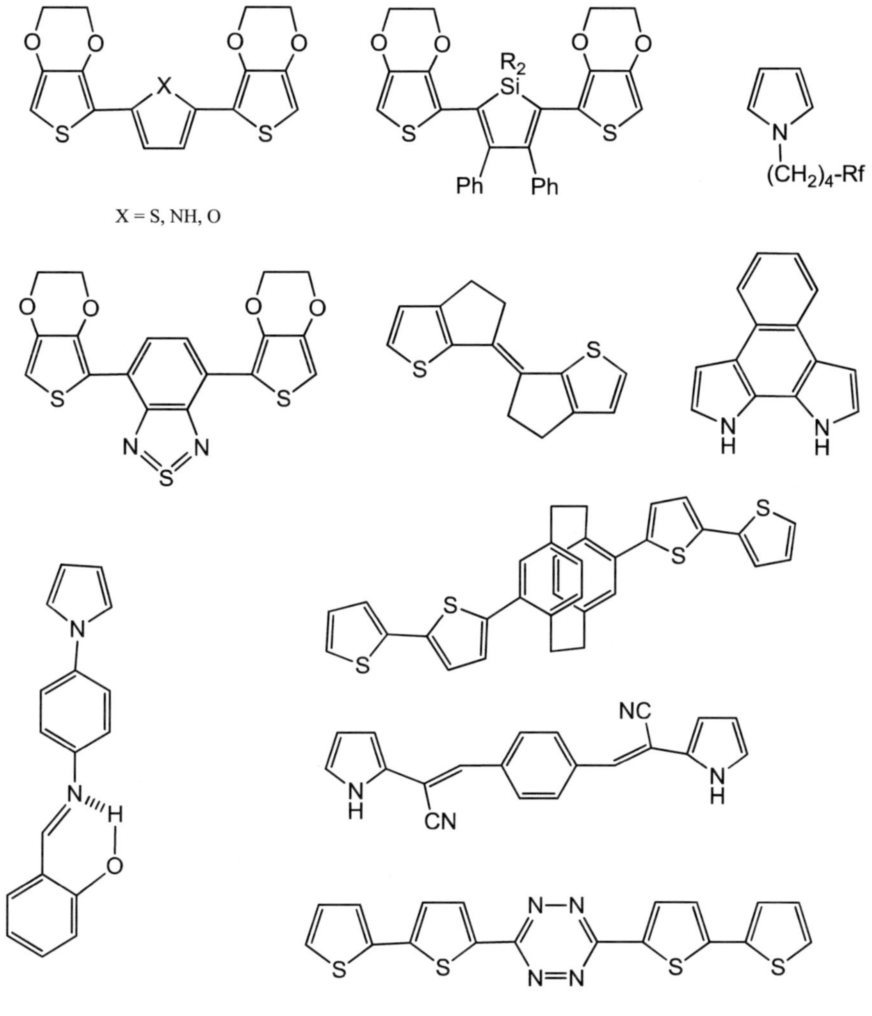

X = S, NH, O

Chart 1. A selection exemplifying original monomers giving polymers of interest by electro-polymerization (R = alkyl chain ; Rf = perfluorinated alkyl chain).

New functional polymers were also prepared, with an extended range of new and interesting additional functionalities, like ferrocene [70], TTF [51], catenanes [71], biotin [72], quinones [73-74] or side chains bringing on a desirable property like liquid crystal behavior [75], solution processability [76], polyelectrolyte nature [77-78], adhesion [79], etc… although all this is now more easily achieved by insertion in the main chain through chemical polymerization using the new organometallic coupling reactions that have emerged later on. A very interesting and to some extent little investigated topic is the preparation of fluorinated polypyrrole or polythiophenes. Since the seminal work of Roncali and Lemaire [44] other works have appeared. Indeed the preparation of poly(difluoropyrrole) has been reported by us [59] but, although the polymer displays very interesting features like a very high redox potential and a high doping level, its low stability, especially to residual water, precludes its use in devices. However, perfluoromethyl derivatives could be an alternative, as described in a recent theoretical analysis [80]. Meanwhile, Nicolas et al. have resumed work in the field, reporting the superhydrophobic and lipophobic properties of a series of polymers issued from perfluoroalkylpyrroles and EDOT [81-82] (see also Chart 1, monomer extreme upright).

Chart 1 shows a restricted selection including some very classical and some of the most recent new monomers published and electropolymerized (of course an extensive list is quite beyond the topic of this chapter, and the reader interested can look at the articles previously quoted and the references quoted therein).

A last field of interest worth signaling in this area is the search for higher dimensional architectures in electropolymerized conducting polymers, than the classical linear conjugated chain. This can be achieved through polymerization (or copolymerization) of branched or "star-like" monomers. Actually, since the first publication of the electropolymerization of tetra(terthienyl)silane [54] there has been a constant publication flux dealing with the electropolymerization of higher symmetry monomers, the most widespread consisting in 3-fold type.

While benzene most of the time introduces a barrier for electronic conductivity, on the other hand the s-triazine ring for example is particularly well adapted to this goal, since not only it provides the required symmetry, but in addition it has a sometimes desirable electron withdrawing character and synthesis of many derivatives is straightforward from cyanuryl chloride [83-84]. We have also described thiophenic analogues of violet crystal, where the node is a stable carbocation and which are fully conjugated [58]. This research field is still quite active, and one has to cite the beautiful work of Moutet and coworkers, who managed to prepare electrochemically cyclic polypyrroles of various sizes directly from pyrrole or classical pyrrole derivatives, using carefully controlled cyclization conditions with multidentate anions in the electrolyte [85-87].

b) Other Monomers

Many monomers have been subjected to electrochemical polymerization, but here will be only mentioned what seems to be the most significant examples: carbazoles, fluorenes, siloles, phospholes and finally salens. Polycarbazoles are a very interesting class of materials, however, like polyaniline, they are much preferably prepared by chemical coupling [88]. However, electrochemical polymerization, like with many other polymers, allows coating substrates with various shapes [89] and also to some extent leads to nanostructuration.

Polyfluorenes are a very interesting case, because chemical and electrochemical polymerization are both possible and especially, electrochemical polymerization is relatively easy; like with all other monomers, this introduces the advantage of avoiding expensive synthesis of adequate halogenated or boronate precursors, while on the other hand the electrochemically synthesized material presents more defects. Many examples can be found in the literature [90], sometimes including other electroactive moieties, most owing to the group of Rault-Berthelot [91-92]. A very nice feature with fluorenes is the easy access to spirofluorenes [93], and higher analogues, allowing the preparation of a wide range of superstructured polymers [94-95].

Siloles [96-97] and especially phospholes [98-99] containing polymers are also particularly interesting, and have shown outstanding properties, especially in view of photovoltaic applications, as demonstrated by Réau and coworkers [98]. However, neither silole nor phosphole does electropolymerize alone (although 1,1'-bisphosphole was very recently shown to electropolymerize [100]). These moieties are always linked (usually at the α positions) to more easily polymerizable rings, most of the time pyrroles or thiophenes [99]. Finally, an interesting class of monomers, which almost exclusively polymerize through electrooxidation, is the one of salens (Figure 10). These very classical organometallic phenolic tetracoordinated complexes have been found to electropolymerize independently first by Goldsby [101], then by Bedioui [102] and us [42,103]. Actually, a very interesting feature of these compounds is the possibility to exchange the metal and perform catalytic reactions [104]. The Freire's group has also developed some polymers in this direction [105-106].

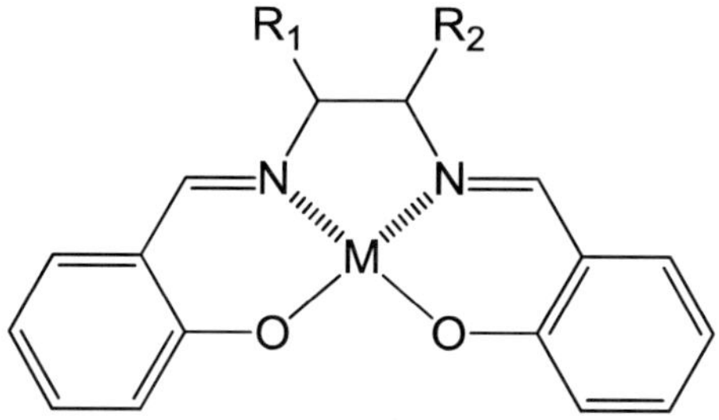

M = Ni, Cu, Sn, Co, Fe, Zn....

Figure 10. Generic formula of a salen complex; polymerization occurs whatever the metal by oxidative coupling on the para position vs. the oxygen atoms.

2.2. New Electrolytic Media

Actually, over the last 20 years, only three noticeable new domains concerning new synthetic methodologies, besides standard electropolymerization in organic media, have been explored, with different success. The first and oldest is water emulsion polymerization. The second domain is the use of boron trifluoride etherate as an additive for polymerization. Finally, the last noticeable extension of the classical electropolymerization is the polymerization in room temperature ionic liquids (RTIL).

a) Polymerization in Water Emulsions

The incentive for these investigations was obvious since water is a much cheaper and cleaner solvent than the usual organic solvents (acetonitrile, dichloromethane, propylene carbonate ...) classically used for the electrosynthesis of conducting polymers. However, at the exception of pyrrole which is slightly soluble in water, none of the classical monomers that can be easily polymerized is soluble in water; in addition, water is a nucleophilic solvent which can easily attack the electrogenerated cation-radicals, especially the reactive ones (like thiophenic cation-radicals e.g.) and inhibit the polymerization. Therefore some tricks have to be used in order to perform this type of electropolymerization. Although one report describes electropolymerization in unstable emulsions mechanically prepared [107], almost all the time

[108] a surfactant agent is necessary to obtain a stable emulsion, and therefore perform the electrochemical synthesis.

However, the different techniques developed in water do not seem to produce polymers much superior to the ones obtained in organic solutions [109], except in the case of EDOT [110-111], which is a relatively low potential monomer. Therefore its cation-radical is more tolerant to the presence of water and polymerization proceeds more easily. In addition, lowering of the oxidation potentials has been observed in the presence of anionic surfactants, as a probable result of the interactions between the surfactant anions and the EDOT cation-radicals [110].

b) Polymerization with the Addition of BF_3-Etherate

The use of boron trifluoride etherate for thiophene electropolymerization has been reported for the first time in 1995 by Shi and coworkers [112]. At this stage, this unusual and highly acidic reagent used pure as a solvent was shown to considerably improve the conductivity (around 50 S/cm against 1-10 for conventional polythiophene) but especially led to polythiophene with an exceptional tensile strength comparable to aluminum! This work has triggered the use of BF_3-etherate for the polymerization of numerous monomers like benzene [113] or furan [114]. A very strange feature is that the use or the addition of BF_3-etherate to the electrolytic medium considerably lowers the oxidation potential of hardly oxidizable monomers like thiophene or their halogenated derivatives for example [115-116]. The reason for this –complexation of the monomer and decrease of the aromaticity–[117] does not sound very convincing, since usually electron withdrawal to an organic substrate of any kind usually makes it harder to oxidize. Anyhow, the use of this medium has allowed electropolymerizations that were impossible to perform before its arrival on the scene. However, and somewhat bizarrely, there are no reports on the use of other Lewis acids in the polymerization of heterocycles. This absence is rather puzzling and tends to show that, in our opinion, the role of BF_3-etherate might be much more complicated and less understood than already proposed.

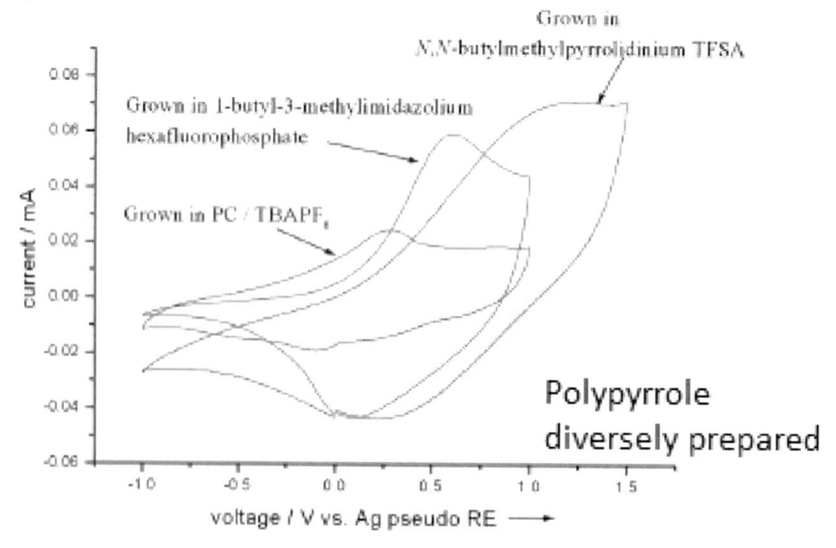

Figure 11. Very different behaviors from PPy films grown conventionally and in various ionic liquids. (adapted from ref [118]).

c) Ionic Liquids (RTIL)

The use of room temperature ionic liquids for the synthesis of conducting polymers has been pioneered by Wallace [118] et al., who in their first paper on the topic in Science, demonstrated both the synthesis, and especially the long term stability to cycling of several conducting polymers including polypyrrole. They demonstrated that very robust films could be made and cycled much longer than in standard organic or aqueous electrolytes.

The Wallace's group published several papers [108,119] and was followed by others. Meanwhile, Hapiot et al. published recently a complete approach of electrochemical mechanisms in RTILs [120] showing that indeed electrochemical processes can be strongly different, especially from the kinetic point of view. In the field of polymerization, this may lead to important differences, as Wallace pointed out. Figure 11 shows the difference between conventional and RTIL grown polymers.

Actually, and besides the obvious fact that polymers much more resistant to cycling can be prepared, for the rest the influence of RTILs compared to conventional media is not very clear. There are statements that smoother and more electroactive films can be prepared [108], however the conditions used in RTIL, and especially the concentration of monomer, are higher than in conventional polymerization, which renders the comparison more difficult [120]. More conductive, free-standing films of polyselenophene can be prepared in RTIL [121]. In some RTILs (but not all !) the redox system of PEDOT is split into two parts; this has been attributed to the coexistence of two different morphologies in the polymer, one compact film close to the electrode, switchable at a low potential, and one less dense film, switchable at a higher potential [34]. Finally, PEDOT prepared in RTILs has been shown to display better performances when used in interpenetrated network actuators [122].

d) Sonochemical electropolymerization

Actually these last years, a novel way to exert a better control on the polymerization of conducting polymers was the use of sonochemical conditions [123]. This has been performed in the case of EDOT [124] and pyrrole [125] by the group of Hihn and also by Reyman in the case of polythiophene [126]. Usually the polymer deposited under ultrasounds has a more compact morphology. The use of ultrasounds, in specific conditions of wavelength and power, sometimes allows also defining a more focused polymerization zone, and has even been used as a way to micropattern the deposition of the polymer on a copper substrate (figure 12) [127].

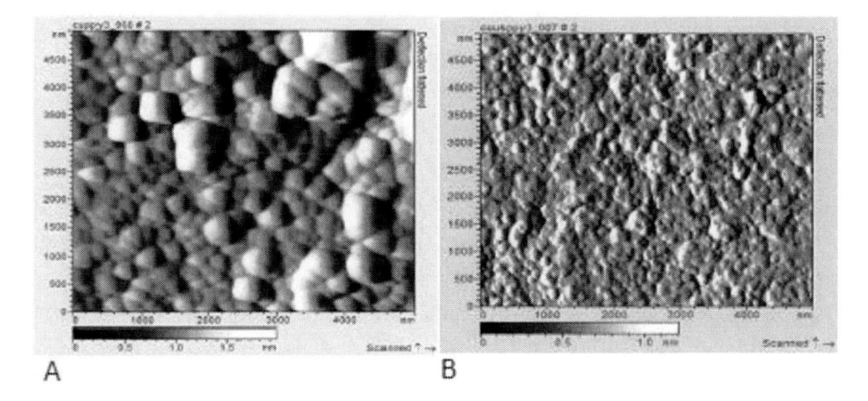

Figure 12. (Continued).

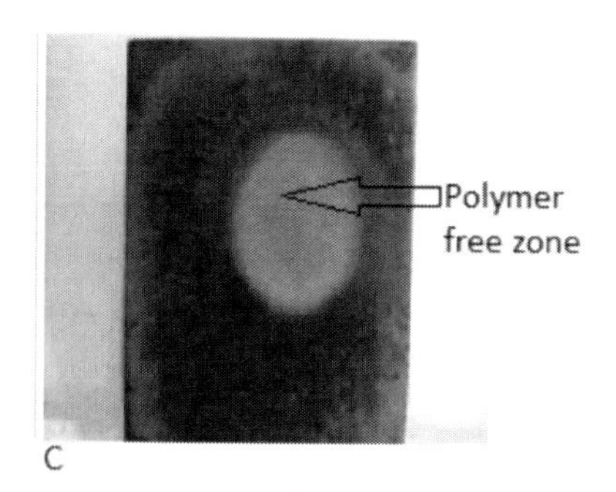

Figure 12. Atomic force microscopy (AFM) images of polypyrrole electrosynthesized by chronoamperometry in sodium salicylate (0.1M) aqueous solution. A) without ultrasound and B) with ultrasound irradiation (500 kHz, 25 W). C)Photography of a copper substrate obtained after ultrasound masking against polymer deposition. The central area (about 3 mm wide) which was irradiated by the focused ultrasound beam (750 kHz; 25W) during electropolymerization is polymer free (adapted from ref.[127]).

2.3. Nano-Objects

In the last decade, a great effort was made to develop new morphologies of conducting polymers, especially at the nano scale. A recent review of the subject lists the various methods encountered with their own advantages and drawbacks [128]. Historically, the first methods used to generate nanowires were based on the template approach, either hard (alumina matrix, porous silicon, polycarbonate …) or soft templates (micelles of surfactants, liquid crystals …). Most of the time these methods involved chemical polymerization but in some cases electropolymerization can also be employed. The pioneering work in the field of electropolymerization in porous membranes comes from the Martin's group, who developed the polymerization within pores of anodized aluminum oxide membranes [129-130]. Based on this concept, Lee et al. obtained PEDOT nanotube arrays by electrochemical polymerization after dissolution of the porous template sputter coated on an ITO working electrode, another ITO acting as the counter electrode (see Figure 13) [131]. These nanostructures exhibit enhanced switching rates compared to bulk polymers, due to limited diffusion paths for counter-ions [132]. However, this synthesis method requires template dissolution and this may lead to nanostructure collapse and degradation of the properties. A way to cope with this issue is either to use polymeric scaffolds as templates that are easier to remove, or to use self-degrading templates. Block copolymer templates using PC, PS, PMMA[6] or other soluble polymers have been coated on electrodes to generate similar nanowire arrays [133]. Electropolymerization allows controlling the height, diameter and in some cases morphology of the nano-objects.

Soft template methods use self-assembly of amphiphilic molecules to make spherical or cylindrical micelles (or reverse micelles depending on the solvent) around which polymerization occurs. They are associated to chemical oxidative polymerization, as these nanostructures are formed in bulk solutions rather than on a conductive surface. The use of special dopants acting as structuring agents (like naphthalene sulfonic acids) has opened the

[6] PolyCarbonate, PolyStyrene, PolyMethylMethAcrylate.

way to new "template free" syntheses [134], with interesting adaptation to electrochemical techniques when electrogenerated gas bubbles are also involved as "templates"[7][135].

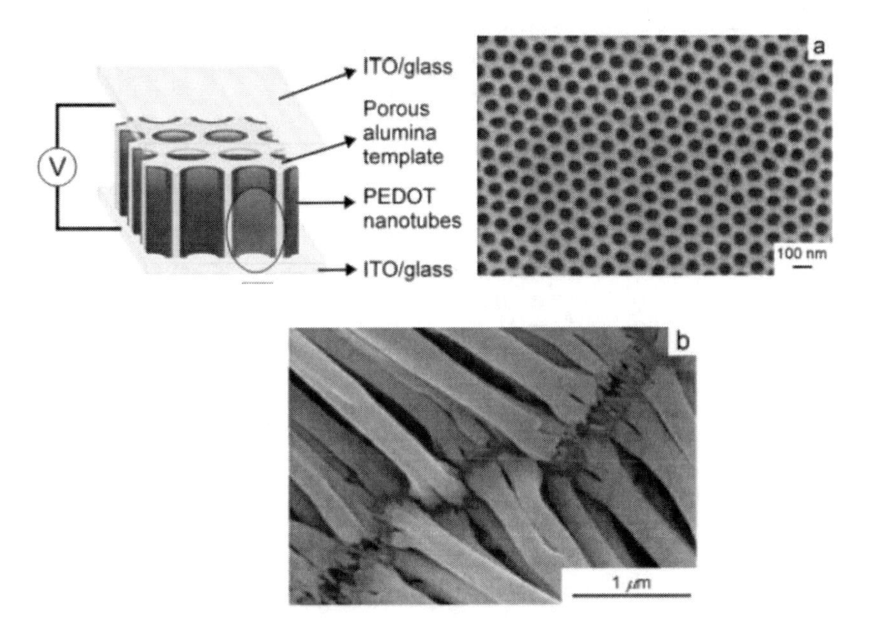

Figure 13. left : Scheme of the set-up for electropolymerization of EDOT within pores of alumina template ; right : SEM pictures of the template (a) and PEDOT arrays after template dissolution (b) (reproduced from ref. [131] with permission, Copyright Wiley).

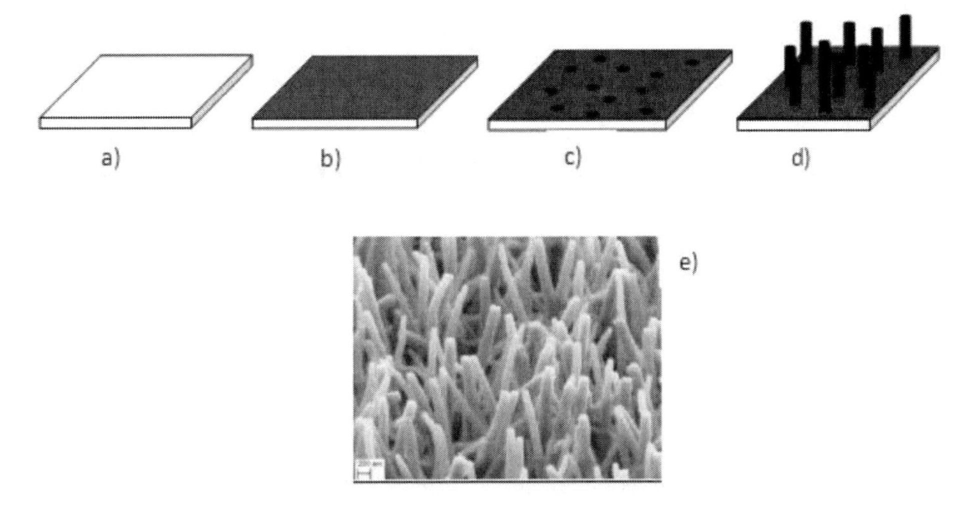

Figure 14. Template free electrochemical method to make conducting polymer nanorods : a) initial electrode surface ; b) first layer of conducting polymer ; c) generation of oxygen gas bubbles on the overoxidized polymer ; d) further polymerization on conductive spots after gas evolution; e) SEM pictures of the final PPy nanorods (reproduced with permission from ref. [136]).

[7] Actually, these methods are called 'template free' as they do not need external template agents, although they use structuring agents coming from the dopant or from electrogenerated species.

In a recent paper, Debiemme-Chouvy took benefit of these gas bubbles template to obtain PPy nanowires by a pure electrochemical method [136]. The various steps of the mechanism are illustrated in figure 14.

The polymerization is performed at constant potential in aqueous solution using an excess of weak acid (HA) as the main dopant and a small amount of salt (X^+Y^-) as a minor dopant. Initially, the electropolymerization leads to a first layer but polymerization stops when A^- is consumed by the reaction with the protons generated by the polymerization. As the potential remains constant, the electrochemical reaction changes, and water oxidation relays the monomer oxidation, leading to hydroxyl radicals which overoxidize the polymer first layer, resulting in electrode passivation. A small amount of hydroxyl radicals are able to dimerize in hydrogen peroxide and then in oxygen gas, making bubbles on the surface that block the passivation. Nanometer-size spots of non passivated surface are thus obtained that allow further polymerization with the remaining small amount of dopant (Y^-). The length and density of the conducting polymer nanorods are controlled by the electrode potential and polymerization time.

The overall process can be summarized by the following equations:

$$n(Py) \rightarrow (Py)_n + 2ne^- + 2nH^+ \quad \text{(r1)}$$

$$H^+ + A^- \rightarrow HA \quad \text{(r2)}$$

~~$$(Py)_n + n\delta A^- \rightarrow \left(Py^{\delta+}, \delta A^-\right)_n + n\delta e^- \quad \text{(r3)}$$~~

$$H_2O \overset{k_1}{\rightarrow} OH \cdot + e^- + H^+ \quad \text{(r4)}$$

$$(Py)_n + xOH \cdot \rightarrow (Py - OH)_n + xOH^- \quad \text{(r5)}$$

$$2OH \cdot \overset{k_2}{\rightarrow} H_2O_2 \rightarrow O_{2(g)} + 2e^- + 2H^+ \quad \text{(r6)}$$

$$n(Py) + n\delta Y^- \rightarrow \left(Py^{\delta+}, \delta Y^-\right)_n + (2+\delta)ne^- + 2nH^+ \quad \text{(r7)}$$

(r1) features the initial polymerization step, with consumption of the weak base A^- by the generated protons (r2). This impedes further doping (r3) and polymerization stops. Water oxidation takes place generating hydroxyl radicals (r4) that can overoxidize the first PPy layer (r5) resulting in a passivation of the electrode surface. In parallel, the dimerization of these radicals into hydrogen peroxide, being further oxidized into oxygen gas (r6), is likely to occur, albeit with a slower kinetics ($k_2 < k_1$). Finally PPy growth can take place again where oxygen bubbles have been generated, the surface remaining active at these places (r7). Other interesting examples of template-free electropolymerization processes leading to nano-size structures can be found in the literature [137-138]. One can mention that phosphate ions are able to induce nanofibrillar growth for PPy through hydrogen bonding, provided high concentrations of phosphate vs. perchlorate dopant and moderate oxidation potential are used as experimental conditions (see figure 15).

PANI was also successfully electropolymerized under the form of nano-objects in moderate acidic and galvanostatic conditions. According to the current density used, nanospheres or nanobelts are generated on the electrode surface (see figure 16).

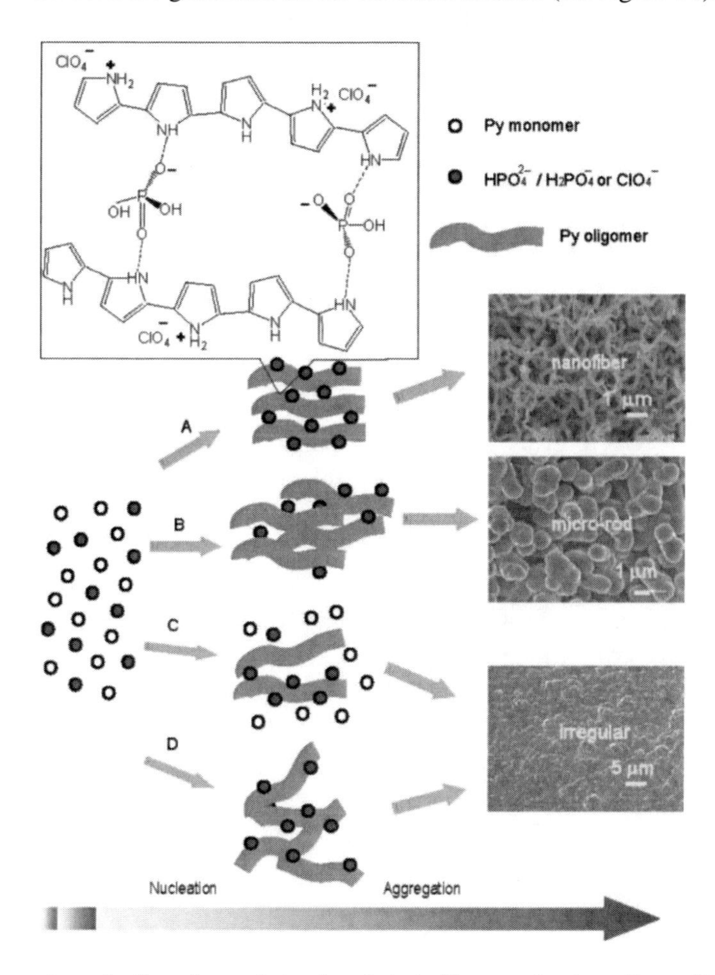

Figure 15. Electropolymerization of pyrrole in phosphate buffer : mechanism of nanofiber formation (Reproduced from ref. [137] with permission).

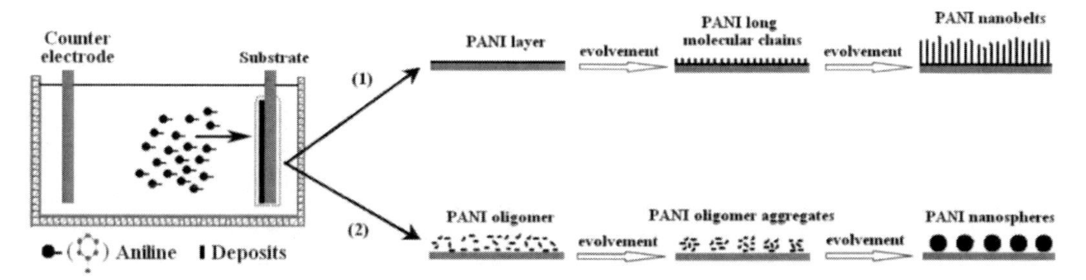

Figure 16. Electrodeposition of PANI nanostructures under galvanostatic conditions : (1) low current density (0.5 $mA.cm^{-2}$) ; (2) high current density (5 $mA.cm^{-2}$). (Reproduced with permission from ref. [138]).

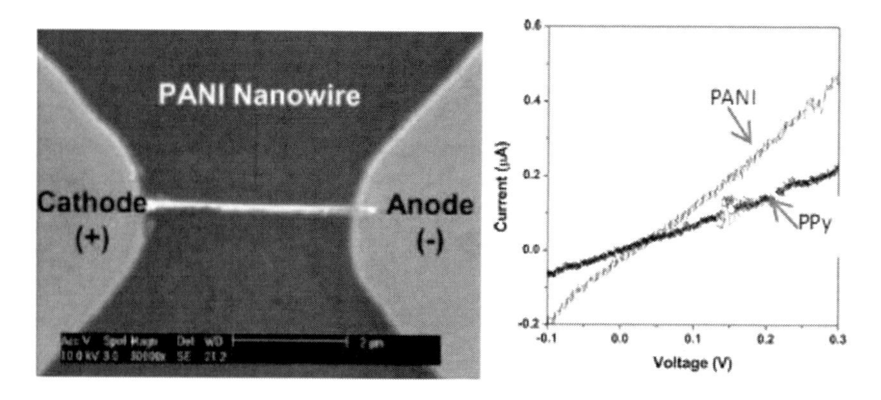

Figure 17. left) SEM picture of a single PANI nanowire bridging the gap between two electrodes ; right) I-V ohmic trace for PANI and PPy nanowire (Reproduced with permission from ref. [139]).

These various examples show that nanostructures of various shapes can be easily obtained with conducting polymers without external templates and using electrodeposition processes. The electrochemical parameters used for the synthesis (current density, potential, electrodeposition time) allow controlling the size and morphology. These template-free synthesized nanosystems usually exhibit higher conductivity than counterparts obtained from templates, and their high specific areas are essential for high capacitance and electrocatalytic properties. Other applications in nanoelectronics can be envisaged, when single nano-objects are generated between two electrodes [139]. Figure 17 shows an example of PPy and PANI single nanowire galvanostatically electrodeposited between two electrodes.

2.4. Nanocomposites with Metal and Carbonaceous Materials

a) Metal Nanoparticles

Among the various examples of composites made of conducting polymers with metal nanoparticles (NP), only a few involves electrochemical techniques in the synthetic pathway. Liu and Chiang proposed a nice way to design core-shell Au-PPy nanocomposites in two steps [140]: first, Au NP stabilized by tetrachloroaurate ions are generated by electrochemical oxidation-reduction cycles on a gold substrate in a potassium chloride solution. Then pyrrole is added and autopolymerization occurs on the Au NP surface upon oxidation of pyrrole by $AuCl_4^-$. Gold NPs deposited on a PANI film have also been reported as new materials with electrocatalytic properties. After electropolymerization of the monomer on a titanium surface, galvanostatic deposition of Au from $AuCN_2^-$ dissolved in a citrate buffer is performed [141]. A recent paper describes the one-step formation of Au-PPy nanocomposite from cyanoaurate solution by applying oxidation-reduction cycles. In that case micrometer size Au platelets deposited on a PPy film can be obtained by this full electrochemical process [142]. In a similar way, Cu NP-PPy nanocomposites were obtained by electropolymerization of pyrrole followed by pulse potentiostatic reduction of copper chloride in aqueous solution. According to the charge involved in this latter step, copper clusters of various sizes in the few hundreds of nanometer range could be synthesized [143]. Although less numerous than the chemical routes in presence of templates or in biphasic conditions, electrochemical methods may reveal an easy and versatile way to obtain such nanocomposites.

b) Carbonaceous Nanomaterials

The main purpose for combining conducting polymers with carbonaceous materials is the development of new materials for energy storage, especially supercapacitors. During the past decade, there has been a great interest in using carbon nanotubes (CNT) to make either composite films with conducting polymers or to act as conducting substrates for coating thin layers of conducting polymers on their surface. The various synthesis methods of CNT-ECP composites and their use for supercapacitors were recently reviewed [144]. As CNT are good electrical conductors, there is no impediment to the electropolymerization at their surface; the main issue is to fix them on a conducting support for macroscopic handling. In the first example of CNT-ECP nanocomposite published, this was performed using a conducting paint [145]. Later on, several examples appeared with well-aligned arrays of CNT allowing the obtention of coaxial wires [146] with either PPy [147-148] or PANI [146]. The CNT can be deposited on the electrode surface under other forms like buckypaper or solvent cast films, while the use of a cavity microelectrode enables electropolymerization directly on the CNT surface without any electrode deposition process [149]. Efforts were also concentrated on coating CNT with ultrathin ECP layers, with the aim of improving the conductivity by decreasing the contact resistance. This was mainly achieved by vapor phase polymerization [150-151] or chemical polymerization in aqueous solution controlling the monomer concentration [151].

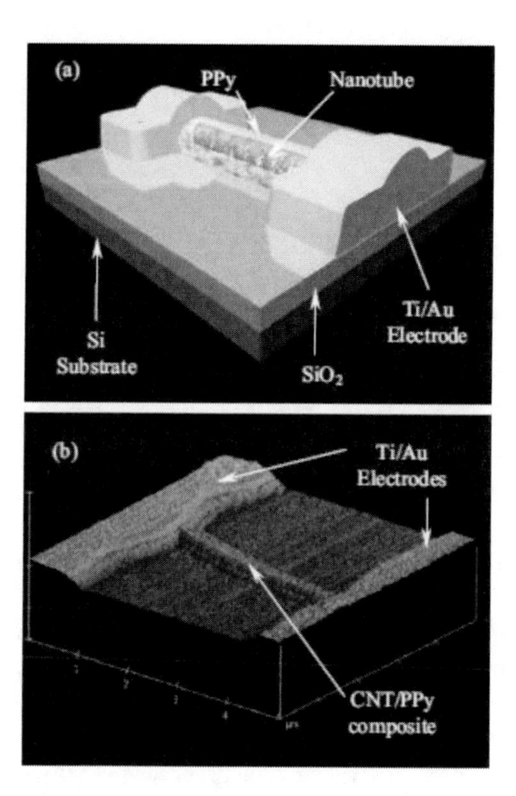

Figure 18. Schematic representation (a) and AFM picture (b) of a single carbon nanotube/PPy hybrid between two electrodes (Reproduced from ref. [152] with permission).

Finally coating of individual single wall carbon nanotubes with a thin layer of PPy was achieved in a source-drain configuration on Si/SiO_2 substrate (Figure 18) [152]. These nano-devices exhibited unexpected suppressed conductance after PPy electrodeposition.

Of course alternative protocols to make composite films of CNT and ECP simply consist in co-electrodeposition of both nanomaterials from solutions [153]. This usually requires oxidative pre-treatment of the CNTs or the use of surfactants since CNTs are usually not soluble enough in the polar media used for electrodeposition [154-156]. Composites films of that type were prepared with each of the classical ECPs, namely PPy [153], PANI [157-158] and PEDOT [159]. Nanocomposites with PTh were also electrosynthesized in that case for photovoltaic applications [160].When CNTs are negatively surface charged, they may be incorporated in the polymer films also as dopants, with consequences on the electrochemical properties of the composite that may differ from the pure conducting polymer ones. For example, the CV is more capacitive and electron delocalization is found higher in the composites [158].

In the last five years, the enthusiasm for composites with CNT has gradually shifted towards graphene materials. The first composites were prepared by chemical polymerization after dispersion of the graphene material in the polymerization medium. Quickly after, electropolymerization techniques appeared as alternative ways to give transparent conductive graphene-ECP films [161-162]. Similarly to CNTs, graphene sheets can be incorporated as dopants when functionalized by negatively charged groups [163]. The targeted applications for these composites are mainly supercapacitors [164-167]. It is clear that the search for nanocomposites with tailored and controlled design will be the aim of many investigations in the future and electrochemical techniques can find a place in that query.

3. Applications

While the fundamental works like the search for low bandgap polymers has receded, applications are the main driving force nowadays to prepare new conducting polymers. The first historical application of conducting polymers, and especially PPy, has been accumulators [168]. While research in the area has been extremely active in the 80's and even in the 90's [169], it is more or less finished by now. The reason is probably the success of new electrode materials for energy storage like nano-divided metal oxides or chalcogenides, which now outrank most conducting polymers [170], although PANI batteries are still fabricated. However, the field of high-energy supercapacitors has meanwhile emerged as a prolongation of this application. The use of especially PPy and PEDOT derivatives for high energy capacitors continues to be the focus of several works [171]. Similarly the relatively ancient development of conducting polymers based modified electrodes has much decreased, at the exception of the biosensors area [172]. Electrochromism is still an important issue, but concerns now mainly PEDOT and its derived materials [173].

Fortunately, some other applications have emerged in the 90's, and continue to develop, while very new and original applications also continue to appear from time to time. Among the classical applications that appeared more than a decade ago, especially one met a considerable development, which is the search for efficient anti-corrosion coatings. In this paragraph, we will successively describe the state-of-art and recent advances in the fields of energy storage, actuators, biosensors, anti-corrosion coatings and drug delivery.

3.1. Batteries and Supercapacitors

As previously mentioned, the research for standard batteries from conducting polymers has almost stopped, at the exception of the realization of nano-devices [174]. On the other hand, some polymers for use in the preparation of supercapacitors are still in use. For example, pulse polymerization has been shown to lead to a more performing polymer [175]. The search for better contacts in many types of accumulators has triggered technical studies, but PPy is one of the best conducting link between the active matter of a device and the metal electrode, especially if the metal is divided. However, PPy (or PEDOT) [171] is almost never used alone, but most of the time in association with low-dimensional carbon materials, like nanotubes, graphene or fullerenes (see §2.4) [176]. Actually in this approach, the conducting polymer not only contributes to the energy storage, but, more importantly, creates a conducting link between the active carbon moieties. Indeed it is known that contacting as well as processing new carbon materials like the above cited, is often difficult or even impossible, and therefore the conductivity of composite materials often lies well below the one of each single component [149].

On the other hand, many composites like conducting polymers/polyoxometallates and others have been used towards such purpose [177]. It seems that large size dopant ions improve the porosity of the composite and therefore its performances in supercapacitors [178]. The use of conducting polymers in fuel cells has also been attempted and recently reviewed [179].

Some results are encouraging, because conducting polymers can be electropolymerized and create a link between the catalyst particles, and frequently also present proton conduction. However, it seems unfortunately unlikely that this polymer could efficiently resist the harsh conditions of a fuel cell over its life time.

3.2. Actuators

Since the pioneering work of Otero [180], Inganas [181], and Smela [182], the field of conducting polymers, and especially PPy based actuators has shrunk a little bit, but recent publications continue to appear. The principle of a performing actuator is represented in figure 19. Two separate conducting polymer films are deposited on both sides of a supple insulating substrate as represented on the left drawing. Both films are connected to an electrochemical device, which can control the redox state of the film; when it gets oxidized, it swells and curves the actuator in the opposite direction. Both films can be controlled separately, which allows a huge deformation span for this type of actuator.

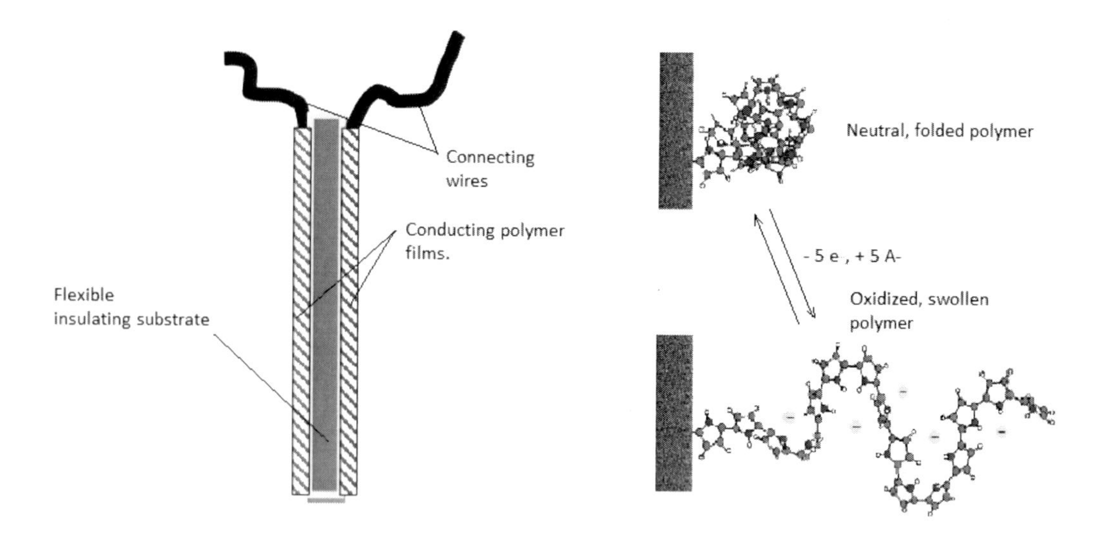

Figure 19. Left: principle of a two-films conducting polymer actuator. Right : drawing representing the electrochemical unfolding of the film at the molecular level.

More simple devices have also been described in the past, involving just a flexible metal foil acting as the electrode, and a CP film. Such actuator can also bend, but have at most a twice smaller deformation span.

While nice progresses have been made in the field up to the early 2000's, now it seems that the performances of the devices tend to reach a limit; however some progress continues to be observed concerning the motion kinetics of the actuators, which by now encompasses the 20 Hz range [183]. It should be noticed that actuators can be used the other way round, as stress sensors, measuring the variation of the redox potential as a function of the applied deformation to the system. In this case, since the deformation span is small, the most simple kind of device is used, that is, a single polymer film deposited (but as adherent as possible) on a flexible conducting substrate.

3.3. Anticorrosion

Application of conducting polymers for anticorrosion protection emerged quite later on, and roughly dates back from the end of the 90's [184-185], although earlier patents in the field might have appeared slightly earlier. Actually, the fact that the action mechanism of conducting polymers as corrosion protecting agents is quite counter-intuitive (why an oxidant should protect a metal against oxidative corrosion?) probably explains this situation. It has been recognized even earlier that the fact that the polymer is redox active allows defining the electrochemical potential at the metal/polymer interface and therefore bends importantly the Fermi level at this interface. The working principle is illustrated in Figure 20.

This process is fundamentally different from the classical anti-corrosion agents which behave, either as a sacrificial electrode, or simply like a physical barrier, and are rarely properly understood and explained in the literature [184,186].

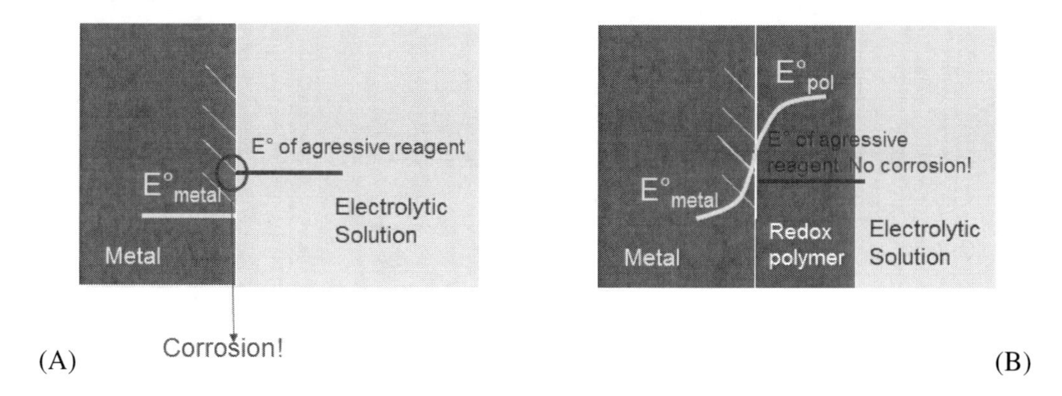

Figure 20. Scheme depicting the action of a conducting polymer deposited on an oxidizable substrate. Left) No protection.Right) The polymer makes the metal surface potential bend toward higher values and corrosion can be stopped.

At the beginning however, the coatings were too coarse and too porous, allowing the penetration of corroding agents which often exfoliate the metal/polymer interface, leading to a fast protection loss. The relative frailty of PPy, which is easily submitted to chemical degradation, or undoped by atmospheric agents, was also a drawback for this type of applications. However, quickly researchers learned to protect PPy with a strong mechanical barrier. Sol-gel coatings, PMMA and related polymers proved particularly efficient under this respect, as noticed by several works [184,187-188]. Along a parallel path, many works aimed and succeeded at polymerizing pyrrole onto several types of oxidizable metals. By now, electropolymerization techniques have been developed for almost every corrodible metal, including of course iron [189-193] but also more reactive metals like zinc [186], lead and especially aluminum [187-188,194-195], this latter representing a special goal due to the extreme need in aeronautics industry. The subject has been reviewed recently [196].

Meanwhile, the adherence of the polymer layer has been considerably improved onto the metal, thus increasing the duration of the corrosion protection because of a more efficient and long-lasting electrical contact between the metal and the polymer. In this area also, many type of composites, specifically aimed at increasing the overall robustness of the polymer in view of anticorrosion applications have been tried with various success, like for example soluble PEDOT additives in paints [197].

3.4. Biosensors and Related Materials

The use of PPy for biosensors has started in the 90's [198], most of the time with the glucose oxidase wiring as the main goal. The subject was reviewed by Cosnier a few years later [199].The conducting polymer in this respect acts as the mechanical holder of the enzyme, and insures the overall conductivity of the entire modified electrode. However, an additional redox mediator was almost all the time necessary in these first generation enzyme electrodes, because the active site of the enzyme, especially in the case of glucose oxidase, is much too far from the conducting polymer for it to act as the electron source or sink.

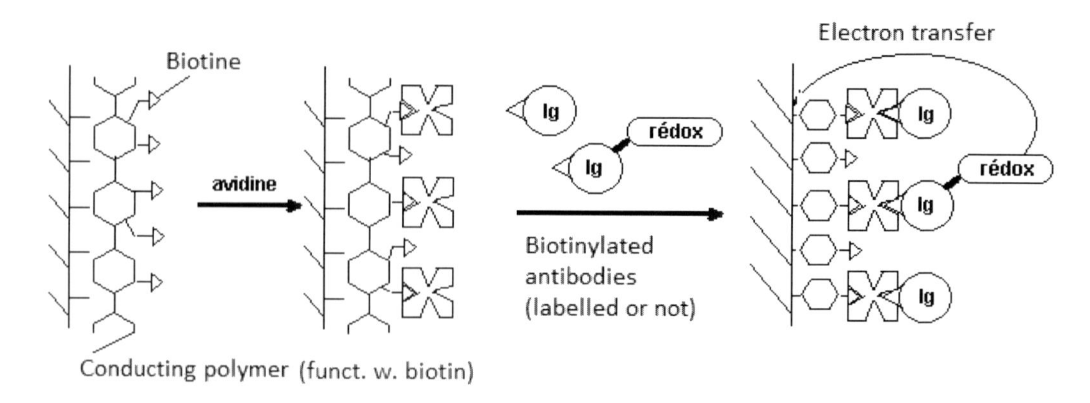

Figure 21. Scheme and functioning of a biotinylated conducting polymer biosensor.

Further works have aimed at improving the enzyme-polymer-electrode contact by using modern objects, like for examples nanotubes or Pt nanoparticles [200]. The efficiency of conducting polymers based biosensors has also been improved by nanostructuration, and Wang et al. have shown that electropolymerization can be successfully used in that direction [201]. Also recently Au nanoparticles embedded in a modified electrode made of electropolymerized PANI nanospheres have been shown to bind efficiently DNA for sensing applications [202]. While in first published works PPy, and to a lesser extent PANI, were the preferred polymers used for biosensors, very recently PEDOT has supplanted these "historical" polymers, as clearly explained in two excellent recent reviews in the field [48,203].

Later on, many enzymes have been integrated to electropolymerized conducting polymers, like peroxidases [204-205], superoxide dismutase [206], as non-limitative examples. The main way to attach bioactive systems to conducting polymers is certainly post-functionalization through the biotin-avidin system (Figure 21) [207] The extremely high affinity constant ($K = 10^{15}$) of the biotin-avidin binding is the main reason for the extremely frequent choice of this otherwise simple system.

3.5. Drug Delivery through Conducting Polymers Switches.

Due to the possibility of switching under electrochemical control, conducting polymers have been recently intensively studied for drug delivery [208]; development of this approach has increased over the last ten years. The drug delivery can be performed through two main paths: 1) Expelling of a charge drug, for example during the polymer reduction. 2) Release of the drug to electro triggered polymer swelling. For example, a PPy coated micro electrode (Figure 22) has been used to deliver neurotrophin to neurons.

Many examples of charged drug delivery by various types of conducting polymers, albeit essentially PPy, have been presented in the literature, for example the delivery of the cationic chlorpromazine drug [209]. Anionic drugs can be even more easily delivered, since doped PPy is a positively charged polymer [210].

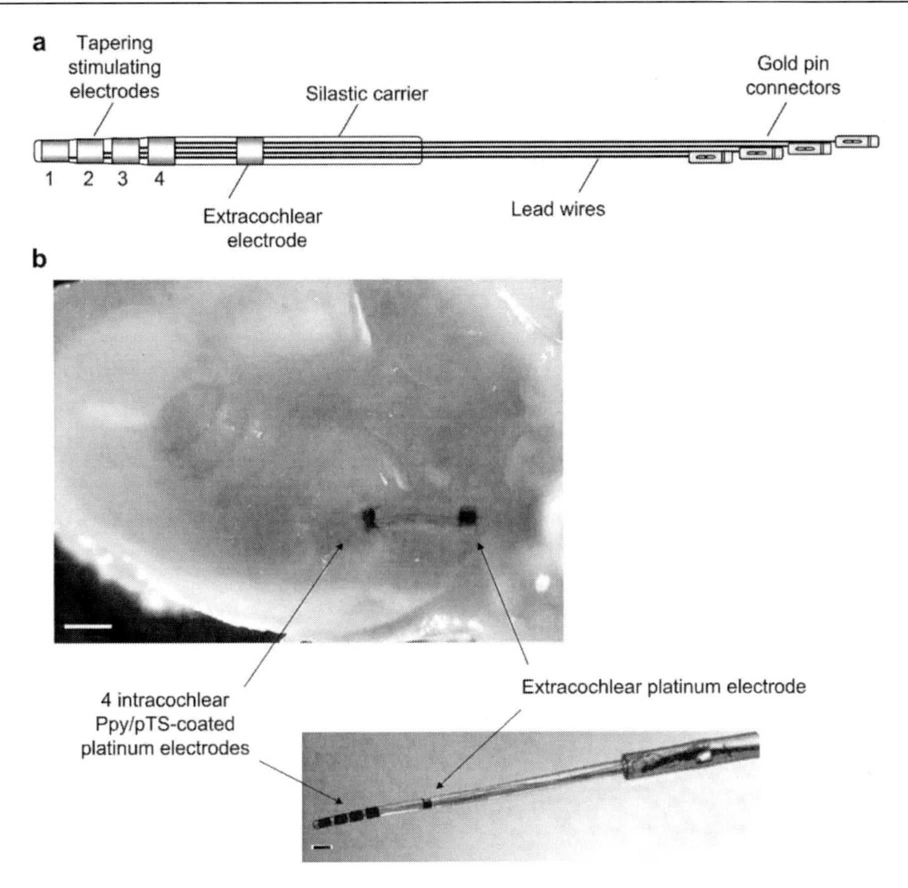

Figure 22. Four-ring platinum electrode array for implantation in guinea pigs (GPs). (a) Scheme of the electrode array showing the four active electrodes individually wired for stimulation as electrode pairs and an extracochlear electrode as a marker for insertion depth. (Diagram not to scale). (b) An electrode array coated with PPy/pTS implanted into a GP cochlea. The fourth electrode can be seen protruding from the cochleostomy in this example. The fifth uncoated platinum extracochlear electrode is also visible (reproduced with permission from ref. [208]).

This recent, but fast expanding field represents probably one of the most promising applications to come related to electropolymerized conducting polymers in a near future.

CONCLUSION AND FUTURE

As can be observed from the content of this chapter, research on electropolymerized conducting polymers is still active. However, one can notice that, while the interest for the fundamental properties and functioning of the polymer has considerably decreased, on the other hand the field of composites synthesis and more over their applications has taken the relay. While improvement of traditional properties like conductivity is less and less regarded, the electrochemical properties of conducting polymers and derived products are still the root of several applications, and even the scope of fundamental studies like in the interaction with plasmonics [211-213] or electrofluorochromism [214].

It is interesting to remark that the quest for new monomers has quite receded while the 'historical' polymers, polypyrrole, polythiophene (and related PEDOT) and to a lesser extent polyaniline continue to attract the largest part of the works in the field.

However, in most of the specified studies, an application is clearly identified and presented. Most of the time, improvements deal with the creation of new composites. Still more often, it also consists into the nanostructuration of the polymer.

Despite more chemistry than electrochemistry is involved in this latter field, yet electrodeposition of nanostructured polymers is still going at a high pace.

Among applications, clearly bio-related materials and anticorrosion protection are by far the most important, and almost the only ones in the most recent works.

We believe that, while the use of conducting polymers against corrosion has begun, and is certainly a hot spot, bio-related materials are still in the infancy, apart in the field of biosensors. Problems related to the relatively high aggressiveness of the biologic fluids are still often leading to confinement of the works in that direction to ex-vivo demonstration, and further breakthrough is still needed in this area.

REFERENCES

[1] Chiang, C. K.; Fincher, C. R., Jr.; Park, Y. W.; Heeger, A. J.; Shirakawa, H.; Louis, E. J.; Gau, S. C.; MacDiarmid, A. G. *Phys. Rev. Lett.* 1977, *39*, 1098.

[2] Audebert, P.; Miomandre, F. In *Handbook of conducting polymers, 3rd edition*; Reynolds, J.R., Skotheim T., Eds.; CRC Press: Boca Raton, 2007.

[3] Audebert, P., Andrieux, C. P., Hapiot, P. and Savéant, J. M. *J. Am. Chem. Soc.*1990, *112*, 2439.

[4] Andrieux, C. P.; Audebert, P.; Hapiot, P.; Savéant, J. M. *J. Phys. Chem.*1991, *95*, 10158.

[5] Garcia, P.; Pernaut, J. M.; Hapiot, P.; Wintgens, V.; Valat, P.; Garnier, F.; Delabouglise, D. *J. Phys. Chem.*1993, *97*, 513.

[6] Heinze, J.; John, H.; Dietrich, M.; Tschuncky, P. *Synth. Met.*2001, *119*, 49.

[7] Meerholz, K.; Heinze, J. *Electrochim. Acta* 1996, *41*, 1839.

[8] Audebert, P.; Bidan, G. *J. Electroanal.Chem.*1985, *190*, 125.

[9] Ruiz, V.; Colina, Á.; Heras, A.; López-Palacios, J. *Electrochim. Acta* 2004, *50*, 59.

[10] Sanchez, J. V.; Diaz, R.; Herrasti, P.; Ocon, P. *Polym. J.*2001, *33*, 514.

[11] Mondal, S. K.; Prasad, K. R.; Munichandraiah, N. *Synth. Met.* 2005, *148*, 275.

[12] Zotti, G.; Schiavon, G.; Berlin, A.; Pagani, G. *Adv. Mat.*1993, *5*, 551.

[13] Cai, Z.; Martin, C. R. *J. Electroanal. Chem.*1991, *300*, 35.

[14] Feldberg, S. W. *J. Am. Chem. Soc.*1984, *106*, 4671.

[15] Tanguy, J.; Pron, A.; Zagorska, M.; Kulszewiczbajer, I. *Synth. Met.*1991, *45*, 81.

[16] Tanguy, J.; Baudoin, J. L.; Chao, F.; Costa, M. *Electrochim. Acta* 1992, *37*, 1417.

[17] Ulgut, B.; Grose, J. E.; Kiya, Y.; Ralph, D. C.; Abruña, H. D. *Appl. Surf. Sci.*2009, *256*, 1304.

[18] Visy, C.; Lakatos, M.; Szucs, A.; Novak, M. *Electrochim. Acta* 1997, *42*, 651.

[19] Heinze, J.; Fontana-Uribe, B. A.; Ludwig, S. *Chem. Rev.*2010, *110*, 4724.

[20] Dietrich, M.; Heinze, J. *Synth. Met.* 1991, *41-43*, 503.

[21] Neudeck, A.; Petr, A.; Dunsch, L. *Synth. Met.* 1999, *107*, 143.
[22] Brédas, J. L.; Street, G. B. *Acc. Chem. Res.*1985, *18*, 309.
[23] John, R.; Wallace, G. G. *J. Electroanal. Chem.*1993, *354*, 145.
[24] Vilas-Boas, M.; Freire, C.; de Castro, B.; Christensen, P. A.; Hillman, A. R. *Chem. Eur. J.* 2001, *7*, 139.
[25] Vorotyntsev, M. A.; Vieil, E.; Heinze, J. *J. Electroanal. Chem.* 1998, *450*, 121.
[26] Zotti, G. *Synth. Met.* 1998, *97*, 267.
[27] Odin, C.; Nechtschein, M. *Phys. Rev.Lett.*1991, *67*, 1114.
[28] Odin, C.; Nechtschein, M. *Synth. Met.*1991, *44*, 177.
[29] Odin, C.; Nechtschein, M.; Hapiot, P. *Synth.Met.*1992, *47*, 329.
[30] Otero, T. F.; Grande, H.; Rodriguez, J. *J. Phys. Chem.B* 1997, *101*, 8525.
[31] Otero, T. F.; Boyano, I. *J. Phys. Chem.B* 2003, *107*, 6730.
[32] Otero, T. F.; Grande, H.; Rodriguez, J. *Synth. Met.* 1997, *85*, 1077.
[33] Heinze, J.; Rasche, A. *J. Solid State Electrochem.* 2006, *10*, 148.
[34] Randriamahazaka, H.; Plesse, C.; Teyssie, D.; Chevrot, C. *Electrochem. Commun.* 2003, *5*, 613.
[35] Otero, T. F.; Santos, F. *Electrochim. Acta* 2008, *53*, 3166.
[36] West, B. J.; Otero, T. F.; Shapiro, B.; Smela, E. *J. Phys. Chem. B* 2009, *113*, 1277.
[37] Genies, E. M.; Noel, P. *Synth. Met.*1992, *46*, 285.
[38] Audebert, P.; Bidan, G. *J. Electroanal. Chem.*1985, *190*, 129.
[39] Deronzier, A.; Moutet, J.-C. *Acc. Chem. Res.*1985, *22*, 249.
[40] Marrec, P.; Dano, C.; Gueguen-Simonet, N.; Simonet, J. *Synth. Met.* 1997, *89*, 171.
[41] Marrec, P.; Simonet, J. *J. Electroanal. Chem.*1998, *459*, 35.
[42] Aubert, P. H.; Neudeck, A.; Dunsch, L.; Audebert, P.; Capdevielle, P.; Maumy, M. *J. Electroanal. Chem.*1999, *470*, 77.
[43] Lemaire, M.; Delabouglise, D.; Garreau, R.; Guy, A.; Roncali, J. *J. Chem. Soc.-Chem. Commun.*1988, 658.
[44] Buchner, W.; Garreau, R.; Roncali, J.; Lemaire, M. *J. Fluor. Chem.* 1992, *59*, 301.
[45] Brisset, H.; Thobiegautier, C.; Gorgues, A.; Jubault, M.; Roncali, J. *J. Chem. Soc.-Chem. Commun.*1994, 1305.
[46] Mangeney, C.; Lacroix, J. C.; Chane-Ching, K. I.; Jouini, M.; Vilain, F.; Ammar, S.; Lacaze, P. C. *Chem.Eur.J.* 2001, *7*, 5029.
[47] Zong, K.; Groenendaal, L.; Reynolds, J. R. *Tet. Lett.*2006, *47*, 3521.
[48] Walczak, R. M.; Reynolds, J. R. *Adv. Mater.* 2006, *18*, 1121.
[49] Thompson, B. C.; Abboud, K. A.; Reynolds, J. R.; Nakatani, K.; Audebert, P. *New J. Chem.* 2005, *29*, 1128.
[50] Zotti, G.; Zecchin, S.; Schiavon, G.; Vercelli, B.; Berlin, A.; Dalcanale, E.; Groenendaal, L. *Chem. Mater.* 2003, *15*, 4642.
[51] Zotti, G.; Zecchin, S.; Schiavon, G.; Berlin, A.; Huchet, L.; Roncali, J. *J. Electroanal. Chem.* 2001, *504*, 64.
[52] Zotti, G.; Zecchin, S.; Schiavon, G.; Vercelli, B.; Berlin, A.; Grimoldi, S. *Macromol. Chem. Phys.* 2004, *205*, 2026.
[53] Divisia-Blohorn, B.; Genoud, F.; Borel, C.; Bidan, G.; Kern, J. M.; Sauvage, J. P. *J. Phys. Chem. B* 2003, *107*, 5126.
[54] Roncali, J.; Thobiegautier, C.; Brisset, H.; Favart, J. F.; Guy, A. *J. Electroanal. Chem.*1995, *381*, 257.

[55] Doussot, J.; Guy, A.; Roncali, J. *Tet.Lett.* 1999, *40*, 8811.

[56] Blanchard, P.; Cappon, A.; Levillain, E.; Nicolas, Y.; Frere, P.; Roncali, J. *Org. Lett.* 2002, *4*, 607.

[57] Leriche, P.; Blanchard, P.; Frere, P.; Levillain, E.; Mahon, G.; Roncali, J. *Chem. Commun.* 2006, 275.

[58] Audebert, P.; Guyard, L.; Cherioux, F. *Journal of the Chemical Society, Chem. Commun.* 1998, 2225.

[59] Audebert, P.; Miomandre, F.; Di Magno, S. G.; Smirnov, V. V.; Hapiot, P. *Chem. Mater.* 2000, *12*, 2025.

[60] Roux, S.; Audebert, P.; Pagetti, J.; Roche, M. *New J. Chem.* 2000, *24*, 877.

[61] Roux, S.; Audebert, P.; Pagetti, J.; Roche, M. *New J. Chem.* 2000, *24*, 885.

[62] Roznyatovskiy, V. V.; Roznyatovskaya, N. V.; Weyrauch, H.; Pinkwart, K.; Tubke, J.; Sessler, J. L. *J. Org. Chem.* 2010, *75*, 8355.

[63] Beaujuge, P. M.; Reynolds, J. R. *Chem. Rev.* 2010, *110*, 268.

[64] Lima, A.; Schottland, P.; Sadki, S.; Chevrot, C. *Synth. Met.* 1998, *93*, 33.

[65] Naitoh, S. *Synth. Met.*1987, *18*, 237.

[66] Audebert, P.; Miomandre, F. In *Conducting Polymers Handbook*; Reynolds, J. R., Skotheim, T., Eds.; Springer: New York, 2007.

[67] Simitzis, J.; Soulis, S.; Triantou, D. *J. Appl. Polym. Sci.*2012, *125*, 1928.

[68] Meng, H.; Wudl, F. *Macromolecules* 2001, *34*, 1810.

[69] Loganathan, K.; Cammisa, E. G.; Myron, B. D.; Pickup, P. G. *Chem. Mater* .2003, *15*, 1918.

[70] Brisset, H.; Navarro, A. E.; Moggia, F.; Jousselme, B.; Blanchard, P.; Roncali, J. *J. Electroanal. Chem.* 2007, *603*, 149.

[71] Kern, J. M.; Sauvage, J. P.; Bidan, G.; Divisia-Blohorn, B. *J. Polym. Sci. Pol. Chem.* 2003, *41*, 3470.

[72] Dupont-Filliard, A.; Billon, M.; Bidan, G.; Guillerez, S. *Electroanalysis* 2004, *16*, 667.

[73] Audebert, P.; Bidan, G.; Lapkowski, M. *J. Chem. Soc.-Chem. Commun.*1986, 887.

[74] Andrieux, C. P.; Audebert, P. *J. Electroanal. Chem.*1989, *261*, 443.

[75] Hosseini, S. H.; Mohammadi, M. *Mater. Sci. Eng. C-Biomim. Supramol. Syst.* 2009, *29*, 1503.

[76] Steckler, T. T.; Zhang, X.; Hwang, J.; Honeyager, R.; Ohira, S.; Zhang, X. H.; Grant, A.; Ellinger, S.; Odom, S. A.; Sweat, D.; Tanner, D. B.; Rinzler, A. G.; Barlow, S.; Bredas, J. L.; Kippelen, B.; Marder, S. R.; Reynolds, J. R. *J. Am. Chem. Soc.* 2009, *131*, 2824.

[77] Zotti, G.; Zecchin, S.; Schiavon, G.; Berlin, A. *Chem. Mater.* 2002, *14*, 3607.

[78] Zotti, G.; Zecchin, S.; Schiavon, G.; Groenendaal, L. *Macromol. Chem. Phys.* 2002, *203*, 1958.

[79] Yin, X.; Li, Y.; Li, Y.; Zhu, Y.; Tang, X.; Zheng, H.; Zhu, D. *Tetrahedron* 2009, *65, 8373.*

[80] Nikoofard, H.; Sabzyan, H. *J. Fluor. Chem.* 2007, *128*, 668.

[81] Nicolas, M. *J. Coll. Int. Sci.* 2010, *343*, 608.

[82] Darmanin, T.; Nicolas, M.; Guittard, F. *Langmuir* 2008, *24*, 9739.

[83] Cherioux, F.; Guyard, L.; Audebert, P. *Chem. Commun.*1998, 2225.

[84] Leriche, P.; Piron, F.; Ripaud, E.; Frere, P.; Allain, M.; Roncali, J. *Tet.Lett.* 2009, *50*, 5673.

[85] Bui, T. T.; Iordache, A.; Chen, Z. R.; Roznyatovskiy, V. V.; Saint-Aman, E.; Lim, J. M.; Lee, B. S.; Ghosh, S.; Moutet, J. C.; Sessler, J. L.; Kim, D.; Bucher, C. *Chem. Eur. J.* 2012, *18*, 5853.
[86] Buda, M.; Iordache, A.; Bucher, C.; Moutet, J. C.; Royal, G.; Saint-Aman, E.; Sessler, J. L. *Chem. Eur. J.* 2010, *16*, 6810.
[87] Bucher, C.; Devillers, C. H.; Moutet, J. C.; Pecaut, J.; Sessler, J. L. *Chem. Commun.* 2006, 3891.
[88] Morin, J. F.; Leclerc, M.; Ades, D.; Siove, A. *Macromol. Rapid Commun.* 2005, *26*, 761.
[89] Chevrot, C.; Ngbilo, E.; Kham, K.; Sadki, S. *Synth. Met.*1996, *81*, 201.
[90] Cihaner, A.; Tirkes, S.; Onal, A. M. *J. Electroanal. Chem.* 2004, *568*, 151.
[91] Rault-Berthelot, J.; Raoult, E.; Le Floch, F. *J. Electroanal. Chem.*2003, *546*, 29.
[92] Rault-Berthelot, J.; Raoult, E.; Pilard, J. F.; Aoun, R.; Le Floch, F. *Electrochem. Commun.* 2001, *3*, 91.
[93] Poriel, C.; Ferrand, Y.; Le Maux, P.; Paul-Roth, C.; Simonneaux, G.; Rault-Berthelot, J. *J. Electroanal. Chem.* 2005, *583*, 92.
[94] Poriel, C.; Liang, J. J.; Rault-Berthelot, J.; Barriere, F.; Cocherel, N.; Slawin, A. M. Z.; Horhant, D.; Virboul, M.; Alcaraz, G.; Audebrand, N.; Vignau, L.; Huby, N.; Wantz, G.; Hirsch, L. *Chem. Eur. J.* 2007, *13*, 10055.
[95] Cocherel, N.; Poriel, C.; Vignau, L.; Bergamini, J. F.; Rault-Berthelot, J. *Org. Lett.* 2010, *12*, 452.
[96] DiCarmine, P. M.; Wang, X.; Pagenkopf, B. L.; Semenikhin, O. A. *Electrochem. Commun.* 2008, *10*, 229.
[97] Byers, J. C.; DiCarmine, P. M.; Moustafa, M. M. A.; Wang, X.; Pagenkopf, B. L.; Semenikhin, O. A. *J. Phys. Chem. B* 2009, *113*, 15715.
[98] Baumgartner, T.; Reau, R. *Chem. Rev.* 2006, *106*, 4681.
[99] de Talance, V. L.; Hissler, M.; Zhang, L. Z.; Karpati, T.; Nyulaszi, L.; Caras-Quintero, D.; Bauerle, P.; Reau, R. *Chem. Commun.* 2008, 2200.
[100] Fadhel, O.; Benko, Z.; Gras, M.; Deborde, V.; Joly, D.; Lescop, C.; Nyulaszi, L.; Hissler, M.; Reau, R. *Chem. Eur. J.* 2010, *16*, 11340.
[101] Hoferkamp, L. A.; Goldsby, K. A. *Chem. Mater.* 1989, *1*, 348.
[102] Bedioui, F.; Labbe, E.; Gutierrezgranados, S.; Devynck, J. *J. Electroanal. Chem.* 1991, *301*, 267.
[103] Audebert, P.; Capdevielle, P.; Maumy, M. *New J. Chem.* 1992, *16*, 697.
[104] Miomandre, F.; Audebert, P.; Maumy, M.; Uhl, L. *J. Electroanal. Chem.* 2001, *516*, 66.
[105] Vilas-Boas, M.; Santos, I. C.; Henderson, M. J.; Freire, C.; Hillman, A. R.; Vieil, E. *Langmuir* 2003, *19*, 7460.
[106] Branco, A.; Pinheiro, C.; Fonseca, J.; Tedim, J.; Carneiro, A.; Parola, A. J.; Freire, C.; Pina, F. *Electrochem. Solid State Lett.* 2010, *13*, J114.
[107] Asami, R.; Atobe, M.; Fuchigami, T. *J. Am. Chem. Soc.* 2005, *127*, 13160.
[108] Pringle, J. M.; Efthimiadis, J.; Howlett, P. C.; MacFarlane, D. R.; Chaplin, A. B.; Hall, S. B.; Officer, D. L.; Wallace, G. G.; Forsyth, M. *Polymer* 2004, *45*, 1447.
[109] Sadki, S.; Schottland, P.; Brodie, N.; Sabouraud, G. *Chem. Soc. Rev.* 2000, *29*, 283.
[110] Schweiss, R.; Lübben, J. F.; Johannsmann, D.; Knoll, W. *Electrochim.Acta* 2005, *50*, 2849.

[111] Sakmeche, N.; Aeiyach, S.; Aaron, J.-J.; Jouini, M.; Lacroix, J. C.; Lacaze, P.-C. *Langmuir* 1999, *15*, 2566.
[112] Shi, G.; Jin, S.; Xue, G.; Li, C. *Science* 1995, *267*, 994.
[113] Shi, G.; Li, C.; Liang, Y. *Adv. Mater.* 1999, *11*, 1145.
[114] Alakhras, F.; Holze, R. *Synth. Met.* 2007, *157*, 109.
[115] Zhou, L.; Xue, G. *Synth. Met.* 1997, *87*, 193.
[116] Xu, J.; Shi, G.; Xu, Z.; Chen, F.; Hong, X. *J.Electroanal.Chem.* 2001, *514*, 16.
[117] Jin, S.; Xue, G. *Macromol.* 1997, *30*, 5753.
[118] Lu, W.; Fadeev, A. G.; Qi, B. H.; Smela, E.; Mattes, B. R.; Ding, J.; Spinks, G. M.; Mazurkiewicz, J.; Zhou, D. Z.; Wallace, G. G.; MacFarlane, D. R.; Forsyth, S. A.; Forsyth, M. *Science* 2002, *297*, 983.
[119] Zhou, D. Z.; Spinks, G. M.; Wallace, G. G.; Tiyapiboonchaiya, C.; MacFarlane, D. R.; Forsyth, M.; Sun, J. Z. *Electrochim. Acta* 2003, *48*, 2355.
[120] Hapiot, P.; Lagrost, C. *Chem. Rev.* 2008, *108*, 2238.
[121] Dong, B.; Xing, Y. H.; Xu, J. K.; Zheng, L. Q.; Hou, J.; Zhao, F. *Electrochim. Acta* 2008, *53*, 5745.
[122] Plesse, C.; Vidal, F.; Randriamahazaka, H.; Teyssie, D.; Chevrot, C. *Polymer* 2005, *46*, 7771.
[123] Compton, R. G.; Eklund, J. C.; Marken, F. *Electroanalysis* 1997, *9*, 509.
[124] Taouil, A. E.; Lallemand, F.; Hihn, J. Y.; Melot, J. M.; Blondeau-Patissier, V.; Lakard, B. *Ultrason. Sonochem.* 2011, *18*, 140.
[125] Taouil, A. E.; Lallemand, F.; Hihn, J. Y.; Blondeau-Patissier, V. *Ultrason. Sonochem.* 2011, *18*, 907.
[126] Reyman, D.; Guereca, E.; Herrasti, P. *Ultrason. Sonochem.* 2007, *14*, 653.
[127] Taouil, A. E.; Lallemand, F.; Hallez, L.; Hihn, J. Y. *Electrochim. Acta* 2010, *55*, 9137.
[128] Long, Y. Z.; Li, M. M.; Gu, C. Z.; Wan, M. X.; Duvail, J. L.; Liu, Z. W.; Fan, Z. Y. *Progr. Polym. Sci.* 2011, *36*, 1415.
[129] Penner, R. M.; Martin, C. R. *J. Electrochem. Soc.* 1986, *133*, 2206.
[130] Martin, C. R. *Acc. Chem. Res.* 1995, *28*, 61.
[131] Cho, S. I.; Kwon, W. J.; Choi, S. J.; Kim, P.; Park, S. A.; Kim, J.; Son, S. J.; Xiao, R.; Kim, S. H.; Lee, S. B. *Adv. Mater.* 2005, *17*, 171.
[132] Cho, S. I.; Lee, S. B. *Acc. Chem. Res.* 2008, *41*, 699.
[133] Lee, J. I.; Cho, S. H.; Park, S.-M.; Kim, J. K.; Kim, J. K.; Yu, J.-W.; Kim, Y. C.; Russell, T. P. *Nano Lett.* 2008, *8*, 2315.
[134] Wan, M. X. *Adv. Mater.* 2008, *20*, 2926.
[135] Gupta, S. *Appl. Phys. Lett.* 2006, *88*.
[136] Debiemme-Chouvy, C. *Electrochem. Commun.* 2009, *11*, 298.
[137] Zang, J.; Li, C. M.; Bao, S.-J.; Cui, X.; Bao, Q.; Sun, C. Q. *Macromol.* 2008, *41*, 7053.
[138] Li, G.-R.; Feng, Z.-P.; Zhong, J.-H.; Wang, Z.-L.; Tong, Y.-X. *Macromol.* 2010, *43*, 2178.
[139] Ramanathan, K.; Bangar, M. A.; Yun, M.; Chen, W.; Mulchandani, A.; Myung, N. V. *Nano Lett.* 2004, *4*, 1237.
[140] Liu, Y. C.; Chuang, T. C. *J. Phys. Chem. B* 2003, *107*, 12383.
[141] Hosseini, M.; Momeni, M. M.; Faraji, M. *J. Mater. Sci* .2010, *45*, 2365.
[142] Rapecki, T.; Donten, M.; Stojek, Z. *Electrochem. Commun.* 2010, *12*, 624.

[143] Cioffi, N.; Torsi, L.; Losito, I.; Di Franco, C.; De Bari, I.; Chiavarone, L.; Scamarcio, G.; Tsakova, V.; Sabbatini, L.; Zambonin, P. G. *J. Mater. Chem.* 2001, *11*, 1434.
[144] Peng, C.; Zhang, S. W.; Jewell, D.; Chen, G. Z. *Progr. Nat. Sci. Mater. Int.* 2008, *18*, 777.
[145] Downs, C.; Nugent, J.; Ajayan, P. M.; Duquette, D. J.; Santhanam, K. S. V. *Adv. Mater.* 1999, *11*, 1028.
[146] Gao, M.; Huang, S.; Dai, L.; Wallace, G.; Gao, R.; Wang, Z. *Angew. Chem. Int. Ed.* 2000, *39*, 3664.
[147] Chen, J. H.; Huang, Z. P.; Wang, D. Z.; Yang, S. X.; Li, W. Z.; Wen, J. G.; Ren, Z. F. *Synth.Met.*2001, *125*, 289.
[148] Chen, J. H.; Huang, Z. P.; Wang, D. Z.; Yang, S. X.; Wen, J. G.; Ren, Z. R. *Appl. Phys. Mater. Sci.and Process.* 2001, *73*, 129.
[149] Bozlar, M.; Miomandre, F.; Bai, J. B. *Carbon* 2009, *47*, 80.
[150] Laforgue, A.; Robitaille, L. *Chem. Mater.* 2010, *22*, 2474.
[151] Pumera, M.; Smid, B.; Peng, X. S.; Golberg, D.; Tang, J.; Ichinose, I. *Chem. Eur. J.* 2007, *13*, 7644.
[152] Liu, X.; Ly, J.; Han, S.; Zhang, D.; Requicha, A.; Thompson, M. E.; Zhou, C. *Adv. Mater.* 2005, *17*, 2727.
[153] Chen, G. Z.; Shaffer, M. S. P.; Coleby, D.; Dixon, G.; Zhou, W. Z.; Fray, D. J.; Windle, A. H. *Adv. Mater.* 2000, *12*, 522.
[154] Zhang, X. T.; Zhang, J.; Wang, R. M.; Zhu, T.; Liu, Z. F. *Chemphyschem* 2004, *5*, 998.
[155] Han, G. Y.; Yuan, J. Y.; Shi, G. Q.; Wei, F. *Thin Solid Films* 2005, *474*, 64.
[156] Zhang, X. T.; Zhang, J.; Liu, Z. F. *Carbon* 2005, *43*, 2186.
[157] Feng, W.; Bai, X. D.; Lian, Y. Q.; Liang, J.; Wang, X. G.; Yoshino, K. *Carbon* 2003, *41*, 1551.
[158] Wu, M. Q.; Snook, G. A.; Gupta, V.; Shaffer, M.; Fray, D. J.; Chen, G. Z. *J. Mater. Chem.* 2005, *15*, 2297.
[159] Peng, C.; Snook, G. A.; Fray, D. J.; Shaffer, M. S. P.; Chen, G. Z. *Chem. Commun.* 2006, 4629.
[160] Patel, R. J.; Tighe, T. B.; Ivanov, I. N.; Hickner, M. A. *J. Polym. Sci. B-Polym. Phys.* 2011, *49*, 1269.
[161] Feng, X. M.; Li, R. M.; Ma, Y. W.; Chen, R. F.; Shi, N. E.; Fan, Q. L.; Huang, W. *Adv.Funct. Mater.* 2011, *21*, 2989.
[162] Si, P.; Ding, S. J.; Lou, X. W.; Kim, D. H. *Rsc Advances* 2011, *1*, 1271.
[163] Liu, A. R.; Li, C.; Bai, H.; Shi, G. Q. *J. Phys. Chem.C* 2010, *114*, 22783.
[164] Bose, S.; Kim, N. H.; Kuila, T.; Lau, K. T.; Lee, J. H. *Nanotechnology* 2011, *22*.
[165] Davies, A.; Audette, P.; Farrow, B.; Hassan, F.; Chen, Z. W.; Choi, J. Y.; Yu, A. P. *J. Phys. Chem. C* 2011, *115*, 17612.
[166] Zhang, D. C.; Zhang, X.; Chen, Y.; Yu, P.; Wang, C. H.; Ma, Y. W. *J. Power Sources* 2011, *196*, 5990.
[167] Alvi, F.; Ram, M. K.; Basnayaka, P. A.; Stefanakos, E.; Goswami, Y.; Kumar, A. *Electrochim. Acta* 2011, *56*, 9406.
[168] Novak, P.; Muller, K.; Santhanam, K. S. V.; Haas, O. *Chem. Rev.* 1997, *97*, 207.
[169] Baumgarten, M.; Huber, W.; Mullen, K. *Adv. Phys. Org. Chem.* 1993, *28*, 1.
[170] Tarascon, J. M.; Recham, N.; Armand, M.; Chotard, J. N.; Barpanda, P.; Walker, W.; Dupont, L. *Chem. Mater.* 2010, *22*, 724.

[171] Ertas, M.; Walczak, R. M.; Das, R. K.; Rinzler, A. G.; Reynolds, J. R. *Chem. Mater.* 2012, *24*, 433.
[172] Cosnier, S. *Electroanalysis* 2005, *17*, 1701.
[173] Mortimer, R. J.; Dyer, A. L.; Reynolds, J. R. *Displays* 2006, *27*, 2.
[174] Lu, X. F.; Zhang, W. J.; Wang, C.; Wen, T. C.; Wei, Y. *Prog. Polym. Sci.* 2011, *36*, 671.
[175] Sharma, R. K.; Rastogi, A. C.; Desu, S. B. *Electrochem. Commun.* 2008, *10*, 268.
[176] Snook, G. A.; Kao, P.; Best, A. S. *J. Power Sources* 2011, *196*, 1.
[177] Karnicka, K.; Chojak, M.; Miecznikowski, K.; Skunik, M.; Baranowska, B.; Kolary, A.; Piranska, A.; Palys, B.; Adamczyk, L.; Kulesza, P. J. *Bioelectrochemistry* 2005, *66*, 79.
[178] Suppes, G. M.; Cameron, C. G.; Freund, M. S. *J. Electrochem. Soc.* 2010, *157*, A1030.
[179] Antolini, E.; Gonzalez, E. R. *Appl. Catal. A-Gen.* 2009, *365*, 1.
[180] Otero, T. F.; Sanchez, J. J.; Martinez, J. G. *J. Phys. Chem. B* 2012, *116*, 5279.
[181] Jager, E. W. H.; Smela, E.; Inganas, O. *Science* 2000, *290*, 1540.
[182] Smela, E. *Adv. Mater.* 2003, *15*, 481.
[183] Skaarup, S.; Bay, L.; West, K. *Synth. Met.* 2007, *157*, 323.
[184] Roux, S.; Audebert, P.; Pagetti, J.; Roche, M. *J.Mater. Chem.* 2001, *11*, 1.
[185] Sitaram, S. P.; Stoffer, J. O.; Okeefe, T. J. *J. Coat. Technol.* 1997, *69*, 65.
[186] Petitjean, J.; Aeiyach, S.; Lacroix, J. C.; Lacaze, P. C. *J. Electroanal.Chem.* 1999, *478*, 92.
[187] Seegmiller, J. C.; da Silva, J. E. P.; Buttry, D. A.; de Torresi, S. I. C.; Torresi, R. M. *J. Electrochem. Soc.* 2005, *152*, B45.
[188] Zubillaga, O.; Cano, F. J.; Azkarate, I.; Molchan, I. S.; Thompson, G. E.; Skeldon, P. *Surf. Coat. Technol.* 2009, *203*, 1494.
[189] Zhu, R.; Li, G.; Huang, G. *Materials and Corrosion-Werkstoffe Und Korrosion* 2009, *60*, 34.
[190] Narayanasamy, B.; Rajendran, S. *Progr. Org. Coat.* 2010, *67*, 246.
[191] Pekmez, N. O.; Abaci, E.; Cinkilli, K.; Yagan, A. *Progr. Org. Coat.* 2009, *65*, 462.
[192] Karpakam, V.; Kamaraj, K.; Sathiyanarayanan, S. *J. Electrochem. Soc.* 2011, *158*, C416.
[193] Solmaz, R.; Sahin, E. A.; Kardas, G. *React.and Funct. Polym.* 2011, *71*, 1148.
[194] He, J.; Gelling, V. J.; Tallman, D. E.; Bierwagen, G. P.; Wallace, G. G. *J. Electrochem. Soc.* 2000, *147*, 3667.
[195] Rizzi, M.; Trueba, M.; Trasatti, S. P. *Synth. Met.* 2011, *161*, 23.
[196] Sorensen, P. A.; Kiil, S.; Dam-Johansen, K.; Weinell, C. E. *J. Coat. Technol. Res.* 2009, *6*, 135.
[197] Liesa, F.; Ocampo, C.; Aleman, C.; Armelin, E.; Oliver, R.; Estrany, F. *J.Appl.Polym. Sci.* 2006, *102*, 1592.
[198] Cosnier, S.; Innocent, C. *J. Electroanal. Chem.* 1992, *328*, 361.
[199] Cosnier, S. *Anal. Bioanal. Chem.* 2003, *377*, 507.
[200] Wang, A. F.; Ye, X. Y.; He, P. G.; Fang, Y. Z. *Electroanalysis* 2007, *19*, 1603.
[201] Konyushenko, E. N.; Stejskal, J.; Trchova, M.; Hradil, J.; Kovarova, J.; Prokes, J.; Cieslar, M.; Hwang, J. Y.; Chen, K. H.; Sapurina, I. *Polymer* 2006, *47*, 5715.
[202] Wang, X. X.; Yang, T.; Li, X. A.; Jiao, K. *Biosens. Bioelectron.* 2011, *26*, 2953.
[203] Rozlosnik, N. *Anal. Bioanal. Chem.* 2009, *395*, 637.
[204] Hamid, M.; Khalil ur, R. *Food Chem.* 2009, *115*, 1177.

[205] Rahman, A.; Park, D. S.; Shim, Y. B. *Biosens. Bioelectron.* 2004, *19*, 1565.
[206] Descroix, S.; Bedioui, F. *Electroanalysis* 2001, *13*, 524.
[207] Nita, II; Abu-Rabeah, K.; Tencaliec, A. M.; Cosnier, S.; Marks, R. S. *Synth. Met.* 2009, *159*, 1117.
[208] Richardson R.T., Wise A.K., Thompson B.C., Flynn B.O., Atkinson P.J., Fretwell N.J., Fallon J.B., Wallace G.G., Shepherd R.K., Clark G.M., O'Leary S.J., *Biomaterials* 2009, 30, 2614.
[209] Hepel, M.; Mahdavi, F. *Microchem.J.* 1997, *56*, 54.
[210] Svirskis, D.; Travas-Sejdic, J.; Rodgers, A.; Garg, S. *J. Control. Release* 2010, *146*, 6.
[211] Stockhausen, V.; Martin, P.; Ghilane, J.; Leroux, Y.; Randriamahazaka, H.; Grand, J.; Felidj, N.; Lacroix, J. C. *J. Am. Chem. Soc.* 2010, *132*, 10224.
[212] Leroux, Y.; Eang, E.; Fave, C.; Trippe, G.; Lacroix, J. C. *Electrochem. Commun.* 2007, *9*, 1258.
[213] Leroux, Y. R.; Lacroix, J. C.; Chane-Ching, K. I.; Fave, C.; Felidj, N.; Levi, G.; Aubard, J.; Krenn, J. R.; Hohenau, A. *J. Am. Chem. Soc.* 2005, *127*, 16022.
[214] Kim, Y.; Kim, J.; You, J.; Kim, E. *Mol. Cryst. Liq. Cryst.* 2011, *538*, 39.

In: Conducting Polymers
Editor: Luiz Carlos Pimentel Almeida
ISBN: 978-1-62618-119-9

Chapter 4

SYNTHESIS, CHARACTERIZATION AND APPLICATIONS OF CONDUCTING POLYMERS AND THEIR NANOCOMPOSITES

Jasper Chiguma, Edwin Johnson, Preya Shah, Jessica Rivera, Natalya Gornopolskaya, Eliud Mushibe and Wayne E. Jones, Jr.*

Materials Science and Engineering Program and Department of Chemistry, Binghamton University, Binghamton, NY, US

ABSTRACT

Traditionally, polymers have always been classified into two broad categories which are insulating and intrinsically conducting polymers. Among the insulating polymers are thermoplastics and thermosets. Intrinsically conducting polymers, such as polypyrrole, polyaniline, poly (3, 4-ethylenedioxythiophene) and polythiophene, can be prepared from their respective monomers electrochemically or by chemical oxidation of the monomers. Composite materials have recently been investigated which impart electrical conductivity to otherwise insulating materials. In recent years, materials at the nanoscale such as single-walled and multiwalled carbon nanotubes, graphite fibers and graphene have increasingly been used as fillers for this purpose. It has been envisioned that these properties can be imparted to the insulating polymer matrix once these materials are incorporated. In this research article, we explore recent advances in conducting polymer nanocomposites that were synthesized using both insulating and intrinsically conducting polymers as matrices. These nanocomposites have potential applications as thermal or electrical interface materials and other applications where both light weight and strength are required.

---

* E-mail address:wjones@binghamton.edu

Introduction

This chapter explores the synthesis, characterization and applications of both conducting and insulating polymers and their conducting nanocomposites. Nearly all aspects of human endeavor in the modern world depend on polymer-based materials hence the demand for polymers in their original or modified form is almost unlimited. With the advent of nanoscience, came the idea of incorporating nanomaterials such as nanoparticles, nanotubes and nanorods into polymers in order to induce properties such as thermal and electrical conductivity as well as mechanical strength. The blending of polymer with nanomaterials results in what are called nanocomposites. For inherently insulating polymers, the introduction of conducting nanomaterials makes them electrically and thermally conducting, thereby making them potential electronic materials. For inherently conducting polymers that could be used as electronic materials without further modifications, the introduction of nanomaterials, enhances their potential application as electronic materials by making them more electrically and thermally conductive. Mechanically, these nanocomposites find use in spacecraft, the automobile industry and electronics where light weight and strength are increasingly becoming important.

1. Synthesis and Characterization of Conducting Nanocomposites Based on Insulating Polymers PMMA Nanocomposites

Composite materials containing carbon nanotubes have been proposed as components of numerous electronic assemblies. We have been investigating the change in electrical conductivities of multiwalled carbon nanotube (MWCNT) loaded composites for poly (methyl methacrylate) (PMMA). Our interest in these MWCNT containing polymer composites stems from their envisioned use as interface materials in electronic packages (IMs). Since CNT have been reported to have high electrical conductivities and a high aspect ratio, their presence in polymer matrices should improve the electrical, thermal, and mechanical properties of insulating polymers. The nanocomposites were prepared by in-situ chemical polymerization in the presence of pristine MWCNT or pre-oxidized MWCNT. Composites were also prepared by ex-situ. Characterization of the nanocomposites was done using FTIR, Four-point probe and FESEM. Data on the mechanical strength of the composites were determined using the Instron Tensile Testing Machine as well as the Nanoindenter. Electrical conductivity of the nanocomposites was found to increase with wt % loading of MWCNT. The use of oxidized MWCNT resulted in higher electrical conductivity in the MMA nanocomposites.

PMMA is a versatile polymer whose applications abound. It finds use in a lot of imaging and non-imaging applications in microelectronics. Some uses of PMMA are high resolution positive resist for direct write e-beam, x-ray and deep UV microlithographic processes, protective coating for wafer thinning, bonding adhesive, and sacrificial layer [1].

Carbon nanotubes are characterized by unique properties that make them attractive for use in advanced technology. Small diameter, high surface area and high aspect ratio are some of the most important properties of carbon nanotubes that make them attractive for different technological applications. [2] They have been used as tips for AFM, cells for hydrogen storage [3], nanotransistors [4], electrodes for electrochemical applications [5], sensors of biological molecules [6], as well as catalysts. [7] [8] In engineering a lot of attention has been paid to developing polymer composites in which carbon nanotubes are incorporated. The carbon nanotubes are expected to serve as reinforcement within the polymer matrix, thereby imparting improved mechanical properties. [9] Some of the mechanical properties that have been found to improve significantly from the incorporation of carbon nanotubes in polymers are modulus, stiffness and flexibility. When incorporated in polymer matrix, the resultant nanocomposites are known to have high thermal stability, making them ideal for aerospace applications. The electronics industry stands to benefit from carbon nanotubes because of their potential use as semiconductors capable of dissipating heat. [10] This application whereby carbon nanotube filled polymers are used as thermal interface materials is the focus of present work.

While there is a lot of interest in carbon nanotubes as polymer filler materials, the properties of the resultant polymer/carbon nanotube nanocomposites, is dependent on a number of factors. Among the most important factors are the intrinsic properties of the carbon nanotubes which are determined by the carbon nanotube production method and the post-production treatment in order to facilitate their dispersion in solvents. [11] Different carbon nanotube production methods result in different degrees of graphitization which significantly influences their electron conductivity. [12] Further treatment, such as functionalization, depending on the method used, can seriously impair the integrity of the carbon nanotubes, which in turn compromises the properties of the nanocomposites. However, oxidation and functionalization of carbon nanotubes, has been used to improve interfacial interaction between the carbon nanotubes and the polymer matrix. [13] [14].

Experimental

Preparation of H2SO4/HNO3 Oxidized MWCNTs

0.50 g of pristine Nanocyl®-7000 MWCNTs was added to a mixture of 50 ml concentrated HNO3 and 150 ml concentrated sulfuric acid (H2SO4). The mixture was placed in a bath sonicator until the MWCNTs were uniformly dispersed. The suspension was refluxed in an oil bath with constant stirring. The temperature was slowly raised from 90 °C to 133 °C, and then allowed to continue refluxing for 2 h. As the reaction proceeded, a brown gas was observed which escaped at the top of the condenser. To collect the oxidized MWCNTs, the mixture was diluted with distilled water and centrifuged until the MWCNTs sedimented. Centrifugation was performed several times until a neutral pH was obtained [15].

Preparation of KMnO4 Oxidized MWCNTs

0.12 g MWCNTs and 25 mL dichloromethane (CH2Cl2) were added to a 100 mL flask and the mixture was bath sonicated for 30 minutes. 1.0 g Aliquat 336 (a phase transfer agent) from Aldrich was added to the dispersion. 5 g of powdered KMnO4 were added in small portions over 2h followed by 5 mL of acetic acid. The mixture was then stirred vigorously overnight at room temperature followed by filtering, washing with concentrated hydrochloric acid (HCl) and water and dried to get oxidized MWCNT (o-MWCNTs).

Preparation of In-Situ PMMA-MWCNT Nanocomposites

Methyl methacrylate monomer (MMA) was doubly distilled prior to use. MMA contains the compound monomethylether hydroquinone (MEHQ). MEHQ is an inhibitor compound that is added to MMA to prevent decomposition. MMA and MEHQ have different boiling points of 100°C and 243°C, respectively. It was therefore possible to distill out the MMA leaving the MEHQ behind. To make the composite 0.002 grams of pristine Nanocyl 7000 MWCNTs were added to 1 gram of MMA in a three necked flask. A small amount of 2, 2'-Azo-bis-isobutyrylnitrile (AIBN) was added to act as a free radical vinyl polymerization initiator. The flask was placed in a Branson 5150 bath sonicator to disperse the MWCNTs in the MMA. After sonication, the reaction flask was briefly purged with nitrogen gas. The flask was then stoppered, and heated in an oil bath with constant stirring at 90 oC for 2 h. As the MMA polymerized to PMMA, it solidified and the MWCNTs were evenly distributed in the solid. The flask was then allowed to cool before the PMMA nanocomposite was removed from the flask. The above procedure was repeated using 0.004, 0.006, 0.008, and 0.01 g of pristine MWCNTs. This resulted in samples of 0.2, 0.4, 0.6, 0.8, and 1.0 percent by weight of MWCNTs. The experiment was repeated with identical loadings of H2SO4/HNO3 and KMnO4 pre-oxidized MWCNT. For comparison, PMMA was also prepared without MWCNTs. In order to prepare films, the in-situ prepared PMMA-MWCNT nanocomposites were dissolved in nitromethane (CH3NO2) and the solution cast into films. Dissolution of the nanocomposite was achieved by 2 hours of bath sonication in a Branson 5150 bath sonicator followed by one minute tip sonication using a SONICS VibraCell tip sonicator. Nitromethane was used because it resulted in very uniform dispersion of MWCNTs throughout the PMMA films. Films were also cast for performing FTIR scans. For these films dichloromethane (CH2Cl2) was used as a solvent, because it produced thin, transparent films which were ideal for FTIR. The films did not require any further preparation before taking FTIR spectra.

Preparation of Ex-Situ PMMA-MWCNT Nanocomposites

The weight percentage of carbon nanotubes in the MWCNT/PMMA nanocomposite was calculated using the definition of MWCNT to polymer mass ratio. For example, a 9.1 wt% sample means that the MWCNT and PMMA mass ratio is 2:20 and therefore [2/(2+20)]*100% = 9.1%. [16] PMMA was obtained from Sigma-Aldrich (average = M.W. 350,000) and used as received.

The desired final mass of each sample was set at 0.5 g. PMMA and MWCNTs were weighed to give the desired loading of MWCNTs (0.2 – 1.0 wt %) so that when added together their total mass would be 0.5 g. Each mass of PMMA was dissolved overnight with stirring in 10 mL of nitromethane. The appropriate mass of MWCNTs was then added to the dissolved PMMA. This mixture was sonicated in a Branson 5150 bath sonicator for 2 hours, followed by 2 minutes of sonication by a SONICS VibraCell tip sonicator to ensure uniform MWCNT dispersion. The resulting dispersion was cast in petri dishes and allowed to dry overnight, yielding thin plastic films. This process was repeated to form nanocomposites with loadings of 0.2, 0.4, 0.6, 0.8, and 1.0 wt% of pristine H2SO4/HNO3 and KMnO4 MWCNTs. Ex-situ PMMA/MWCNT nanocomposites with H2SO4/HNO3 oxidized MWCNTs were prepared using THF as the solvent instead of nitromethane. Nitromethane was not able to adequately disperse the acid oxidized MWCNTs. THF was found to produce stable dispersions of the acid oxidized MWCNTs.

Mechanical Properties

Films were cut into strips of dimensions 1.5 inches by 0.25 inches. The average thickness of the strips was 0.0134 cm. The instrument used for mechanical strength testing was an Instron Tensile Testing Machine. During measurement the gauge length was set at one inch. The films were pulled from both ends until failure. Five strips were used for every sample to insure reliability of the results. Mechanical information was also obtained by nanoindentation. During nanoindentation, from four selected regions, four indentations were made on each area. The load was varies from 1000 μN to 4000 μN using an open loop. A Berkovich tip, 250 nm in diameter was used.

Results and Discussion

Figure 1 shows the FTIR spectra of pristine, H2SO4/HNO3 and KMnO4 MWCNTs. In order obtain these spectra the carbon nanotubes were dispersed in KBr, pulverized and pressed into transparent pellets. The band at 3000-2800 cm-1 on the spectrum of pristine MWCNTs corresponds to –CH stretching, a result of possible defects from manufacture or handling. The feature at around 1720 cm-1 on the spectra of both oxidized MWCNT corresponds to the carboxylic stretching (COOH). Features around 1610-1550 cm-1, 1075-1010 cm-1, 3000-2800 cm-1 are the absorption peaks of COO- asymmetric stretching, -OH in primary alcohol, -CH stretching and –COO- stretching respectively. The presence of these bands is indication that oxidation of the MWCNT had taken place [17] [18].

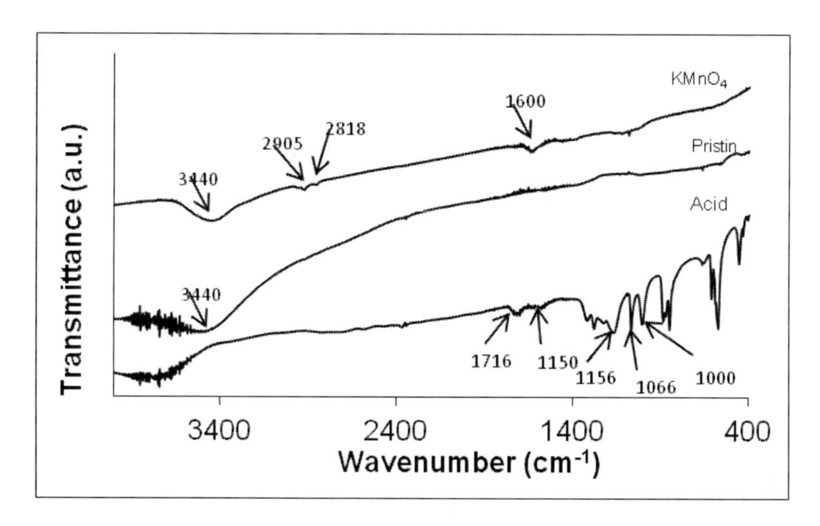

Figure 1. FT-IR spectra of MWCNTs.

Characterization of PMMA-MWCNT Nanocomposites

FTIR

Figure 2 shows the FTIR spectra of PMMA that was prepared by the polymerization of the monomer MMA and the commercially available PMMA. The two spectra are identical showing that the polymerization process that we used was able to yield PMMA.

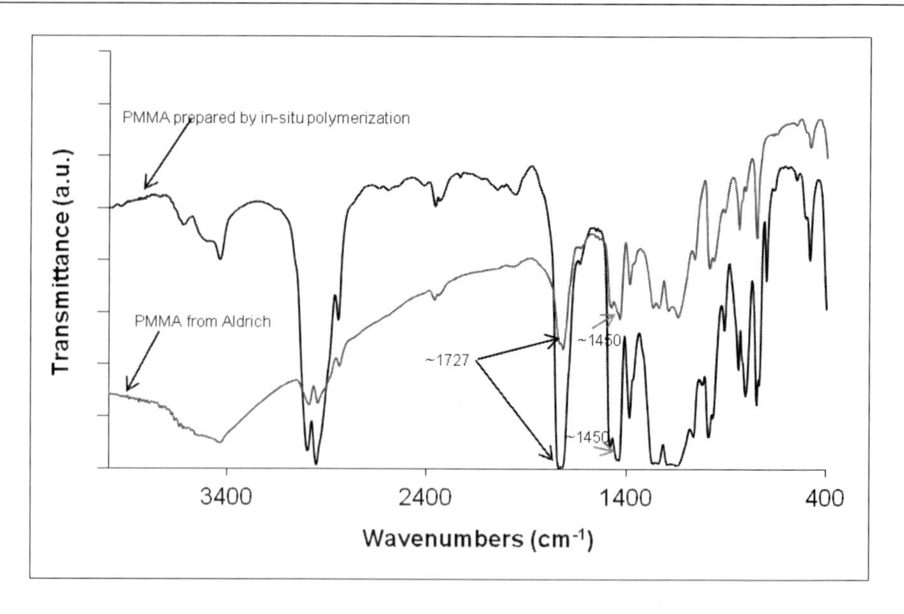

Figure 2. FT-IR spectra of in-situ prepared PMMA and PMMA from Aldrich.

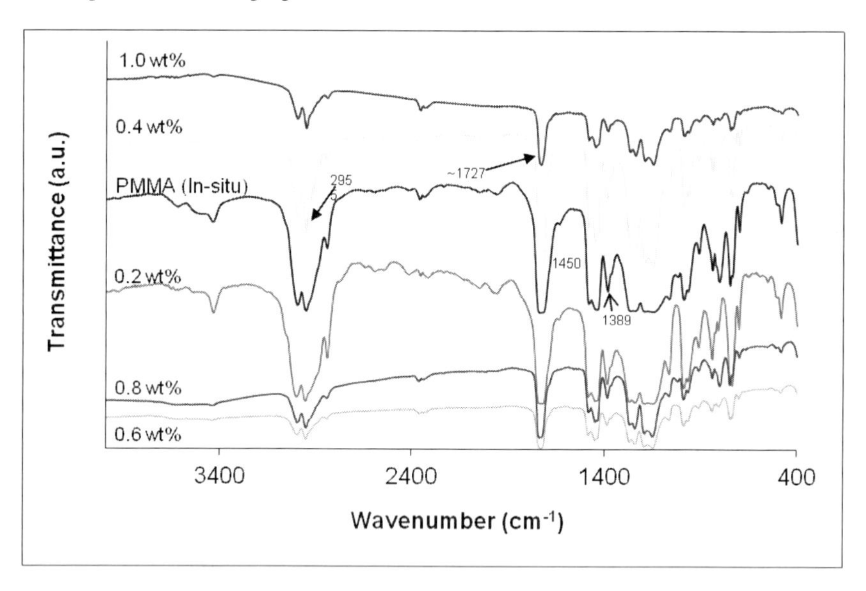

Figure 3. FT-IR spectra of PMMA nanocomposites loaded with pristine MWCNTs.

After verifying the successful synthesis of PMMA, nanocomposites were synthesized and the spectra are shown in Figures 3, 4, and 5 for PMMA loaded with pristine, H2SO4/HNO3 and KMnO4 oxidized MWCNTs respectively. All the spectra are similar except for a few changes in the spectra of the nanocomposites. The features that are similar verify the presence of PMMA. The fingerprint characteristic vibration bands of PMMA appear at 1727 cm-1(C=O) and 1450 cm-1(C-O). The bands at 3000 and 2900 cm-1 correspond to the C-H stretching of the methyl group (CH3) while the bands at 1300 and 1450 cm-1 are associated with C-H symmetric and asymmetric stretching modes, respectively.

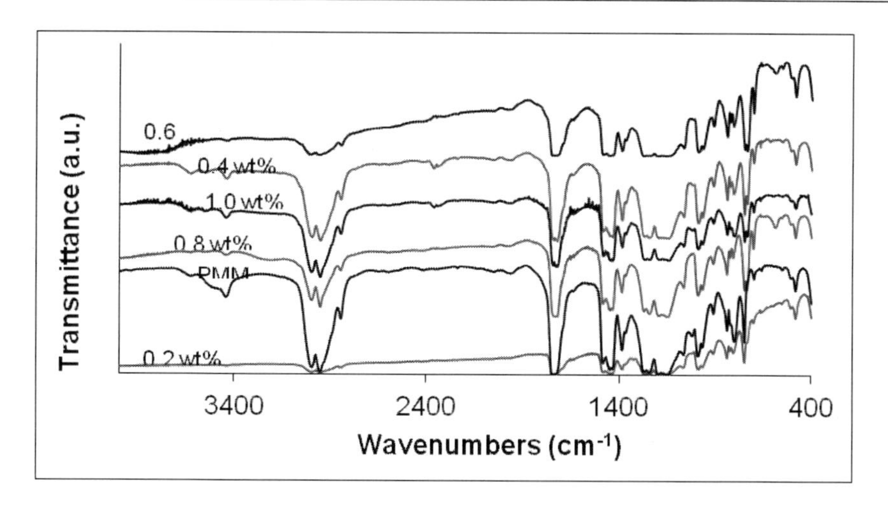

Figure 4. FT-IR spectra of PMMA nanocomposites loaded with H2SO4/HNO3 oxidized MWCNTs.

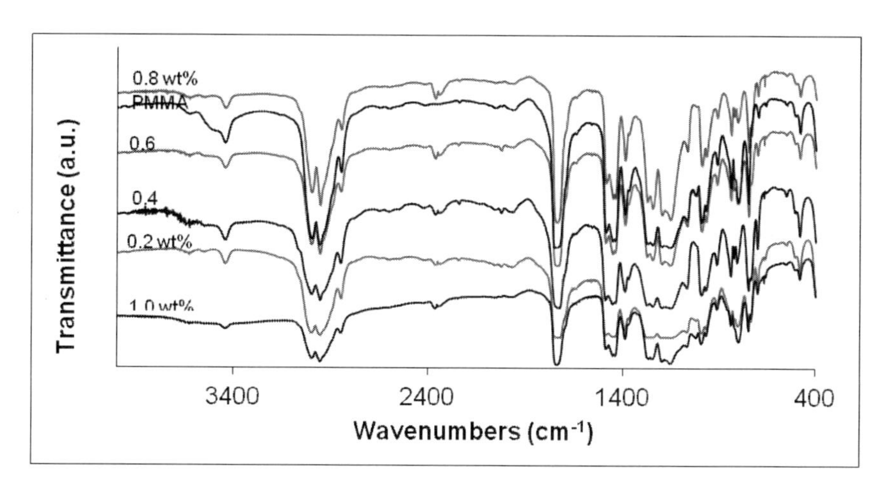

Figure 5. FT-IR spectra of PMMA nanocomposites loaded with KMnO4 oxidized MWCNTs.

Figure 6. Dispersions of Nanocyl 7000 MWCNT in (a) nitromethane (b) tetrahydrofuran (c) N-methyl pyrrolidone.

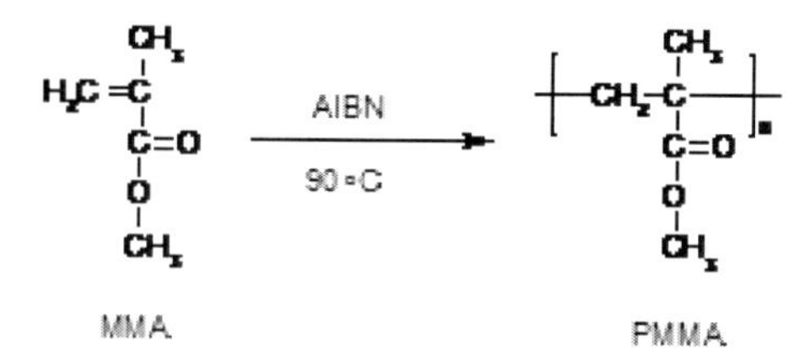

Figure 7. Polymerization of MMA monomer to PMMA.

The 1240 cm-1 band is assigned to torsion of the methylene (CH2) and the 1150 cm-1 band corresponds to vibration of the ester group C-O, while C-C stretching bands are at 1000 and 800 cm-1. [19].

Some researchers have reported the formation of band at 1650 cm-1 in PMMA nanocomposites. They say the band originates from the formation of a new C-C bond between PMMA and MWCNTs during the polymerization because the π bonds get opened in the process thereby allowing MWCNTs to link to PMMA.

However, we did not observe that band in our spectra. Instead there is a band in PMMA at 1624 cm-1 which decreases in size as the carbon nanotube loading increases and finally disappears at 1 wt% loading of MWCNTs. The 1624 cm-1 band may be associated with the COO- group in PMMA which gradually disappears as the PMMA links with the MWCNTs.

While dichloromethane was found suitable for the preparation of FTIR films from the nanocomposites, nitromethane was selected for the preparation of films for electrical conductivity and mechanical testing. The suitable solvent had to be able to dissolve the PMMA and at the same time disperse the carbon nanotubes. Solvents were selected based on the solubility parameters of several solvents which were tested for the fabrication of PMMA-MWCNT nanocomposite films. [20]

Solvents that were found to disperse pristine MWCNT uniformly were nitromethane (NM), tetrahydrofuran (THF), and N-methyl pyrrolidone (NMP). Figure 6 shows stable dispersions of MWCNT in NM (a), THF (b), and NMP (c). These dispersions were prepared by bath sonication for 90 minutes. This photograph was taken after the dispersions were left to stand still and undisturbed for 24 hours.

Figure 7 shows the polymerization of the monomer MMA to PMMA in the presence of AIBN. It has been suggested that using this process with pristine carbon nanotubes results in the initiator opening the π bonds of the MWCNTs. [21] An important consequence of the opening of the π bonds is the formation of strong covalent bonds between the carbon nanotubes and the polymer.

Four-Point Probe Electrical Conductivity

Figure 8(a) shows the electrical conductivity of PMMA nanocomposites loaded with pristine as well as H2SO4/HNO3 and KMn04 oxidized MWCNTs. The electrical conductivity percolation threshold for this nanocomposite was observed at below 0.2 wt% loading of pristine MWCNTs. Steady increase in electrical conductivity was seen up to 1.0 wt% of MWCNTs. Figure 8(b) and (c) show an initial increase in conductivity with percolation threshold of below 0.2 wt% acid-oxidized MWCNTs as well as KMnO4 oxidized MWCNTs. Conductivity then remains relatively constant as the loadings increase from 0.2 wt% to 1.0 wt% of acid-oxidized MWCNTs.

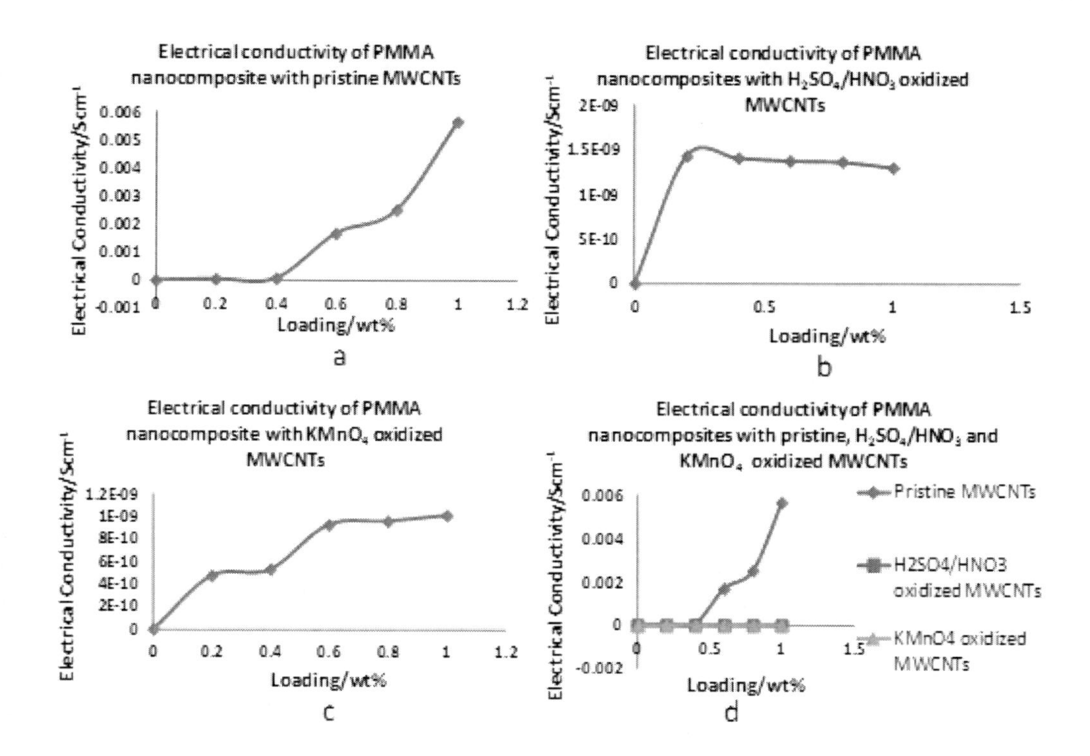

Figure 8. Electrical conductivity of PMMA nanocomposite with (a) pristine MWCNTs (b) H2SO4/HNO3 (c) KMnO4 oxidized MWCNTs (d) plots (a), (b) and (c) together.

While the nanocomposites prepared by all the types of carbon nanotubes showed increased electrical conductivity with increase in the percentage loading of carbon nanotubes, there was a marked difference in the conductivity between the pristine and functionalized carbon nanotubes.

PMMA/pristine-MWCNTs nanocomposites showed significantly higher electrical conductivity while the PMMA/oxidized carbon nanotube nanocomposites electrical conductivity was significantly low probably due to damage during functionalization. [22].

Mechanical Properties

Data on tensile strength, elongation at break, modulus and toughness of the in-situ prepared PMMA nanocomposites were obtained and plotted. Literature reports suggest a significant improvement in electrical, thermal and mechanical properties with increase in carbon nanotube loading. [23] [24] Figure 9 (a), (b), (c) and (d) show the plots of tensile strength, elongation at break, toughness and modulus respectively. The mechanical properties of PMMA nanocomposites that contain KMnO4 oxidized MWCNTs seem to decrease the most. The nanocomposite that contained acid oxidized MWCNTs seemed to exhibit the best mechanical properties in agreement with the results of other researchers. [25] As the carbon nanotube loading increased, the electrical conductivity generally increased up to a certain point, however, the mechanical properties in some cases decrease.

It can be seen that the highest electrical conductivities do not necessarily yield better mechanical properties. The nanocomposites containing acid oxidized MWCNTs yielded better mechanical properties. This might be due to enhanced polymer-oxidized carbon nanotube interaction. However pristine MWCNTs exhibited the highest electrical conductivity, probably because the integrity and intrinsic properties of the carbon nanotubes were maintained.

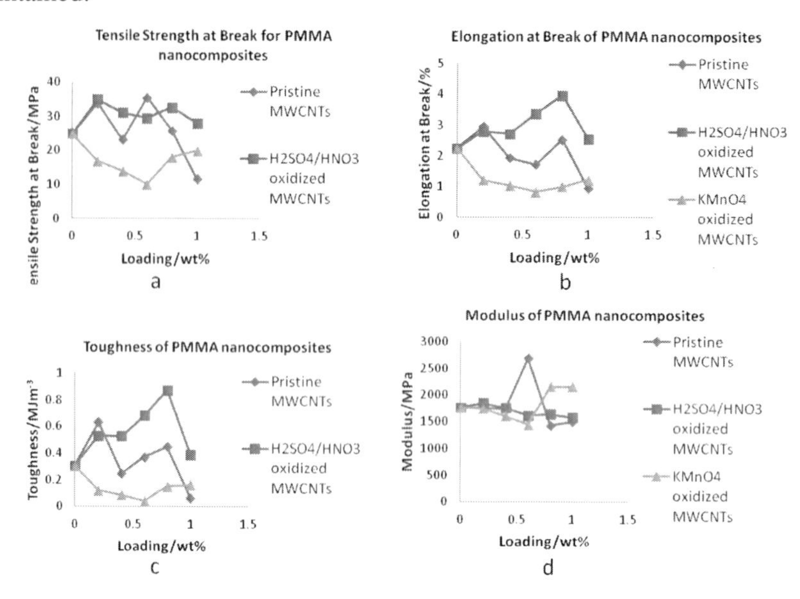

Figure 9. Mechanical properties of PMMA-MWCNT nanocomposites.

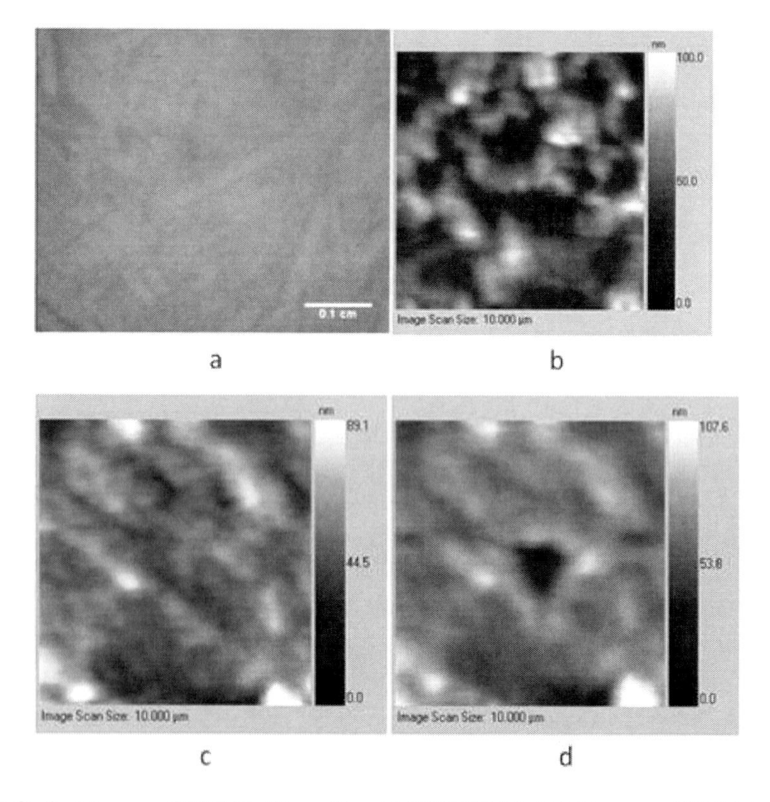

Figure 10. (a) Optical image of PMMA film without MWCNT (b) AFM image of PMMA film without MWCNT (c) AFM image of PMMA/ pristine MWCNT nanocomposite before nanoindentation, (d) AFM image of PMMA/ pristine MWCNT nanocomposite after nanoindentation.

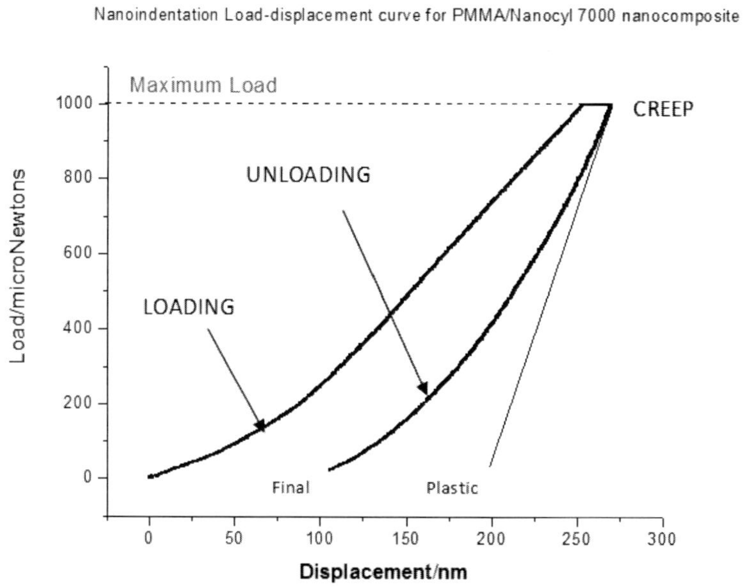

Figure 11. Nanoindentation load-displacement curve for the PMMA/pristine Nanocyl 7000 MWCNT nanocomposite.

The mechanical properties of nanocomposites loaded with 0.2 wt% pristine MWCNTs were evaluated using nanoindentation. Figure 10(a) and (b) show the optical and AFM images of the PMMA film without carbon nanotubes respectively. Figures 10(c) and (d) show the AFM pictures of the PMMA nanocomposite loaded with 0.2 wt% pristine MWCNT before and after indentation respectively. The indentation mark which is triangular in shape.

Figure 11 shows the resulting nanoindentation load-displacement curve for the PMMA/pristine Nanocyl 7000 MWCNTs.

Young's modulus was calculated based on the Equation 1 [26].

$$\frac{1}{E_r} = \frac{(1-v^2)}{E} + \frac{(1-v_i^{\,2})}{E_i} \qquad (1)$$

E_r is the reduced modulus measured from the indenter and it can be called the indentation modulus, E_i is the indenter modulus, which has a value of 1140 GPa for a diamond tip, E is the elastic modulus of the sample material, vi is the Poisson's ratio of sample material with a value of 0.5 and v_i is the Poisson's ratio for the diamond indenter which has a value of 0.07.

Figure 12 shows the plots of indenter and Young's moduli. While the indenter modulus increases with MWCNT loading, the Young's modulus decreases continuously. The modulus for this composite obtained from the Instron Tensile Testing Instrument showed an increase at 0.6 wt% before decreasing.

This difference between the two moduli may be attributed to nanoindentation only providing surface modulus whereas the Instron Tensile Testing Instrument provides bulk modulus. [27].

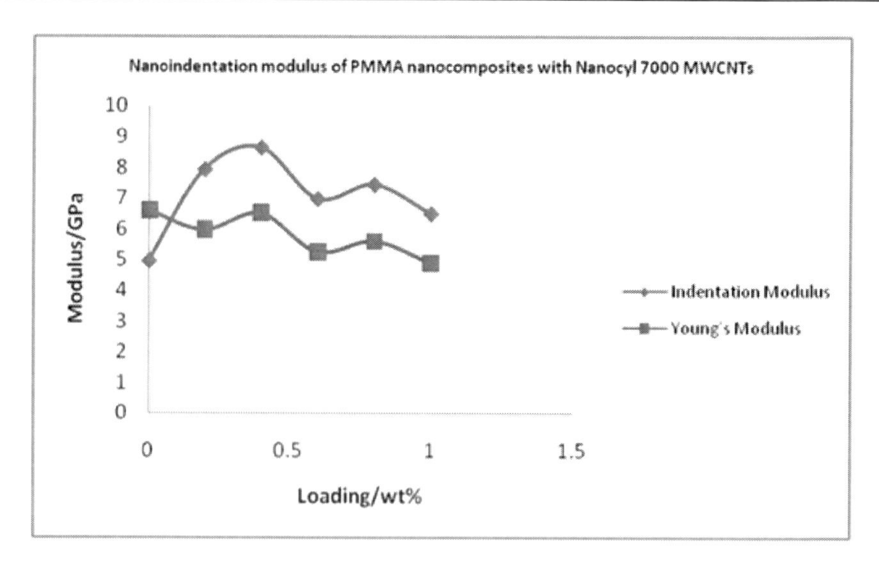

Figure 12. Plot of Nanoindenter modulus and Young's modulus as a function of percentage loading of MWCNT.

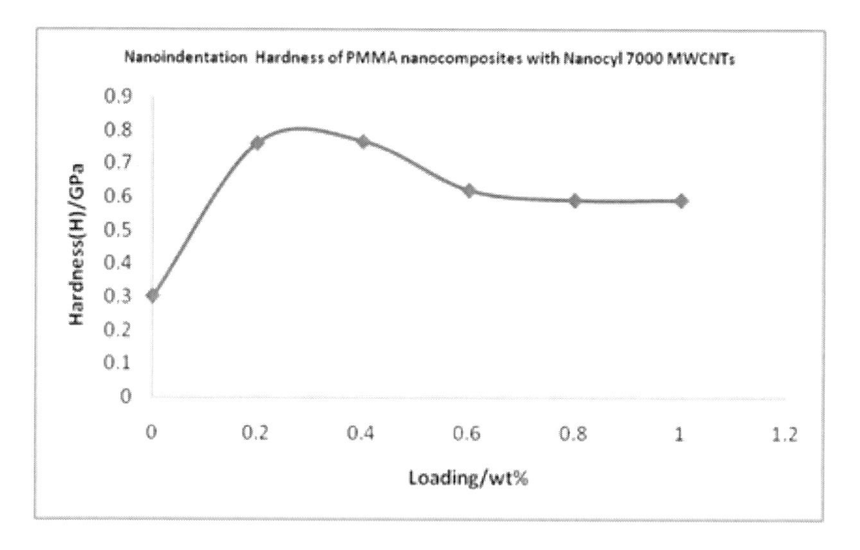

Figure 13. Plot of Hardness as a function of percentage loading of MWCNTs.

Figure 13 shows an increase in the hardness with a maximum at percentage loadings of between 0.2 and 0.4 wt%. Above a percentage loading of 0.4 wt% the hardness decreases. This shows that there is a critical value of carbon nanotube loading at which good mechanical properties can be achieved after which further addition of carbon nanotubes results in the deterioration of the mechanical properties.

Besides Nanocyl 7000 MWCNTs, other carbon nanotubes were used. These carbon nanotubes had been produce by the combustion method. The carbon nanotubes were MWCNTs like the Nanocyl 7000, except that they had been manufacture by a different method. We set out to explore the electrical and thermal properties of PMMA nanocomposites in which these MWCNTs had been used as fillers. The nanocomposites were prepared by the ex-situ method described earlier.

Electrical Conductivity

Figure 14 shows the electrical conductivity plot for PMMA nanocomposite loaded with Nanocyl 7000 MWCNTs and prepared by the ex-situ method. It can be seen from the plot in Figure 14 that the electrical percolation threshold for this nanocomposite is below 1 wt%. The electrical conductivity rises to a maximum below 3 wt% after which it seems to level off. This behavior, as we have seen before, confirms the significant effect Nanocyl 7000 MWCNTs have on the electrical conductivity of PMMA.

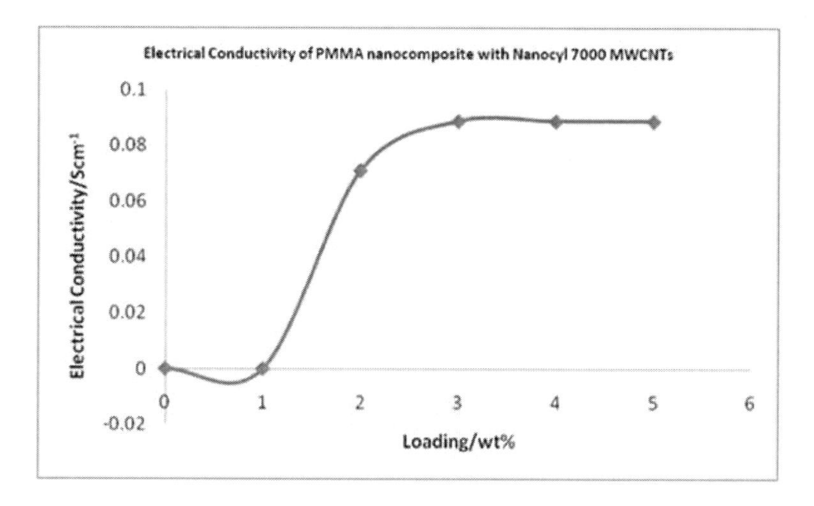

Figure 14. Plot of the electrical conductivity of PMMA nanocomposites loaded with Nanocyl 7000 MWCNTs.

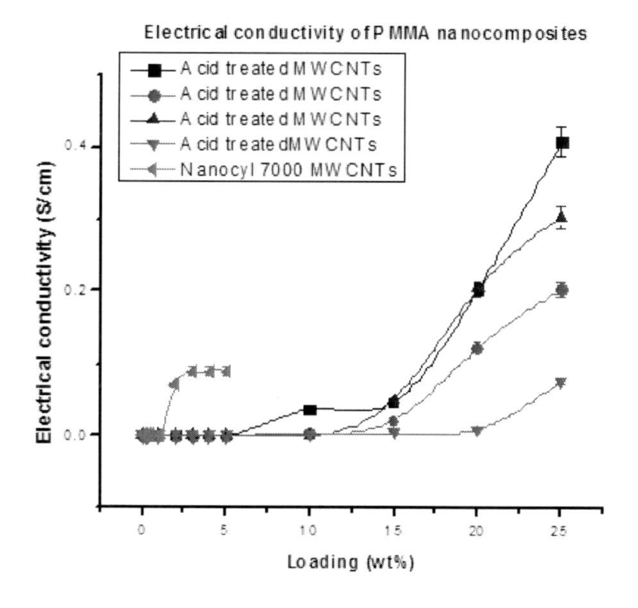

Figure 15. Electrical conductivity of PMMA nanocomposites loaded with MWCNT in the 5-25 wt% range.

In the 1-5 wt% percent loading, no measurable electrical conductivity was recorded for PMMA nanocomposites prepared in the presence of the other MWCNTs that were prepared by the combustion method. However, we were able to record data on their mechanical properties, such as tensile strength at break, elongation at break and modulus. PMMA nanocomposites loaded with the MWCNTs that did not show measurable electrical conductivity in the 1-5 wt% range. So, nanocomposites were prepared again but with percentage loadings in the 5-25 wt% loading range. In this percentage loading range, all the nanocomposites with MWCNTs showed measurable electrical conductivity.

It can be seen in Figure 15 that in loading range 0-5 wt%, no measurable electrical conductivity is observed in PMMA nanocomposites made with all the MWCNTs that were manufactured by the combustion process. This shows a huge difference in the performance of Nanocyl 7000 MWCNTs, which below 1 wt%, shows electrical percolation. The combustion manufactured MWCNTs filled PMMA nanocomposites were also compared among themselves. It can be seen that measurable electrical percolation starts at a percentage loading of 5 wt%. Besides showing higher electrical conductivity between 5 and 10 wt%, the highest electrical conductivity was achieved at a maximum loading of 25 wt%. In these composites the loading of carbon nanotubes was stopped at 25 wt% because, any further increase in the amount of MWCNTs resulted in poor dispersion.

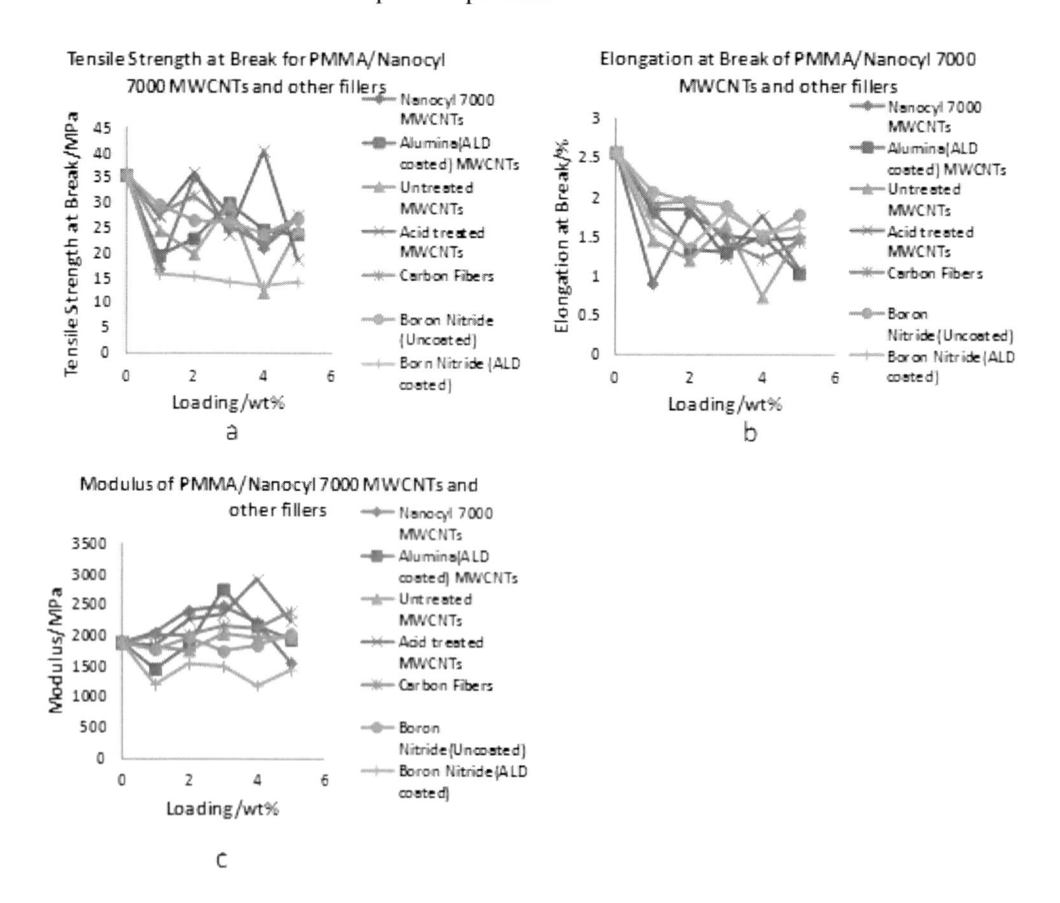

Figure 16. Plot of (a) Tensile strength at break (b) Elongation at break and (c) Modulus of PMMA nanocomposites loaded with Nanocyl 7000 MWCNTs and other fillers.

Mechanical Properties

In Figure 16, the nanocomposites show a decrease in tensile strength at break as well as elongation at break at low percent loadings of 1-3 wt%. However there is an increasing trend in the modulus with increasing loading of the fillers. The elongation at break for all the nanocomposites decreases as the filler percentage weight is increased. [28] The data on the tensile strength at break, elongation at break and modulus show that the modulus of these nanocomposites, increase at the expense of the tensile strength at break and elongation at break.

2. SYNTHESIS AND CHARACTERIZATION OF POLYIMIDE NANOCOMPOSITES BY *IN-SITU* POLYMERIZATION

Experimental

The PMDA-ODA polyimide was synthesized by reacting 0.5006g (0.0025 mol) ODA and 0.5455g (0.0025 mol PMDA in 15 ml of dry DMAc. [29] For the preparation of the PMDA-ODA nanocomposites with MWCNT, 1-5 wt% loadings of carbon nanotubes were added to a solution of DMAc in a three neck flask and sonicated for 1 hr in a bath sonicator. After 1hr the carbon nanotubes were found to be well dispersed. A stir bar was then added and the three neck flask transferred to a hot plate with a magnetic stirrer. While stirring the ODA was added and allowed to dissolve before PMDA could be added. In order for the reaction to take place under inert atmosphere, the three neck flask was purged with nitrogen and a balloon filled with nitrogen attached. The reaction was left to proceed at room temperature overnight under nitrogen with vigorous magnetic stirring. Films were obtained by casting the PMDA-ODA polyamic acid in petri dishes. In order to avoid blistering during thermal curing, the cast solution was put under vacuum at room temperature in order to drive out air pockets. Upon drying, the films were peeled off the petri dishes and heated in a convection oven at 100, 200, and 260 ^{o}C. The temperature was maintained at each temperature for 1 hour. After the final temperature treatment, the oven was allowed to cool before the samples could be withdrawn.

Results and Discussion

Polyimide with significant commercial importance Kapton was synthesized from pyromellitic dianhydride (PMDA) and 4, 4'-oxydianiline (ODA). The process by which the polyimide is formed from the two monomers is represented by the scheme in Figure 38. In the presence of a dipolar aprotic solvent such as DMAc, a nucleophilic attack by the amino group on the carbonyl carbon of the anhydride group will occur when the dianhydride is added to the diamine at room temperature. During the synthesis of the polyimide, the amine was added first followed by the dianhydride. That ensured that the dianhydride that was added second reacted faster with the diamine than with any water that might have existed. [30] The nucleophilic attack process is illustrated by the scheme in Figure 17.

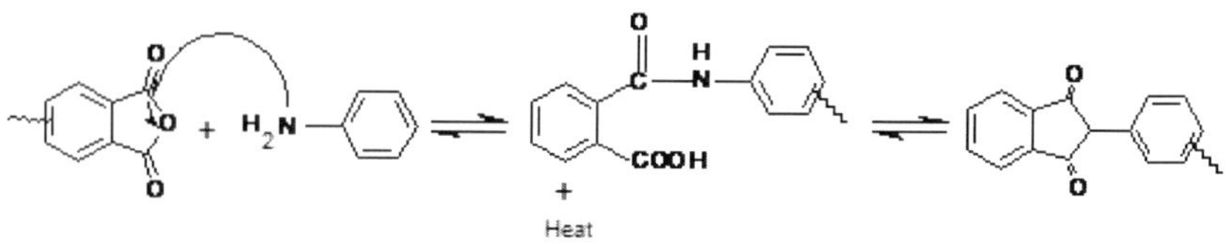

Figure 17. Reaction mechanism of aromatic imide formation.

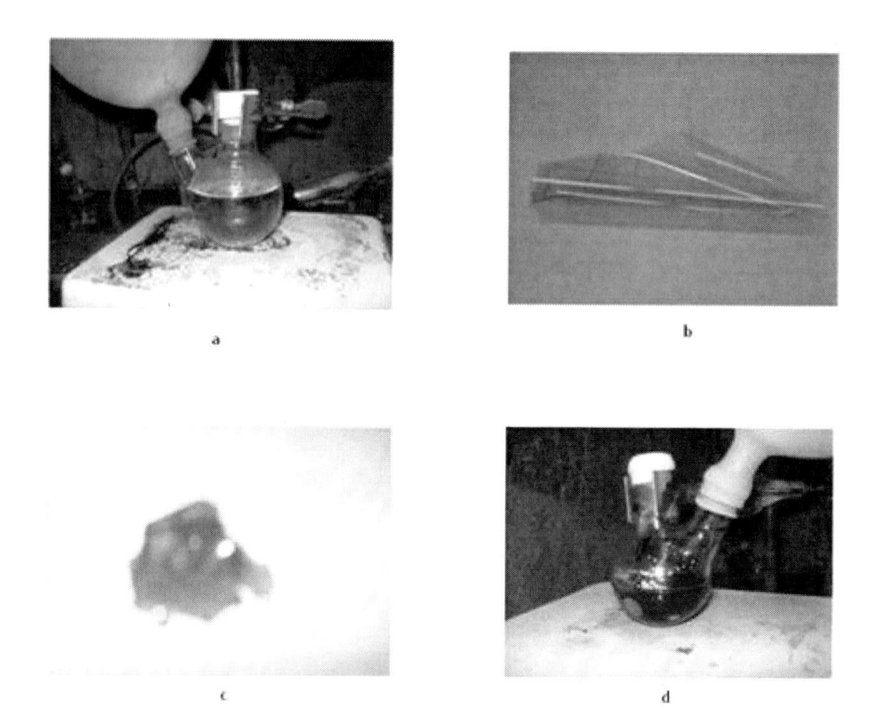

Figure 18. (a) Polyamic acid formed from PMDA and ODA under nitrogen atmosphere (b) Thin flexible polyamic acid film (c) Polyamic acid film after curing at 300 °C (d) Polyamic acid with MWCNT under nitrogen.

To ensure an inert atmosphere during the polymerization of the polyimide, a nitrogen atmosphere was used as shown in Figure 18(a). The polyamic acid solution as well as the thin films that were obtained after casting the solution on glass slides was found to be yellow. However, after curing which thermal imidization the films turned orange as shown in Figure 18(c).

In order to verify the successful synthesis of the polyimide from PMDA and ODA monomers, IR spectra of the polyamic acid film were compared before and after curing. Figure 41 shows the IR spectra of polyamic acid film and the polyimide film obtained after curing the polyamic acid film. The successful formation of polyimide from PMDA and ODA is evidence by the presence of IR bands at 730, 1370, 1720 and 1780 cm-1 as shown in Figure 19 [31]. Smooth films with well dispersed filler materials were obtained and tested for electrical conductivity. It was surprising to find out that although the filler materials were well dispersed in the nanocomposites, none of them exhibited measurable electrical conductivity except Nanocyl 7000 MWCNT filled nanocomposites.

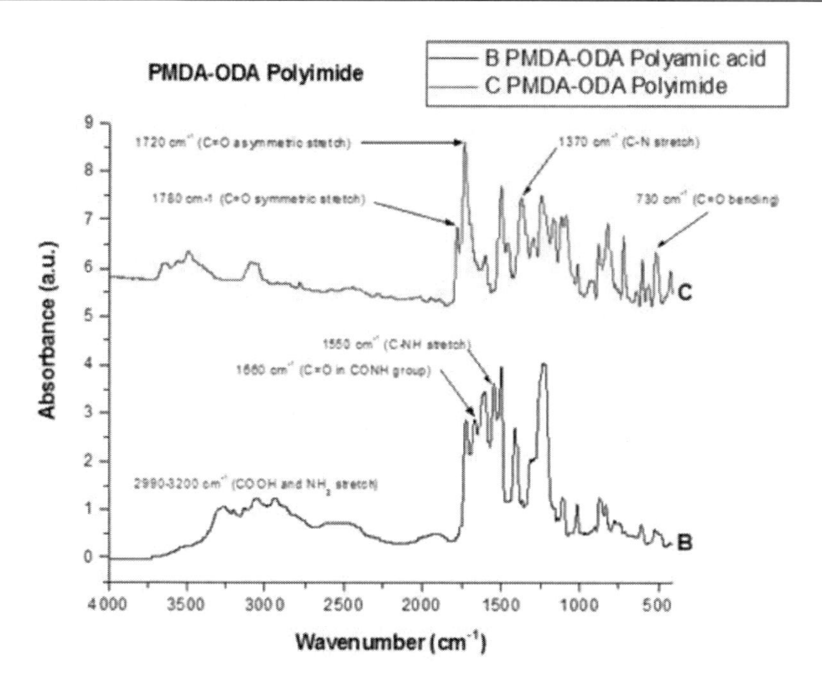

Figure 19. FTIR spectra of (B) Polyamic acid thin film (C) Polyimide thin film obtained after curing at 260 °C.

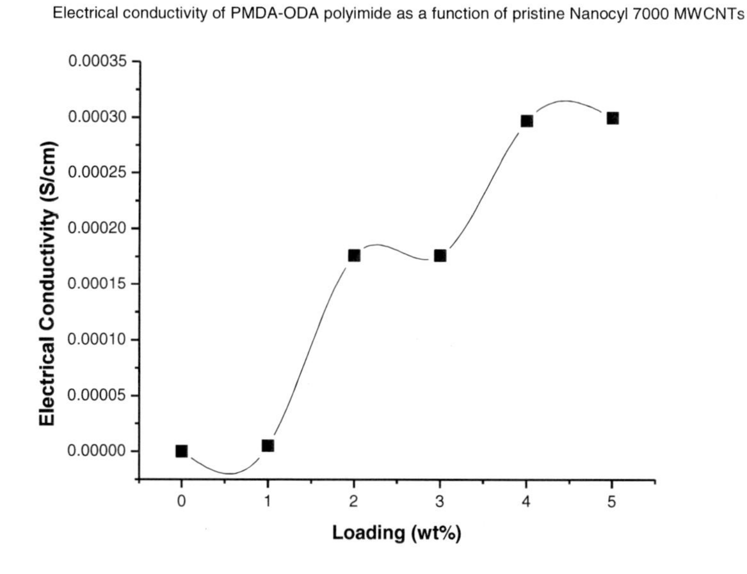

Figure 20. Plot of the electrical conductivity of PMDA-ODA/Nanocyl 7000 MWCNT nanocomposite.

Figure 20 shows the plot of the electrical conductivity of PMDA-ODA nanocomposite loaded with Nanocyl 7000 MWCNT in the range 1-5 wt%.

It can be seen that even though the electrical conductivity is low, electrical conductivity percolation is achieved at low percentage loading of carbon nanotubes.

It can be seen from Figure 21 that the tensile strength at break and the modulus both increase significantly between 1 and 3 wt% loading of fillers after which there is a decrease. The general decrease in the tensile strength and modulus at higher loadings may be due increased brittleness as the carbon nanotube loading increases. [32] This effect is even more evident in the elongation at break. The elongation at break of the nanocomposites with most of the fillers, start to drop at the first percentage loading of 1 wt%.

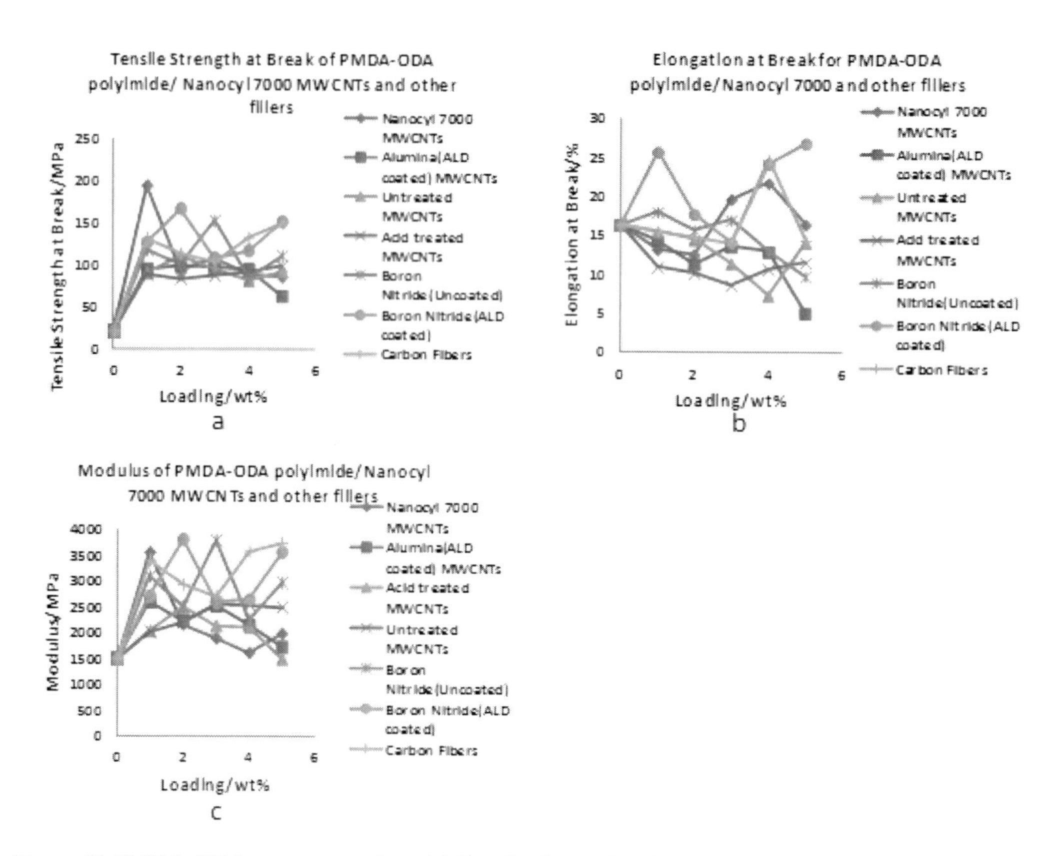

Figure 21.PMDA-ODA nanocomposites, (a) Tensile Strength at Break of at Break, (b) Elongation at Break, (c) Modulus.

3. Polycarbonate (PC), Poly (Vinyl Chloride) (PVC), Poly (Acrylonitrile-Co-Butadiene-Co-Styrene) (ABS), (ABS-PC), (ABS-PVC) and (ABS-PC-PVC) Nanocomposites

Introduction

In this section we report the synthesis of nanocomposites of polycarbonate, poly (vinyl chloride), poly (acrylonitrile-co-butadiene-co-styrene) and their alloys (ABS-PC), (ABS-PVC) and (ABS-PC-PVC) with multiwalled carbon nanotubes. Electrical, and mechanical properties were obtained from nanocomposites that were prepared by solution casting. The percentage loadings of carbon nanotubes in the nanocomposites ranged from 1 to 25 wt%.

Polycarbonate (PC), poly (vinyl chloride) (PVC), poly (acrylonitrile-co-butadiene-co-styrene) (ABS) are among the six leading resins that find extensive use in industry, the other three being polyphenylene-based resins, polyethylene and polypropylene. [33] PC is a homogeneous single phase polymer. It consists of two methyl side groups in addition to phenyl groups on the molecular chain. The phenyl groups make the PC molecule stiff and this stiffness is increased by the presence of the methyl groups on the sides. The phenyl groups in each molecule tend to interact with phenyl groups in the other molecules thereby restricting mobility. [34] This lack of mobility has the advantage of making polycarbonate a good thermal resistor. Polycarbonate does not have a significant crystalline structure as a result of its inflexibility and lack of movement. An important consequence of this lack of crystallinity or its amorphous nature is that it a transparent polymer. [35] It is a polymer that finds extensive use in engineering.

ABS is a heterogeneous two-phase terpolymer which consists of a dispersed rubbery phase made up of polybutadiene (PB) rubber grafted with styreneacrylonitrile (SAN) and then dispersed in a continuous plastic phase of more styreneacrylonitrile (SAN). ABS finds extensive industrial applications in the vehicle, computer and general equipment industry. This wide application is a consequence of the outstanding mechanical properties, processability and recyclability of this polymer. [36]

PVC is a polymer with diverse applications in industry. When plasticized, it becomes flexible for floor applications while when it is unplasticized it becomes suitable for use in building. Window frames are also constructed from the unplasticized PVC. [29] Because of the importance of these polymers we set out to investigate the effect of nanomaterial fillers would have on their electrical, and mechanical properties.

Initially all these materials were used to make PC, PVC and ABS nanocomposites with percentage loadings of 1-5 wt%. Based on initial electrical conductivity measurements some of the MWCNT fillers were reused at higher loadings. Due to the supplier's continued effort to improve the performance of the MWCNT produced by the combustion process, two more batches of MWCNT were supplied. Nanocomposites of PC, PVC and ABS with these two sets of MWCNT were also prepared. Due to the importance of the alloys of these three polymers in industrial applications we felt motivated to also investigate the electrical and mechanical properties of the alloys ABS-PC, ABS-PVC and ABS-PC-PVC with Nanocyl 7000 MWCNTs and acid treated MWCNTs manufactured by the combustion process. All the alloy nanocomposites had a Nanocyl 7000 MWCNT loading of 1-5 wt% and a 5-25 wt% loading of the acid treated MWCNTs manufactured by the combustion process.

Experimental

Poly (bisphenyl-A) carbonate, Poly(vinyl chloride) and poly(acrylonitrile-co-butadiene-co-styrene) were purchased from Aldrich. Filler materials for the preparation of nanocomposites were supplied by Nanodynamics Buffalo. To prepare polymer solutions of 1g of each of the polymers was dissolved by magnetic stirring in 15 ml of chloroform, tetrahydrofuran and 1,2-dichloroethane respectively. Two types of mulitwalled carbon nanotubes (MWCNT) were used. Nanocyl 7000 produced by chemical vapor deposition (CVD) and Nanodynamics MWCNTs that were produced by a combustion process of ethylene in air. Prior to mixing the MWCNT with the polymer solutions, they were dispersed in 10 ml of the solvent that was used to dissolve the polymer. Dipersion of the MWCNT in solvents was achieved by bath sonication for an hour. The percentage loadings of the MWCNT were 1-5 wt% for Nanocyl 7000 and 5-25 wt% for the MWCNTs produced by the combustion process.

The polymer solutions were then mixed with the dispersed MWCNT and bath sonicated for 30 minutes. The polymer solutions with the MWCNT were then left to mix overnight by stirring gently using magnetic stir bars. Before casting in Teflon cups the polymer solutions with MWCNT were tip-sonicated for 2 minutes. The solutions were left to dry at room temperature. ABS-PC, ABS-PVC alloy nanocomposites were prepared by mixing equal amounts 7.5 ml of ABS with 7.5 ml of PC and PVC solutions prepared following the previous described procedure. The polymer alloy was mixed by magnetic stirring for 1 hour. Carbon nanotubes were dispersed for 1 hour by bath sonication in a mixture of 5 ml of 1, 2-dichloroethane and 5ml chloroform for ABS-PC and the same volumes of 1, 2-dichloroethane and tetrahydrofuran for ABS-PVC. The dispersed carbon nanotubes were then mixed with the polymer blend.

The mixture was further bath sonicated for 1 hour and left to mix overnight by magnetic stirring. Before casting, the polymer alloy/carbon nanotube mixtures were tip-sonicated for 2 minutes. The ABS-PC-PVC alloy polymer solutions were prepared by mixing 5 ml of each of the polymer solutions prepared following the procedure described above. The polymer alloy mixtures were mixed by magnetic stirring for 1 hour. Carbon nanotubes were dispersed in 9 ml solvent mixture made up of 3 ml of 1, 2-dichlorethane, 3 ml of chloroform and 3 ml of tetrahydrofuran by bath sonicating for 1 hour. The ABS-PC-PVC polymer blend was then mixed with the dispersed carbon nanotubes and bath sonicated for 1 hour followed by magnetic stirring overnight.

Tip-sonication was done for 2 minutes before film casting. The morphology of the films were obtained using FESEM SUPRA 55, electrical conductivity data were obtained using the Four-point probe, thermal and Tensile measurements were made using the Instron Tensile Testing Machine.

Results and Discussion

Electrical Conductivity

Of all the initial PC, PVC and ABS nanocomposites that were prepared with loadings of 1-5 wt%, only Nanocyl 7000 MWCNT loaded nanocomposites showed measurable electrical conductivity. The electrical conductivity of the PC, PVC and ABS nanocomposites with Nanocyl 7000 MWCNTs are shown in Figures 22 (a), (b), and (c) respectively. It can be seen that the electrical percolation threshold in the PC, PVC and ABS nanocomposites loaded with Nanocyl 7000 MWCNT is below 5 wt%. There is also variation in the carbon nanotube percentage weight at which the percolation threshold occurs for each polymer nanocomposite. This may be attributed to the difference in the properties of the polymers such as crystallinity and the nature of interactions the polymer has with the carbon nanotubes [37].

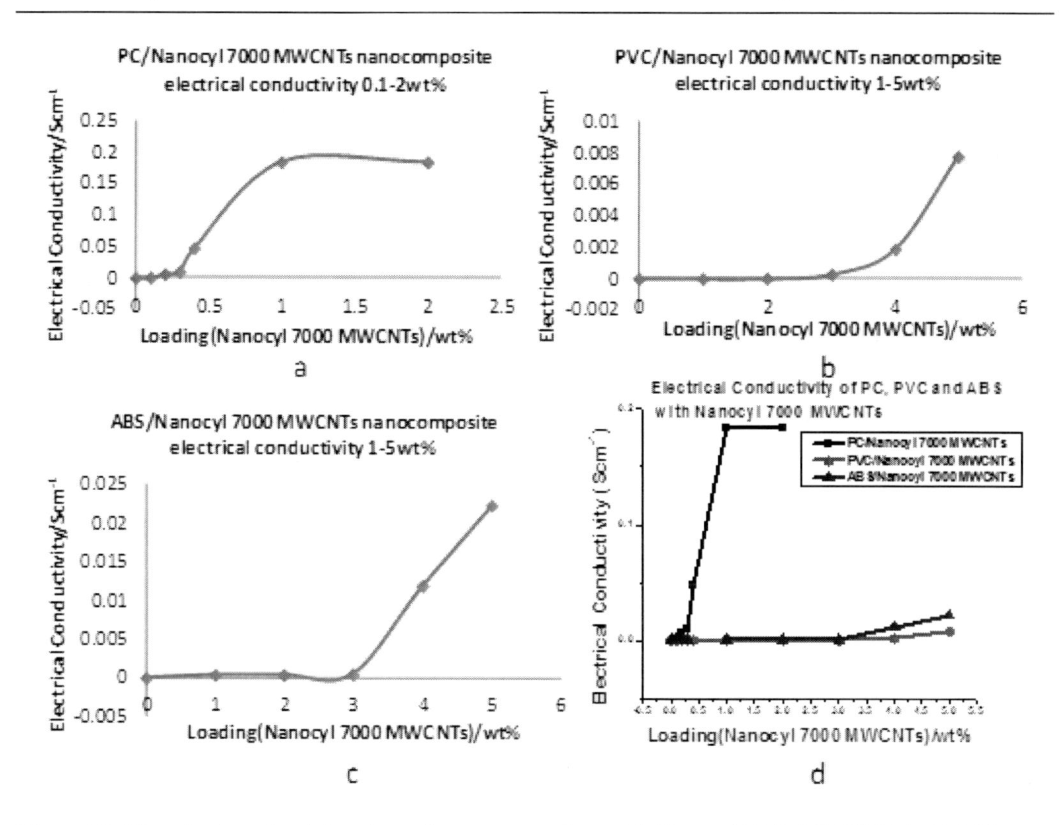

Figure 22. Electrical conductivity plots of nanocomposites with Nanocyl 7000 MWCNT (a) PC (b) PVC and (c) ABS (d) PC, PVC and ABS plots together.

In order to further investigate and understand the electrical conductivity of the nanocomposites loaded with MWCNTs prepared by the combustion process, nanocomposites of PC, PVC and ABS were synthesized with 5-25 wt% MWCNT. Figure 23(a), (b), (c) and (d) show the plots of the electrical conductivities that were obtained for 5-25 wt% loading of MWCNTs in PC, PVC and ABS.

Although we had not managed to record electrical conductivity readings at low loadings of MWCNTs manufactured by the combustion process, the plots indicate that indeed these carbon nanotubes cause electrical conductivity in the nanocomposites starting at percentage loadings around 5 wt%. Solution processing was possible up to 25 wt%, after which dispersing the MWCNTs became difficult and the polymer solution-carbon nanotube mixture became too viscous. Figure 24(a), (b), (c) and (d) show the electrical conductivity plots that were obtained using acid treated MWCNTs.

The acid treated MWCNT based nanocomposites showed an electrical percolation threshold of greater than 5 wt%. Compared to Nanocyl 7000 MWCNT loaded nanocomposites, it was found that even at high percentage loadings such as 25 wt%, the acid treated and untreated combustion process produced MWCNTs loaded nanocomposites did not achieve electrical conductivity of 0.2 Scm-1 which was attained by Nanocyl 7000 MWCNT filled PC nanocomposites at 2 wt%.

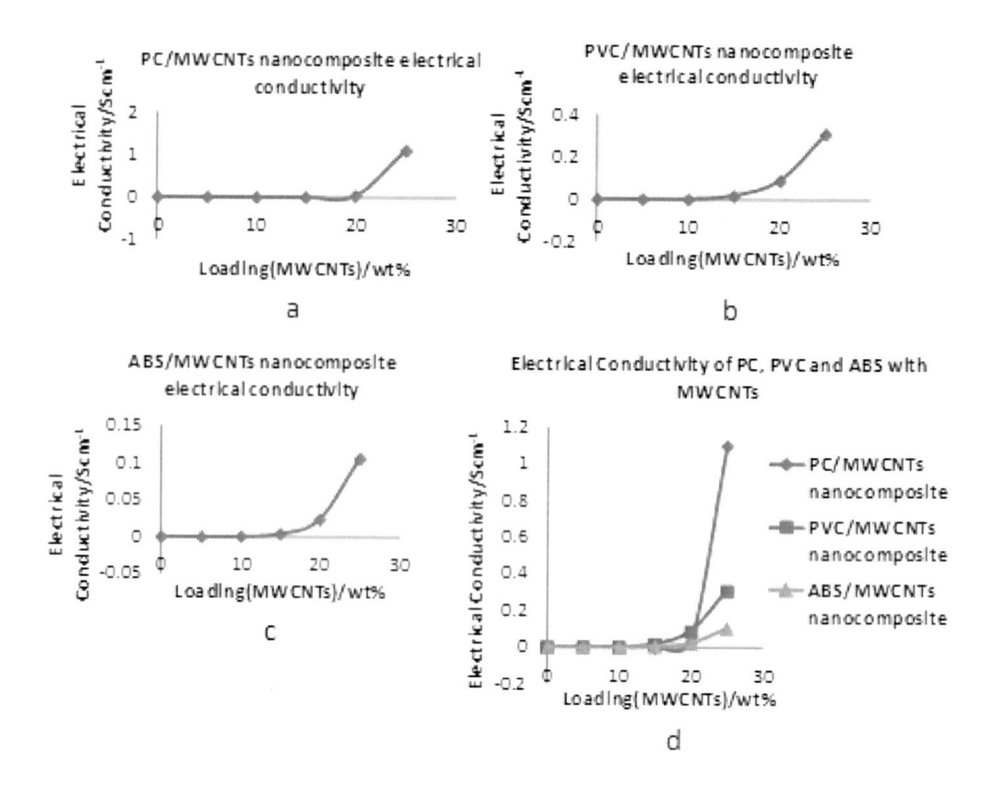

Figure 23. Electrical conductivity plots of nanocomposites with MWCNTs (a) PC (b) PVC (c) ABS (d) PC, PVC and ABS plots together.

The percolation threshold for Nanocyl 7000 filled nanocomposites was below 5 wt% and going as low as 0.1 wt% in the PC nanocomposite. This could be attributed to the fact that these MWCNT are coated with alumina which may be shielding their electrical properties. However, alumina coated carbon nanotubes have been found to increase electrical conductivity in composites. [38].

An analysis of the plots in Figure 25 shows that polycarbonate when loaded with combustion process manufactured carbon nanotubes, results in better electrical conductivity compared to PVC and ABS.

This may be attributed to the high degree of crystallinity which is not present in PVC and ABS. Only PC-Nanocyl 7000 MWCNTs nanocomposites showed an increase in elongation at break at a percentage loading of 1 wt%. The rest of the PC-filler nanocomposites showed a continued decrease in elongation at break as the nanofiller loading increased. The decrease in elongation at break may be attributed to increased stiffness and as result they become more brittle. [39].

Figure 28 shows the tensile measurements for PVC nanocomposites loaded with the different filler materials. It can be seen that the modulus of the PVC nanocomposites filled with Nanocyl 7000 MWCNTs, tend to decrease after a slight rise at 1 wt% loading. Carbon fiber and Boron Nitride (both coated and uncoated) filled PVC nanocomposites showed a steady increase in modulus.

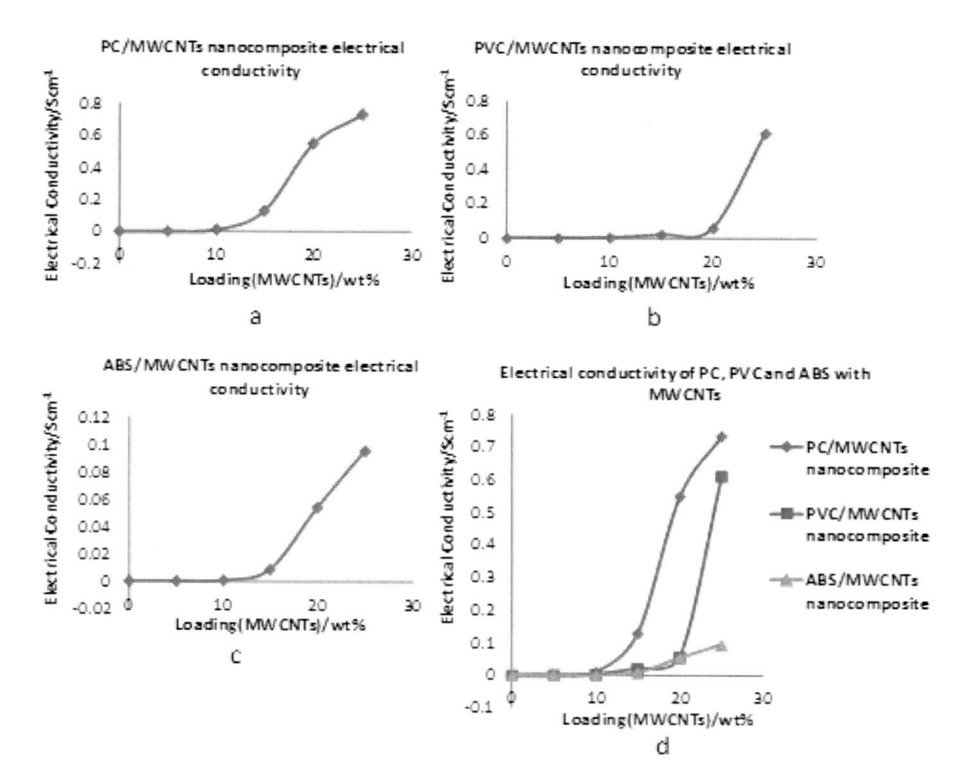

Figure 24. Electrical conductivity plots of nanocomposites with Nitiric Acid treated MWCNTs (a) PC (b) PVC and (c) ABS (d) PC, PVC and ABS plots together.

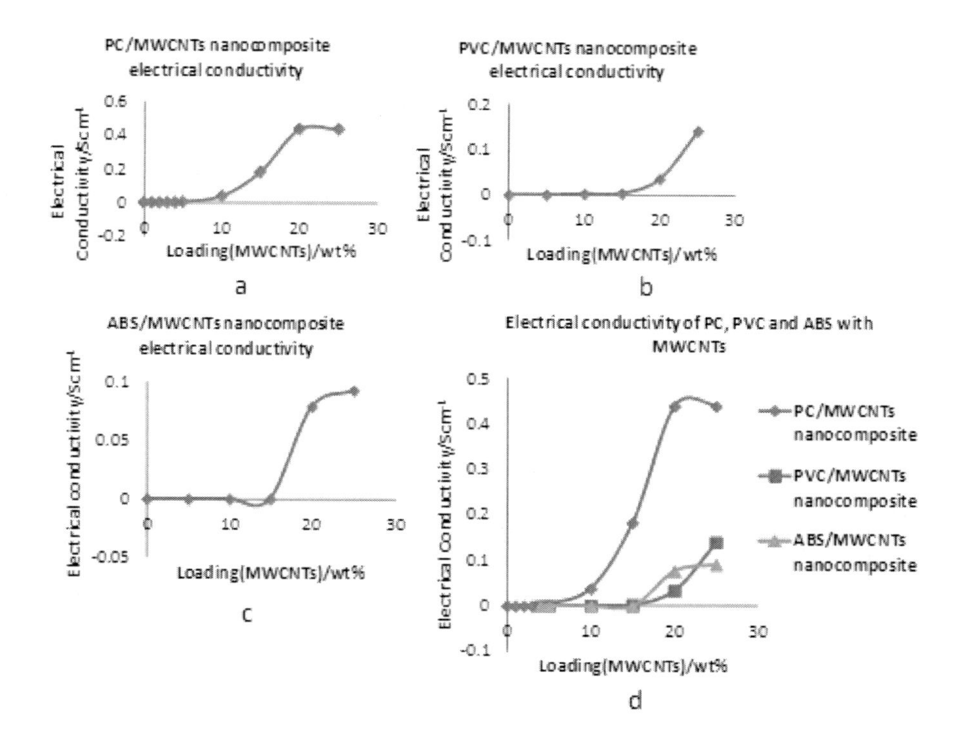

Figure 25. Electrical conductivity plots of nanocomposites with MWCNTs (a) PC (b) PVC and (c) ABS (d) PC, PVC and ABS plots together.

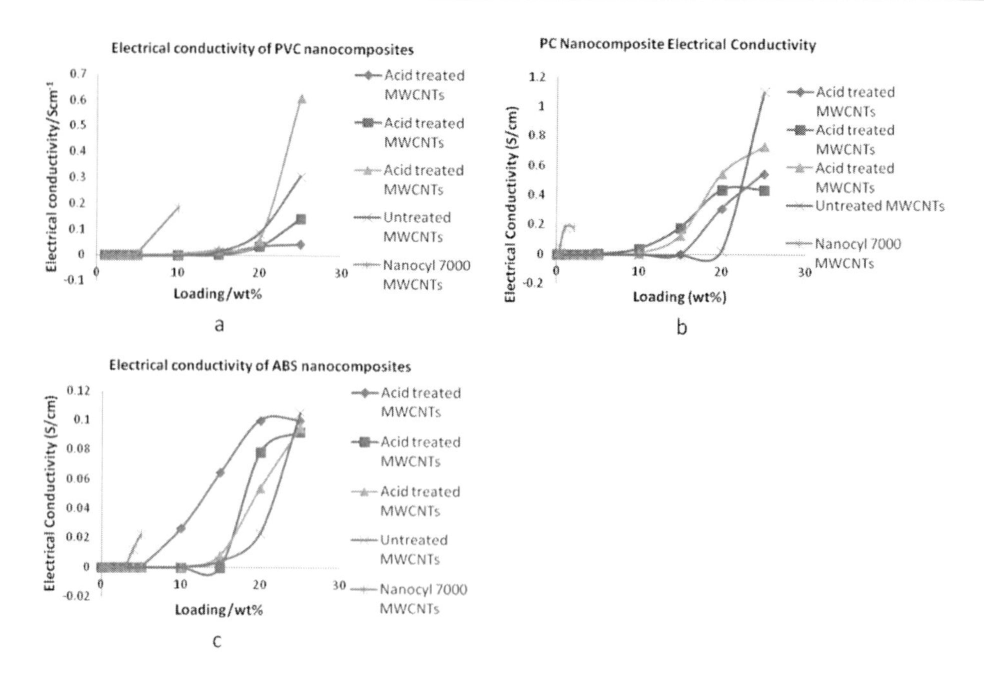

Figure 26. Electrical conductivity of (a) PVC (b) PC and (c) ABS nanoposites loaded with different types of MWCNTs.

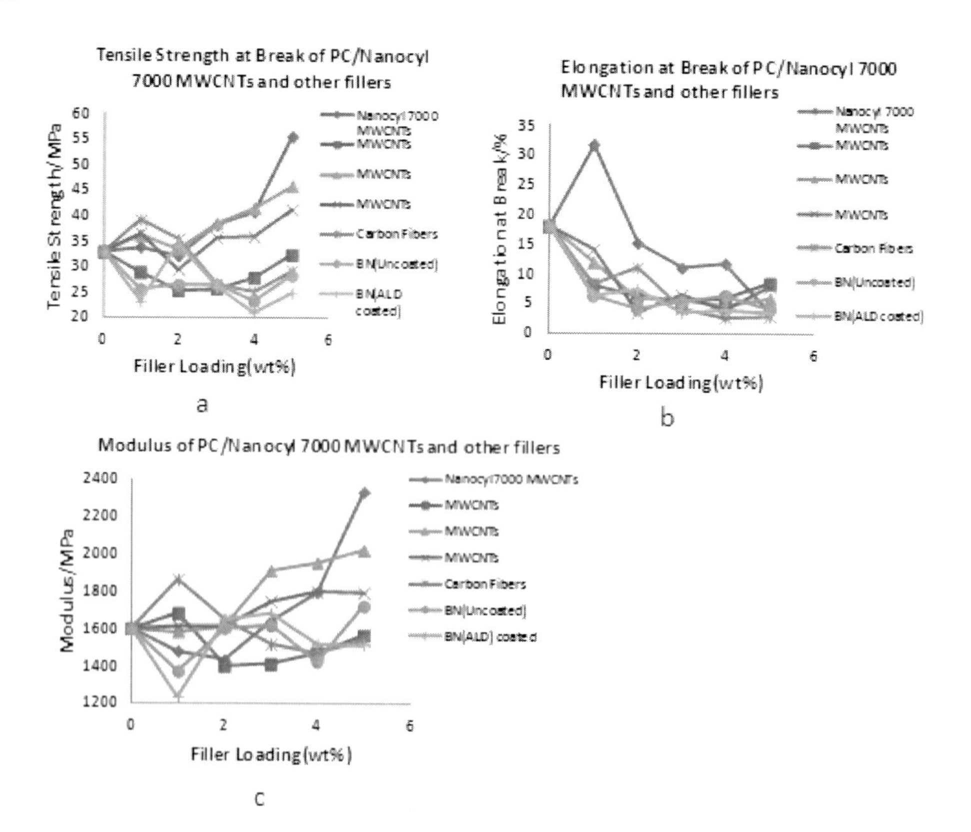

Figure 27. Tensile properties of PC nanocomposites with Fillers (a) Tensile Strength at Break (b) Elongation at Break (c) Modulus.

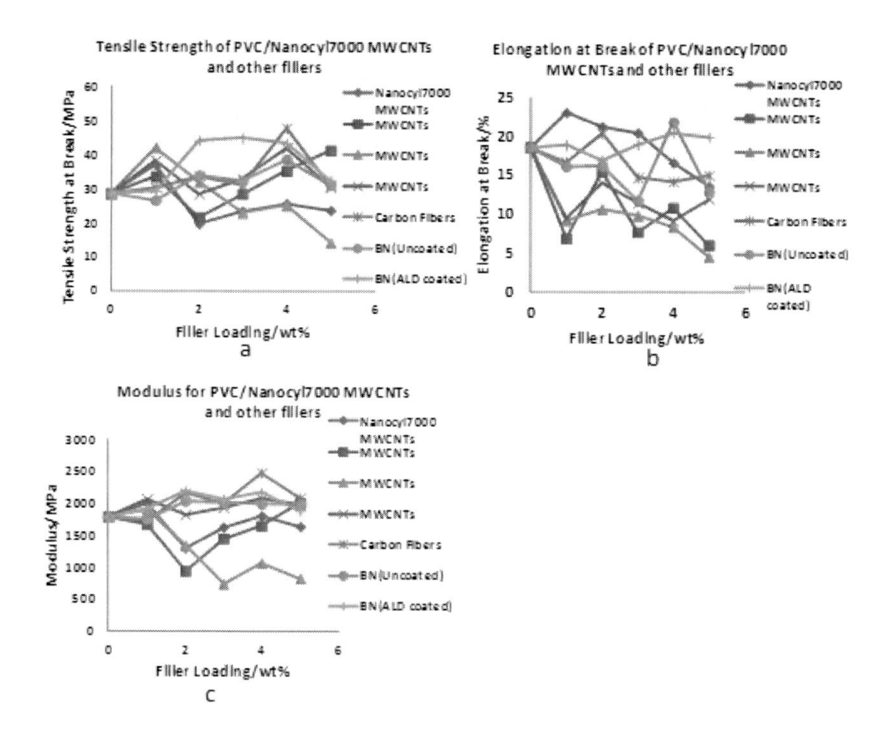

Figure 28. Tensile properties of PVC nanocomposites with Fillers (a) Tensile Strength at Break (b) Elongation at Break (c) Modulus.

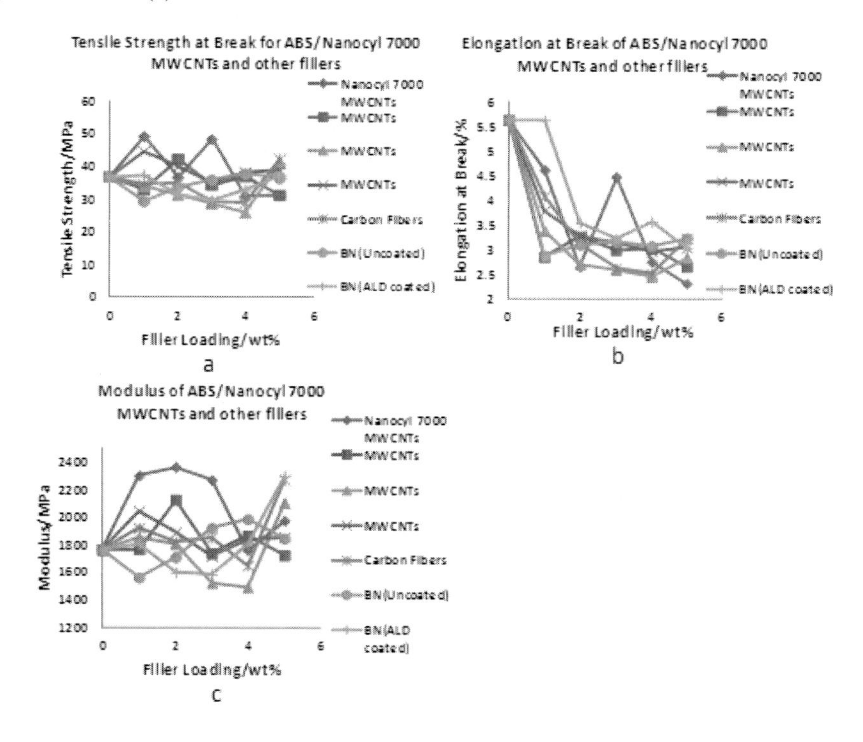

Figure 29. Tensile properties of ABS nanocomposites with Fillers (a) Tensile Strength at Break (b) Elongation at Break (c) Modulus.

The nanocomposite tensile strength for most of the fillers including Nanocyl 7000 MWCNTs, and Carbon fiber filled PVC nanocomposites show an increase at 1 wt% loading. The elongation at break for the PVC nanocomposites filled with most of the fillers showed a decrease except for Nanocyl 7000 MWCNTs and Boron Nitride filled PVC nanocomposites which show an increase at 1 wt%. Figure 29 shows the tensile properties of ABS nanocomposites filled with the same fillers used to make PC and PVC nanocomposites. It can be seen that the moduli of ABS filled with all the fillers increase at a filler loading of 1 wt% except for one and coated Boron Nitride.

The tensile strength of ABS nanocomposites filled with Nanocyl 7000 MWCNT and coated Boron Nitride is seen to increase at 1 wt% loading and the rest of the nanocomposites show a decrease. The elongation at break for ABS nanocompostes filled with all the fillers was found to decrease with increasing percentage loading of fillers.

Figure 30 shows the electrical conductivity of an ABS-PC alloy loaded with acid treated MWCNTs. This was the latest set of carbon nanotubes supplied by Nanodynamics. In this plot, where the range of percentage loading of carbon nanotubes was 5-25 wt%, it can be seen that significant electrical conductivity percolation is not seen until between 5 and 10 wt%. The electrical conductivity of this alloy nanocomposite achieves a peak at around 20 wt% in the 5-25 wt% range.

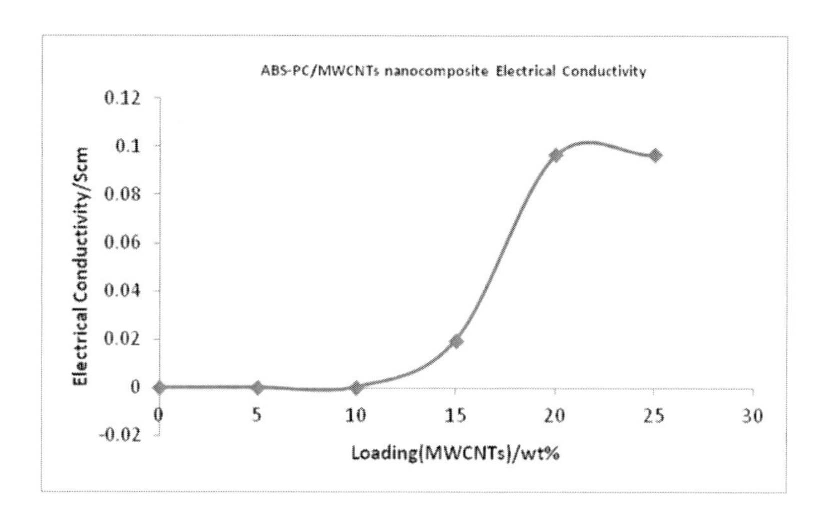

Figure 30. Electrical conductivity of ABS-PC loaded with MWCNTs.

Figure 31 shows a plot of the same alloy, ABS-PC, filled with Nanocyl 7000 MWCNT. It can be seen that the ABS-PC/Nanocyl 7000 MWCNT nanocomposite achieves electrical conductivity percolation at below 1 wt% and at 5 wt% the electrical conductivity is much higher than that achieved by ABS-PVC/ MWCNT nanocomposite at 25 wt%.

Given that both Nanocyl 7000 and the other MWCNT dispersed well in both the solvents that were used as well as the polymer alloy solutions, this difference in electrical conductivity may be attributed to the production process. It is likely that the carbon nanotubes produced by the combustion process whereby ethylene is burned in air tend to produce carbon nanotubes that have defects and the problem is exacerbated by mechanical manipulation during the processing of the nanocomposites. Most importantly the combustion process may produce carbon nanotubes that are more likely to be armophous due to lack of complete graphitization.

The graphitization process is likely to be achieved in the chemical vapor deposition (CVD) process that is used to manufacture Nanocyl 7000 multiwalled carbon nanotubes.

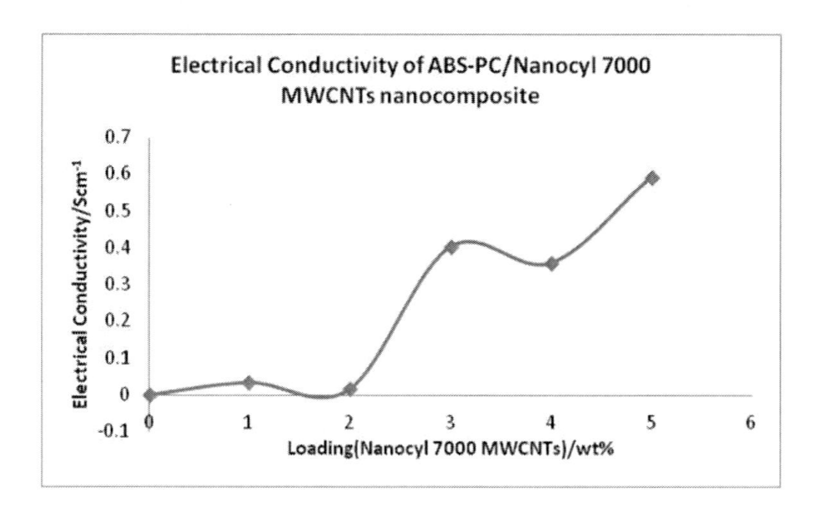

Figure 31. Electrical conductivity of ABS-PC loaded with Nanocyl 7000 MWCNTs.

Figure 32(a) and (b) show the electrical conductivity plot of ABS-PVC alloy nanocomposites with Nanodynamics MWCNTs and Nanocyl 7000 multiwalled carbon nanotubes respectively.

Figure 32(a) shows that the electrical percolation threshold for the ABS-PVC/acid treated MWCNT nanocomposite is between 5 wt% and 10 wt%. For the ABS-PVC/Nanocyl 7000 MWCNT nanocomposite the onset of electrical conductivity is between 2 wt% and 3 wt%. What is interesting though is that within the same range of filler loading, the ABS-PC/Nanocyl 7000 nanocomposite has a much higher electrical conductivities at each loading in the 1-5 wt% range.

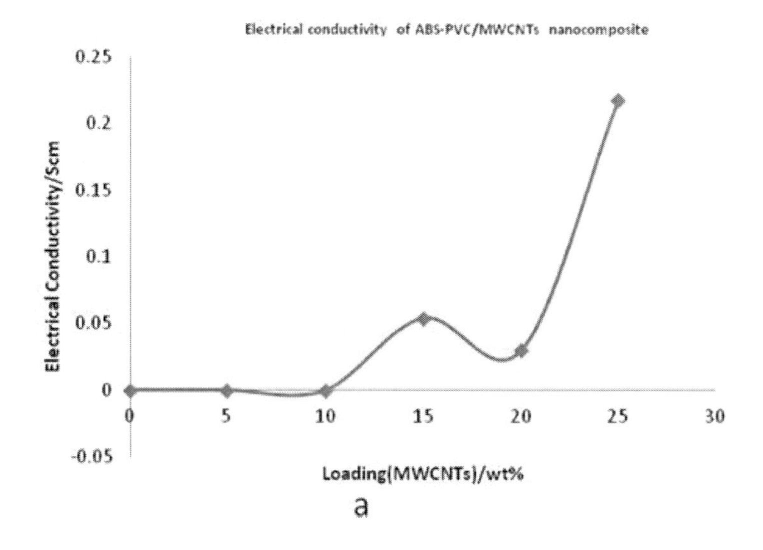

Figure 32. (Continued).

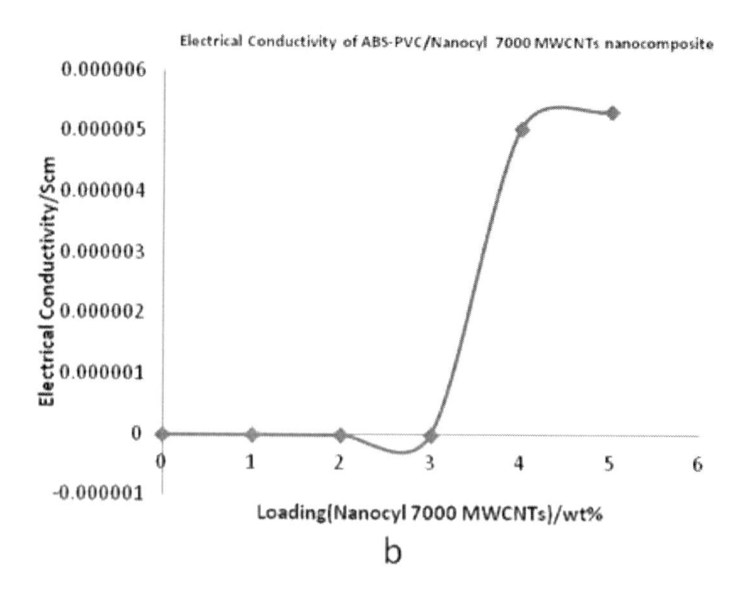

Figure 32. Electrical conductivity of ABS-PVC loaded with (a) Acid treated MWCNT and (b) Nanocyl 7000 MWCNT.

The low conductivity of the ABS-PVC/Nanocyl 7000 MWCNT nanocomposite may be a consequence of the different type of interaction the alloy has with the two types of carbon nanotubes since they are manufactured by different processes and that Nanocyl 7000 MWCNT are pristine while the other MWCNT are post processed. [40].

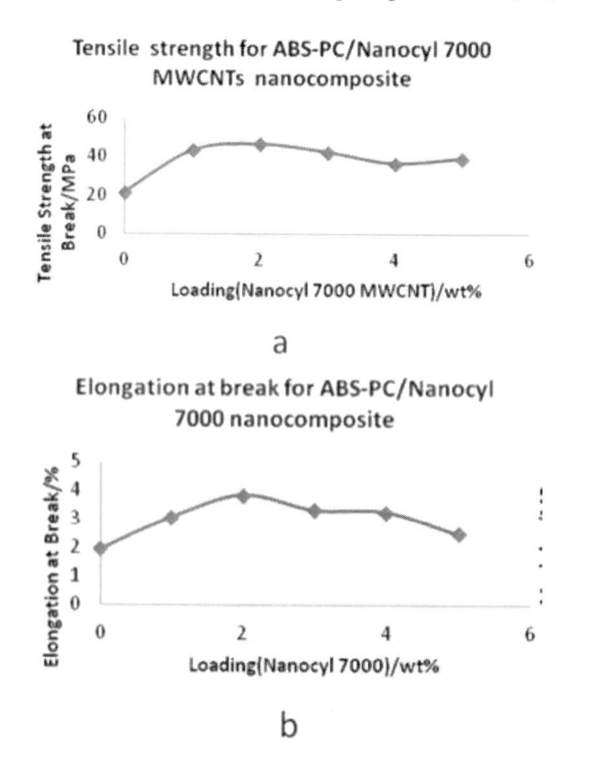

Figure 33. (Continued).

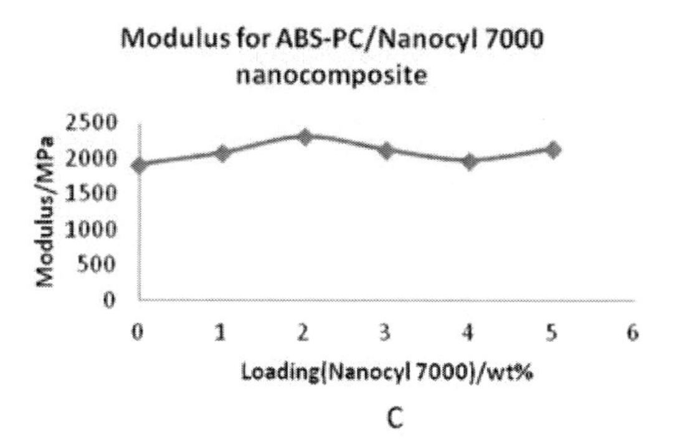

Figure 33. Tensile measurements of ABS-PC loaded with Nanocyl 7000 MWCNT (a) Tensile Strength at Break (b) Elongation at Break (c) Modulus.

Data on the mechanical properties of these two alloys were collected and plotted as shown in Figure 33. Figure 33(a), (b), and (c) show the tensile strength at break, elongation at break and modulus of ABS-PC alloy nanocomposites loaded with Nanocyl 7000 MWCNT. These three properties show a rise between 1 and 2 wt%.

However with increased loading of carbon nanotubes, these mechanical properties start to decrease. It can be seen however, that once the modulus increases between 1 and 2 wt%, it does not decrease significantly, but seems to hold steady with increased carbon nanotube loading.

Figures 34(a), (b) and (c) show the tensile strength at break, elongation at break and modulus of ABS-PVC alloy nanocomposite loaded with nanocyl 7000 MWCNT in the range 1-5 wt% respectively, the same range in which ABS-PC was loaded. The tensile strength at break and modulus show a rise at 1 wt% and then holds steady with increased loading of carbon nanotubes. However, the elongation at break decreases after 1 wt%.

Figure 35 (a) and (b) show the tensile strength at break and the modulus that have been plotted together for comparison. When the mechanical properties of two alloy nanocomposites, ABS-PC and ABS-PVC loaded with Nanocyl 7000 MWCNT are plotted together, it can be seen that ABS-PC nanocomposite has better tensile strength at break and modulus in the 1-5 wt% loading range.

In contrast to the pattern seen for the tensile strength at break and the modulus, the ABS-PVC nanocomposite with Nanocyl 7000 MWCNT, shows better elongation at break compared to the ABS-PC nanocomposite loaded with the same carbon nanotubes.

The tensile measurements of ABS-PC alloy with acid treated MWCNT as filler material are shown in Figure 36. Acid treated MWCNTs were loaded in the range 5-25 wt%. The tensile strength at break and the elongation at break are seen to increase gradually up to 10 wt% after which they start to go down. However, the modulus increases within the whole 5-25 wt% loading range.

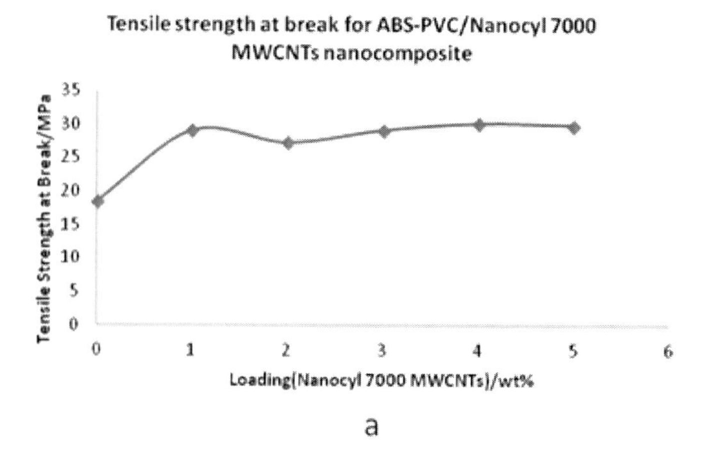

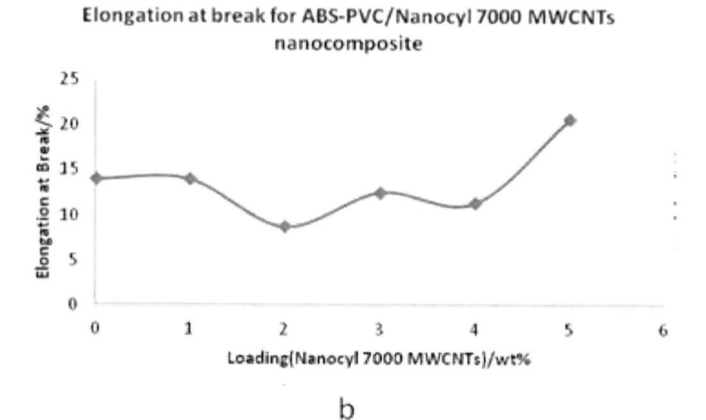

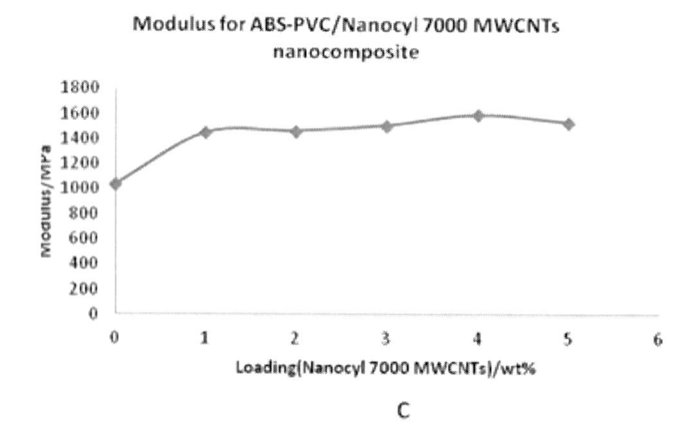

Figure 34. Tensile measurements of ABS-PVC loaded with Nanocyl 7000 MWCNTs (a)Tensile Strength at Break (b) Elongation at Break (c) Modulus.

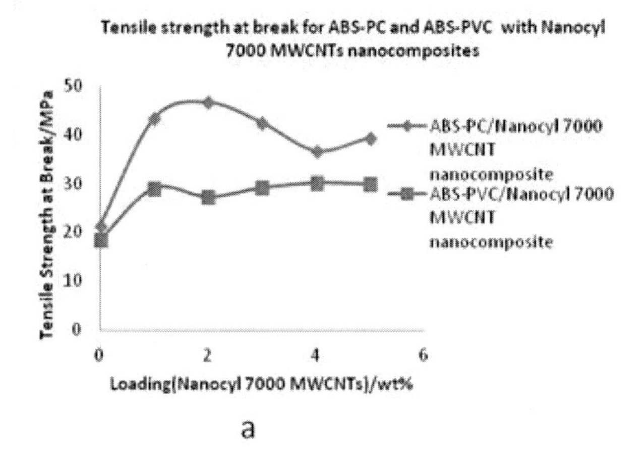

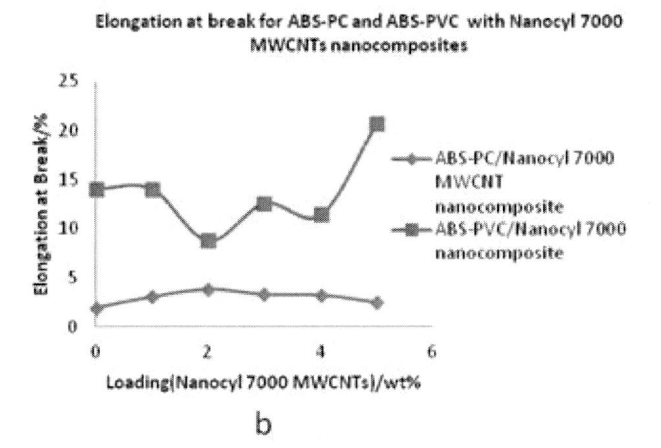

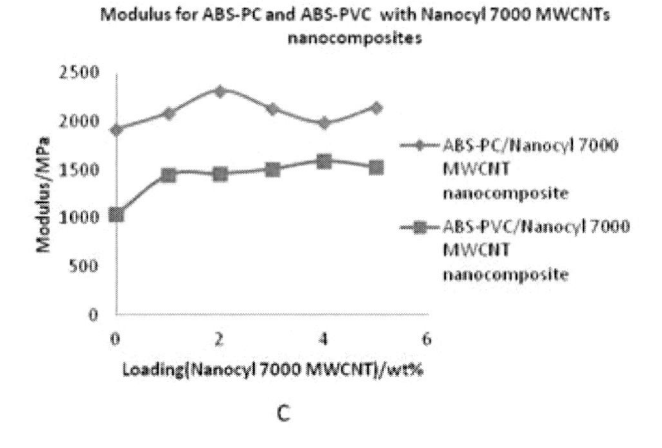

Figure 35. Tensile measurements of ABS-PC and ABS-PVC loaded with Nanocyl 7000 MWCNT (a) Tensile Strength at Break (b) Elongation at Break (c) Modulus.

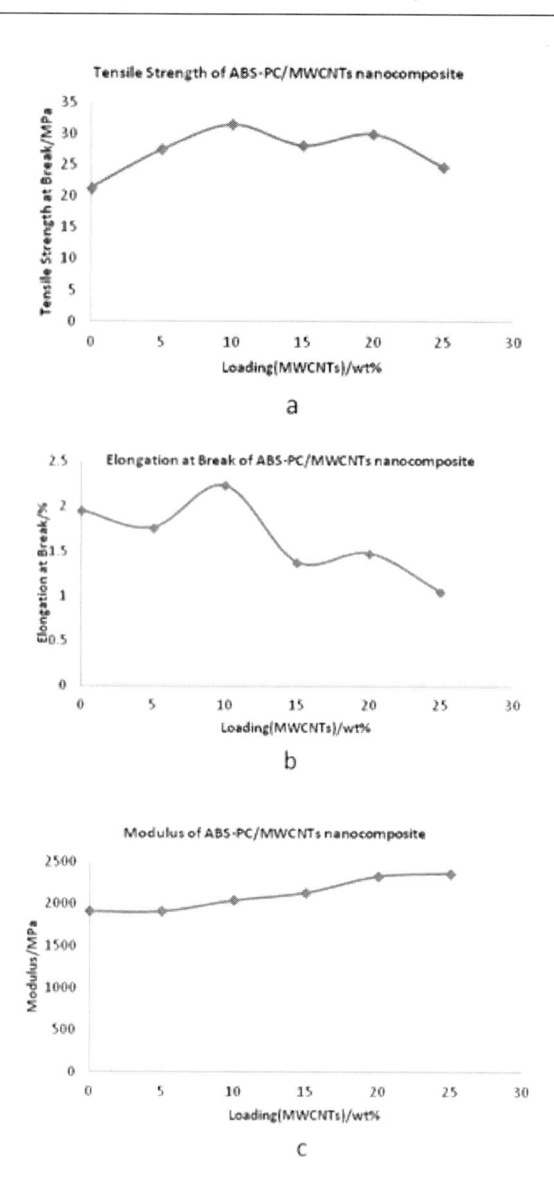

Figure 36. Tensile measurements of ABS-PC loaded with acid treated MWCNT (a) Tensile Strength at Break (b) Elongation at Break (c) Modulus.

Figure 37(a) and (d), show the tensile strength at break, and modulus of ABS-PVC alloy nanocomposite loaded with acid treated MWCNT respectively. The carbon nanotube loading was in the range of 5-25 wt%.

As can be seen the tensile strength at break and the modulus both increase with carbon nanotube loading up to a percentage loading of 10 wt%, after which they both seem to level off.

Although ABS-PVC/acid treated MWCNTs nanocomposite tensile strength at break and the modulus both increase with increase in carbon nanotube loading, the elongation at break decreases drastically as shown in Figure 37(b).

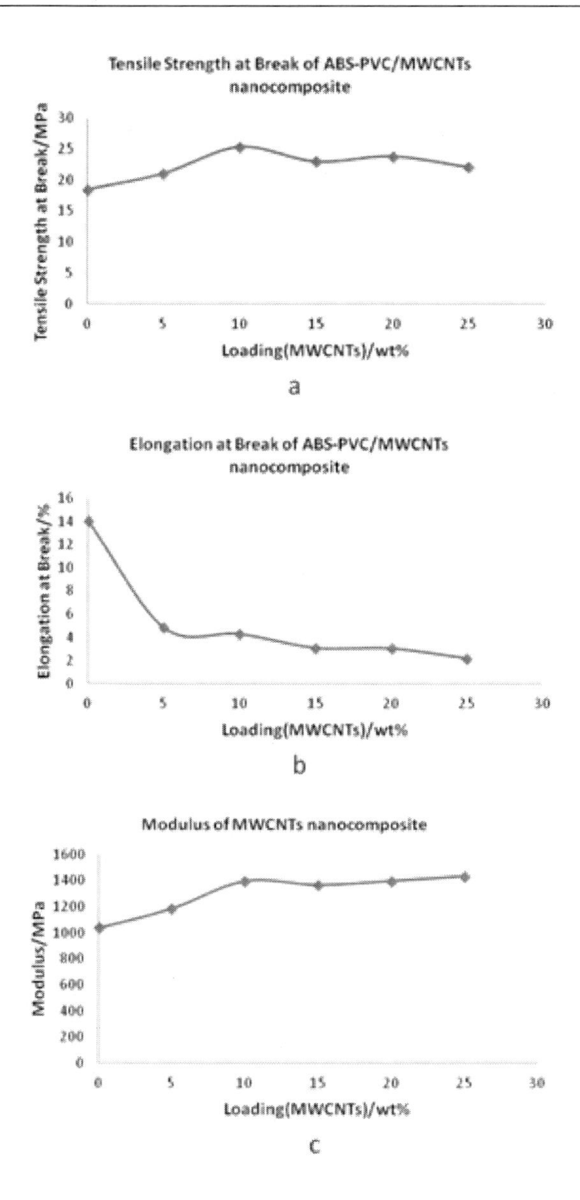

Figure 37. Tensile measurements of ABS-PC loaded with acid treated MWCNT (a) Tensile Strength at Break (b) Elongation at Break (c) Modulus.

In order to compare the mechanical properties of ABS-PC and ABS-PVC with acid treated MWCNT, their tensile strength at break, elongation at break and modulus were plotted together. Figure 38(a) shows that both alloys show an increase in tensile strength at break and have a peak at 10 wt% loading. The ABS-PC alloy nanocomposite shows better modulus performance in the entire loading range of 5-25 wt% as shown in Figure 38(c). Although the elongation at break values are higher for ABS-PVC, Figure 38(b) shows a rapidly decreasing trend as the carbon nanotube loading increases, while the ABS-PC holds steady. This better performance in mechanical properties was also manifested in ABS-PC loaded with Nanocyl 7000 MWCNT.

The plots in Figure 39 show the same pattern that was revealed when the two alloys were loaded with Nanocyl 7000 MWCNTs. The ABS-PVC nanocomposite shows better performance than the ABS-PC. However, the ABS-PVC shows a continuous decrease in elongation at break as the carbon nanotube loading increases while for ABS-PC, even though the elongation at break values are lower than those of the ABS-PVC alloy nanocomposite, the trend seems to hold steady with no drastic decrease.

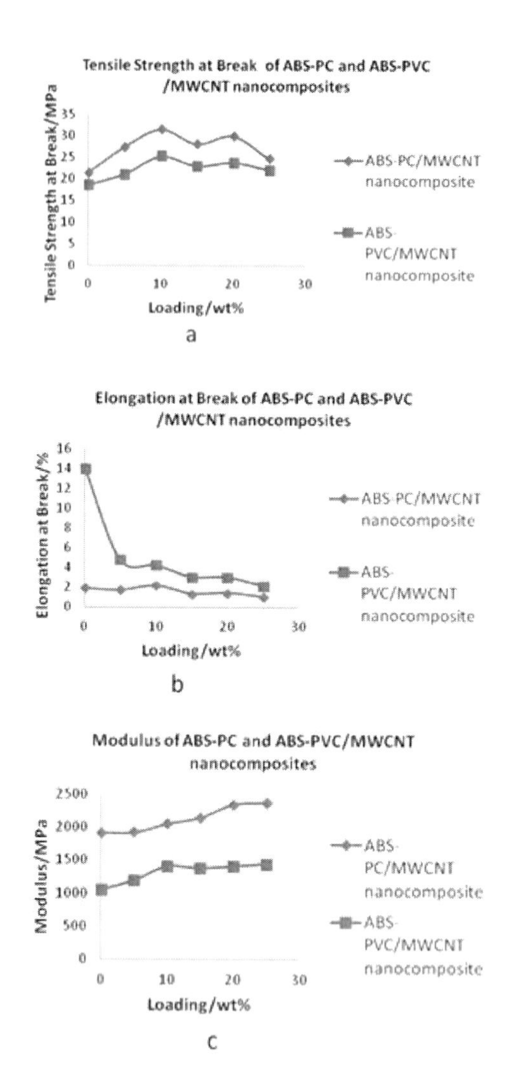

Figure 38. ABS-PC and ABS-PVC loaded with MWCNT (a) Tensile strength at break (b) Elongation at Break (c) Modulus.

Figure 39 shows plots of the ABS-PC-PVC alloy with Nanocyl 7000 MWCNT and acid treated MWCNTs as fillers respectively. While ABS-PC-PVC/Nanocyl 7000 MWCNT nanocomposite shows an electrical conductivity percolation threshold between 2 wt% and 3 wt%, ABS-PC-PVC/acid treated MWCNT nanocomposite shows an onset of electrical conductivity between 10 wt% and 15 wt%.

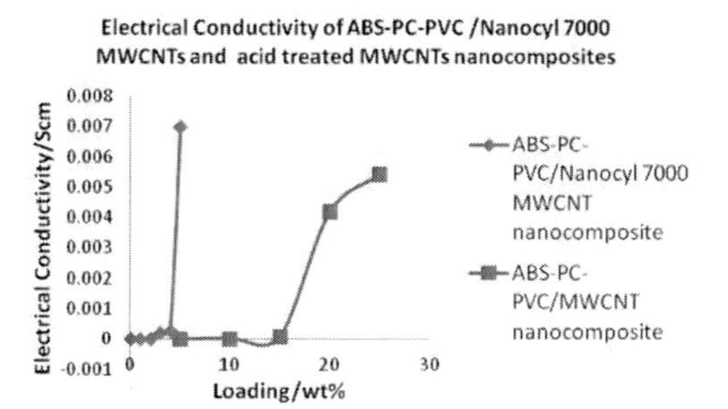

Figure 39. Electrical conductivity of ABS-PC –PVC loaded with Nanocyl 7000 MWCNT and Acid treated MWCNTs plotted together for comparison.

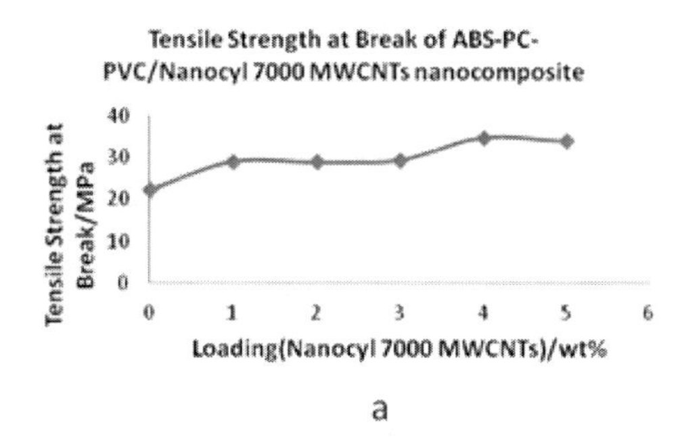

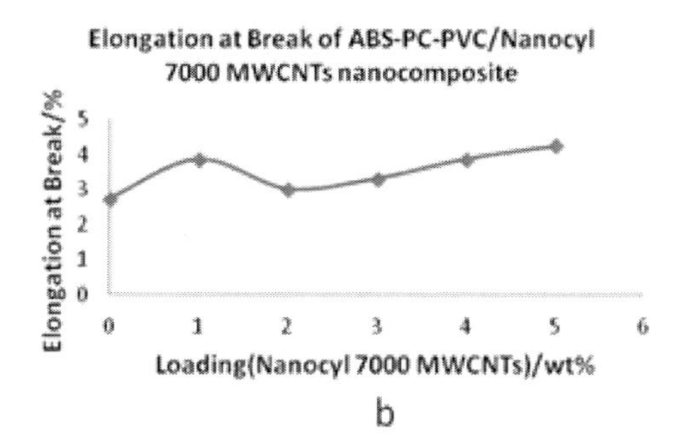

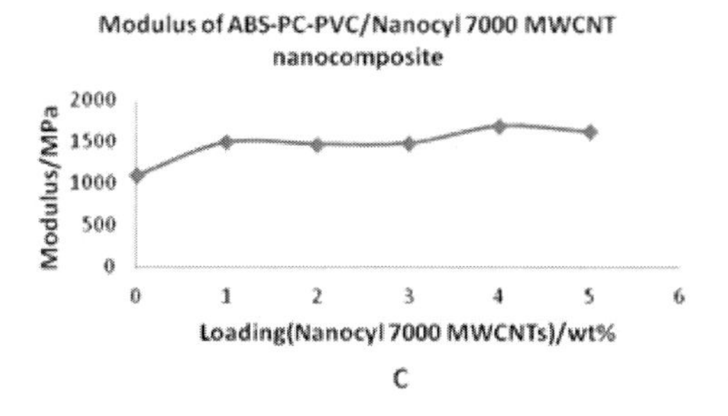

Figure 40. ABS-PC –PVC loaded with Nanocyl 7000 MWCNTs (a) Tensile Strength at Break (b) Elongation at Break (c) Modulus.

The plots of ABS-PC-PVC/Nanocyl 7000 MWCNT nanocomposite and ABS-PC-PVC/MWCNT nanocomposite are plotted together for comparison.

The tensile measurements of ABS-PC-PVC alloy loaded with Nanocyl 7000 MWCNT and acid treated MWCNT were investigated as well. Figure 40(a), (b) and (c) show the tensile strength at break, the elongation at break and the modulus of ABS-PC-PVC alloy nanocomposites loaded with Nanocyl 7000 MWCNT.

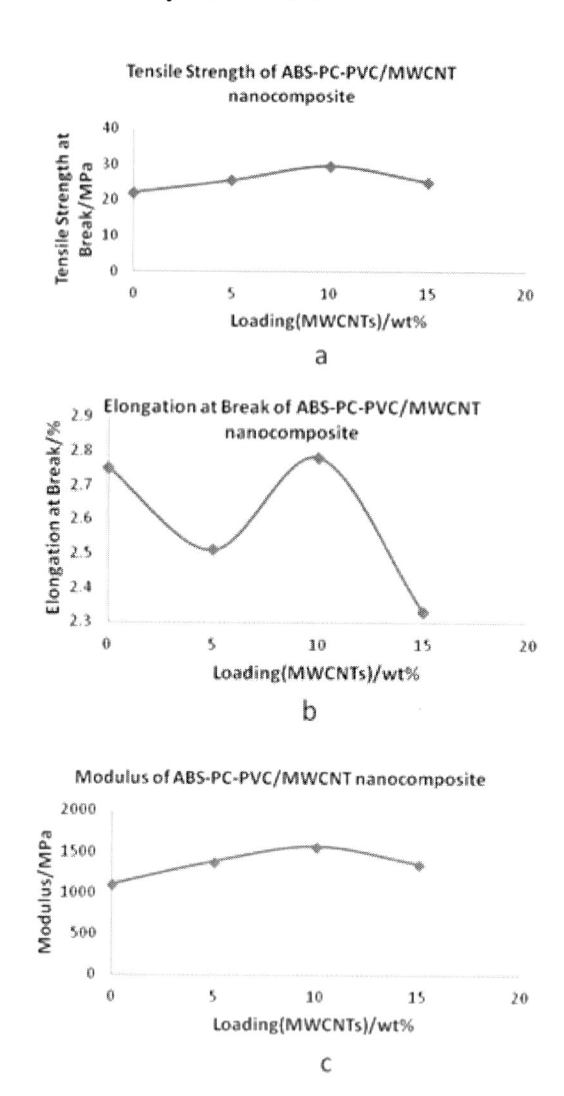

Figure 41. ABS-PC –PVC loaded with acid treated MWCNT tensile strength at break.

Both the tensile strength at break and modulus show an increase with increasing loading of carbon nanotubes. After showing a distinct increase at 1 wt%, the elongation at break drops before it increases gradually up to 5 wt%.

The tensile measurements of ABS-PC-PVC loaded with acid treated MWCNT were taken and plotted. Figure 41(a), (b) and (c) show the tensile strength at break, elongation at break and modulus of this ABS-PC-PVC/acid treated MWCNT nanocomposite respectively.

It can be seen in that the tensile strength at break and the modulus increase as the carbon nanotube loading increases. Although the elongation at break initially decreases, it rises significantly at 10 wt% before it falls.

All the parameters, tensile strength at break, elongation at break and modulus show a maximum peak at a percentage loading of 10 wt%. These mechanical properties of ABS-PC-PVC loaded with the two types of carbon nanotubes were plotted together for comparison. The graphs are shown in Figure 42.

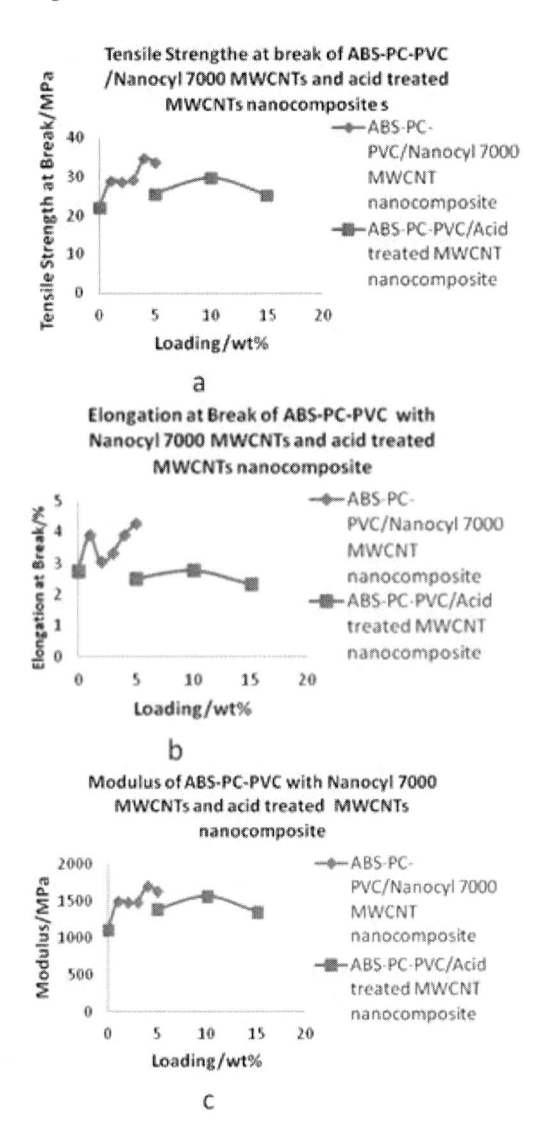

Figure 42. ABS-PC –PVC loaded with acid treated MWCNT and Nanocyl 7000 MWCNT Tensile Strength at Break plotted together.

A comparison of the plots in Figure 42 clearly shows that while the ABS-PC - PVC/Nanocyl 7000 MWCNT nanocomposite, which can be successfully prepared with percentage loadings of 1-5 wt%, has both electrical and mechanical properties, the ABS-PC-PVC/acid treated MWCNT nanocomposite did not manifest any electrical conductivity in the 1-5 wt% loading. As a result, the electrical and mechanical properties were only measured in the 5-25 wt% range.

ABS is a blend of polybutadiene and styrene acrylonitrile (SAN). The name ABS denotes the three monomers from which the polymer is made. These are acrylonitrile, butadiene and styrene. [41] The structures of ABS, PVC, PC and PVC with SAN of ABS are shown in Figure 43.

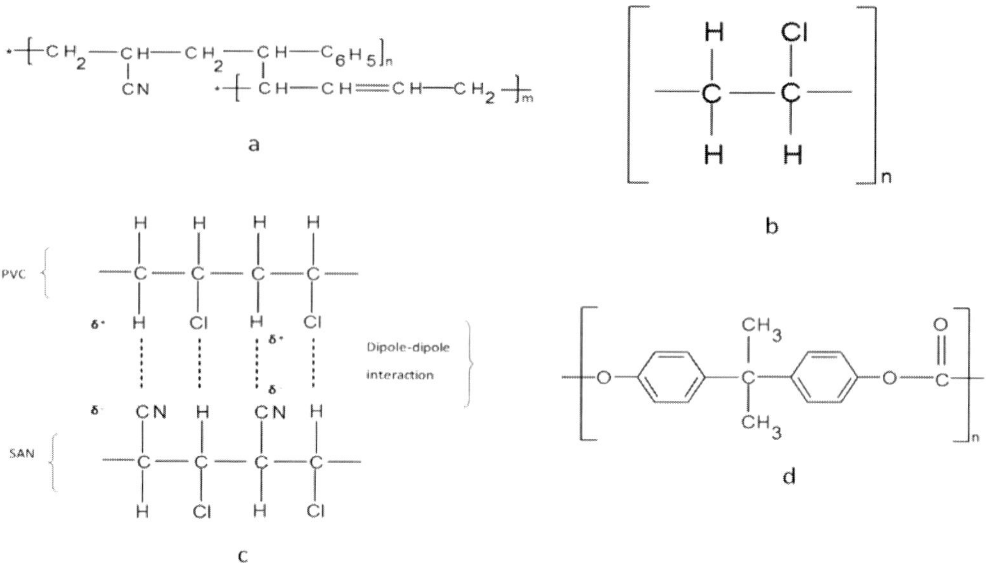

Figure 43. Chemical structure of (a) ABS (b) PVC (c) PVC with SAN of ABS and (d) Poy(bisphenol carbonate).

ABS has the advantage of being able to be blended with other polymers to further improve its properties. [42] One of the best candidates for blending with ABS is PVC which is cheap and its applications abound. A material with good flame resistance, toughness and flame resistance results when PVC is blended with ABS. [43] [44] [45] [46] As result this blend finds applications in construction, electrical and vehicle industry.

Crazing, shear yielding and cavitation are known as the major toughening mechanisms of polymers. Figure 43(c) shows the possible interaction between these two polymers. Crazing is the deformation mechanism when the entanglement density of the polymer is below a critical value. At higher entanglement densities both crazing and shear take place, whereas above an upper critical density, disentanglement becomes very difficult, thereby suppressing crazing and only shear deformation can be observed. [47] It is possible that the presence of carbon nanotubes can aid in the toughening of the ABS-PVC alloy.

Polycarbonate, shown in Figure 40(d) is another polymer that forms a useful alloy with ABS. The π system of the polycarbonate is likely to react with the SAN of the ABS. Carbon nanotubes in these polymer systems are expected to use their π systems to interact with the polymer system.

4. Synthesis and Characterization of Conducting Nanocomposites based on Intrinsically Conducting Polymers

Introduction

Since the discovery of carbon nanotubes by Iijima in 1991, [48] there has been an increasing research into their possible applications. It was found that carbon nanotubes have large surface area and high aspect ratio. [49] Among the growing areas of research is in electronics applications. However, using individual or bundles of carbon nanotubes without any support is difficult. To solve this problem, insulating and conducting polymer matrice have been used as media in which the carbon nanotubes can be embedded with the hope that the excellent properties of the carbon nanotubes will be transferred to the polymer matrix. [50] Incorporating carbon nanotubes into these polymer matrice has its own challenges. During production, carbon nanotubes come out in bundles. It is therefore important to have them separated into individual carbon nanotubes in order to achieve isotropic thermal and electrical conduction. [51] A number of methods which include sonication, cationic and nonionic surfactants, as well as functionalization have been used to disperse carbon nanotubes in polymer matrice. The use of surfactants and functionalization often allow the carbon nanotubes to disperse homogeneously in polymer solution [52].

In this work we used functionalization as a way to enhance the interaction of polyaniline with MWCNTs in nanocomposites. The process of oxidizing carbon nanotubes creates defects. The creation of defects leads to the destruction and conversion of sp2 bonds into sp3 bonds, which serve to bridge adjacent graphene shells and chemically functionalize broken –C=C- bonds by oxidation. It has been reported that intershell cross-linking in MWCNTs also results from oxidation. Functionalization and defect creation are likely to increase the number of scattering centers, thus reducing the electrical conductivity. However, the increase in conductivity due to cross-linking offsets the effect of scattering. Despite the decreasing of carbon nanotubes due increased sp3 bonds, the formation of -COOH and –C=O moeties at dangling bonds can serve as means to chemically anchor other nanostructures to the carbon nanotubes [53] [54] [55] [56].

In order to understand the dual effect on electrical conductivity of incorporating carbon nanotubes and functionalizing them before incorporating them in conducting polymers, we need to look at how charge is transported in conjugated polymers. Charge transport in conjugated polymers can involve many processes such as (1) conduction along the polymer backbone (2) hopping across chains due to inter-chain interactions (3) Tunneling between conducting segments that are separated by less conducting regions. Polymer properties are improved by carbon nanotubes in that they (i) induce additional structural ordering of the polymer (ii) improve the compactness and conjugation or chain length (iii) induce higher delocalization of charges in the nanocomposite (iiii) induce thermal stability and (iv) induce charge carrier mobility [57] [58].

The use of para-phenylenediamine results in MWCNTs with attached phenylamino groups which facilitates the dispersion of MWCNTs in hydrochloric acid. During polymerization, the phenylamino groups will be converted into polyaniline and therefore there is no need to worry about removing it as an impurity.

Commercialized products of polyaniline include 3V-coin shaped batteries by Brigetone-Seiko, antistatic layers in computer disks by Hitachi-Maxwell, camouflage by Milliken and Co., dispersable polyaniline powder version-jointly developed by Allied Signal, Americhem and Zipping Kessler, electrostat loudspeaker-0.1 μA polyamine on 6 μm polyester film and Incoblend-electro static dissipation which IBM is utilizing as an antistatic component carrier (e.g. computer chips). [59].

Polypyrrole is one of the intrinsically conducting polymers such as polyaniline, polythiophene, and poly (3, 4-ethylenedioxythiophene), polyisothianaphthene, polyparaphenylene vinylene, polyparaphenylene sulfide, polyparaphenylene and polyacetylene. It is characterized by p-type doping and has a conductivity of 40-200 S/cm in the doped state. Polypyrrole requires a single step to be produced in the conducting form. It is surprisingly tolerant of preparation conditions, particularly given the intermediacy of cationic species of cationic species which in principle could be attacked by nucleophiles. It can even be prepared in water and is stable in both air and water in its conducting form.

While a lot of work has been done in the synthesizing of CNTs containing nanocomposites, challenges still abound. Among them are, polymer-CNT interfacial resistance, CNT-CNT junction resistance, polymer particle-polymer particle resistance. An understanding of the way conducting polymers conduct is important in order to solve the problems mentioned above. Figure 44 illustrates how charge is transported in a conducting polymer like polypyrrole [60]. These problems are more pronounced in non-conducting polymer nanocomposites. To overcome these obstacles researchers have used different methods such as oxidizing and functionalizing the CNTs as well as using surfactants. Among some of the oxidizing reagents that have been used by Satishkumar are concentrated HNO3, concentrated H2SO4, aqua regia, superacid HF/BF3, aqueous OSO4, OSO4-NaIO4. [61].

Another reagent that has been used is a mixture of nitric acid and sulfuric acid. However, besides oxidizing the CNTs, this method tends to chop them into short pieces. [62] In this work we chose to oxidize MWCNTs with potassium permanganate in the presence of a phase transfer catalyst. During the oxidation process potassium permanganate is reduced according to equation 2.

$$MnO_4^- + 8H^+ + 5e^- \longrightarrow Mn^{2+} + 4H_2O \qquad (2)$$

The phase transfer catalyst (FTC) serves to extract potassium permanganate from the solid phase to an organic solvent. This method has been reported to realize 65-70 % functionalized carbon nanotubes, percentages believed to be a lot higher compared to any other method used to date [63].

In this work we report the preparation of polypyrrole/multiwalled carbon nanotube (PPy/MWCNT) nanocomposites as well polypyrrole/boron nitride (PPy/BN) nanocomposites in the presence of two surfactants, namely cetyltrimethylammonium bromide (CTAB) which is cationic and polyoxyethylene(9)nonylphenyl ether, branched (IGEPAL-CO-630) which is nonionic. The structures of these two surfactants are shown in Figure 45.

Figure 44. Schematic of conduction pathway in a conducting polymer (A) intrachain; (B) interchain; (C) interparticle.

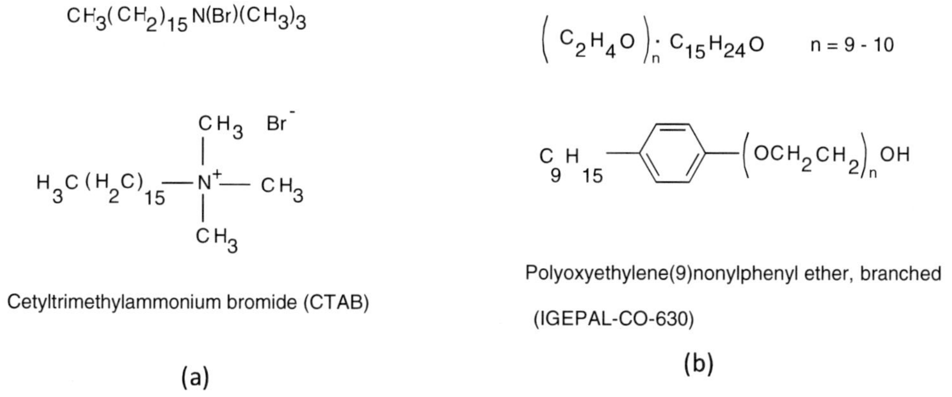

$CH_3(CH_2)_{15}N(Br)(CH_3)_3$

$H_3C(H_2C)_{15}-N^+(CH_3)_3 \; Br^-$

Cetyltrimethylammonium bromide (CTAB)

(a)

$(C_2H_4O)_n \cdot C_{15}H_{24}O$ n = 9 - 10

$C_9H_{15}-C_6H_4-(OCH_2CH_2)_nOH$

Polyoxyethylene(9)nonylphenyl ether, branched

(IGEPAL-CO-630)

(b)

Figure 45. Structure of surfactants (a) CTAB (b) IGEPAL-CO-630.

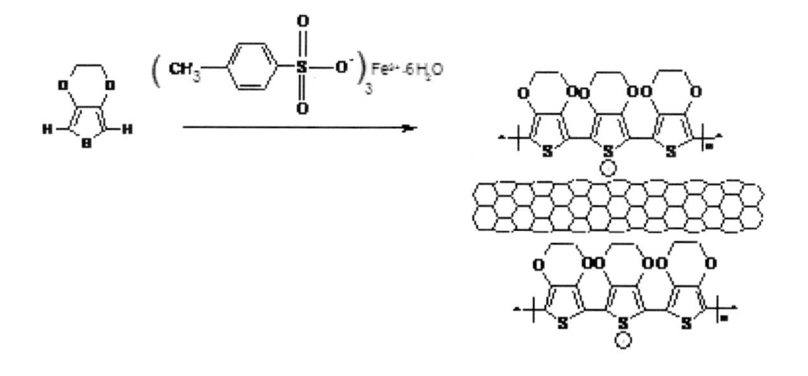

Figure 46. Interaction of PEDOT with carbon nanotubes.

Poly (3, 4-ethylenedioxythiophene), commonly known by the name PEDOT, is a conducting organic polymer (COP). It is synthesized by in situ chemical oxidative polymerization of the monomer 3, 4-ethylenedioxythiophene (EDOT). Much of the research done on PEDOT has been stimulated by its special characteristics that are not common to other COPs such as polypyrrole, polyaniline and polythiophene. A small band gap and the characteristic transparent light sky-blue film are the most important and distinct properties of PEDOT. [64] [65].

The main broad area where PEDOT has found potential applications is in electronics. Applications include electrochromics, supercapacitors, antistatic and electrostatic coatings, light-emitting diodes, photovoltaic cells and sensors among others. [66] In some of these applications the tuning of PEDOT properties is essential and this has resulted in researchers introducing different kinds of dopants in the backbone in order to aid in charge conduction.

Poly (3, 4-ethylenedioxythiophene) that was synthesized in the presence of multiwalled carbon nanotubes (PEDOT/MWCNTs) for use as thermal interface materials (TIMs) in electronics packaging. Figure 46 shows the proposed mechanism by which EDOT polymerizes to form PEDOT. In the presence of carbon nanotubes or other fillers, a PEDOT film is expected to be deposited on the surface. In the case of carbon nanotubes charge transfer is expected to take place between the PEDOT and the carbon nanotubes.

Experimental

Synthesis and Characterization of Polyaniline Nanocomposites with Para-Phenylenediamine Functionalized Multiwalled Carbon Nanotubes

The functionalization of the MWCNTs was started by oxidizing them using potassium permanganate ($KMnO_4$) and a phase transfer catalyst (PTC). Pristine MWCNTs (0.12 g) and 25 ml of dichloromethane (CH_2Cl_2) were mixed in a 100 ml flask and the suspension was sonicated for 30 minutes. A phase transfer catalyst, Aliquat 336 (1.0 g) was added. Over a period of 2 h, 5 g of powdered $KMnO_4$, were added. To the mixture was added 5 ml of acetic acid (CH_3COOH) were added followed by stirring the mixture overnight at room temperature. To obtain the oxidized MWCNTs (o-MWCNTs) the mixture was filtered, washed with concentrated hydrochloric acid (HCl) and deionized water and then dried.

For comparison, polyaniline/MWCNT nanocomposites were prepared using pristine, oxidized and functionalized MWCNTs. In a typical synthesis, 100 mg of MWCNT were dispersed in 10 ml HCl and sonicated for 30 minutes to disperse the carbon nanotubes, after which 0.5 ml of aniline was added and sonication continued. Oxidative chemical polymerization of aniline was done by slowly adding while stirring, a solution of 1.14 g of ammonium persulfate(($(NH_4)_2S_2O_8$)) dissolved in 10 ml of water. After 5 h of reaction at room temperature, the product was filtered and washed with HCl, followed by washing with deionized water and acetone. The nanocomposites were then left to dry.

Surface morphology was analyzed using Field Emission Scanning Electron Microscopy (FESEM) type SUPRA-55 SEM. FTIR was recorded with a FTS 40 Pro system on KBr pellets. TEM was recorded using a Jeol TEM 2010. Conductivities measurements on compressed pellets were taken using a Four-point probe equipped with a Keithley 220 Programmable current source and a Keithley 182 sensitive voltmeter.

Synthesis of Surfactant Assisted Polypyrrole Nanocomposites

Nanocyl 7000 MWCNTs oxidation using potassium permanganate MWCNTs was carried out as follows: 0.12 g MWCNTs and 25 mL dichloromethane (CH_2Cl_2) were added to a 100 mL flask and the mixture was sonicated in a in water bath for 30 minutes. 1.0 g

Aliquat 336 (a phase transfer agent) from Aldrich was added. 5 g powdered KMn04 were added in small portions over 2h. 5 mL was also added. The mixture was then stirred vigorously overnight at room temperature followed by centrifugation, washing with concentrated hydrochloric acid (HCl) and water and dried to get oxidized MWCNT (o-MWCNTs). Before use the monomer, pyrrole from Aldrich was distilled to get a clear solution. 0.124 g of CTAB or 0.210 g of IGEPAL-CO-630 and 6 mg of p-MWCNTs/o-MWCNTs were added to 31 mL of DI water. The mixture was sonicated in water bath for at least 2h. Well-dispersed solutions were obtained and then cooled to 0-5 ^{o}C. The monomer, pyrrole and the oxidant ammonium persulfate (NH4)2S2O8 were also pre-cooled at the same temperature. 0.06 mL of pyrrole was then added to the cooled, well dispersed solution followed by 0.204 g of (NH4)2S2O8 in 6.25 mL of de-ionized water. After bath sonication for 2 minutes, the reaction mixture was allowed to stand at 0-5 ^{o}C for 24 hours. The black precipitate was filtered and washed with distilled water and methanol and then dried at room temperature. [67] PPy nanocomposites were synthesized with p-MWCNTs and o-MWCNTs loadings of 0, 5, 10, 15, 20 and 25 wt%. The same procedure was used to prepare the nanocomposites that were loaded with the other fillers. Surface morphology was analyzed using a Field Emission Scanning Electron Microscopy (FESEM) type SUPRA-55. FTIR was recorded with a FTS 40 Pro system on KBr pellets. TEM was recorded using a Jeol TEM 2010. Conductivities were measured on compressed pellets were taken using a Four-point probe equipped with a Keithley 220 Programmable current source and a Keithley 182 sensitive voltmeter.

Synthesis of Poly (3, 4-Ethylenedioxythiophene) Nanocomposites with Multiwalled Carbon Nanotubes Prepared by In-Situ Polymerization

All reagents were obtained from Aldrich. Before use, EDOT from Aldrich was distilled to give a clear liquid. Fe(OTs)3.6H2O, imidazole and 1-pentanol were used as received. Both Fe(OTs)3.6H2O and imidazole were dissolved separately in 1-pentanol at 100 ^{o}C in an oil bath with stirring for 2 minutes to give 1.75 mol/L and 0.5 mol/L respectively. The dissolved imidazole was mixed with 1 mol/L solution of EDOT. A specific amount of MWCNTs was added to the mixture followed by bath sonication for 30 minutes to disperse the MWCNTs. After sonication the Fe(OTs)3.6H2O solution was added to the dispersion. The orange solution was dropped on a glass slide and the slide placed in an oven for 2 minutes after which the films turned sky-blue. The rest of the solution was left to polymerize overnight at room temperature. The product was then vacuum filtered, washed with methanol and left to dry at room temperature. Using this procedure, three set of experiments were carried out. In the first experiment the amounts of Fe(OTs)3.6H2O and EDOT were fixed and the amount of imidazole varied. The second experiment was carried out in the absence of imidazole, and the amounts of Fe(OTs)3.6H2O and EDOT were held constant, while the amounts of MWCNTs were varied. The third experiment was carried out in the presence of constant amounts of Fe(OTs)3.6H2O, imidazole and EDOT while varying the amounts of MWCNTs. Thickness of the film on glass was determined using AFM and resistivity measurements were measured using the Four-point probe. FESEM was used study the surface morphology of the bulk composites. Bulk conductivities were determined from hydraulic press compressed pellets. Images were also taken using TEM. Spectra were obtained using FTIR. For comparison EDOT was polymerized in acetonitrile using iron (III) chloride hexahydrate as oxidant without using any imidazole. PEDOT nanocomposites were also prepared by oxidizing 0.1M

EDOT solution in acetonitrile with 0.2 M FeC13.6H2O. Percentage weights of MWCNTs ranging from 1-5wt% were added to 5 ml of EDOT solution and bath sonicated until dispersion was achieved then 5 ml of FeC13.6H2O were added. The mixture was left to stir overnight.

Results and Discussion

Polyaniline Nanocomposites with Para- Phenylenediamine Functionalized Multiwalled Carbon Nanotubes

Figure 47 shows the dispersion of p-MWCNTs and o-MWCNTs in HCl after sonication and sitting for 30 minutes.

Figure 47. Stability of MWCNTs in HCl after sonication (a) p-MWCNTs (b) o-MWCNTs.

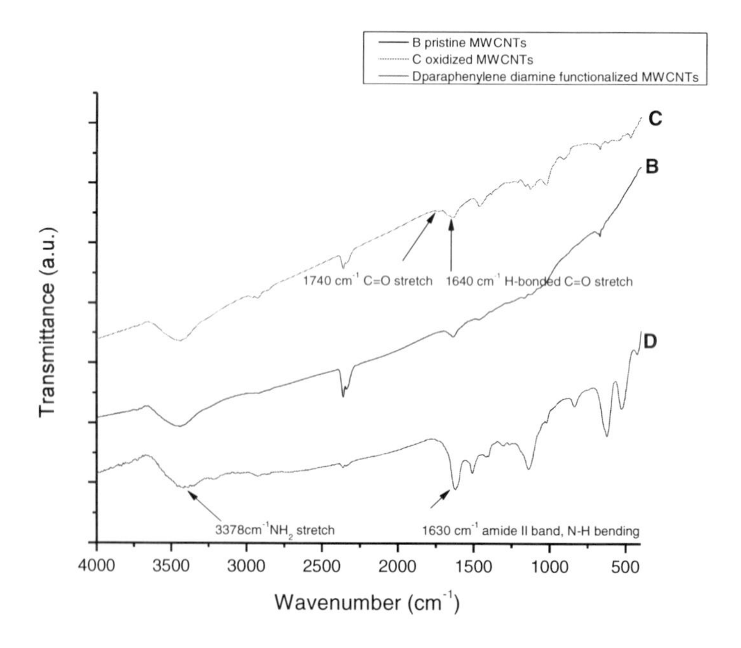

Figure 48. FTIR spectra of (b) p-MWCNTs (b) o-MWCNTs (c) f-MWCNTs.

It can be seen from Figure 47(b) that oxidation of carbon nanotubes facilitates their dispersion in HCl. The MWCNTs that are not oxidized tend to agglomerate immediately after sonication due to their high surface energy. [68] The oxidation of MWCNTs results in the attachment of carboxylic and carbonyl groups on the walls of the nanotubes. These functional groups cause electrostatic repulsion between the carbon nanotubes that results in the lowering of their surface energy and hence dispersion. [69].

Figure 48 shows the FTIR spectra of pristine and oxidized and functionalized multiwalled carbon nanotubes. In Figure 48(c), the oxididized MWCNTs show the presence of bands around 1740 cm-1 and 1640 cm-1 characteristic of C=O stretch and H bonded C=O stretch. [70].

Figure 48(d) shows the spectrum of paraphenylenediamine functionalized MWCNTs. The presence of the band at 1630 cm-1 confirms the presence amide II band, N-H bending. The conversion of the carboxylic groups of the oxidized MWCNTs to the acyl group and subsequently to the amino group.

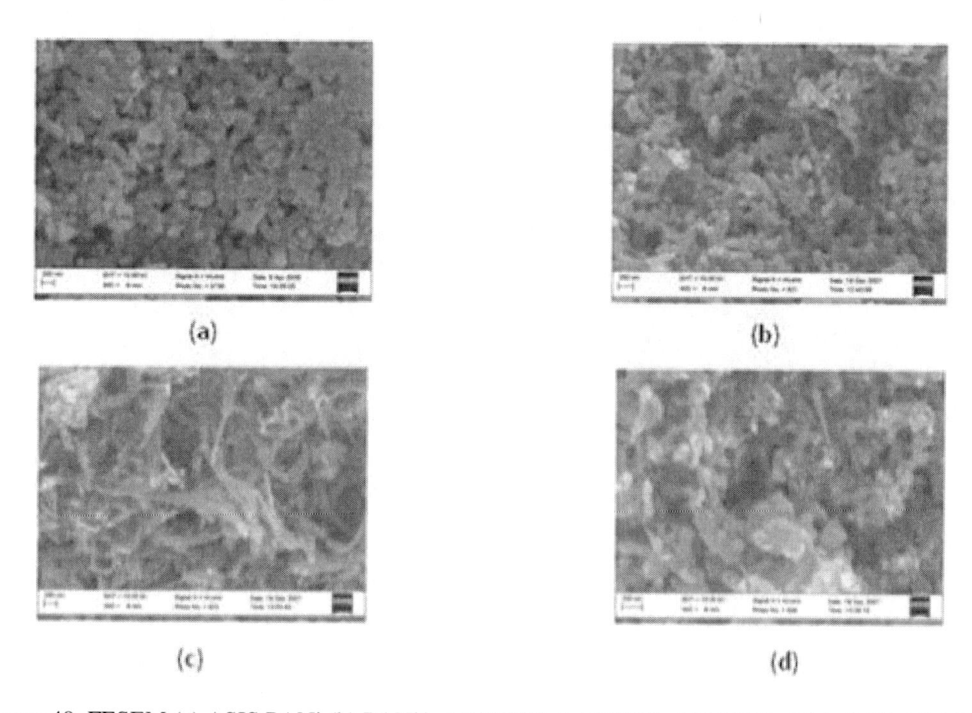

Figure 49. FESEM (a) ASIS PANi (b) PANi/p-MWCNTs (c) PANi/o-MWCNTs (d) PANi/f-MWCNT.

The presence of NH2 groups on the surface of the carbon nanotubes which tend to protonate resulting in the formation of NH3+ ensures electrostatic repulsion between them and the HCl which facilitites dispersion. [71] Figure 49 shows the FESEM of the polyaniline nanocomposites that were prepared in the presence of pristine, oxidized and paraphenylenediamine functionalized MWCNTs. Figure 49(a) shows the presence of nanorods/nanotubes indicating the formation of PANI nanorods/nanotubes which could participate in the conduction process. No MWCNTs were added in the fabrication of this nanocomposite. This confirms the phenomenon that was observed in the preparation of polypyrrole nanocomposites. Nanorods/nanotubes were observed in polypyrrole without any carbon nanotubes. This suggests that the elongated structures that can be seen in Figures

49(b), (c) and (d) are composed of polyaniline nanorods/nanotubes and the MWCNTs that were added as fillers. It is not clear whether the presence of the MWCNTs inhibits the formation of polyaniline nanorods/nanotubes. If that is the case, then the elongated structures in Figures 49(b, c and d) are MWCNTs only.

The nature of the structures that were observed by FESEM was further made clear when the nanocomposites were analyzed using TEM. Indeed nanorods/nanotubes can be seen in Figure 50(a) which is the TEM of polyaniline without any MWCNTs.

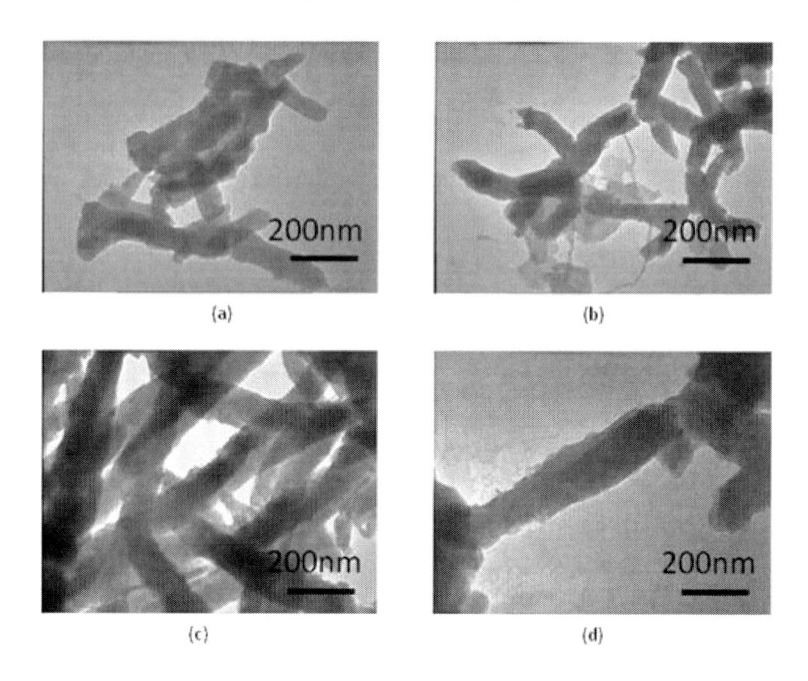

Figure 50. TEM (a) ASIS PANi, (b) PANi/p-MWCNTs, (c) PANi/o-MWCNTs, (d) PANi/f-MWCNT.

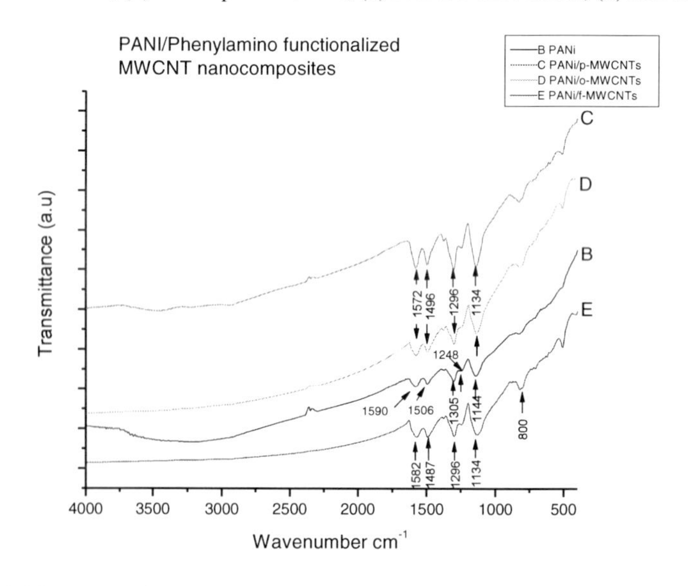

Figure 51. FTIR (B) ASIS PANi, (C) PANi/p-MWCNTs, (D) PANi/o-MWCNTs, (E) PANi/f-MWCNT.

The TEM of the nanocomposites loaded with oxidized MWCNTs look thicker and darker compared to the nanocomposite with pristine MWCNTs. The same is observed for the nanocomposites loaded with paraphenylenediamine functionalized carbon nanotubes. It has been suggested that the increase in thickness is a result of better attachment of polyaniline to the walls of the MWCNTs which have carboxylic and amino groups on the surface.

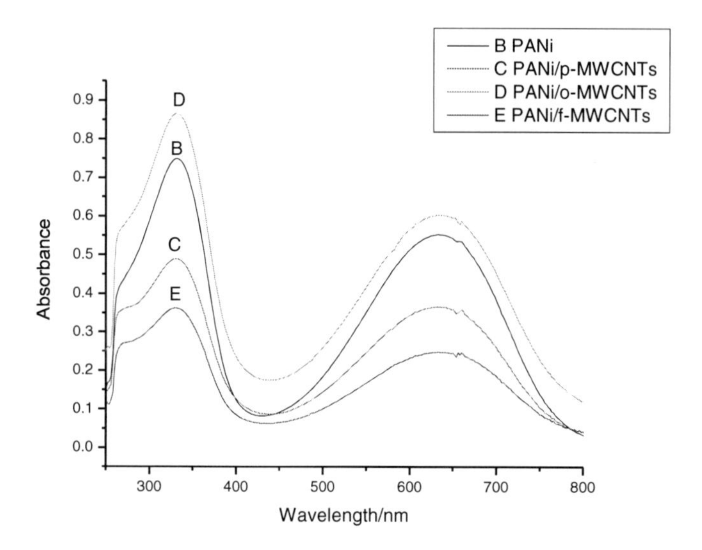

Figure 52. UV-vis (B) ASIS PANi, (C) PANi/p-MWCNTs, (D PANi/o-MWCNTs, (E) PANi/f-MWCNT.

Figure 51 shows the FTIR spectra of PANi as well as the nanocomposites that were prepared in the presence of pristine, oxidized and paraphenylenediamine. The peaks at around 1580 cm-1 shifted from 1590 cm-1 corresponds to the quinoid ring C=N stretching mode and the peak at around 1487 cm-1 shifted from around 1506 cm-1 is associated with C=C of the benzoid ring. The peaks at 1305 cm-1 and 1248 cm-1 are due to the C-N stretching mode of the benzoid ring while the peak at 1144 cm-1 corresponds to the quinoid ring of doped PANi.

The peak at 820 cm-1 corresponds to the out of plane C-H bending of 1, 4 disubstituted benzenoid rings. [72] [73] It can be seen that the IR spectra of MWCNTs loaded polyaniline has all the characteristics of polyaniline without MWCNTs. It appears that the positions of the IR bands are independent of the presence of carbon nanotubes. All the IR spectra are the same which seems to imply that the presence of MWCNT does not affect the backbone structure.

Figure 52 shows the UV-vis spectra of PANi before and after being loaded with MWCNTs. The spectra exhibit peaks at 340 and 638 nm. These two bands represent the transition and charge-transfer excitation-like transition bands of the emeraldine base form of PANI respectively. After polymerization, both the polyaniline with and without carbon nanotubes were expected to show peaks at 430 and around 830 nm, but these bands were not observed. Generally peaks at 340, 430 and a broader one at 830 nm would correspond to the electronic transitions from the valence band to the polaron band characteristic of the doped emeraldine oxidation state of PANi. [74] [75] These are normally used to confirm the oxidation state PANi. It is not clear what effect the MWCNTs have on the electronic transition of polyaniline. If they were playing the role of a dopant like certain sulfonated surfactants then a broad band at around 800 nm should have appeared.

To further confirm the existence of MWCNTs in the nanocomposites, XRD was used. The XRD diffractograms are shown in Figure 53. PANi with no MWCNTs exhibits a characteristic peak at around 25o, which indicate that the PANi is to some extent crystalline. Although MWCNTs also have a peak at 25o like PANi, they show a distinct peak at 45o that cannot be seen in PANi. [76] [77] [78].

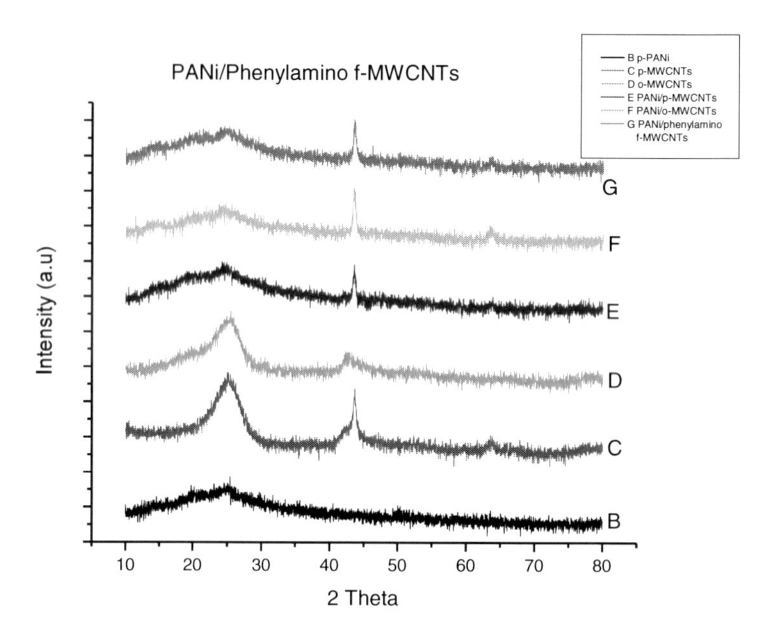

Figure 53. XRD diffractograms of (B) ASIS PANi, (C) pristine MWCNTs (D) oxidized MWCNTs (E) PANi/p-MWCNTs (F) PANi/o-MWCNTs (G) PANi/f-MWCNTs.

Table 1. Electrical Conductivity of PANi nanocomposites

Sample	Current (amps)	Voltage (volts)	Resistivity	Conductivity (Scm^{-1})
PANi EB without MWCNT	1.00×10^{-6}	17.34746	1,572,498.725	6.35×10^{-7}
PANi-*p*-MWCNT	1.00×10^{-6}	0.000167	3.027616575	0.330
PANi-*o*-MWCNT	1.00×10^{-6}	0.000005	0.679854021	1.47
PANi-f-MWCNT	1.00×10^{-6}	0.000001	0.317265209	3.15

The peak at 45° appears in all the nanocomposites in which MWCNTs were used as fillers, which is a clear indication of the presence of MWCNTs. In the nanocomposites it was expected that the peak at 25° would increase in intensity due to the superimposition of PANi and MWCNTs. Instead we did not see much difference in the intensity of this peak between PANi and MWCNTs loaded PANi. [79] Apart from having a low concentration of MWCNTs in the nanocomposites, one reason why we do not see an increase in the peak at 25°, is that the MWCNTs may act as an impurity which interferes with the growth of PANi crystallites. It is can also be noticed that even after oxidation, the characteristic peaks of MWCNTs were still present indicating that to a greater extend the integrity of the MWCNTs was preserved.

Table 1 shows the electrical conductivity of the samples that were prepared starting with undoped PANi that was prepared by stirring PANi without MWCNTs in 3wt% of NH4OH for 2 h. All the conductivity measurements were done on compressed pellets. The results show a marked increase in electrical conductivity with PANi loaded with functionalized MWCNTs showing the highest electrical conductivity of 3.15 Scm-1.

Electrical Properties of Surfactant Assisted Synthesized Polypyrrole Nanocomposites

Figure 54 shows the effect of the surfactants CTAB and IGEPAL-CO-630 on the dispersion of pristine Nanocyl 7000 carbon nanotubes (p-MWCNTs) in water. It can be seen in Figure 54(e) that the pristine carbon nanotubes settled immediately after sonication. However, when the surfactants were used, the carbon nanotube suspensions that were obtained remained stable for a long time as shown in Figure 54(e). Adsorption at interface and self-accumulation into supramolecular structures are the two important features that characterize surfactants. Once the adsorption of surfactant molecules on particle surfaces occur, self-organization of the surfactant into micelles takes place above a critical micelle concentration [80].

Figure 54. (a) p-MWCNTs/H_2O before sonication (b) p-MWCNTs/H_2O after 2 h sonication (c) p-MWCNTs/H_2O/CTAB and p-MWCNTs/H2O/IGEPAL-CO-630 before sonication (d) p-MWCNTs/H_2O/CTAB and p-MWCNTs/H2O/IGEPAL-CO-630 after sonication (e) Sonicated samples after sitting for 30 minutes.

Flat PPy/MWCNT nanocomposite discs were obtained after filtration, washing with DI water followed by methanol. Figure 55 shows the SEM pictures of PPy/MWCNTs that were obtained in the presence of the surfactant CTAB.

It can be seen from Figure 55(b) that when polypyrrole is prepared in the presence of CTAB, nanorods are formed which look like carbon nanotubes. In the absence of CTAB these nanorods were not observed. These nanostructures can be clearly seen in the TEM images shown in Figure 56(a). It can be seen that the nanostructures are actually nanotubes.

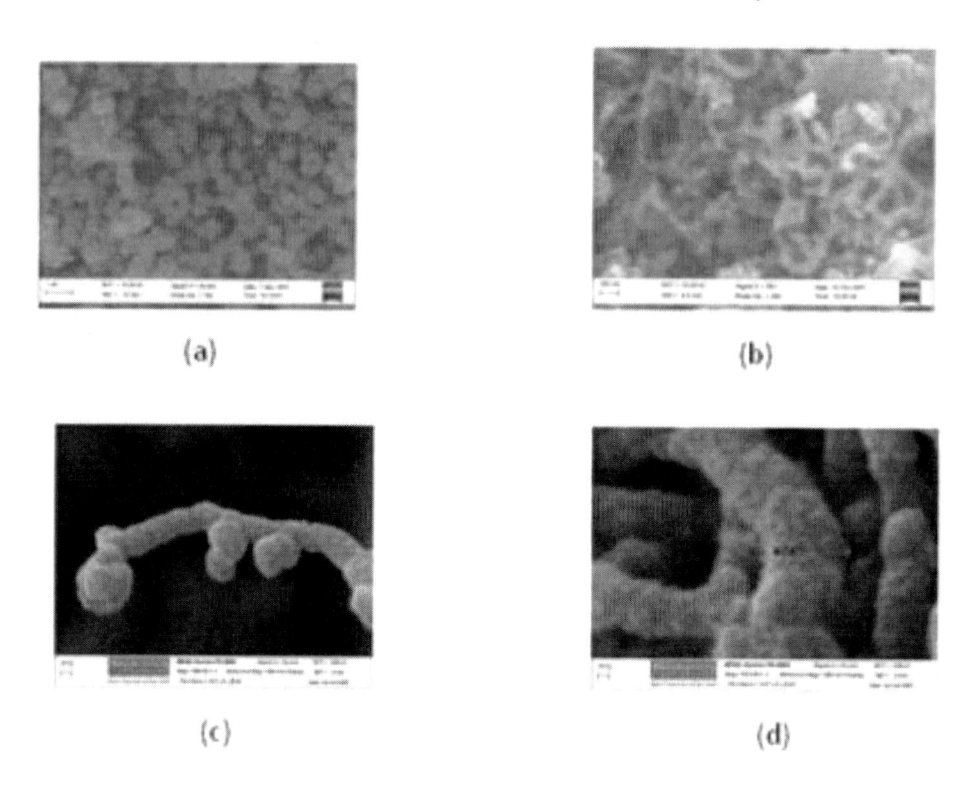

Figure 55. FESEM (a) PPy without CTAB (b) PPy with CTAB (c), (d) PPy with CTAB/20 wt% p-MWCNTs.

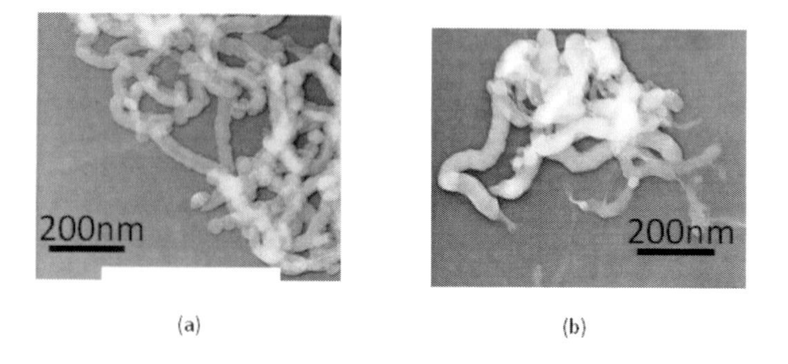

Figure 56. TEM, PPy prepared in the presence of (a) CTAB (b) CTAB/20 wt% p-MWCNT.

Figure 56(b) shows the TEM image that was obtained for the nanocomposite in the presence of carbon nanotubes. The image shows the presence of a thin polypyrrole coating on the surface of the carbon nanotubes.

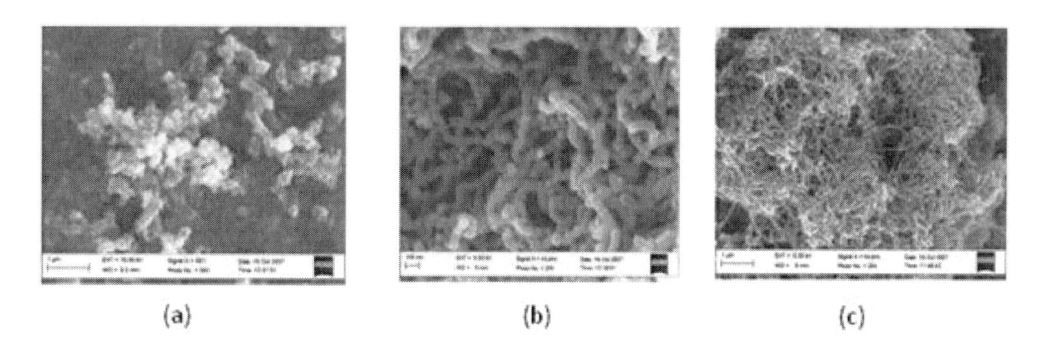

Figure 57. FESEM, PPy prepared in the presence of (a) IGEPAL-CO-630 (b), (c) IGEPAL-CO-630 /20 % p-MWCNTs different magnification.

The polypyrrole nanocomposite therefore consists of polypyrrole nanotubes as well as the filler multiwalled carbon nanotubes which are coated by a thin layer of polypyrrole. The growth of PPy nanotubes in the presence of surfactants is an example of template free synthesis of conducting polymer nanotubes which has been observed before.

When the nonionic surfactant IGEPAL-CO-630 was added during the polymerization of polypyrrole, nanotubes were also observed. However, the nanotubes were branched unlike those that were observed in the presence of CTAB. The FESEM images of PPy that was obtained in the presence of IGEPAL-CO-630 and in the presence of IGEPAL-CO-630 and MWCNTs are shown in Figure 57. The TEM images shown in Figure 58(a) clearly show the branched nature of the PPy that was prepared in the presence of the surfactant IGEPAL-CO-630. It is also interesting to note the presence of knot-like features on the nanotubes. The presence of the knot-like structures may be due to the nonionic surfactant which is branched. It is assumed that when nonionic surfactants are used, amphiphilic macromolecular chains of the surfactants self-assemble into sphere-like micelles without interference from the ionic oxidizing agent. The sphere-like micelles spatially separate and organize hydrophobic pyrrole monomers into the micelles through hydrophobic interactions. [81] In the presence of carbon nanotubes, a thin coating of polypyrrole is deposited on the surface [82] and wraps them uniformly as shown in Figure 58(b).

Figure 58. TEM, PPy prepared in the presence of (a) IGEPAL-CO-630 (b) IGEPAL-CO-630/20 wt% p-MWCNT.

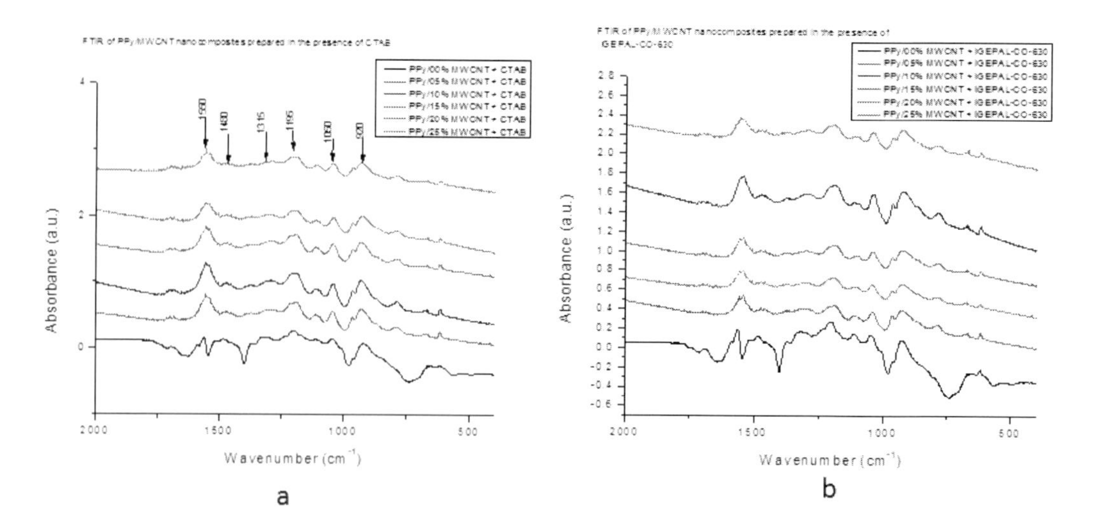

Figure 59. FTIR of (a) PPy/CTAB/p-MWCNTs (b) PPy/IGEPAL-CO-630/p-MWCNTs.

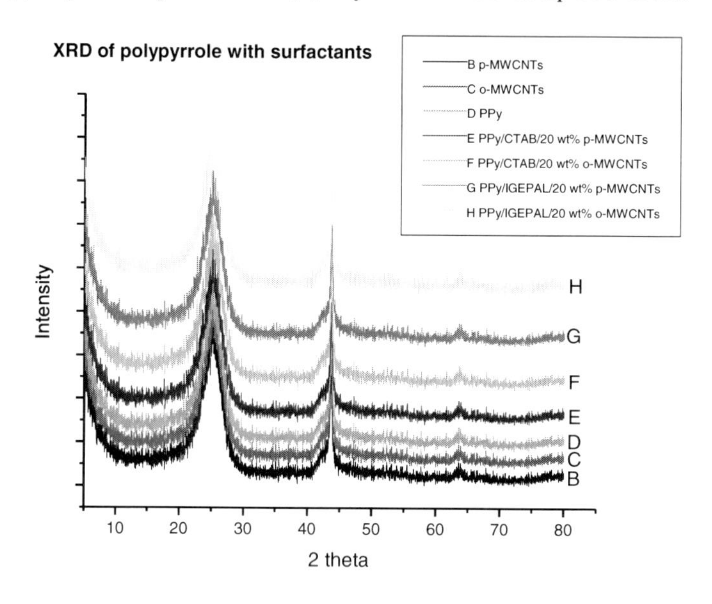

Figure 60. XRD diffractograms of MWCNTs, PPy and nanocomposites.

Figure 59(a) shows the FITR spectrum of the PPy/p-MWCNT composites prepared in the presence of CTAB with different percent loadings of carbon nanotubes. The bands indicated in the spectrum can be assigned as follows: 1195 – 920 cm-1(doping state of polypyrrole), 1050 cm-1(C-H deformation), 1315 cm-1(C-N stretching vibration), 1560 cm-1(Asymmetric ring stretching mode), 1480 cm-1 (symmetric ring stretching mode) . [83] The same bands are observed when PPy is polymerized in the presence of IGEPAL-CO-630 as shown in Figure 59(b).

Figure 60 shows the XRD diffractograms of MWCNTs, PPy and PPy/MWCNTs. Characteristic peaks of MWCNTs were observed at around 25o and 45o corresponding to the graphite-like structure. [84] These peaks were also seen in oxidized MWCNTs indicating that their integrity was not compromised by the oxidation process. It was expected that the diffractogram of PPy would not show the peaks that were seen in the MWCNTs due to its possible amorphous structure. Instead the peaks around 25o and 45o were seen in PPy without MWCNTs. It is possible that this may be to the presence of the nanorods/nanotubes that were formed during polymerization which could be having a structure similar to that of MWCNTs. Since the PPy show crystallinity, the coating of the MWCNTs with a thin layer of PPy does not seem to induce any significant additional crystallinity. Figure 61 shows a plot of the electrical conductivity of PPy nanocomposites in the presence of surfactants and p-MWCNTs.

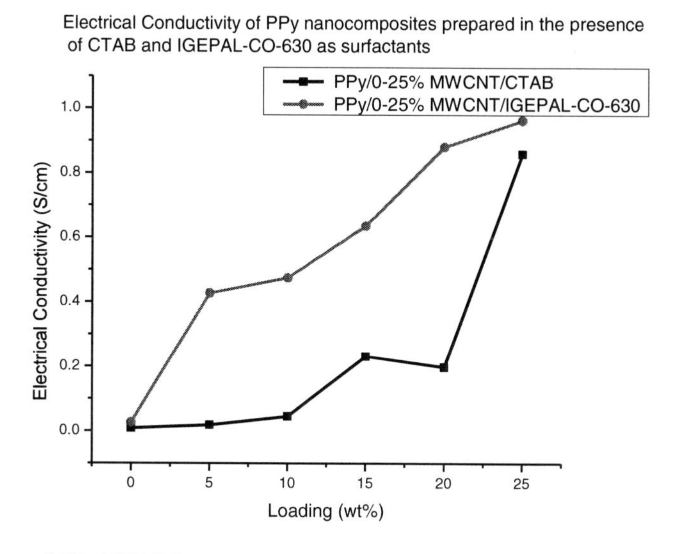

Figure 61. Conductivty of PPy/CTAB/p-MWCNTs and PPy/IGEPAL-CO-630/p-MWCNTs.

It can be seen that in both cases the percolation threshold is below 10 wt% and way below 5 wt% for polypyrrole nanocomposites in the presence of IGEPAL-CO-630. Probably, IGEPAL-CO-630 is better at directing the growth of nanorods/tubes compared to CTAB. [85] Electrical conductivity measurements were also taken for nanocomposites that were prepared in the presence KMnO4 oxidized MWCNTs (o-MWCNTs). Figure 62 shows the FTIR spectra of p-MWCNTs and the o-MWCNTs.

The presence of IR peaks at 1730 cm-1 and 1630 cm-1 confirms the presence of C=O groups which indicate that the carbon nanotubes were indeed oxidized. The other bands that are important are 3440 cm-1, 1053 cm-1 and 670 that correspond to OH-stretch, CH2-O-H and –C=C-stretch respectively [86].

When the electrical conductivities of PPy nanocomposites loaded with KMnO4 oxidized MWCNTs that were prepared in the presence of CTAB and IGEPAL-CO-630 were plotted together as shown in Figure 63, it was seen that the electrical conductivity increased for the nanocomposites in which CTAB was used. Probably CTAB is better at dispersing the carbon nanotubes. It can also be seen that after oxidizing the carbon nanotubes the electrical conductivities increased significantly before decreasing after reaching a maximum at around 15 wt% MWCNT loading. Improved dispersion of the oxidized MWCNTs and better MWCNT/polymer interfacial interaction may be responsible for the significant improvement in electrical conductivity. [87]

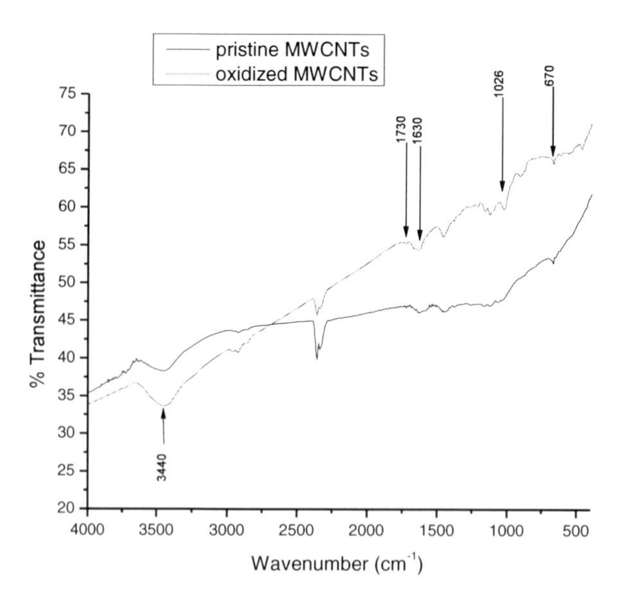

Figure 62. FTIR (a) p-MWCNT (b) o-MWCNTs.

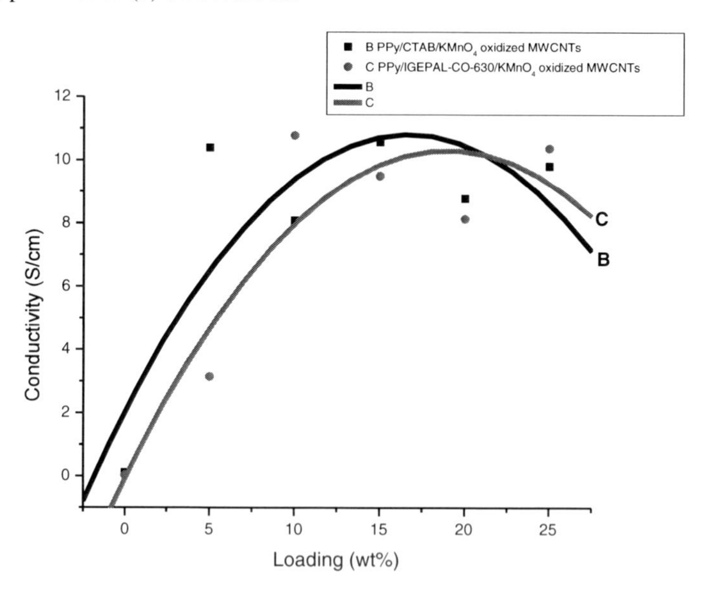

Figure 63. Electrical conductivity of PPy with KMnO4 oxidized MWCNTs prepared in the presence of (B) CTAB (C) IGEPAL-CO-630.

It can also be seen from Figures 64(a) and 64(b) that the nanocomposite that was prepared in the presence of KMnO4 oxidized MWCNTs had higher electrical conductivity than the one that was prepared with pristine MWCNTs. The improvement in electrical conductivity of the nanocomposite can be seen as a result of the improved wetting of the MWCNTs which was carried out under conditions that caused no damage to the MWCNTs as well as generating no impurities. [88]

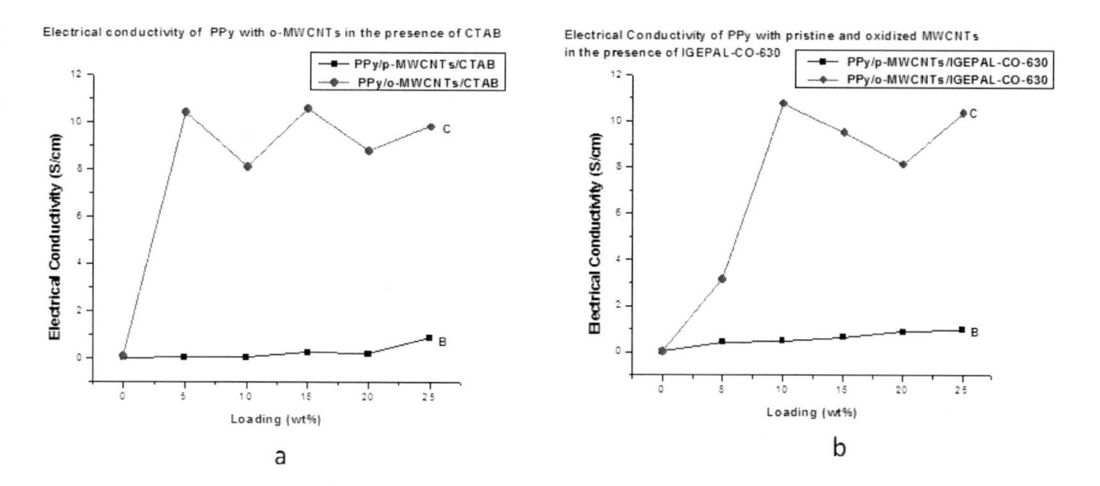

Figure 64. Electrical conductivity of PPy nanocomposites prepared with (a) CTAB (b) IGEPAL-CO-630.

Since CTAB has been known to disperse carbon nanotubes and get washed out we chose to use it for the dispersion of nanomaterials in the rest of our experiments. [89] After dispersing the carbon nanotubes and washing out the CTAB, it means the surfactant was not a contaminant in the nanocomposite. Pyrrole monomers then insert themselves between the surfactant and carbon nanotubes due to the π-π interaction between them. Upon the oxidation of pyrrole by S2O82- the CTAB is regenerated. [90] Figure 65 shows the electrical conductivity plots of PPy nanocomposites loaded with different filler materials. It can be seen in Figure 65 that the PPy nanocomposite that is loaded with Nanocyl 7000 MWCNTs produced the highest conductivity at 5 wt% loading. This may be attributed to the better properties of this type of MWCNTs that is produced by the CVD method. Carbon nanotubes produced by this method are known to have better graphitization. [91] The Nanodynamics carbon nanotubes are produced by the combustion method which results in less graphitization and are likely to have more defects [92].

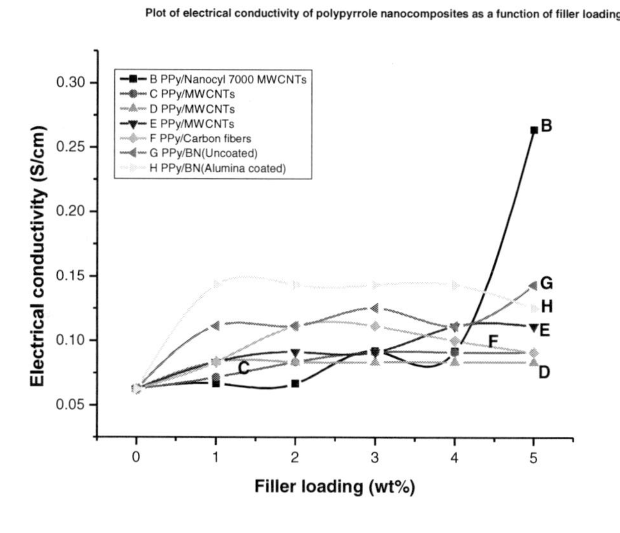

Figure 65. Electrical conductivity of PPy different fillers prepared in the presence of CTAB.

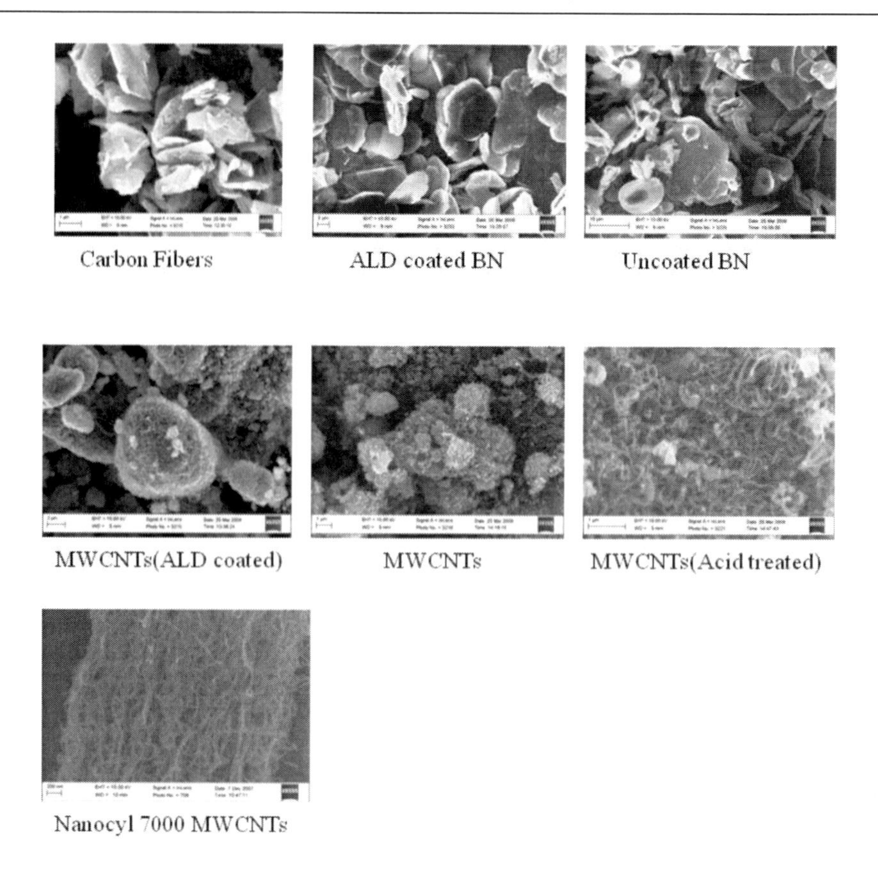

Figure 66. FESEM images that were used as fillers in the preparation of PPy nanocomposites.

Figure 65 shows that PPy loaded with boron nitride and alumina coated boron nitride achieved lower electrical percolation threshold. The lowest percolation threshold was achieved by the nanocomposite in which alumina coated boron nitride was used as filler. No significant improvement in the electrical conductivity of PPy was observed when carbon fibers, and the other Nanodynamics MWCNTs were used. The highest electrical conductivity was achieved by the Nanocyl 7000 MWCNTs filled PPy at between 4 and 5 wt% loading. Figure 66 shows the FESEM images of the materials that were used as fillers in the preparation PPy nanocomposites.

Poly (3, 4-Ethylenedioxythiophene) Nanocomposites with Multiwalled Carbon Nanotubes Prepared by *In-Situ* Polymerization

It was found that beyond a certain concentration of imidazole, formation of PEDOT could not take place. After using a 2 mol/L solution of imidazole, this concentration was doubled, trebled and quadrupled and used during the polymerization reaction. After heating for 2 minutes at 110 oC to initiate the polymerization, only the sample in which 2 mol/L solution of imidazole had been used produced a sky-blue film. Figure 67(a) (left to right) shows the three samples in which the concentration of imidazole was doubled, trebled and quadrupled. The corresponding films after heating are shown in Figure 67(b) (second, third and fourth respectively). It can be seen that these films remain pretty much orange after heating in sharp contrast with the first film which turned blue for which the imidazole concentration was 2 mol/L.

Figure 67(d) shows a series in which the amount of Fe(OTs)3.6H2O, and EDOT were kept constant and the amounts of imidazole were 0, 0.5, 1.0, 1.5 and 2.0 mol. It was found that the sample with no imidazole (first vial in Figure 67(d)), turned sky blue immediately, yet the rest of them only became sky blue after heating.

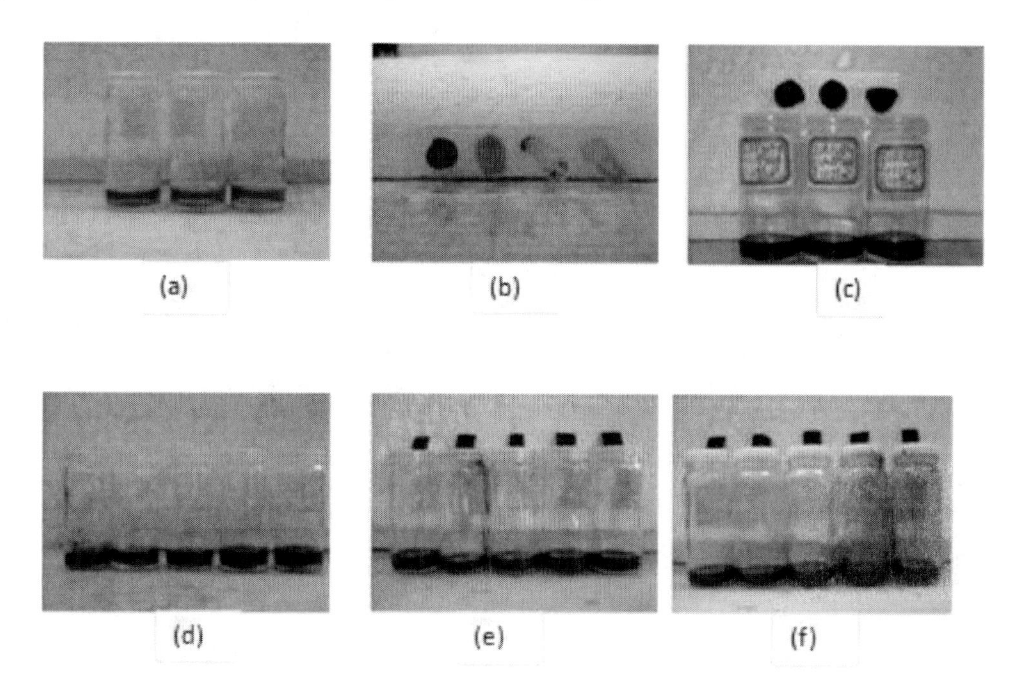

Figure 67. Imidazole amounts in samples composites (a) 4.0, 6.0, 8.0 mol (b) 2.0, 4.0, 6.0, 8.0 mol(c) 0.5, 1.0, 1.5 mol (d) 0, 0.5, 1.0, 1.5, 2.0 mol (e) and (f) all 0.5 mol.

Figure 67(e) shows the results of samples that were prepared with a constant amount (0.5 mol) of imidazole and constant amount (1.75 mol) of Fe(OTs)3.6H2O but the amounts of MWCNTs were 0, 0.001g (0.7 wt%), 0.002g (1.4 wt%), 0.0003g (2.1 wt%) and 0.004g (2.7 wt%). Figure 67(f) is a series that was prepared under the same conditions as series (e) except that no imidazole was used. Shown on top of the vials in Figure 67(c, e and f) are the glass slides with PEDOT and PEDOT/MWCNTs nanocomposite films. The bulk samples in the vials that did not polymerize immediately only polymerized after heating on glass slides. The bulk PEDOT went on to polymerize, forming blue solutions after being left still overnight.

It has been reported that increasing the amount of imidazole increases the conductivity and transparencry of PEDOT films. Imidazole is expected to coordinate with the Fe(OTs), thereby quenching the monomeric EDOT radicals. The overall effect of the imidazole is to push the polymerization kinetics towards the formation of longer PEDOT chains which in turn allow for the better orbital delocalization and consequently better electrical conductivity. [24] Figure 68 shows the FESEM of the bulk nanocomposites that were prepared.

Figure 69(a) and (b) shows the TEM images of PEDOT and PEDOT/MWCNTs nanocomposite respectively. Figure 69(a) shows that nanorods were formed during the formation of PEDOT.

(a)

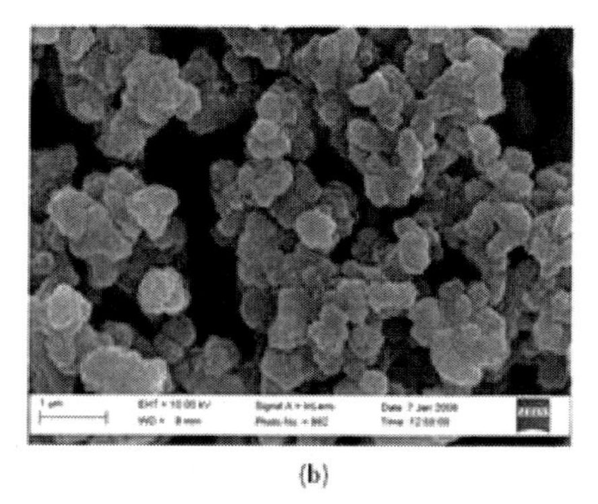

(b)

Figure 68. FESEM images of PEDOT (a) with imidazole and no MWNTs (b) with imidazole and 0.7 wt% MWCNTs.

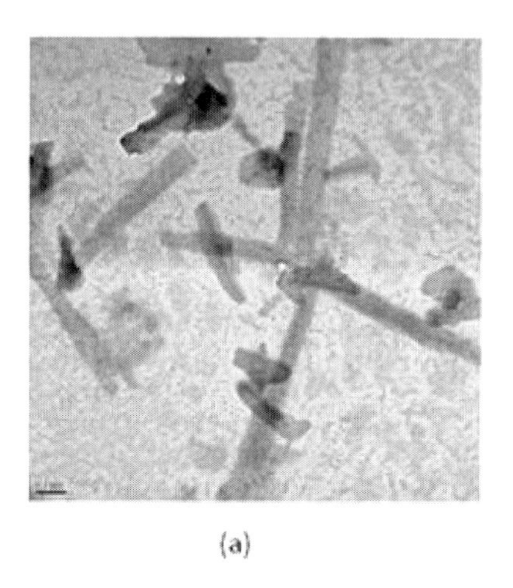

(a)

(b)

Figure 69. TEM images of PEDOT (a) with imidazole and no MWNTs (b) with imidazole and 0.7 wt% MWCNTs.

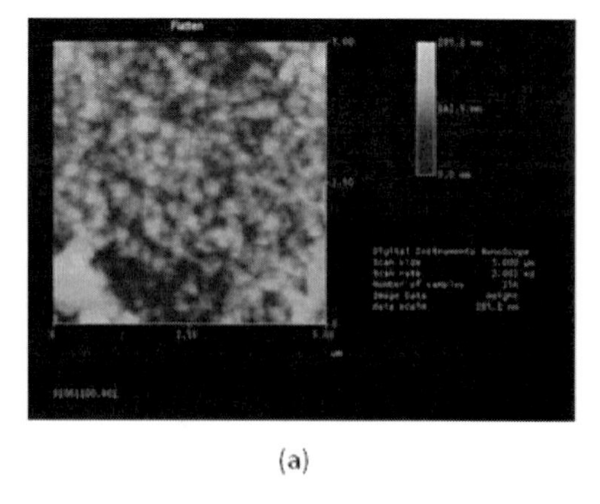

(a)

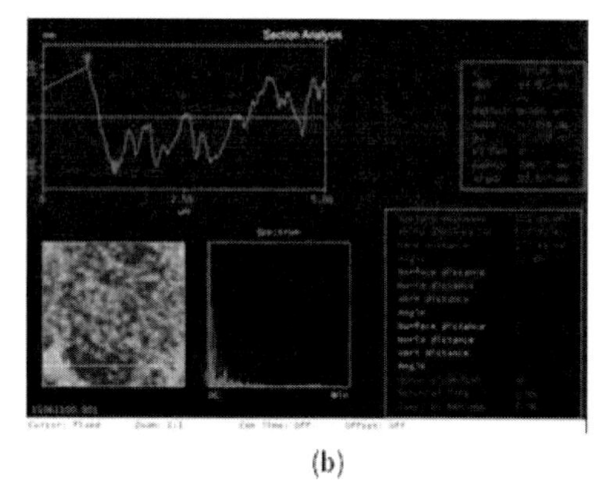

(b)

Figure 70. (a) TM AFM images of PEDOT/MWCNT composite thin films on glass (b) Section analysis of the nanocomposite film.

Figure 71. FESEM images PEDOT/MWCNTs prepared in the absence of imidazole (a) 0, (b) 0.001g (0.70 wt%).

The formation of these structures has been observed before in PEDOT, PANi and PPy. [93] Figure 69(b) shows the presence of many elongated structures. Since the PEDOT without MWCNTs showed the presence of nanorods, the elongated structures in the PEDOT nanocomposite should be a mixture of PEDOT nanorods and MWCNTs.

The thickness of the PEDOT/MWCNTs nanocomposite that was deposited on glass was measured using AFM. Figure 70(a) shows the AFM picture of the nanocomposite film on glass and Figure 70(b) shows the section analysis from which the thickness of the film was read and found to be 213.48 nm. While the film with imidazole reaches a high electrical conductivity of 2.67 x 10-6 Scm-1 at 1.4 wt%, the one without imidazole shows electrical conductivity of 1.28 x 10-6 Scm-1 at the same loading. Figure 71 shows FESEM of PEDOT/MWCNTs nanocomposites prepared in the absence of imidazole.

In the absence of imidazole, the bulk PEDOT nanocomposite did not show significantly high electrical conductivity. At the highest percent loading of 2.7 wt% the electrical conductivity of 1.26 x 10-9 Scm-1 was achieved which is still very low.

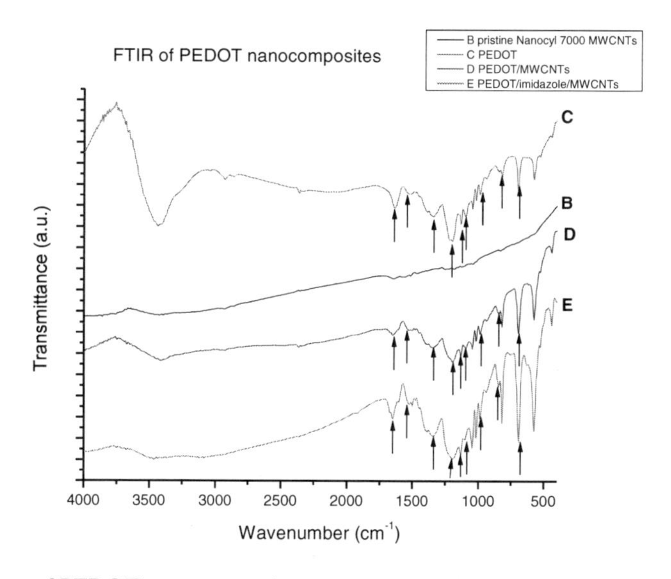

Figure 72. FTIR spectra of PEDOT nanocomposites.

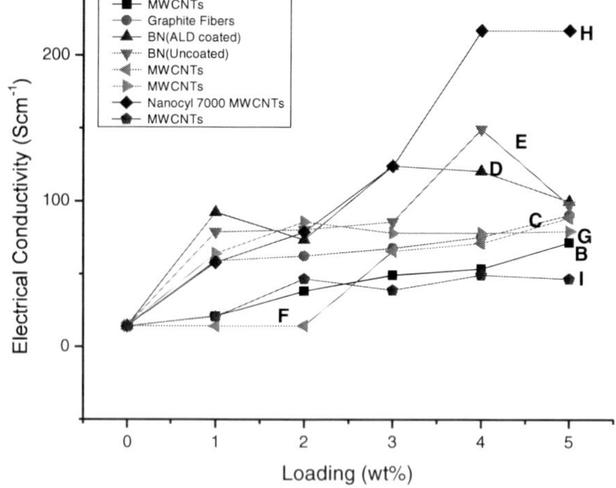

Figure 73. Electrical conductivity of PEDOT nanocomposites prepared from 0.1 M EDOT and 0.2 M FeCl3.6H2O.

Figure 72 shows the FTIR spectra of pristine MWCNTs (B), PEDOT (C), PEDOT with MWCNTs (D) and PEDOT with MWCNTs in the presence of imidazole (E). The spectrum of pristine MWCNTs shows no significant peaks. The peak at 1640 cm-1 is associated with the doping level of PEDOT. Some researchers have assigned it to a C=C bond whose position depend on the doping level of the polymer. The peaks at 1338 and 1540 cm-1 are due to C-C or C=C stretching of quinoidal structure of the thiophene ring and the ring stretching of thiophene ring respectively. The peaks at around 1208-1089 cm-1 are due to stretching in the alkylenedioxy group. The peaks around 987, 815 and 682 cm-1 originate from the C-S bond stretching in the thiophene ring. The peak round 1043 cm-1 can be assigned to the –SO3H group which proves that the resulting PEDOT is in the doped state [94] [95] [96].

Figure 73 shows the electrical conductivity results of the PEDOT nanocomposites that were prepared from 0.1 M PEDOT solution and 0.2 M FeCl3.6H2O solution in acetonitrile. The reaction was left to run for overnight before vacuum filtration and drying. Electrical conductivity measurements were obtained from compressed pellets. It can be seen that the PEDOT nanocomposite loaded with Nanocyl 7000 MWCNTs achieved the highest electrical conductivity of 217 Scm-1. In general, this method of preparing PEDOT nanocomposites yield high conductivity values with all the fillers that were used although PEDOT with Nanocyl 7000 MWCNTs stood out.

References

[1] M. Na, S.W. Rhee, *Organic Electronics*, 7, 205 (2006).

[2] L. Zhu, J. Xu, Y. Xiu, Y. Sun, D. Hess, and C. P. Wong, *Carbon,*44, 253 (2006).

[3] H. M. Cheng, Q. H. Yang, and C. Liu, "Hydrogen storage in carbon nanotubes,", 39, 1447 (2001).

[4] W. Bong et al.,*Applied Physics Letters*, 79, 3696(2001).

[5] M. Pumera, *Nanoscale Research Letters*, 2,87(2007).
[6] F. Picaud, C. Girardet, *Surface Science*, 602, 235 (2008).
[7] H. Catalysis, "OMb," Society, no. 1, pp. 7935-7936, 1994.
[8] I. Eswaramoorthi, V. Sundaramurthy, N. Das, A. K. Dalai, J. Adjaye, *Applied Catalysis A: General*, 339, 187 (2008).
[9] Y. X. Zhou, P. X. Wu, Z. -Y. Cheng, J. Ingram, S. Jeelani, *eXPRESS Polymer Letters*,2, 40 (2008).
[10] N. Mingo, D. A. Broido, *Physical Review Letters*, 93, 1 (2004).
[11] M. Endo, T. Hayashi, Y.-A. Kim, *Pure App. Chem*, 78, 1703 (2006).
[12] M. Urbán, Z. Kónya, D. Méhn, J. Zhu, I. Kiricsi, *PhysChemComm*,5, 138 (2002).
[13] A. A. Koval'chuk, V. G. Shevchenko, A. N. Shchegolikhin, P. M. Nedorezova, A. N. Klyamkina, A. M. Aladyshev, Macromolecules, 41, 7536 (2008).
[14] C. V. Santos, A. L. M. Hernández, R. Ruoff, and V. M. Castaño, *Chem. Mater.*, 15, 4470 (2003).
[15] Y. Li et al., *Nanotechnology*, 15, 1645 (2004).
[16] Y. Long, Z. Chen, X. Zhang, J. Zhang, and Z. Liu, *Journal of Physics D: Applied Physics*, 37, 1965 (2004).
[17] J. Kathi, K. Y. Rhee, *Journal of Materials Science*,43, 33 (2007).
[18] T. Ramanathan, F. T. Fisher, R. S. Ruoff, L. C. Brinson, *Chem. Mater.*,17, 1290 (2005).
[19] S. Ahmad, S. Ahmad, S. A. Agnihotry,*Materials Science,* 30, 31 (2007).
[20] J. Liu, T. Liu, and S. Kumar, *Polymer,* 46, 3419 (2005).
[21] J.-H. Du, J. Bai, H.-M. Cheng, *Polymer*, 1, no. 5, pp. 253-273, 2007.
[22] Z. Yang et al., *Journal of Materials Science*, 42, 9447 (2007).
[23] D. K. Pradhan, *eXPRESS Polymer Letters*, 2, 630 (2008).
[24] J. -P Salvetat et al. *Advanced Materials*, 11, 161 (1999).
[25] J. Kim, D. Seong, T. Kang, J. Youn, *Carbon*, 44, 1898 (2006).
[26] M. Ohring, *Materials Science of Thin Films. Deposition and Structure*, San Diego, (2002).
[27] T.-H. Fang and W.-J. Chang, *Polymer*, 35, 595 (2004).
[28] Y. S. Song, J. R. Youn, *Carbon*, 43, 1378 (2005).
[29] S.-M. Yuen, C.-C. M. Ma, C.-L. Chiang, Y.-Y. Lin, C.-C. Teng, *Polymer*,45, 3349 (2007).
[30] T. J. Shin, B. Lee, H. S. Youn, K.-bong Lee, and M. Ree, *Langmuir*, 17, 7842 (2001).
[31] H. H. So, J. W. Cho, N. G. Sahoo, *European Polymer Journal*, 43, 3750 (2007).
[32] C. Blanco, S. P. Appleyard, B. Rand, 205, 21 (2002).
[33] J. Wu, C. Xiao, A. F. Yee, C. A. Klug, J. Schaefer, *Polymer*, 39, 1730 (2001).
[34] J. P. F. Inberg, R. J. Gaymans, *Polymer*,43, 3767(2002).
[35] T. Nishikawa, K. Ogi, T. Tanaka, Y. Okano, and I. Taketa, *Advanced Composite Materials*, 16, 1 (2007).
[36] A. Marcilla, S. García, J. C. G. -Quesada, 71, 457 (2004).
[37] G. Yamamoto, M. Omori, T. Hashida, H. Kimura,*Nanotechnology*, 19, 457 (2008).
[38] A. Fakhru, M. A. Atieh, N. Girun, T. G. Chuah, *Young*, 75, 496 (2006).
[39] T. Watanabe, H. Wang, Y. Yamakawa, M. Yoshimura, *Carbon*, 44, 799 (2006).
[40] R. Balart, D. Garcı, M. D. Salvador, J. Lo, *European Polymer Journal*, 41, 2150 (2005).
[41] G. Wildes, H. Keskkula, D. R. Paul, *Polymer*,40, 7089 (1999).

[42] A. Gawade, A. V. Lodha, P. S. Joshi, *Journal of Macromolecular Science, Part B,*47, 201 (2008).
[43] P. P. Lizymol, S. Thomas,*PolymerDegradation and Stability*, 57, 187 (1997).
[44] L. F. Lu, D. Price, G. J. Milnes, P. Carty, S. White, *Polymer*, 64, 601 (1999).
[45] P. Carty, S. White, *Polymer*, 47, 305 (1995).
[46] R. Jurk, M. Saphiannikova, J. Fritzsche, H. Lorenz, *Polymer*, 49, 5276 (2008).
[47] C. Zhou et al., *Journal of Polymer Science Part B: Polymer Physics*, 44, 687 (2006).
[48] G. Editorial, *Carbon*, 44,1621 (2006).
[49] J. N. Coleman, U. Khan, W. J. Blau, Y. K. Gun, *Carbon*, 44, 1624 (2006).
[50] B. Philip, J. Xie, A. Chandrasekhar, J. Abraham,V. K. Varadan, *Smart Materials and Structures*, 13, 295 (2004).
[51] F. Du, J. Fischer, K. Winey, *Physical Review B*, 72, 1 (2005).
[52] H. Chen et al., *Nanotechnology*, 18, 415 (2007).
[53] S. Agrawal, M. S. Raghuveer, H. Li, and G. Ramanath, *Applied Physics Letters*, 90, 193 (2007).
[54] R. H. Telling, C. P. Ewels, A. a El-Barbary, M. I. Heggie, *Nature Materials*,2, 333 (2003).
[55] M. S. Raghuveer et al., *Chem. Mater.*, 18, 1390 (2006).
[56] M. S. Raghuveer, a. Kumar, M. J. Frederick, G. P. Louie, P. G. Ganesan, G. Ramanath, *Advanced Materials*, 18,547 (2006).
[57] W. Feng, X. D. Bai, Y. Q. Lian, J. Liang, X. G. Wang, K. Yoshino, *Carbon*, 41, 1551 (2003).
[58] B. Philip, J. Xie, J. K. Abraham, V. K. Varadan, *Smart Materials and Structures*, 13, N105 (2004).
[59] D. C. Trivedi, Handbook of Organic Conductive Molecules and Polymers: Conductive Polymers: Synthesis and Electrical Properties. Chichester (1997).
[60] D. Walton, P. Lorimer, *Polymers*, New York (2000).
[61] B. C. Satishkumar, A. Govindaraj, J. Mofokeng, G. N. Subbanna,*J. Phys. B: At. Mol. Opt. Phys.*,29, 4925 (1996).
[62] J. Zhang et al., *The Journal of Physical Chemistry B*, 107, 3712 (2003).
[63] N. Zhang, J. Xie, V. K. Varadan, *Smart Mater. Struct.*, 11, 962 (2002).
[64] M. G. Han, S. H. Foulger, *Chem.Commun*, 19, 2154 (2004).
[65] B. W.-Jensen, K. West, *Macromolecules*, 37, 4538 (2004).
[66] T. L. Truong et al., *Thin Solid Films*, 516, 6020 (2008).
[67] R. Sainz et al., *Nanotechnology*, 16, S150 (2005).
[68] D. Hohnholz, A. G. MacDiarmid, D. M. Sarno, W. E. J. Junior, *Chem.Commun.*, 23, 2444 (2001).
[69] K. A. Narh, A.-T. Agwedicham, L. Jallo, *Powder Technology*, 186, 206 (2008).
[70] Y. Si, E. T. Samulski, *Nano Lett.*,8, 1679(2008).
[71] E. Akalin, S. Akyuz, *Vibrational Spectroscopy*,22, 3 (2000).
[72] E. N. Konyushenko, J. Stejskal, M. Trchová, N. V. Blinova, P. Holler, *Synthetic Metals*,158, 927 (2008).
[73] H. Xia, J. Narayanan, D. Cheng, C. Xiao, X. Liu, H. S. O. Chan, *The journal of physical chemistry. B*, 109, 12677 (2005).
[74] B. Su, Y. Tong, J. Bai, Z. Lei, *Indian Journal of Chemistry*, 46, 595 (2007).
[75] F. Yan, G. Xue, *J. Mater. Chem.*,9, 3035 (1999).

[76] L. Tarachiwin, P. Kiattibutr, L. Ruangchuay, A. Sirivat, J. Schwank, *Synthetic Metals*, 129, 303 (2002).
[77] J. P. Pouget, M. E.Jozefowzicz, A. J. Epstein, X. Tang, A. G. MacDiarmid, *Macromolecules*, 24, 779(1991).
[78] X. Sui, Y. Chu, S. Xing, C. Liu, *Materials Letters*, 58, 1255 (2004).
[79] J. Xu, P. Yao, Y. Wang, F. He, Y. Wu, *Journal of Materials Science: Materials in Electronics*, 20, 517 (2008).
[80] H. Guo, H. Zhu, H. Lin, J. Zhang, *Materials Letters*, 62, 3919 (2008).
[81] L. Zhang, M. Wan, *Macromolecules*, 13, 750 (2002).
[82] X. Zhang, J. Zhang, W. Song, Z. Liu, *The journal of physical chemistry. B*, 110, 1158 (2006).
[83] M. Pumera, B. Smíd, X. Peng, D. Golberg, J. Tang, I. Ichinose, *Chemistry - A European Journal*, 13, 7644 (2007).
[84] T. Dai, X. Yang, Y. Lu, *Nanotechnology*, 17, 3028 (2006).
[85] T.-M. Wu, S.-H. Lin, *Polymer*, 6449 (2006).
[86] J. Ouyang, Y. Li, *Polymer*, 38, 3997 (1997).
[87] L. Wang, D. M. Xing, H. M. Zhang, H. M. Yu, Y. H. Liu, *Journal of Power Sources*, 176, 270 (2008).
[88] K. Y. Yan, Q. Z. Xue, Q. B. Zheng, L. Z. Hao, *Nanotechnology*, 18, 255 (2007).
[89] Y. Kim, T. Shin, H. Choi, J. Kwon, Y. Chung, H. Yoon, *Carbon*, 43, 23 (2005).
[90] M. F. Islam, E. Rojas, D. M. Bergey, A. T. Johnson, A. G. Yodh, *Nano Letters*, 3, 269 (2003).
[91] J.-F. Liu, W. A. Ducker, *Langmuir*, 16, 3467 (2000).
[92] V. M. Castano, *Composite Interfaces*, 11, 567 (2005).
[93] J. U. I. H. Chen, C.-An Dai, W.-Yen Chiu, *Polymer*, 1662 (2007).
[94] J. Jang, M. Chang, H. Yoon, *Advanced Materials*, 17, 1616 (2005).
[95] J. W. Choi, M. G. Han, S. Y. Kim, S. G. Oh, S. S. Im, *Synthetic Metals*, 141, 293 (2004).
[96] W. Feng et al., *J. Phy. Condens. Matter.*,19, 1 (2007).

In: Conducting Polymers
Editor: Luiz Carlos Pimentel Almeida
ISBN: 978-1-62618-119-9

Chapter 5

SYNTHESIS, CHARACTERIZATION AND APPLICATIONS OF CONDUCTING POLYMER-CLAY NANOCOMPOSITES

Gustavo Morari do Nascimento*
Universidade Federal do ABC, Centro de Ciências Naturais e Humanas
(CCNH)-São Paulo, Santo André, Brazil

ABSTRACT

In recent years many efforts have been paid in the direction of the synthesis of polymers with improved thermal, mechanical and electrical properties. In order to acquire this purpose, clays have been largely employed as inorganic support to improve the polymeric properties. Mainly smectite clays are used in the polymer intercalation. The structure of clay minerals of the smectite group consists of a central sheet of $MO_4(OH)_2$ octahedral symmetrically bound to two MO_4 tetrahedral sheets producing layers designated T:O:T. The octahedral sites are occupied by ions such as aluminum, magnesium, and iron, while the tetrahedral centers accommodate silicon and aluminum. The negative T:O:T individual layers assume a parallel orientation, and the electric charge is neutralized by the presence of exchangeable hydrated positive ions in the interlayer space. The main purpose for performing the polymer confinement in porous frameworks, such as clays, is to work at the molecular level in order to create specific supramolecular arrangements of its chains with a possible improvement of the polymer bulk properties, making attractive their technological application. Several polymers have been synthesized in the nanospace of the clay interlayer and although significant progress has been made in developing nanocomposites, until now it has been very difficult to predict and control their properties. In this chapter this amazing area of polymer-clay nanocomposites, mainly focused on in the Polyaniline-clay nanocomposites, will be reviewed concerning the state-or-art results of synthesis, spectroscopic characterization and applications. Previous and new results obtained by our group, using spectroscopic techniques will be considered. Special attention will be given in the role of the clay in the control of the electrical, thermal and mechanical properties and also on the morphological

* E-mail: gustavo.morari@ufabc.edu.br

aspects of the polymeric material. The main goal of this work is to contribute in the rationalization of some important results obtained in the open area of polymer-clay nanocomposites.

1. General Aspects

1.1. Conducting Polymers

During the 60s it was synthesized the first materials with values of conductivity above 1 $S.cm^{-1}$, named as Krogman salts. These salts are square planar complexes of PtX_4 or IrX_4 (X = CN^- or $C_2O_4^{-2}$). It can form structures of chains, which conductivity reaches to metallic values when exposed to oxidizing Br_2 vapors.[1] The oxidation of molecular chains causes a reduction of the distances between the atoms of the metals, allowing the filling of the conduction band formed by d_{zz} orbitals of neighboring metal atoms in the chain. As a result, the conductivity reaches values as higher as 10^{-7} to 10^{-2} $S.cm^{-1}$, an increase of 10^5 times. Charge-transfer complex forms the second class of molecular substances with high electrical conductivity. Tetrathiofulvalene (TTF) and Tetracyanoquinodimethane (TCNQ) and its derivatives are the most known charge-transfer system [1] (see Figure 1.1).

The intrinsically conducting polymers (ICPs), more commonly known as "synthetic metals", forms the third class of molecular conductors. Initially, the study of ICPs was very complicated by their insolubility, infusibility and instability in the air. In the early 70s, Shirakawa and Ikeda [2] showed the possibility to prepare more stable films of semiconducting poly(acetylene). This discovery has not generated much interest until 1977, when MacDiarmid et al. [3-5] found that, when the poly(acetylene) is doped with Lewis's acid (or base), it is possible to increase the conductivity by 13 orders of magnitude.

Since this initial discovery in 1977 the development of the conducting polymer field has continued to accelerate at an unexpectedly rapid rate. This development has been stimulated not only by the fundamental synthetic novelty and importance but mainly because this field is a cross-disciplinary section of investigators- chemists, electrochemists, experimental and theoretical physicists and electronic and electrical engineers, due to the higher potential technological applications.

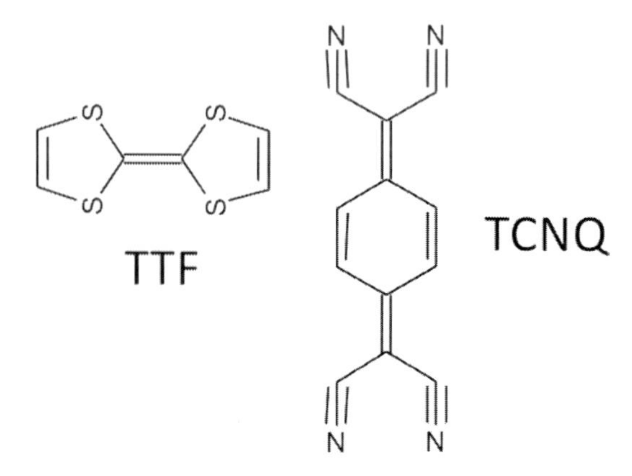

Figure 1.1. Chemical representation of TTF and TCNQ molecules.

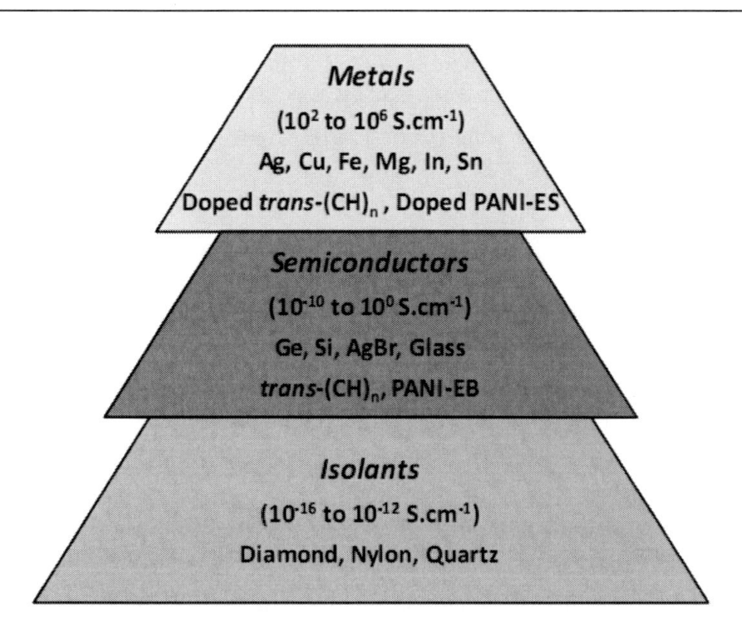

Figure 1.2. Conductivity of the ICPs compared with different materials.[5]

The concept of doping is unique and has central importance, because it is what differentiates the conducting polymers from all other types of polymers.[6-8] During the doping process, an insulating or semiconducting organic polymer with low conductivity, typically ranging from 10^{-10} to 10^{-5} S.cm^{-1}, is converted into a polymer which shows conductivity in a "metallic" regime (ca. 1-10^4 S.cm^{-1}). The addition of non-stoichiometric chemical species in quantities commonly low (≤10%), results in dramatic changes in electrical, magnetical, optical and the structural properties of the polymer (see Figure 1.2). The doping term is used for conducting polymers in analogy to the doping process of crystalline inorganic semiconductors. However, the amount of dopant used in the doping process of inorganic semiconductors is near to ppm and there is no significative disturbance into their crystalline structure. For doping of the organic polymer, the dopant chemically reacts with the chain, generally forming charged segments into the polymeric backbone, and it causes severe disturbance in the crystalline structure of the polymer.

The doping is reversible, and the polymer can return to its original state without major changes in its structure. In the doped state, the presence of counter ions stabilizes the doped state. By adjusting the level of doping, it is possible to obtain different values of conductivity, ranging from non-doped insulating state to the highly doped or metallic one. All conductive polymers (and their derivatives), for example, poly (*p*-phenylene) (a), poly(*p*-phenylene-vinylene) (b), poly(pyrrole) (c), poly(thiophene) (d), poly(furan) (e), poly(heteroaromatic vinylene) (f, where Y = NH, NR, S, O), poly(aniline) (g), poly(*p*-phenylenediamine) (h), poly(benzidine) (i), and poly(*o*-phenylenediamine) (i), among others, can be doped by p (oxidation) or n (reduction) through chemical and/or electrochemical process [6-8] (see Figure 1.3).

For instance, the *trans*-poly(acetylene) (*trans*-PA) can be "p" doped by oxidation of the polymeric chain with iodine as oxidant agent, see chemical equation 1.1.[3]

$$trans\text{-}(CH)_n + 1.5(ny)I_2 \rightarrow [CH^{+y}(I_3)_y^-]_n \; (y \leq 0{,}07) \quad (1.1.)$$

During this process there is an increase of the conductivity from ca. 10^{-5} S·cm^{-1} to 10^{3} S·cm^{-1}. The "p" doping in the *trans*-PA can also be done by electrochemical anodic oxidation of a *trans*-PA film immerse in a solution of $LiClO_4$ dissolved in propylene carbonate, see chemical equation 1.2.[7]

$$trans\text{-}(CH)_n + (ny)(ClO_4)^- \rightarrow [(CH^{+y})(ClO_4)_y^-]_n + (ny)e^- \ (y \leq 0,1) \quad (1.2.)$$

The "n" doping can also occur in *trans*-PA by reduction with sodium naphthalide (Naft, see chemical equation 1.3.) [3].

$$trans\text{-}(CH)_n + (xy)Na(Naft) \rightarrow [Na_y(CH)^{-y}]_x + Naft \ (y \leq 0,1) \quad (1.3.)$$

Finally, there is another type of doping in which the number of electrons of the polymer is reorganized inside the chain without gain or lose of electrons from external agents; this process is named internal redox process. This is the characteristic doping process of poly(anilines). The poly(aniline) in its emeraldine base form (PANI-EB, the most stable form of PANI) can be converted to the doped form (emeraldine salt form, PANI-ES) by protonation with strong acids (see Figure 1.4.).[4]

The protonation of the imine (and sometimes amine) nitrogens form charged species as radical cations and dications segments. The conductivity of the polymer can be increased by 10 times, reaching to 3 S.cm^{-1}.[4] The doping with protonic acids was observed later for the poly(heteroaromatic vinylene).[8]

The doping process of PANI can also be described by the terminology of the solid state physics. When PANI is protonated and consequently charged species are formed (radical cations, for instance) in the polymer backbone, it is energetically favorable the localization of the charge by the presence of a structural distortion into the chain. This process can be visualized by the presence of localized electronic states into the band gap of energy. These localized states are formed by reduction of energy ($\Delta\varepsilon$) of the empty state of lowest energy (LUMO, "lowest unoccupied molecular orbital") or by the increase of the energy of the full state of highest energy (HOMO, "highest occupied molecular orbital", see Figure 1.5.).

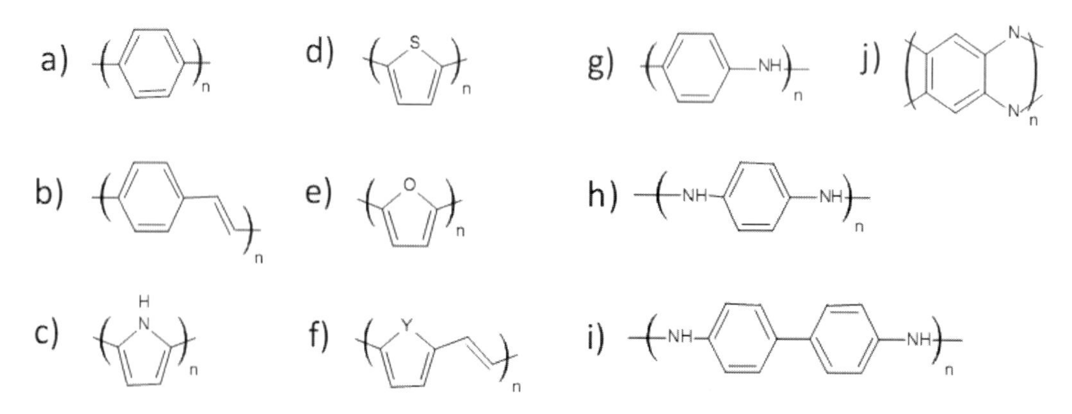

Figure 1.3. Chemical representation of the main classes of the ICPs [5].

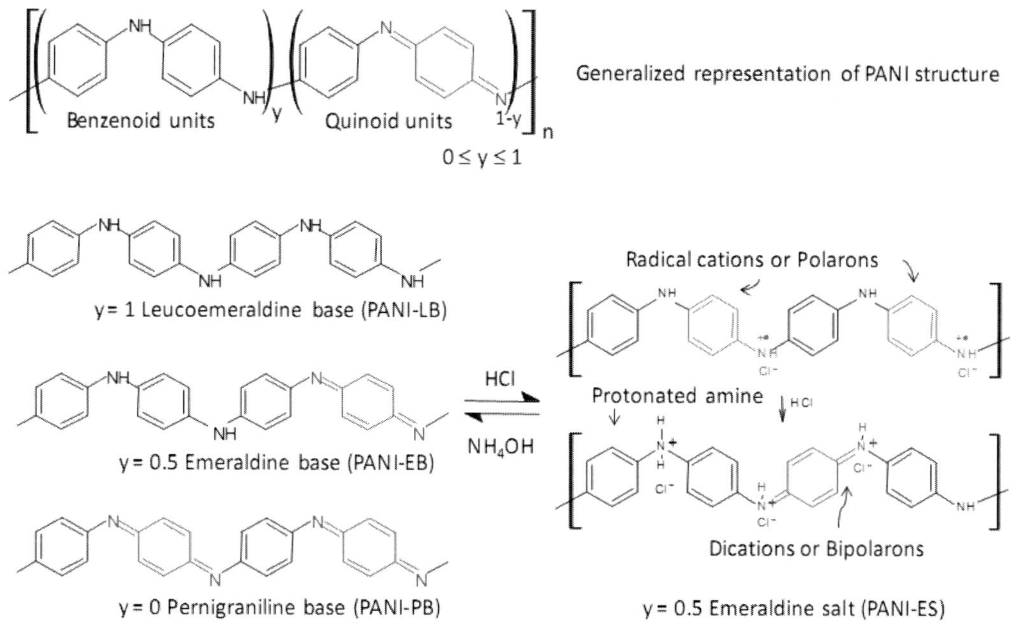

Figure 1.4. Chemical representation of generalized PANI structure and its most common forms.

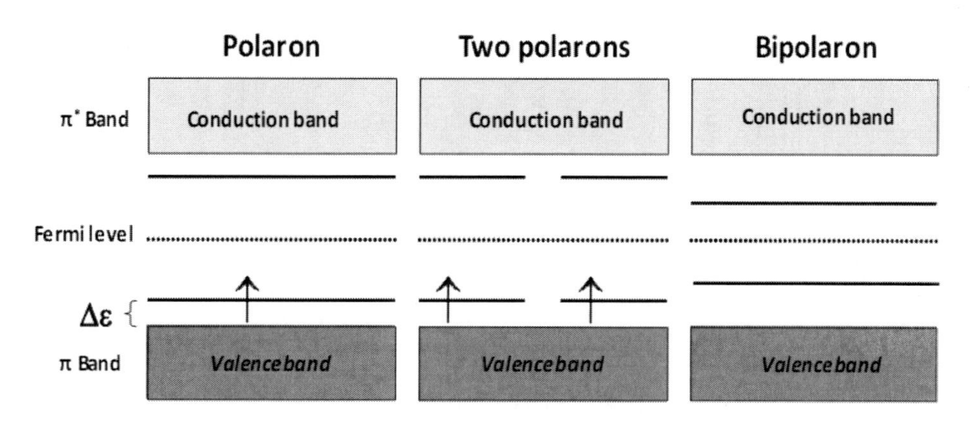

Figure 1.5. Schematic representation of the band structure of a conjugated polymer with a polaron, two polarons or a bipolaron.[9]

If $\Delta\varepsilon$ is near to the distortion energy values (E_{dis}), there is a localization of the charge, this situation is denominated by the solid state physics as polaron.[9] In chemical terminology, polaron is a cation radical (spin$\frac{1}{2}$) associated with the distortion of the structure around the charge and by the presence of localized electronic states into the gap, states called polarons.[9] The charge ability to disturb significantly the chain structure is a consequence of the strong electron-phonon coupling. The $\Delta\varepsilon$ - E_{dis} difference is denominated the polaron binding energy. Calculations using the Hückel theory indicate that the formation the polaron is energetically favorable for all organic conjugated polymer.[10] When a second ionization occurs in the polymer backbone, two polarons or a bipolaron can be formed. In the chemical terminology, bipolaron is defined as a pair of equal charges in a dication associated with a strong distortion in the polymer chain. The band structure of a polymer chain in the presence of a polaron, two polarons or a bipolaron is shown in Figure 1.5.

The E_{dis} for the bipolaron is higher than the polaron, because two near charges cause more distortion in the polymer chain. In addition, the $\Delta\varepsilon$ values for the bipolaron is higher than for the polaron, and the sum of these two factors is the main reason why the formation of bipolaron is more thermodynamically favorable than two polarons. In fact, the formation of bipolaron is higher than the polarons for highly doped polymers.[9,10]

1.2. Polyaniline (PANI)

Polyaniline (PANI) is one of the most important conducting polymers owed to its easy preparation and doping process, environmental stability, and potential use as electrochromic device, as sensor and as corrosion protecting paint. These properties turned PANI attractive to use in solar cells, displays, lightweight battery electrodes, electromagnetic shielding devices, anticorrosion coatings and sensors.

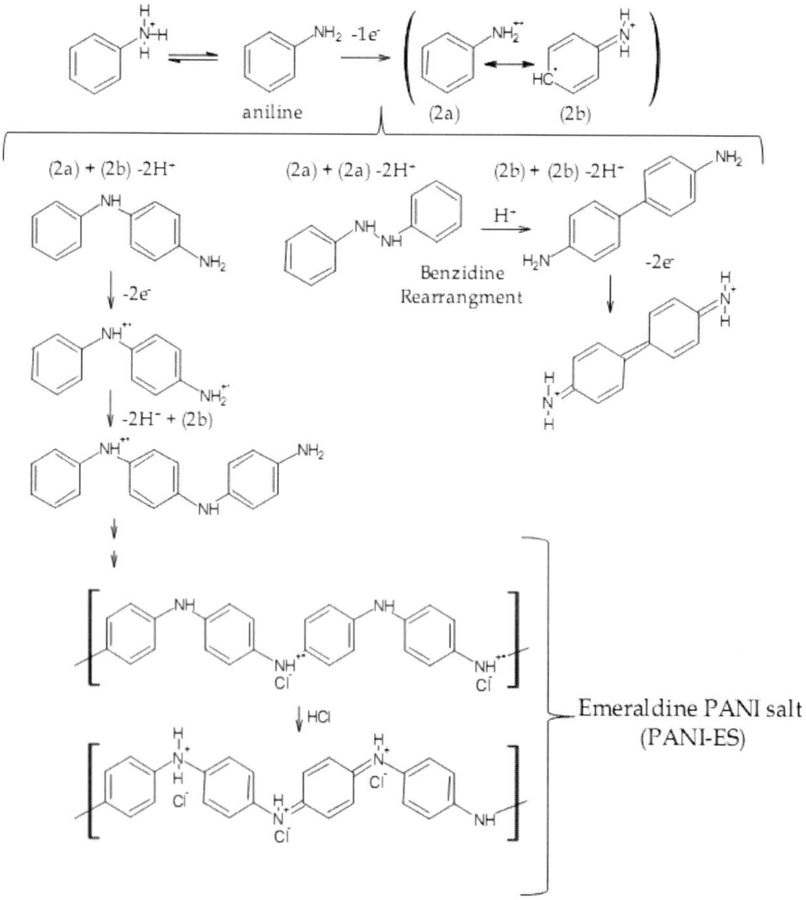

Figure 1.6. Schematic representation of the polymerization steps present during the aniline polymerization. The structure of PANI is shown with radical cations (or polarons) and dications (bipolarons) segments.

The recent research efforts are involved to the control and the enhancement of the bulk properties of PANI, mainly by formation of organized PANI chains in blends, composites and nanofibers.[11-16]

Polyaniline can be synthesized by two main methods, by chemical or by electrochemical polymerization of aniline in acidic media. The chemical oxidation is commonly performed using ammonium persulfate in aqueous acidic media (hydrochloric acid, sulfuric, nitric or perchloric acid) containing aniline. This is the conventional synthetic route of PANI, but one of disadvantages of this route is the presence of excess of oxidant and salts formed during the synthesis, leading to a polymeric sample that is practically insoluble in majority of solvents, making its processing very difficult.[17] During the oxidative polymerization of aniline, the solution becomes progressively colored resulting in a solid dark green. The color of the solution is owing to the presence of soluble oligomers formed by coupling of radical cations of aniline.

The intensity of the color depends on the environment and also the concentration of oxidant.[17] There are many variations of the chemical synthesis of PANI, however there is a certain consensus that there are four main parameters that affect the course of the reaction and the nature of final product, being: (1) nature of the synthetic medium, (2) concentration of the oxidant, (3) duration of the reaction, and (4) temperature of the synthetic medium.[17] The polymerization of aniline was observed as an autocatalytic process.[18,19] Kinetic studies suggest that initial oxidation of aniline leads to the formation of dimeric species, such as *p*-amino-biphenyl-amine, N,N'-Biphenyl-hidrazine, and Benzidine.[20] The oxidation of aniline and its derivatives in strongly acidic media favors the formation of benzidine,[21,22] while in a slightly acid or neutral prevails *p*-amino-Biphenyl-amine. In the other hand, in basic medium the formation of azo bonds, resulting from the head-head coupling is favored. These dimeric species have lower oxidation potential than aniline and are oxidized immediately after its formation (the N,N'-Biphenyl-hidrazine is converted to benzidine through rearrangement that occurs in acid medium,[23] resulting two types of charged quinoid-di-imine species. Afterwards, electrophilic attacks in these species, followed by deprotonation, are responsible for the growth of oligomers with subsequent formation of polymer chains of PANI (see Figure 1.6.).[24-29]

1.3. PANI-Clay Nanocomposites

Among the different types of hosts used in the formation of nanocomposites with PANI undoubtedly lamelar materials are widely employed. The main reason is that the distance between the layers can be modified, facilitating the intercalation of various chemical species. Hosts, such as MoO_3,[30] V_2O_5,[31,32] α-$Zr(HPO_4)_2H_2O$,[33] $HUO_2PO_4.4H_2O$,[34] FeOCl,[35] layered double hydroxide (LDH) [36] and MoS_2 [37] and most frequently clays were used for intercalation of PANI.[38-52]

Among the clay minerals, the most employed are those of the smectite family, whose structure consists of a central sheet containing groups $MO_4(OH)_2$ of octahedral symmetry associated with two tetrahedral sheets (MO_4) producing layers designated T:O:T (see Figure 1.7.).[38] The octahedral sites are occupied by ions of aluminum, iron and magnesium, while the centers accommodate tetrahedrons of silicon and aluminum ions.

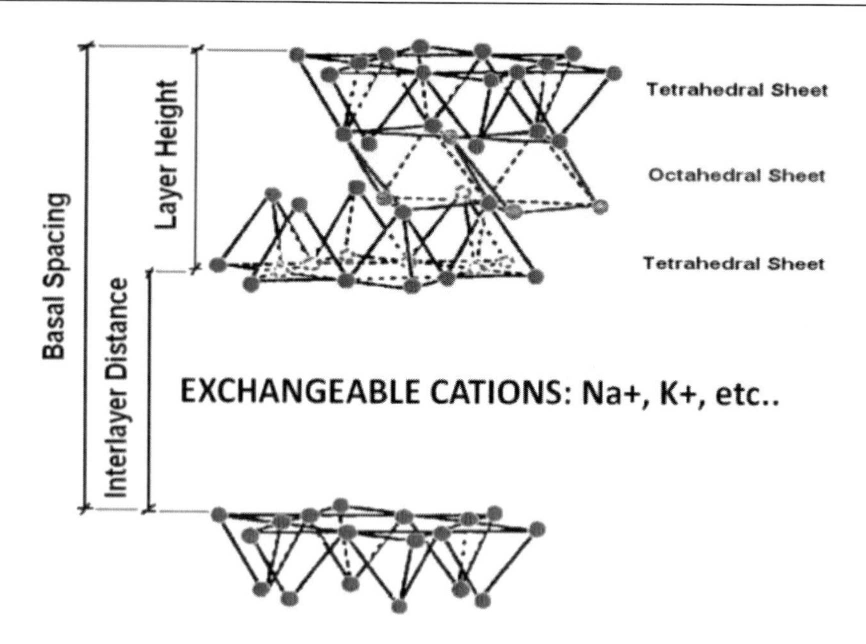

Figure 1.7. Schematic representation of Montmorillonite clay (MMT).

The individual layers of negative T:O:T assume a mutually parallel orientation and the electric charge is neutralized by the presence of hydrated positive ions that are in the interlayer region and can be exchanged by ion exchange.[38] These clays show large surface adsorption and catalytic activity in organic reactions. The montmorillonite (MMT) is the most common smectite clay and the negative charges of the lamellae arise mainly in the layers of octahedral substitution of aluminum by magnesium ions.

The adsorption of aniline on MMT clay has been studied a long time ago, and since them it is well-known that clays have a property to generate colored species by the adsorption of aromatic amines. The best known case is the blue color generated by the adsorption of benzidine (4,4'-diaminobiphenyl) in clay.[39] Among the earlier studies Yariv et al.,[40] Brindley et al.,[41] and Cloos et al.[42] Yariv et al.[43] are reported that films of MMT containing Cu^{2+} ions become black after immersion in aniline, the authors suggest that it is due to the polymerization of monomer. Similar effect was observed for the MMT clay containing ions of Ni, Hg, and Ca. Brindley et al.[44] also reported that the MMT and Hectorite clays containing adsorbed aniline, become black when the films of these materials suffer heating to 110°C in air. Cloos et al.[45] also studied the adsorption of aniline on MMT clay containing ions Cu^{2+} and Fe^{3+}, the authors describe the achievement of a black material containing metal ions of MMT after exposure to aniline. Spectroscopic measurements of electron paramagnetic resonance (EPR) and the infrared (FTIR) showed that copper ions oxidize aniline to produce polymeric species through the formation of radical cations and generating monovalent copper. Soma and Soma[43-47] used the resonance Raman spectroscopy (RR) in the study of oxidation of aromatic compounds (benzene and derivatives) adsorbed on clay, and showed that when vapors of aniline adsorbed on MMT containing Cu^{2+} formed benzidine as dication.

However, when the adsorption of aniline on Cu^{2+} or Fe^{3+}-MMT is made in the liquid phase occur the formation of the polymer. Soma and Soma proposed by the analysis of RR spectrum that the polymer formed was equal to that generated electrochemically (PANI-ES), but the observation of an additional band at 1432 cm^{-1} in the RR spectrum of the polymer-MMT excited at 457.9 nm, led the authors to consider the presence of azo linkages (-N=N-) in the structure of the polymer formed in the MMT clay. Although the study of the interaction of amines with clay is relatively old, just recently (a little over a decade) the amine-clay research moves to the aim to produce conducting polymer-clay nanocomposites. Thus, Giannelis et al.[48] synthesized the PANI intercalated in a synthetic hectorite containing Cu^{2+} ions, the UV-vis-NIR spectrum was very similar to that observed for PANI-EB (see Figure 1.8.), and the polymer was converted to conductive PANI-ES form, simply by exposuring the material HCl vapors (the obtained conductivity was $5{\cdot}10^{-2}$ $S{\cdot}cm^{-1}$).[48] Other work done by Chao et al.[49] reported in 1992 the polymerization of PANI into MMT clay galleries. The intercalation was confirmed by measures of X-ray diffraction, and obtained an interlayer distance was changed from 1.47 to 0.36 nm after the polymerization of aniline. The authors suggested that after polymerization the anilinium rings changed from perpendicular to almost planar position in relation to the clay layers. Absoprtion bands were observed at 420 and 800 nm in the UV-vis-NIR spectrum of the material, which are characteristic of the PANI-ES form. Furthermore, the FTIR spectrum of PANI-MMT showed bands at 1568, 1505, 1311 and 1246 cm^{-1} characteristic of the emeraldine salt, but the bands are shifted to higher frequencies in relation to the FTIR spectrum of PANI-ES. According to the authors this displacement is a consequence of geometric restriction imposed over the aromatic rings. There was a band at 1400 cm^{-1} which was attributed by Chao et al.[49] to the presence of NH_4^+ ions in the clay due to a partial exchange of these ions with anilinium ions intercalted in the clay. The conductivity obtained for this material was approximately 10^{-6} $S.cm^{-1}$.

In another work, Wang et al.[50,51] also obtained PANI-MMT using ammonium persulphate as oxidizing agent, the electronic spectrum of the material obtained is very similar to that obtained for secondary doped PANI-CSA, suggesting that the PANI was obtained in a extended conformation (see Figure 1.9.).

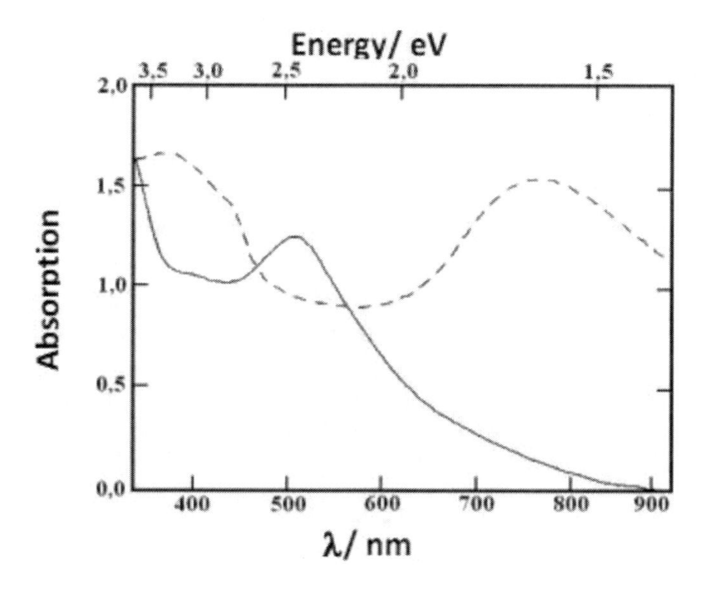

Figura 1.8. Absorption spectra of PANI-hectorite, before (—) and after exposure to the HCl acid vapors (---).[48]

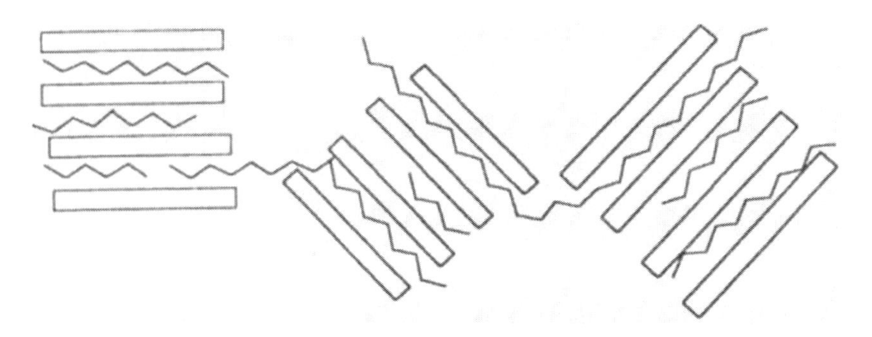

Figure 1.9. Schematic representation of PANI-MMT nanocomposites, according to Wang et al. [50,51]

The formation of PANI-ES was confirmed by the presence of bands at 1489, 1562, and 1311 cm^{-1} in the FTIR spectrum of the material. Despite of the high organization level of PANI chains into the MMT clay, the conductivity of the material, ca. 10^{-3} $S.cm^{-1}$, was not so higher than to those obtained previously. The justification of the authors is that there are few polymeric connections between the particles of clay, which significantly reduces the conductivity observed for the material. Later, other authors reported the synthesis of PANI into MMT clay by the intercalation of anilinium ions into MMT followed by oxidation with ammonium persulphate as a standard method to obtain PANI-MMT nanocomposites.[52-64] The common approach is the use of FTIR spectroscopy to confirm the formation of PANI.

Some studies were performed by the variation of the aniline/clay ratio during the intercalation and it was possible to show the increase of interlayer space and also the amount of intercalated PANI as well the increase of the conductivity of the material.[65] In addition, the morphology of the material were also found recently by Park et al.[57] similar to the morphology of the clay gradually changes to a morphology similar to that observed for free PANI with increase of the aniline/clay ratio.[57]

The synthesis of PANI with clay in a medium containing surfactants (dodecylbenzenesulfonic acid, DBSA, and camphorsulfonic acid, HCSA) was also used.[58-60] The intercalation was confirmed by X-ray diffraction data, being obtained interlayer distances of ~ 1.5 nm and ~ 1.6 nm for composites of PANI-DBSA-MMT and PANI-CSA-MMT, respectively. DC conductivity values for PANI-DBSA-MMT and PANI-CSA-MMT at room temperature were near to 0.3 $S.cm^{-1}$ and 1.0 $S.cm^{-1}$, respectively. The intercalated PANI was also obtained by electrochemical polymerization of aniline, using modified clay electrode[61] graphite electrode modified clay,[62] Pt electrode modified clay,[63] and electrode stainless steel.[64] Inoue et al.[61] were the first authors to report the electrochemical synthesis of PANI nanocomposites with MMT. Inoue et al.[61] used a clay-modified electrode and the intercalation is performed by immersing the electrode in aniline solution. The interlayer distance value of 0.54 nm was obtained for MMT clay after immersion. The electropolymerization was performed in a constant current of 20 $\mu A.cm^{-2}$ and in 2 mol/L of HCl solution. After polymerization, it was observed a reduction of the interlayer distance to 0.34 nm. Cyclic voltammograms obtained for this material showed peaks around 0.2 V (versus saturated calomel electrode / SCE), which are associated with processes of protonation or deprotonation of PANI. It were observed in the first scan, peaks at higher potential values (around 0.5 V), which were attributed to oxidation of residual aniline. Another work using graphite or Pt electrode modified with clay also reported the formation of PANI, confirmed by the profile curves in the voltammograms.[62,63] The oxidation of a suspension of aniline containing MMT clay intercalated with stainless steel electrodes produced a polymer-MMT-valued interlayer distance of 0.51 nm. The FTIR spectrum of the material presented bands at 1579, 1490 and 1311 cm^{-1}, similar to that obtained by Wang et al.[51] in the chemical polymerization of aniline with ammonium persulphate.

2. Structure of PANI-Clay Nanocomposites

2.1. Vibrational Spectroscopy Data: Resonance Raman

Using resonance Raman (RR) spectroscopy it was possible to show that the structure of intercalated PANI is different from the PANI-ES structure,[15,16,66-72] contrarily to that previous observed using only FTIR spectroscopy. The main reason for this difference is that owing to the resonance process between the laser line and the excited vibronic states of the polymer it was possible to search each chromophoric segment, contrarily to the FTIR spectroscopy. Figure 2.1. shows the RR spectra of two PANI-MMT samples and the PANI-ES, using two exciting radiations. The RR spectrum of the PANI-MMT prepared by *in situ* polymerization and excited at 632.8 nm is very different from the PANI-ES spectrum. The bands at 984, 1179, 1201, 1264, 1349, 1395, and 1630 cm^{-1} cannot be associated with any form of PANI, strongly indicating the formation of other kind of chromophoric segment in the intercalated PANI chains. This resonance is supported by the presence of an electronic absorption at 620-670 nm observed in the intercalated PANI-MMT composites.[66,67] At 488.0 nm exciting radiation the RR spectrum of PANI-MMT prepared by *in situ* polymerization shows bands at 1195 and 1618 cm^{-1} that can be assigned to reduced benzenoid segments of PANI-ES. In addition, a band at ca. 1449 cm^{-1} was assigned to νN=N of azo segments. This assignment was further proved by the extensive study of adsorption, intercalation and polymerization of benzidine (4,4'-diaminobiphenyl) in MMT clay [68,73] Another important result is that the RR spectra of PANI-MMT obtained by *ex situ* polymerization (without previous intercalation of anilinium ions) are very similar to those of PANI-ES for both exciting radiations (see Figure 2.1.).

In the spectra of PANI-MMT obtained by *in situ* polymerization, the spectral region near 1100 cm^{-1} is dominated by clay bands in the region from 3700 to 3000 cm^{-1} and at 1630 cm^{-1}, the bands are associated to adsorbed water and structural OH groups of the clay OH.[67,68] There is no significant difference between the IR bands of the polymers obtained by the two routes, being similar to the PANI-ES spectrum. Hence, the use of the IR information and the observation of green color of the PANI-MMT nanocomposites can lead to an incorrectly interpretation that PANI-ES is formed in the clay interlayer, as is frequently observed in the literature. This result is clear evidence that RR spectroscopy is more sensitive than FTIR to characterize this type of systems.

Using several techniques to follow the polymerization of intercalated aniline it was possible to elucidate the structure of the nanostructured PANI chains.[67]

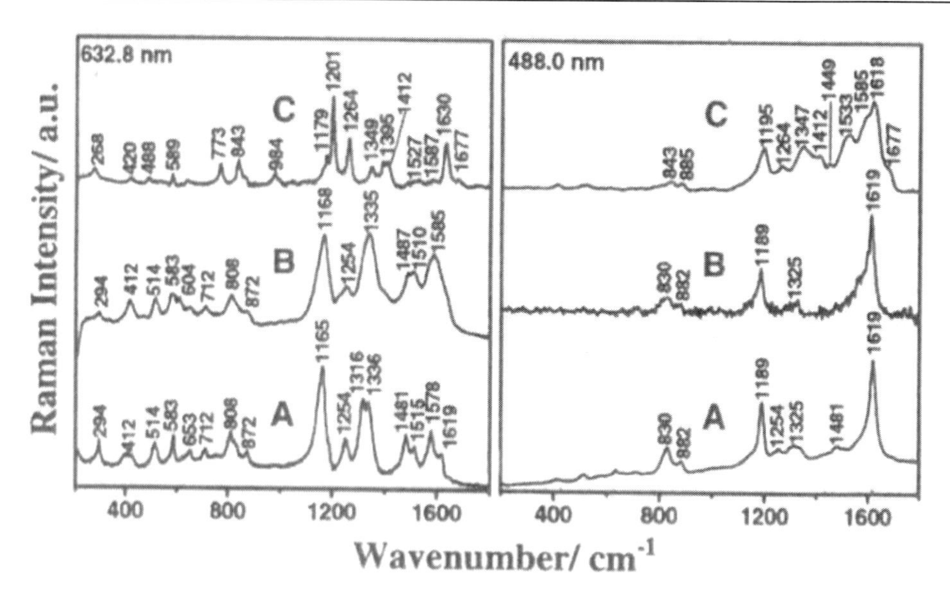

Figure 2.1. Resonance Raman spectra excited at 632.8 nm and 488.0 nm of powdered samples of: PANI-ES (A), PANI-MMT samples prepared from: *ex-situ* polymerization (b) and *in-situ* polymerization (C).[67]

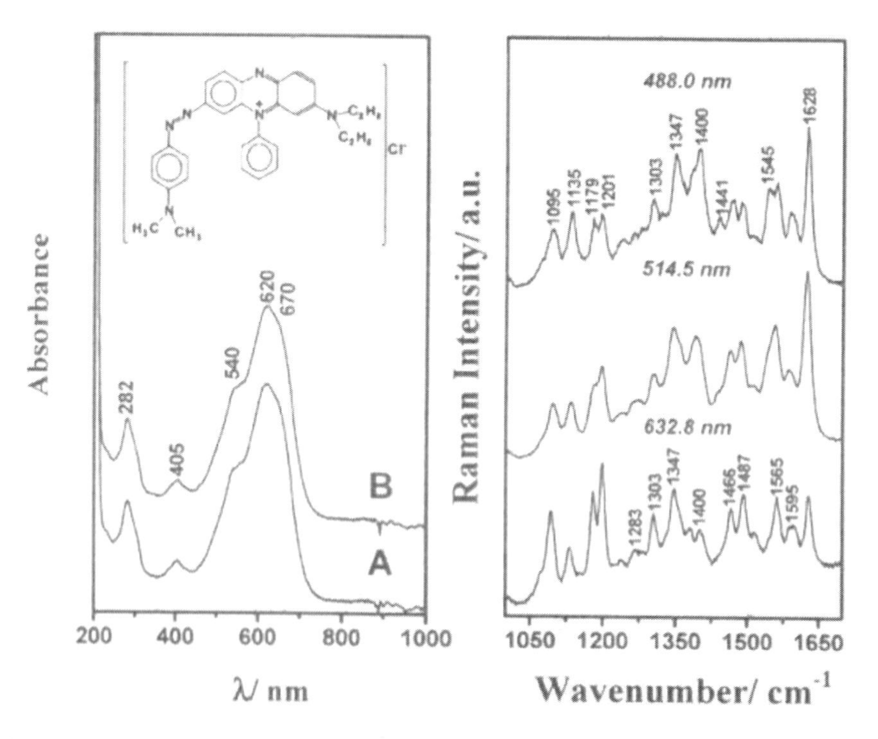

Figure 2.2. UV-vis-NIR spectra of solutions of Janus Green B azo dye in NMP (A), and aqueous acidic solution (B), and Resonance Raman spectra of Janus Green B azo dye in a solid state obtained at indicated exciting radiations. The dye molecular structure is also shown.[67]

The RR spectra obtained at early stages of the polymerization of intercalated anilium ions show the presence of bands of radical cation of PANI-ES (at 1167, 1318/1339, and 1625 cm-

^{1}) and dication (at 1481 and 1582 cm^{-1}) and also bands due to benzidine dication at 1211, 1370, 1455, and 1608 cm^{-1}, confirming that into clay galleries, occurs the head-to-tail and also tail-to-tail coupling between the aniline monomers. In the final stages of the polymerization, mainly the RR spectra obtained at 632.8 nm do not resembles neither the spectrum of radical cation of PANI-ES nor the spectrum of benzidine dication, indicating that another type of chromophore was formed.

At this point, the use of XANES spectroscopy was very important. Through the analysis of several dyes, it was found that the structure of intercalated PANI also has phenazine rings. Taking all these considerations into account, the best standard compound, having azo group together with phenazine or oxazine-like rings in their structure, which give out vibrational and electronic signatures similar to those of the PANI-MMT nanocomposites was the Janus Green B (JGB).

Figure 2.2. displays the UV-vis-NIR spectra of JGB in aqueous solution at pH 0 and in *N*-methylpyrrolidone (NMP) and the RR spectra at different excitation wavelengths of the dye in the solid state. The bands at 620 and 670 nm are similar to those observed for PANI-MMT. Also, the frequencies and relative intensities of the dye bands at 1179, 1201, 1347, 1400, and 1628 cm^{-1} in the RR spectrum at 632.8 nm exciting radiation (Figure 2.2.) are in good agreement with those of PANI-MMT nanocomposites at the same radiation (Figure 2.1.). Therefore, this result indicates that a phenazine-type ring is another kind of segment present in the polymeric structure.

2.2. Electronic Spectroscopy Data: X-Ray Absorption

Henning et al. [74-76] and Hitchcock et al. [77] demonstrated the feasibility of using the XANES technique in the characterization of different oxidation states of nitrogen atoms in the chains of PANI. Henning et al. [74-76] showed that the K edge XANES spectra of nitrogen (NK XANES) of PANI is dominated by transitions 1s $\rightarrow$ $2p\pi^*$ whose energy values and intensities are dependent on the oxidation state of PANI. Figure 2.3. shows the spectrum obtained by NK XANES Henning et al. [74-76] for PANI-EB, and PANI-LB compound used as a standard, 2-hydroxy-3-methoxy-benzilanilina. It is seen that the NK XANES spectrum of PANI-EB is dominated by a band at 397.4 eV, which can be assigned by comparison with the spectrum of the standard compound with the transition 1s $\rightarrow 2p\pi^*$ the imine nitrogen (= N-). This band is not observed in NK XANES of PANI-LB, demonstrating the consistency of the assignment made. The presence of two other bands observed at 400.4 eV and 402.1 eV in NK XANES spectra of PANI-EB, are due to conjugation effects, and were assigned to the transitions 1s $\rightarrow 2p\pi^*$ of delocalized imine and amine nitrogens, respectively.

Henning et al.[74-76] obtained the NK XANES spectra of PANI synthesized electrochemically with different acids and found that the intensity of the band at 397.4 eV, which is dominant in the spectrum of PANI-EB, suffers significant reduction in the spectrum of PANI-ES. In addition, the intensity of this band depends on the degree of protonation of the polymer, figure 2.4. A second effect of the protonation is the complete removal of the band N1s $\rightarrow 2p\pi^*$ (=N-) that is replaced by a new band at 398.8 eV, assigned to the protonated imine nitrogen. The lowest energy, ca. 2.6 eV, indicates that the imine delocalized positive charges stabilize π^* antibonding states. Table 2.1. summarizes the results of characterization NK XANES obtained by Henning et al.[74-76] for PANI.

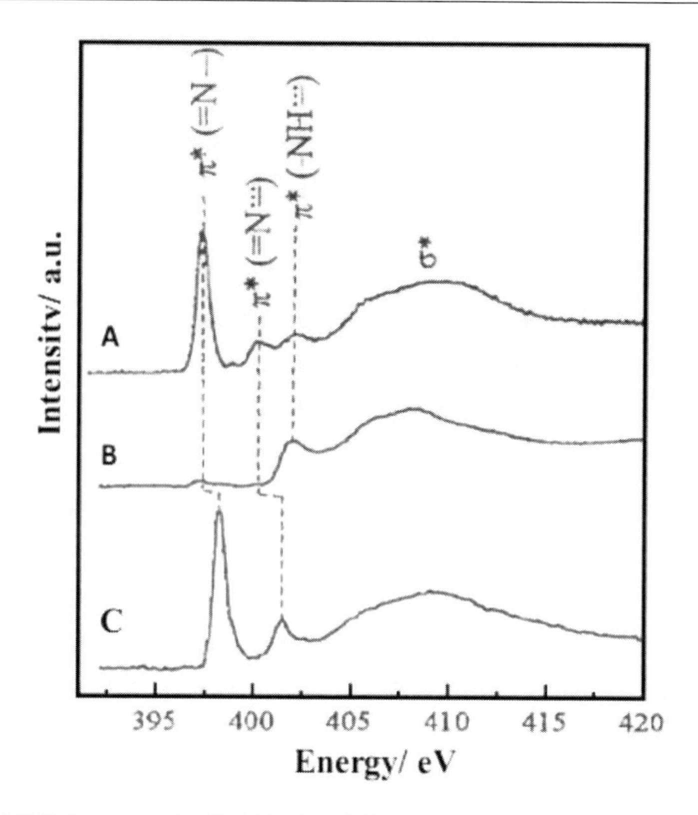

Figure 2.3. N K XANES spectra of: PANI-EB (A), PANI-LB (B) and 2-hydroxyl-3-methoxi-benzylaniline (C).[74]

Table 2.1. N1s→ 2pπ* transitions observed in the N K XANES spectra for some forms of PANI

Sample	Energy/ eV	1s → 2pπ*		
	−N=	$\overset{+\bullet}{-\text{NH}-}$	=N∴	−NH−
PANI-LB	397.2	---	---	402.1
PANI-EB	397.4	---	400.4	402.1
PANI-ES	397.4	398.8	---	402.1

To perform X-ray absorption experiments is required a light source capable to emitting electromagnetic waves almost continuously from UV region to the X-rays or high-energy spectra (λ <0.1 nm). A source capable to accomplishing this is called synchrotron ring. In these accelerators the electrons are circulating in relativistic speeds (near to the speed of light), and under these conditions when they pass through magnetic devices placed in their path, their orbit is disturbed and occurs the emission of a "light" (called synchrotron light). The synchrotron accelerators are obviously expensive devices and are part of National Laboratories where they can be used by different users. Brazil has a facility of this type, being named the National Synchrotron Light Laboratory (LNLS). At this site we performed experiments of X-ray absorption at the edge of Nitrogen, Silicon and SAXS of Polyaniline and its nanocomposites, whose results will be presented in following lines.

Our group was the first to introduce the XANES technique in the study of PANI-clay nanocomposites and also for PANI nanofibers.[15,16,67-70] The spectra of PANI nanocomposites show different profile compared to the "free" PANI, however, through the use of the spectral database previously built (see Table 2.2) it was possible to analyze the new data.

Table 2.2 summarizes all spectral database, the difference between the energy values is little, and however it is possible to correlate the energy value with the type of nitrogen. It should be mentioned that the broad bands are observed at energy values larger than ca. 404 eV, and are assigned to 1s →σ* transitions, whose analysis is more complex, since these bands are much wider than the 1s → π* transitions, hindering the use of these bands in a comparative analysis.

The study of the NK XANES spectra of PANI nanocomposites prove the presence of phenazinic and azo nitrogens in the PANI backbone.[15,16,67-70] The N K XANES spectra of PANI nanofibers reveal that the spectral profile is similar to the PANI-ES, however the signal of polarons are more intense and a shoulder at higher energies indicate the presence of nitrogen with delocalized bond and also protonated phenazinic rings. This is a strong indication of the presence of phenazinic rings in PANI nanofibers, as suggested by our group in several papers[67-70,78,79] and also from other groups.[80,81]

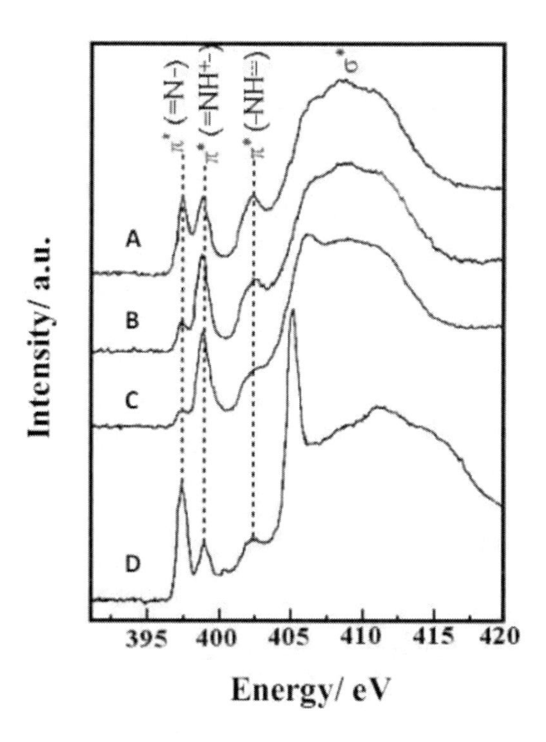

Figure 2.4. N K XANES spectra of PANI-ES films electrochemically synthesized with 0.5 mol/L H_2SO_4 (A), 2.0 mol/L H_2SO_4 (B), 1.0 mol/L $HClO_4$ e 1.0 mol/L HNO_3 (D). The band at 405.1 eV observed in spectrum (D) is attributed to NO_3^- ion.[74]

Table 2.2. Spectral database of 1s → π* transitions observed at N K edge for different compounds having nitrogen atoms bonded in different types of bonds and chemical environments

Samples	Energy/ eV										
	1s → π* transitions at N K edge										
	=N- Quinoid ring	=N- Phenazine or Oxazine like ring	-N=N-	=N- Pyridine like ring	$-\overset{+\bullet}{N}-$	$=\overset{+}{N}-$	=N--	$=\overset{+}{N}-$ Imidazolium ring	-NH-	$-NO_2$	$N_2H_4^{+2}$
PANI-EB	397.7	--	--	--	--	--	400.6	--	402.7	--	--
PANI-PB	397.7	--	--	--	--	--	400.6	--	--	--	--
PANI-ES	--	--	--	--	399.1	--	--	--	402.7	--	--
Phenazine	--	398.2	--	--	--	--	400.6	--	--	--	--
Phenosafranine	--	398.5	--	--	--	399.4	400.7	--	~ 402	--	--
Janus Green B	--	398.3	398.7	--	--	399.6	400.4	--	~ 402	--	--
Nile Blue	--	398.3	--	--	--	399.6	--	--	~ 402	--	--
Oxazine	--	398.3	--	--	--	399.6	--	--	401.8	--	--
4-Amine azobenzene	--	--	398.7	--	--	--	--	--	402.5	--	--
4,4'-Diamine azobenzene	--	--	398.7	--	--	--	--	--	402.0	--	--
Congo red	--	--	398.7	--	--	--	--	--	--	--	--
Methyl Orange	--	--	398.7	--	--	--	--	--	401.5	--	--
Methyl red	--	--	398.8	--	--	--	--	--	--	--	--
Titan yellow	--	--	398.9	--	--	--	400.4	--	--	--	--
Sudan III	--	--	399.0	--	--	--	--	--	--	--	--
Naftol Black	--	--	398.8	--	--	--	--	--	--	--	--
Red Lake C	--	--	399.0	--	--	--	--	--	--	--	--
Hydrazinium chloride	--	--	--	--	--	--	--	--	--	--	404.6
Hydrazinium sulfate	--	--	--	--	--	--	--	--	--	--	403.5
1,10-fenantroline	--	--	--	398.8	--	--	400.9	--	--	--	--
3,5-Dinitrobenzoic acid	--	--	--	--	--	--	--	--	--	403.8	--
N,N'-difenil-1,4-phenylenediamine	--	--	--	--	--	--	--	--	402.0	--	--
Gentian violet	--	--	--	--	--	399.6	--	--	401.8	--	--
[BMIm]Cl*	--	--	--	--	--	--	--	401.9	--	--	--
[HMIm]Br**	--	--	--	--	--	--	--	401.9	--	--	--
[OMIm]Br***	--	--	--	--	--	--	--	401.9	--	--	--

* 1-butyl-3-methylimidazolium chloride, ** 1-hexyl-3-methylimidazolium bromide, *** 1-octyl-3-methylimidazolium bromide.

2.3. Electronic Spectroscopy Data: X-ray Photoelectron Spectroscopy

X-ray photoelectron (XPS) spectroscopy is another valuable technique to study polymeric materials in the solid state. The conduction electrons of a metal, which can be considered as free electrons are not ejected out of the material, because an energy barrier or potential (called work function, Φ) is what prevents this from occurring. Hence, only when the electrons have energy to overcome this barrier is that they can escape from the material. Electrons which are in the deepest levels require additional energy, called binding energy, E_b (binding energy), it corresponding to the difference between the initial energy level and the Fermi level. Thus, the electron will need a power with the minimum value of $E_b + \Phi$ to overcome the potential barrier and escape from the material. Providing energy greater than $E_b + \Phi$ the electron can ejected out with a kinetic energy (K) different from zero (Figure 2.5.). We describe the photoelectric effect by equation 2.1.

$$K = h\nu - E_b - \Phi \qquad (2.1.)$$

From measured values of the kinetic energy of the photoelectron the E_b can be determined using equation 2.1. By E_b values we can identify the chemical elements; this is the basis of Electron Spectroscopy for Chemical Analysis (ESCA) or XPS. Much more information can be obtained in the XPS spectrum than simply identification of the chemical elements. For instance, in determination of the oxidation state of the chemical elements as well the study of electronic states involved in the chemical bonds.

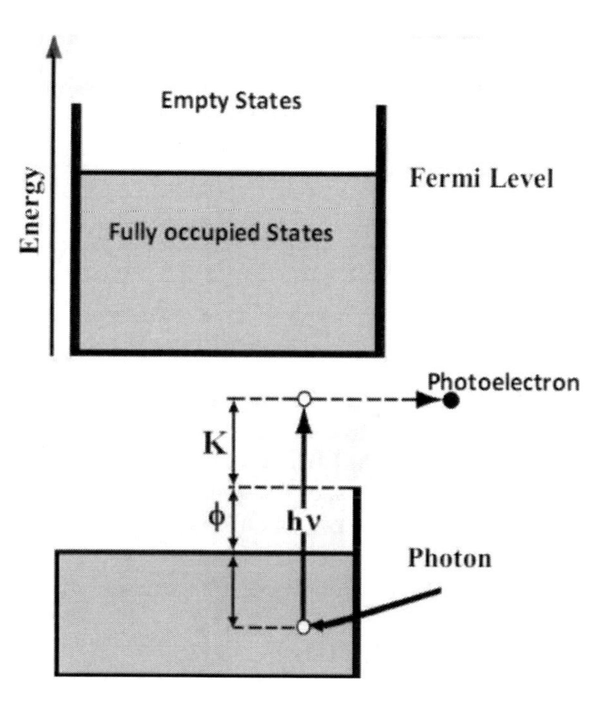

Figure 2.5. Schematic representation of the photoelectron effect in a metal. The minimum energy necessary to eject the electron below the Fermi level is the sum of the $E_b + \Phi$.

The development of the XPS technique was originally difficult due to the high sensitivity of the XPS signal to the surface contamination of the material. This is a consequence that the XPS signal is originated mainly from the photoelectrons of the first atomic layers of the material. Hence, the photoemission spectrum reflects much more the electronic properties of the surface than the whole material. If the surface is contaminated, the XPS analysis of the spectrum will be compromised. Very clean surface and ultra high vacuum conditions are strictly required for the photoemission experiments and only since 1950 the instrumental apparatus was enough for performing photo-emission experiments.

The thickness of the material in which the photoelectron material can pass by without being reabsorbed is called depth of escape. The values of depth of the escape depend on the photoelectron energy and the energy of the incident photon. Thus, the dependence of the depth of escape of the photoelectron with the kinetic energy is similar for different materials and corresponds approximately to a universal curve (Figure 2.6). The curve with the kinetic energy displays minimum from 50 to 200 eV and the depth of escape corresponds to a few atomic plates.

The XPS spectroscopy has been widely used in the literature with the aim to determine the oxidation states and protonation degree of PANI. Tan et al.[83] obtained the XPS spectra for some characteristic forms of PANI. The XPS spectra of PANI-EB was deconvoluted in two main components (bands) centered at 399.3 and 398.1 eV, which was attributed to the amine nitrogen (-NH-) and imines (-N=), respectively. By comparison, with other nitrogen compounds, Tan et al. [83] also found that the band attributed to imine nitrogens gradually disappears with the protonation of PANI, following the mechanism suggested by MacDiarmid for the protonation of PANI. In addition, two bands arise in higher energies in the XPS spectrum and their intensities increase for higher protonation levels. These bands were attributed to the presence of protonated nitrogens in PANI backbone.[83] In literature, authors usually make the deconvolution of the XPS spectrum of PANI using only four components or bands, but we find differences in the allocation of bands of the protonated nitrogen [84-86].

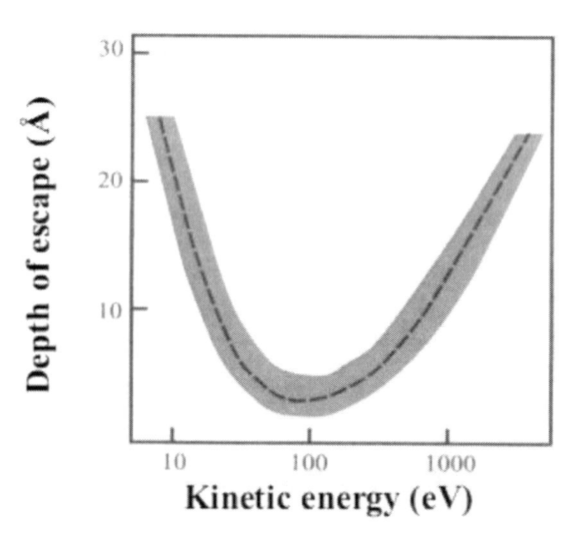

Figure 2.6. Depth of Escape curve for the photoelectron as a function of the kinetic. The dashed area contains all the known experimental points.

Table 2.3. Band values observed in the N 1s XPS spectra of PANI-ES and their respective assignments from different authors

Zeng et al.[100]		Nakajima et al.[101]		Monkman et al.[102]	
Bands/ eV	Atribution	Bands/ eV	Atribution	Bands/ eV	Atribution
398.2	$-N=$	399.0	$-N=$	398.2	$-N=$
399.4	$-NH-$	399.1	$-NH-$	399.5	$-NH-$
400.7	$-\overset{+\bullet}{N}H-$	400.8	$-\overset{+\bullet}{N}H-$	400.8	$-\overset{+}{N}H_2-$
402.6	$-\overset{+}{N}H=$	402.4	$-\overset{+}{N}H_2-$	402.4	$-\overset{+}{N}H=$

Table 2.2 summarizes the main results of the literature and their suggested assignments by different authors for the components of the XPS spectrum of PANI-ES. The XPS spectra of PANI-clay nanocomposites show similar signal from that observed for free doped PANI. It is known that in the XPS signal the majority of the photoelectron signal comes from the first three atomic monolayers, which means that only the polymeric PANI chains on the external surface or on the edge of clay crystal contribute to the XPS signal of PANI-MMT nanocomposite [16,17]. Therefore, the low values of N/Si ratio (0.026) obtained for the PANI-MMT suggest that only a small fraction of polymer is on the external clay surface, which is in agreement with the SEM images.[67] The absence of peaks in the PANI-MMT XPS spectrum which can be assigned to nitrogen atoms of the phenazine-like rings and azo units suggests that these segments are not present in the polymeric chains on the external surface (basal surface and the edges) of the clay. Some reason for the preferential localization of chains, having head-to-tail segments on the basal surface or on the edges of the clay particles, can be the easier access of the oxidizing agent. This observation is consistent with spectroscopic data at the initial stages of An+-MMT polymerization, which show that PANI-ES is formed and phenazine-like rings plus N=N bonds are absent.[67] From the FTIR, resonance Raman, XANES and XPS data, it was possible to state that the intercalated polymeric chain is different from that of PANI-ES. The formation of these new structures, not common in the absence of the clay, could be a consequence of the confinement of the monomers. The An^+ can assume some different orientations inside the gallery height and the electric field between the sheets can change the An^+ electronic distribution. These effects, together with smaller degree of freedom for An^+ imposed by the gallery height (ca. 9.5 Å), are important factors to be considered when trying to rationalize the difference between the free and the confined monomer polymerization mechanisms.

CONCLUSION

The investigation of the electronic and vibrational structure of the conducting polymers-clay nanocomposites through several spectroscopies has been decisive in the study of the formation of the nanostructured conducting polymers, in determination of their structure and also in the study of the interactions between the polymeric chains and the clay hosts. The polymerization inside of porous of a rigid host has been commonly employed to acquire nanostructured conducting polymer chains; it is generally observed shifts in the vibrational frequencies and changes in the UV-vis-NIR spectra of the confined conducting polymer. Hence, this effects are a clearly indication of the electrostatic interactions between the nanostructured chains and the walls of the hosts and also owed to the restricted geometry imposed on the aromatic rings. The force of these interactions is responsible for the formation of a polymeric backbone with a distinct structure to that observed in the free polymer, as observed for poly(aniline). We believe that the study of the structural pattern of the conducting polymer-clay nanocomposites will be decisive for their applications.

ACKNOWLEDGMENTS

This work is dedicated *in memorian* to my grandparents (Maria Ap. M. Morari and Humberto Morari). Many thanks to the National Synchrotron Light Laboratory (LNLS, Campinas-Brazil) for XANES measurements at the N K edge in the SGM beam line.

REFERENCES

[1] Liepins, R.; Ku, C. C. *In: Electrical Properties of Polymers* Inc: C. H. Verlag, Munich, Viena 1987, 256.

[2] Shirakawa, H.; Ikeda, S. *J. Poylm. Sci. Chem.* 1974, *12*, 929. Shirakawa, H.; Ikeda, S. *Polymer* 1971, *2*, 231.

[3] Chiang, C. K.; Fincher Jr, C. R.; Park, Y. W.; Heeger, A. J.; Shirakawa, H.; Louis, E. J.; MacDiarmid, A. G. *Phys. Rev. Lett.* 1977, *39*, 1098.

[4] Chiang, C. K.; Druy, M. A.; Gau, S. C.; Heeger, A. J.; Louis, E. J.; MacDiarmid, A. G. *J. Am. Chem. Soc.* 1978, *100*, 1013.

[5] Shirakawa, H.; Louis, E. J.; MacDiarmid, A. G.; Chiang, C. K.; Heeger, A. J. *J. Chem. Soc.: Chem. Commun.* 1977, 578.

[6] Shirakawa, H. *Angew. Chem. Int. Ed.* 2001, *40*, 2574.

[7] MacDiarmid, A. G. *Angew. Chem. Int. Ed.* 2001, *40*, 2581.

[8] Heeger, A. J. *Angew. Chem. Int. Ed.* 2001, *40*, 2591.

[9] Brédas, J. L.; Street, G. B. *Acc. Chem. Res.* 1985, 18, 309.

[10] Brédas, J. L.; Chance, R. R.; Silbey, R. *Mol. Cryst. Liq. Cryst.* 1981, *77*, 319. Brédas, J. L.; Chance, R. R.; Silbey, R. *Phys. Rev. B.: Condens. Matter.* 1982, *26*, 5843. Brédas, J. L.; Scott, J. C.; Yakushi, K.; Street, G. B. *Phys. Rev. B.: Condens. Matter.* 1984, *30*, 1023.

[11] MacDiarmid, A. G.; Epstein, A. J. *The polyanilines: a novel class of conducting polymers* In: Conducting polymers, emerging technologies, Technical Insights Inc.: New Jersey, 1989, 27.

[12] MacDiarmid, A. G.; Epstein, A. J. *Faraday Discuss. Chem. Soc.*, 1989, *88*, 317.

[13] MacDiarmid, A. G.; Chiang, J. C.; Richter, A. F.; Sonosiri, N. L. D. *In Conducting Polymers*, Alcácer, L., Ed.; Reidel Publications: Dordrecht 1987, 105.
[14] MacDiarmid, A. G.; Epstein, In *Frontiers of Polymers and Advanced Materials;* Prasad, P. N., Eds.; Plenum Press, New York, 1984, 251.
[15] Do Nascimento, G. M.; De Souza, M. A. Spectroscopy of Nanostructured Conducting Polymers, in: A. Eftekhari (Ed.), John Wiley and Sons, 2010, 341.
[16] Do Nascimento, G. M. Spectroscopy of Polyaniline Nanofibers, in: Ashok Kumar. (Org.). Nanofibers. Austria/Croacia: In-tech/Sciyo Book, 2010, 349.
[17] Syed, A. A.; Dineson, M. K. Talanta 1991, *38*, 815.
[18] Wei, Y.; Sun, Y.; Tang, X. *J. Phys. Chem.* 1989, 93, 4878.
[19] Sasaki, K.; Yaka, M.; Yano, J.; Kitani, A.; Kumai, A. *J. Electroanal. Chem.* 1986, *215*, 401.
[20] Mohilmer, D. M.; Adams, R. N.; Argersinger, W. J. *J. Am. Chem. Soc.* 1962, *84*, 3618.
[21] Bacon, J.; Adams, R. N. *J. Am. Chem. Soc.* 1968, *90*, 6596.
[22] Wawzonek, S.; MacIntyre, T. W. *J. Electrochem. Soc.* 1967, *114*, 1025.
[23] Geniés, E. M.; Boyle, A.; Lapkowski, M.; Tsintavis, C. *Synth. Met.* 1990, *36*, 139.
[24] Wei, Y.; Jang, G.-W.; Chan, Ch.-Ch.; Hsuen, K. F.; Hariharan, R.; Patel, S. A.; Whitecar, C. K. *J. Phys. Chem.* 1990, *94*, 7716.
[25] Wei, Y.; Tang, X.; Sun, Y.; Focke, W. W. *J. Polym. Sci.* 1989, *27*, 2385.
[26] Wei, Y.; Harirahan, R.; Patel, S. A. *Macromolecules* 1990, *23*, 758.
[27] Wei, Y.; Hsueh, K. F.; Jang, G. –W. *Polymer* 1994, *35*, 3572.
[28] Doriomedoff, M.; Cristofini, F. H.; De Surville, R.; Jozefowicz, M.; Yu, L. T.; Buvet, R. *J. Chim. Phys.* 1971, *68*, 1055.
[29] Yu, L. T.; Borredon, M. S.; Jozefowicz, M.; Belorgey, G.; Buvet, R. *J. Polym. Sci.* 1987, *10*, 2931.
[30] Bissessur, R.; DeGroot, D. C.; Schindler, J. L.; Kannewurf, C. R.; M. G. Kanatzidis, M. G. *J. Chem. Soc.: Chem. Comumm.* 1993, 687.
[31] Liu, Y.-J.; DeGroot, D. C.; Schindler, J. L.; Kannewurf, C. R.; Kanatzidis, M. G.; *J. Chem. Soc.: Chem. Commumm.* 1993, 593.
[32] Wu, C.-G.; DeGroot, D. C.; Marcy, H. O.; Schindler, J. L.; Kannewurf, C. R.; Liu, Y.-J.; Hirpo, W.; Kanatzidis, M. G. *Chem. Mater.* 1996, *8*, 1992.
[33] Chang, T.; Ho, S.; Chao, K. *J. Phys. Org. Chem.* 1994, *7*, 371.
[34] Kanatzidis, M. G.; Liu, Y. *Inorg. Chem.* 1993, *32*, 2989.
[35] Wu, C.-G.; DeGroot, D. C.; Marcy, H. O.; Schindler, J. L.; Kannewurf, C. R.; Bakas, T.; Papaefthymiou, V.; Hirpo, W.; Yesinowski, J. P.; Liu, Y.-J.; Kanatzidis *J. Am. Chem. Soc.* 1995, *117*, 9229.
[36] Moujahid, E. M.; Dubois, M.; Besse, J.-P.; Leroux, F. *Chem. Mater.* 2002, *14*, 3799.
[37] Wypych, F.; Seefeld, N.; Denicoló, I. *Quimica Nova* 1997, *20*, 356.
[38] Yariv, S.; Michaelian, K. H. "*Structure and Surface Acidity of Clay Minerals*" In *Organo-Clay Complexes and Interactíons;* Yariv, S.; Cross, H., Eds.; Marcel Dekker: New York, 2002, 1.
[39] Hauser, E. A.; Leggett, M. B. *J. Am. Chem. Soc.* 1940, *62*, 1811.
[40] Yariv, S.; Heller, L.; Safer, Z. *Israel J. Chem.* 1968, *6*, 741.
[41] Furukawa, T.; Brindley, G. W. *Clay Clay Miner.* 1973, *21*, 279.
[42] Cloos, P.; Morale, A.; Broers, C.; Badat, C. *Clay Miner.* 1979, *14*, 307.
[43] Soma, Y.; Soma, M. *Clay Miner.* 1988, *23*, 1.

[44] Soma, Y.; Soma, M. Harada, I. *Chem. Phys. Lett.* 1983, *94*, 475.
[45] Soma, Y.; Soma, M.; Harada, I. *J. Phys.Chem.* 1984, *88*, 3034.
[46] Soma, Y.; Soma, M.; Harada, I. *J. Phys. Chem.* 1985, *89*, 738.
[47] Soma, Y.; Soma, M. *Environmental Health Perspectives* 1989, *83*, 205.
[48] Mehrotra, V.; Giannelis, E. P. *Solid State Commun.* 1991, *77*, 155.
[49] Chang, Te-C.; Ho, S.-Y.; Chao, K.-J. *J. Chin. Chem. Soc.* 1992, *39*, 209.
[50] Wu, Q.; Xue, Z.; Qi, Z.; Hung, F. *Polymer* 2000, *41*, 2029.
[51] Wu, Q.; Xue, Z.; Qi, Z.; Wang, F. *Acta Polym. Sin.* 1999, *10*, 551.
[52] Biswas, M.; Ray, S. S. *J. Appl. Polym. Sci.* 2000, *77*, 2948.
[53] Biswas, M.; Ray, S. S. *J. Appl. Polym. Sci.* 2000, *77*, 2948.
[54] Lu, D.; Lu, S.-H.; Char, K.; Kim, J. *Macromol. Rapid Commun.* 2000, *21*, 16.
[55] Yeh, J.-M.; Liou, S.-J.; Lai, C.-Y.; Wu, P.-C.; Tsai, T.-Y. *Chem. Mater.* 2001, *13*, 1131.
[56] Zeng, Q. H.; Wang, D. Z.; Yu, A. B.; Lu, G. Q. *Nanotechnology* 2002, *13*, 549.
[57] Lee, D.; Char, K.; Lee, S. W.; Park, Y. W. *J. Mater. Chem.* 2003, *13*, 2942.
[58] Kim, B. H.; Jung, J. H.; Kim, J. W.; Choi, H. J.; Joo, J. *Synth. Met.* 2001, *117*, 115.
[59] Kim, B. H.; Jung, J. H.; Kim, J. W.; Choi, H. J.; Joo, J. *Synth. Met.* 2001, *121*, 1311.
[60] Kim, B. H.; Jung, J. H.; Joo, J.; Kim, J. W.; Choi, H. J. *J. Korean Phys. Soc.* 2000, *36*, 366.
[61] Inoue, H.; Yoneyama, H. *J. Electroanal. Chem.* 1987, *233*, 291.
[62] Orata, D.; Segar, F. *Reactive and Functional Polymers* 2000, *43*, 305.
[63] Feng, Bo.; Su, Y.; Sang, J.; Kang, K. *J. Mater. Sci. Lett.* 2001, *20*, 293.
[64] Chen, K. H.; Yang, S. M. *Synth. Met.* 2003, *135*, 151.
[65] Jara, P.; Justiniani, M.; Yutronic, N.; Sobrados, I. *J. Incl. Phenom.* 1998, *32*, 1.
[66] Do Nascimento, G. M.; Constantino, V. R. L.; Temperini, M. L. A. *Macromolecules* 2002, *35*, 7535.
[67] Do Nascimento, G. M.; Landers, R.; Constantino, V. R. L.; Temperini, M. L. A. *Macromolecules* 2004, *37*, 9373.
[68] Do Nascimento, G. M.; Constantino, V. R. L.; Temperini, M. L. A. *J. Phys. Chem. B* 2004, *108*, 5564.
[69] Do Nascimento, G. M.; Landers, R.; Constantino, V. R. L.; Temperini, M. L. A. *Polymer* 2006, *47*, 6131.
[70] Do Nascimento, G. M.; Temperini, M. L. A *Eur. Polym. J.* 2008, *44*, 3501.
[71] Do Nascimento, G. M.; Padilha, A. C. M.; Constantino, V. R. L.; Temperini, M. L. A. *Colloids Surf. A: Physicochem. Eng. Aspects* 2008, *318*, 245.
[72] Do Nascimento, G. M.; Temperini, M. L. A *J. Mol. Struct.* 2011, *1002*, 63.
[73] Do Nascimento, G. M.; Barbosa, P. S. M.; Constantino, V. R. L.; Temperini, M. L. A. *Colloids Surf. A: Physicochem. Eng. Aspects* 2006, *289*, 39.
[74] Hennig, C.; Hallmeier, K. H.; Szargan, R. *Synthetic Met.* 1998, 92, 161.
[75] Hennig, C.; Hallmeier, K. H.; Bach, A.; Bender, S.; Franke, R.; Hormes, J.; Szargan, R. *Spectrochim. Acta A* 1996, 52, 1079.
[76] Pavlychev, A. A.; Hallmeier, K. H.; Hennig, C.; Hennig, L.; Szargan, R. *Chem. Phys.* 1995, 201, 547.
[77] Francis, J. T.; Hitchcock, A. P. *J. Phys. Chem.* 1992, 96, 6598.
[78] Do Nascimento, G. M.; Izumi, C. M. S.; Constantino, V. R. L.; Temperini, M. L. A. *Activity report of Brazilian Synchrotron Light Laboratory* 2002-2003, 141.
[79] Do Nascimento, G. M.; Temperini, M. L. A. *Quim. Nova* 2006, *29*, 823.

[80] Trchova, M.; Syedenkova, I.; Konyushenko, E. N.; Stejskal, J.; Holler, P.; Ciric-Marjanovic, G. *J. Phys. Chem. B* 2006, *110*, 9461.
[81] Stejskal, J.; Sapurina, I.; Trchova, M.; Konyushenko, E. M.; Holler, P. *Polymer* 2006, *47*, 8253.
[82] Margaritondo, G. "*Applications of Synchrotron Light*" In Elements of Synchrotron Light for Biology, Chemistry and Medical Research, Oxford University Press, New York, 2002, 72-146.
[83] Tan K. L.; Tan, B. T. G.; Kang, E. T.; Neoh, K. G. *Phys. Rev.* B 1989, *39*, 8070.
[84] Zeng, X-R.; Ko, T.-M. *Polymer* 1998, *39*, 1187.
[85] Nakajima, T.; Harada, M.; Osawa, R.; Kawagoe, T.; Furukawa, Y.; Harada, I. *Macromolecules* 1989, *22*, 2644.
[86] Monkman, A. P.; Stevens, G. C.; Bloor, D.; *J. Phys. D: Appl. Phys.* 1991, *24*, 738.

In: Conducting Polymers
Editor: Luiz Carlos Pimentel Almeida
ISBN: 978-1-62618-119-9

Chapter 6

SYNTHESIS, ELECTRICAL PROPERTIES AND APPLICATIONS OF CONDUCTING POLYMER NANOTUBES AND NANOFIBERS

Yun-Ze Long[1,2,3,*], Bin Sun[1,2], Zhi-Ming Zhang[4], Hong-Di Zhang[1,2] and Chang-Zhi Gu[5]

[1]College of Physics, Qingdao University, Qingdao, China
[2]Key Laboratory of Photonics Materials and Technology in Universities of Shandong (Qingdao University), Qingdao , China
[3]State Key Laboratory Cultivation Base of New Fiber Materials and Modern Textile, Qingdao University, Qingdao, China
[4]College of Chemistry and Chemical Engineering, Ocean University of China, Qingdao, China
[5]Institute of Physics, Chinese Academy of Sciences, Beijing, China

ABSTRACT

This chapter briefly summarizes preparation methods, electrical properties, and potential applications of conducting polymers such as polyaniline (PANI), polypyrrole (PPY) and poly(3,4-ethylenedioxythiophene) (PEDOT) nanotubes/fibers based on our results and some important contributions of other groups. The synthesis approaches include template-free self-assembly, hard physical template-guided synthesis, and electrospinning, *etc*.

Particularly, the size-dependent electrical properties (*e.g*., electrical conductivity, current-voltage (*I-V*) characteristics, magnetoresistance, and nanocontact resistance) of individual nanotubes/fibers are presented. Several potential applications of these nanotubes/fibers in drug release, neural interfaces, field effect transistors, nano-diodes, gas sensors, anticorrosion and hydrogen storage are also discussed.

* E-mail address: yunze.long@163.com.

1. INTRODUCTION

Conducting polymers (CPs) such as polyaniline (PANI), polypyrrole (PPY), and poly(3,4-ethylene dioxythiophene) (PEDOT) are adopted as promising organic semiconductors due to their adjustable electrical conductivities from 10^{-10} to 10^4 S cm^{-1} upon doping, reversible doping/dedoping process, controllable chemical/electrochemical properties and their processability. CPs have potential applications in electronic and optoelectronic devices, chemical and bio-sensors, anti-static and electromagnetic shielding materials, energy harvesting and storage devices, artificial muscles and biomedical materials, *etc.* [1-5] Due to interesting physical/chemical properties and useful applications, one-dimensional (1D) nanostructures have been extensively explored in the last 20 years, including carbon nanotubes, inorganic semiconductor and metallic nanotubes/wires, and conducting polymer nanotubes/fibers, *etc.* [4-6] For example, it is reported that 1D fibrillar morphology can significantly improve the performance of CPs in many conventional applications due to large specific surface area. This leads to faster and more responsive chemical sensors and ultra-fast non-volatile memory devices.

CP nanotubes and nanofibers can be prepared by a variety of approaches like electrospinning, nanoprinting, hard physical template-guided synthesis and soft chemical template synthesis (*e.g.*, interfacial polymerization, template-free method, dilute polymerization, reverse emulsion polymerization). [1-5] Electrical and optical properties of CP nanotubes/fibers have also been studied recently, showing unusual physical and chemical behavior due to the nanosize effects. In order to keep up with this highly active field of nanoscale conducting polymers, therefore, in this chapter, we present a brief summary of recent advances in the synthesis approaches, electronic transport properties and applications of 1D nanostructures of conducting PANI, PPY and PEDOT based on our results and some important contributions of other groups.

2. PREPARATION OF CONDUCTING POLYMER NANOTUBES/FIBERS

CP nanotubes/fibers can be prepared by a variety of approaches such as hard template synthesis [7], interfacial polymerization [8], template-free method [9], dilute polymerization [10], reverse emulsion polymerization [11], electrospinning [1], and lithography technique [12], *etc.* Among the various approaches established, each of the methods has advantages while weakness also exists. In this section, based on our research experience, we only introduce template-free method, hard physical template-guided synthesis and electrospinning, respectively.

2.1. Template-Free Self-Assembly

2.1.1. Pure Conducting Polymer Nanostructures

The template-free method developed by Wan et al. [9, 13, 14] is a simple and versatile self-assembly method without an external hard template. By controlling synthesis conditions such as temperature and molar ratio of monomer to dopants, PANI and PPY nanotubes/wires can be prepared by *in-situ* doping polymerization in the presence of protonic acids as dopants. For example, Zhang et al. reported that PANI and PPY nanostructures (*e.g.*, nanotubes, nanorods or nanofibers) with average diameters of 80-340 nm were synthesized by this template-free self-assembly method in the presence of inorganic acids (*e.g.*, HCl, H_2SO_4, HBF_4, and H_3PO_4) [13, 14] or organic acids (*e.g.*, naphthalene sulfonic acids [15], dicarboxylic acids [16] and poly(3-thiopheneacetic acid) [17]) as dopants, as shown in Figure 1a-c. It is found that the formation yield, morphology (hollow or solid), size, crystalline and electrical properties of the PANI nanostructures are affected by the dopants. Besides, the oxidants such as ammonium persulfate (APS) and $FeCl_3{\cdot}6H_2O$ also have strong influence on the diameter, crystallinity and electrical conductivity of the PANI nanostructures. For instance, highly crystalline PANI nanofibers with 16-23 nm in diameter were successfully prepared by using $FeCl_3{\cdot}6H_2O$ as the oxidant (Figure 1d), the diameter was much smaller than that of the nanostructures oxidized by APS (130 nm). [17, 18] It is proposed that the low redox potential of $FeCl_3$ results in thin nanofibers with ~20 nm in average diameter. [18]

The self-assembled formation mechanism in this approach has also been extensively explored. It is suggested that the micelles were formed by the self-assembly of dopant molecule and/or their monomer salt into a micro/nanostructural intermediate as supermolecular template in the process of forming tubes/wires. [13, 14] Up to now, a variety of PANI micro/nanostructures such as micro/nanotubes [13], nanowires/fibers [18], nanotube junctions and dendrites [15, 20], flower-like structured nanofibers [21] and hollow microspheres [14, 22], have been prepared by the template-free method. For example, the self-assembled PANI nanotubes may aggregate to form a dendritic morphology when the polymerization is performed at a static state possibly resulting from polymer chain interactions including π-π interactions, hydrogen and ionic bonds. [15] In addition, HCl-doped PANI nanofibers were self-assembled by adding inorganic salts (*e.g.*, LiCl, NaCl, $MgCl_2$ and $AlCl_3$) as the additives and aggregated to form chrysanthemum flower-like microstructures.

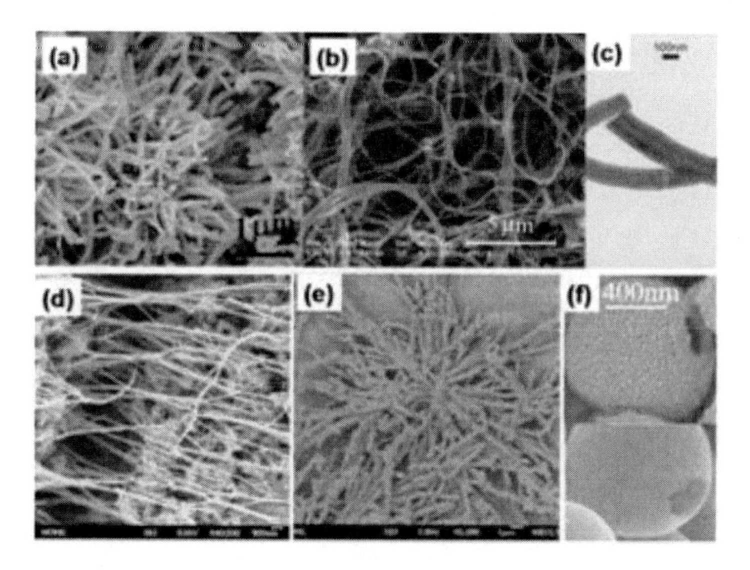

Figure 1. PANI and PPY nanostructures prepared by template-free self-assembly method: (a) PANI nanotubes [19]; (b) PPY nanotubes [19]; (c) TEM image of PANI nanotubes/wires [15]; (d) Ultrathin PANI nanofibers [18]; (e) PANI nanotube dendrites [20]; (f) PANI microspheres [22].

The resultant PANI nanostructures have not only thin diameter (17-23 nm), but also high conductivity (~3.7 S cm^{-1}) at room temperature and high crystallinity due to a low rate of accretion or elongation process of the nanofibers by adding inorganic salts.

2.1.2. Conducting Polymer Nanocomposites

In addition to pure CP nanostructures, various CP nanocomposites have been fabricated by this template-free method. Namely, through using a functional dopant or oxidant (*e.g.*, Fe_3O_4, γ-Fe_2O_3 and TiO_2 nanoparticles, azobenzene sulfuric acid, and $FeCl_3$, *etc.*), (electrical, magnetic, optical, *etc.*) multi-functionalized polymer micro/nanostructures synthesized by this approach can be achieved. [23] For example, PANI nanofibers and nanotubes (180-200 nm in diameter) containing Fe_3O_4 nanoparticles (diameter ~10 nm) and PANI-coated γ-Fe_2O_3 nanoneedles (40-80 nm in diameter and 500-600 nm in length) were prepared by *in situ* doping polymerization in the presence of H_3PO_4 as a dopant. [24-27] PANI/γ-Fe_2O_3 composite nanofibers with diameters of ~20 nm also can be prepared by a chemical one-step method. In this approach, $FeCl_3$ acts as the oxidant either for polymerization of aniline or for preparation of γ-Fe_2O_3 magnets. Besides, it also provides protons produced by the hydrolysis process for doping PANI. The composite nanofibers have a high conductivity and super-paramagnetic properties at room temperature [28]. Particularly, Zhang et al. [29] recently reported an improved chemical one-step method to prepare highly conductive and magnetic PPY/γ-Fe_2O_3 nanospheres with ~80 nm in diameter. In the reaction process involved, $FeCl_3$ acts as an oxidant for the polymerization of pyrrole and as a source of F^{III} for the formation of γ-Fe_2O_3, which also requires the initial presence of Fe^{II}, provided by the addition of $FeCl_2$. The resulting electromagnetic PPY/γ-Fe_2O_3 nanospheres show maximum conductivity of 64.4 S cm^{-1} and saturation magnetization of 4.85 Am^2 kg^{-1}.

2.2. Hard Physical Template-Guided Synthesis

The template-guided synthesis proposed by Martin et al. [7] is an effective technique to synthesize arrays of aligned metal, semiconductor, and polymer micro-/nanotubes or wires with controllable length and diameter within porous materials. Among the various templates, robust porous anodic aluminum oxide and polycarbonate track-etched membrane templates have been demonstrated to be one of the most promising and most commonly used nanoporous materials. By now, PANI [7], PPY [7], and PEDOT [30] nanotubes/wires have been chemically or electrochemically synthesized inside the pores of these membranes, as shown in Figure 2a-b. Their diameters usually range from several tens to several hundreds nanometers. In addition, CdS-PPY heterojunction nanowires [31], multi-segmented Au-PEDOT-Au and Au-PEDOT-PPY-Au nanowires [32], MnO_2/PEDOT [33] and Ni/poly(p-phenylene vinylene) (PPV) [34] coaxial nanowires have also been prepared by the hard template method. The disadvantage of this method is that a post-synthesis process is needed in order to remove the template. However, the Al_2O_3 template can be removed by HF or NaOH solution, and polycarbonate template can be removed by dissolution with a flow of dichloromethane.

Besides these hard templates with channels inside pores, many kinds of pre-existing nanostructures can serve as seeds or templates to synthesize CP nanostructures. For example, V_2O_5 nanofibers [35], MnO_2 nanowires [36], Mn_2O_3 nanofibers [37], poly(styrene-*block*-2-

vinylpyridine) diblock copolymers [38], DNA [39] and tobacco mosaic virus [40] have been used to fabricate PANI, PPY and PEDOT nanofibers.

In addition, aggregated, hollow octahedrons or microspheres of PANI [41, 42] could be synthesized by using octahedral Cu_2O crystals as a template in the presence of H_3PO_4 and ammonium persulfate as a dopant and an oxidant, respectively. PANI/TiO_2 micro/ nanospheres were also prepared successfully by using octahedral Cu_2O as the template. The TiO_2 nanoparticles are uniformly dispersed in PANI. [43] One advantage of these metal oxide templates such as Cu_2O nanocrystals, V_2O_5 nanofibers, MnO_2 nanowires and Mn_2O_3 nanofibers is that they do not need to be removed after polymerization because they can react with the oxidant to form a soluble salt during the polymerization process, and even can serve as oxidant at the same time [35-37, 41].

Furthermore, CP nanostructures can also be prepared by *in-situ* solution or vapor deposition polymerization on the surface of the electrospun polymer nanofibers, which provide a robust and stable template [1, 44, 45]. This approach can keep the porous and fibrous structure of the electrospun nanofibers, and in some cases the template fibers can be removed to obtain pure CP tubes. For instance, Abidian et al. [46-49] recently fabricated PEDOT and PPY nanotubes from biodegradable electrospun nanofibers for controlled drug release and neural interface applications. Long et al. [45] prepared random or aligned poly(methyl methacrylate) (PMMA), polyvinylidene fluoride (PVDF) and polystyrene (PS) nanfibers by electrospinning, then the nanofiber templates were immersed in an aqueous solution containing aniline, ammonium peroxodisulfate as oxidant, and 5-sulfosalicylic acid as dopant. PANI nanostructures were simultaneously deposited on the template surface to form core-shell composite fibers (Figure 2c-d). The nano-branched coaxial PANI fibers show a higher conductivity of about 21 S cm^{-1} at room temperature, and response significantly to low concentration (80 ppb) of NH_3 due to a large specific surface area of nano-branched nanostructure. The sensitivity shows good linear relationship to the NH_3 concentration of ppm level.

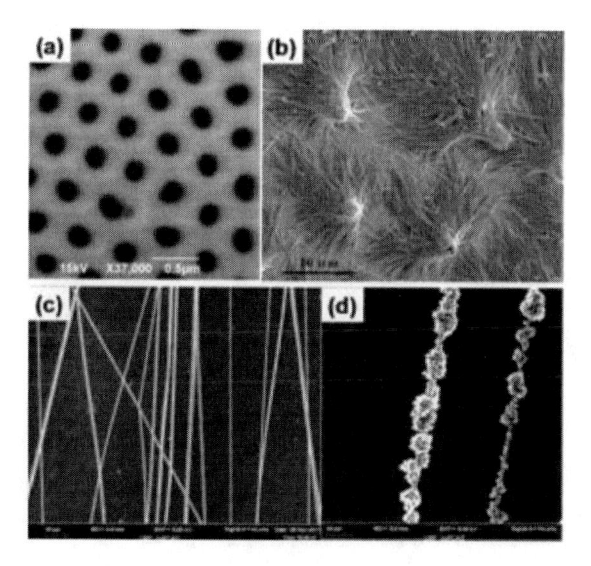

Figure 2. SEM images of (a) porous anodic aluminum oxide template, (b) template-synthesized PEDOT nanowires (polycarbonate track-etched membrane template has been removed) [30]; (c) electrospun PVDF fibers as template, (d) nano-branched PANI grown on the surface of electrospun PVDF fibers.

2.3. Electrospinning

Electrospinning is an efficient, highly versatile and promising technique for fabricating 1D continuous micro/nanoscale fibers [50]. The principle of electrospinning is very simple. Under the action of electric field forces, a charged jet of polymer solution ejects from Taylor cone at the tip of spinneret, and undergoes an instability and elongation/splitting process to become very long and thin. Meanwhile, the solvent evaporates, and ultrafine fibers are randomly deposited on the collector [50-57], as shown in Figure 3a. Generally, the spinneret tip-to-collector distance is usually in the range of 8-20 cm, and a dc high voltage in the range of 7-25 kV is necessary to generate the electrospinning. The principle of electrospinning was first studied by Formhals in the 1930s [51]. And this technology was regained much attention in the 1990s [52]. By now, numerous ultrathin fibers originated from polymer, metal and ceramic have been prepared by electrospinning, and their potential applications in optoelectronics, sensors, catalysis, fiber reinforcement, tissue engineering, drug delivery, *etc.* have also been extensively explored [50, 58-60].

In principle, CPs should be electrospun directly into ultrathin fibers. However, due to the limitations on molecular weight and solubility unsuitable for electrospinning, up to date, only a few conducting polymer solutions with appreciate viscosity have been electrospun directly, including PANI doped with sulfuric acid [1, 52, 61], 2-acrylamido-2-methyl-propane sulfonic acid (AMPSA) [62] or camphorsulfonic acid (CSA) [63] and PPY doped with dodecylbenzene sulfonic acid (DBSA) [64]. For instance, ranging from 10.6 to 19.1 wt%, various concentrations of PANI dissolved in hot sulfuric acid solution could be electrospun into fibers and collected on a coagulation bath, which was made of dilute sulfuric acid ranging from 0 to 30 wt%. Homogeneous PANI fibers with a diameter of 370 nm and a high conductivity of 52.9 S cm^{-1} were prepared [61].

In addition, a popular way is to electrospin PANI and PPY through blending with other electrospinnable polymers such as polyethylene oxide (PEO), PMMA and polystyrene [1].

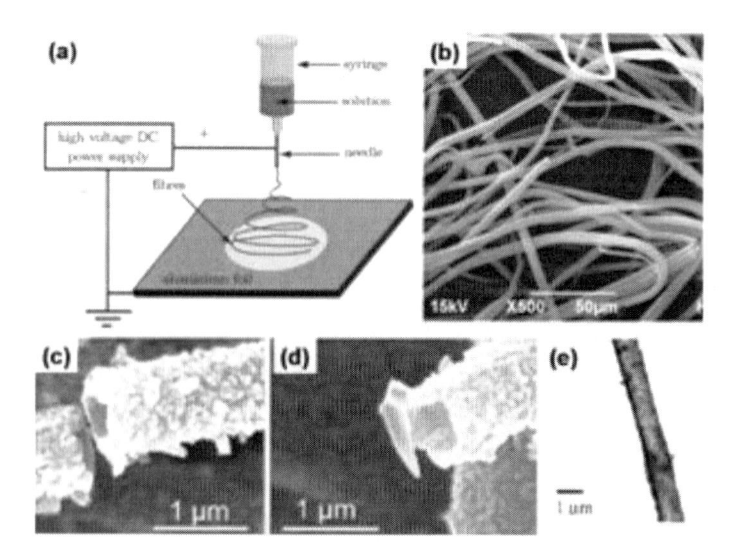

Figure 3. (a) Schematic illustration of electrospinning [56]; (b) SEM image of electrospun PANI/polystyrene composite fibers; (c-d) SEM images of PANI tubes after removal of core fibers [67]; (e) TEM image of PANI tubes [66].

Although the electrospinnable polymers may assist with formation of conducting polymer fibers and also act as a support material for the ultrathin fibers, they thus decrease the conductivity of the as-spun fibers remarkably. For instance, the conductivity is about 0.1 S cm^{-1} when electrospun 50 wt% PANI doped with CSA blended with PEO, dissolved in chloroform [1]. By changing the ratio of PPY to PEO in the blend (25-70 wt%), electrical conductivity (10^{-8}-10^{-5} S cm^{-1}) and average diameter (200-280 nm) of resultant PPY/PEO nanofibers could be controlled [65]. Figure 3b shows the SEM image of electrospun PANI/polystyrene (2:3 by weight) composite fibers by our group.

Electrospun fiber template-assisted synthesis is another efficient strategy, namely, CPs are *in situ* solution or vapor deposition polymerization on the surface of the electrospun nanofiber templates [1, 37, 44, 45, 66]. For example, PANI/poly(*L*-lactide) (PANI/PLA) coaxial fibers were prepared by *in situ* solution polymerization of aniline on the PLA fibers, and the coaxial fibers could be further converted into long PANI tubes after thermal removal of core PLA fibers [66, 67], as shown in Figure 3c-e. In addition, some other approaches such as coaxial electrospinning, electrospinning of CP monomers with subsequent *in situ* polymerization [68] or electrospinning of CP precursors with subsequently heat/solid-state oxidative crosslinking treatment [69] were also reported.

3. Electrical Properties of Individual Conducting Polymer Nanotubes/Fibers

3.1. Electrical Conductivity Measurement

In order to fulfill the potential applications of CP nanotubes and fibers, it is necessary to understand the electronic transport properties of single polymer tubes/fibers. The electrical characterization of single CP nanotubes/fibers has made significant progress during the last decade.

There are several strategies for measuring electrical conductivity: A) The conductivity measurement of single PANI or PPY tube/fiber can be achieved based on a conductive tip of an atom-force microscope (AFM) [70] or two metal microprobes, as shown in Figure 4a. B) A common approach is generally realized by dispersing (or spinning) nanotubes/fibers on patterned micro- or nano-electrodes prepared by photo-lithography or electron-beam lithography, followed by the subsequent searching of nanotubes/fibers just lying on the two or four electrodes only [71] (Figure 4b) . C) Electron- and/or focused-ion beam assisted deposition technique has been employed to attach metal microleads on isolated nanotubes/fibers directly [72], as shown in Figure 4c. D) There are also reports demonstracting a facile technique, *in situ* polymerization and measurement of polymer nanowire arrays between patterned electrodes in channels [73]. All these recent investigations contribute significantly to identifying and understanding the specific electrical behaviour of conjugated polymer nanotubes/fibers in comparison to the bulk materials. For example, the electrical conductivities of compressed pellets made of PANI and PPY nanotubes/fibers prepared by the template-free method are measured in the range of 10^{-2}-10^{0} S cm^{-1} [74, 75]. However, the conductivities of single PANI and PPY nanotube can reach ~30 S cm^{-1} and 70 S cm^{-1}, separately, possibly due to elimination of tube-tube contact resistance [72, 76, 77].

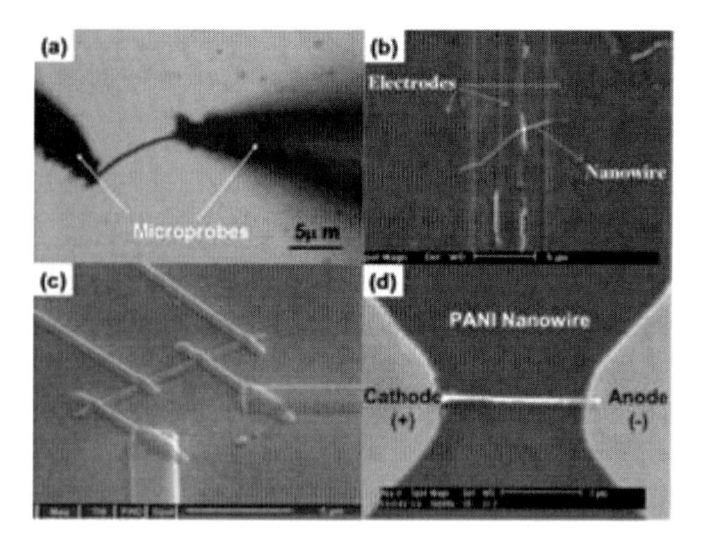

Figure 4. Electrical conductivity measurement of single PANI tube/wire by (a) using two conductive metal microprobes; (b) dispersing tubes/wires on pre-patterned microelectrodes [5]; (c) focused-ion beam assisted deposition of Pt microleads on isolated tube/wire; (d) in situ polymerization and measurement of polymer nanowire between pre-patterned electrodes [73].

3.2. Diameter Dependent Electrical Conductivity

The dependence of room-temperature conductivity on the diameter of the polymer nanotubes/fibers has been widely reported [7, 30]. It was found that the room-temperature conductivities of nanotubes/wires of conducting PPY, PANI and polythiophene derivatives can increase from 10^{-1}-10^{0} to about 10^{3} S cm^{-1} with the decrease of their outer diameters from 1500 to 35 nm. The mechanism responsible for this effect has been ascribed to the enhancement of molecular and supermolecular ordering with the alignment of the polymer chains, as confirmed by studies on polarized infrared absorption spectroscopy and x-ray diffraction [7]. For PEDOT nanowires prepared by the hard template method, the room-temperature conductivities of the nanowires with diameters of 190, 95-100, 35-40, and 20-25 nm are about 11, 30-50, 470-520, and 390-550 S cm^{-1}, respectively, as shown in Figure 5a [78]. For PPY nanotubes prepared by the template-free method, it was found that the PPY tubes with a 560-400 nm outer diameter are poorly conductive and the room-temperature conductivity is only 0.13-0.29 S cm^{-1}. When the outer diameter decreased to 130 nm, the conductivity of the single nanotube increased to 73 S cm^{-1} [77]. Such conductivity dependence on diameter indicates that the polymer tubes/wires prepared by different methods may have similar structural characteristic: the smaller the diameter, the better the polymer chain ordering.

In addition, the resistivity $\rho(T)$ was measured over a wide range of temperatures for the template-free self-assembled PANI and PPY nanotubes/wires [76, 77, 79, 80] and hard template-synthesized PEDOT nanowires with different diameters [78, 81, 82]. It is reflected by the evolution of the corresponding resistivity ratio ρ(10K)/ρ(300K), a useful empirical parameter for quantifying the extent of disorder. The resistivity ratio is found to be 8.6×10^{4} for the 190 nm nanowire, dropping to 15-20 and 1.4 for the 100 nm and 35 nm nanowires, respectively. For the 30 and 25 nm nanowires, it jumps to 2.1×10^{4} and 5×10^{5}.

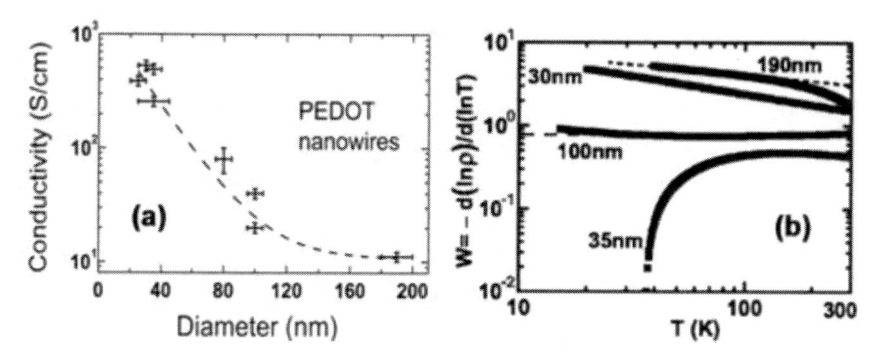

Figure 5. (a) Diameter dependence of room-temperature conductivity of single PEDOT nanowires prepared by hard template-guided synthesis. (b) Log-log plot of the reduced activation energy $W(T)$ for PEDOT nanowires with different diameters [78].

Further insights into the characteristic behavior of $\rho(T)$ and the three conduction regimes can be obtained by considering the reduced activation energy $W(T)$ defined as $W(T)=-d(\ln\rho(T))/d(\ln T)$. In the insulating regime of metal-insulator transition, $W(T)$ has a negative temperature coefficient; in the critical regime, $W(T)$ is nearly temperature independent; and in the metallic regime, $W(T)$ has a positive temperature coefficient. Figure 5b shows the log-log plot of the temperature variation of $W(T)$. The temperature dependences of the four-probe resistivity demonstrate that the PEDOT nanowires with diameter of 190, 95-100, 35-40 and 20-25 nm are lying in the insulating, critical, metallic and insulating regimes of metal-insulator transition, respectively [78]. Furthermore, it is found that the $\rho(T)$ variation of the 190 nm nanowire verify a 3D Mott variable-range hopping law below 120 K. The 100-nm PEDOT nanowire (σ_{RT} ~ 50 S cm^{-1}) falls in the critical regime with $\rho(T)\propto T^{-0.78}$. The 35-40 nm template-prepared PEDOT nanowire (σ_{RT} ~ 490 S/cm) displays a metal-insulator transition at about 32 K, indicating that the nanowire is lying in the metallic regime of the metal-insulator transition [78]. However, for a PEDOT nanowire with a diameter of 20-25 nm, and high room-temperature conductivity (390-550 S cm^{-1}), a very strong temperature dependence ($\rho(10K)/\rho(300K)$ ~10^5) and an insulating behavior at low temperature have been measured. This is possibly due to a confining effect since the value of the diameter (20-25 nm) becomes equal or close to the localization length of electrons (~20 nm).

In such a case, localization of electrons induced by Coulomb interaction or small disorder must be taken into account for explaining this insulating behavior at low temperatures. This result indicates that the electronic transport properties of individual polymer nanowires could be tuned by controlling the diameter.

3.3. Current-Voltage (*I-V*) Characteristics

The current-voltage (*I-V*) characteristics of individual polymer nanowires/tubes have been explored extensively in the past ten years [71, 77, 80, 81]. With lowering temperature, a transition from linear to nonlinear I-V characteristics is usually observed (Figure 6a), and a clear zero bias anomaly (i.e., Coulomb gap-like structure) gradually appears on the differential conductance (d*I*/d*V*) curves (Figure 6b). Similar transition has also been reported in other polymeric and inorganic nanotubes/fibers [83, 84].

Up to now, several theoretical models such as the space-charge limited current [85], fluctuation-induced tunneling [86], Coulomb gap [77], Coulomb blockade [83,87], Lüttinger liquid [88], Wigner crystal [89] models have been considered to explain the conduction mechanism of quasi-one dimensional nanofibers. For example, Kaiser et al. [86] recently proposed a generic expression (extended fluctuation-induced tunneling and thermal excitation model) for the nonlinear *I-V* curves based on numerical calculations for metallic conduction interrupted by small barriers: $G=I/V=G_0\cdot\exp(V/V_0)/\{1+h\ [\exp(V/V_0)-1]\}$, where G_0 is the temperature- dependent zero-bias conductance; V_0 is the voltage scale factor, which strongly depends on the barrier energy; $h=G_0/G_h$ (where $h<1$), which yields a decrease of G below the exponential increase at higher values of voltage V (the saturation of G at a high-field value, G_h, as $V\rightarrow\infty$ given by the expression is an extrapolation of the calculations). Kaiser et al. showed that this expression could give a very good description of the observed nonlinearities in polyacetylene nanofibers, vanadium pentoxide nanofibers, *etc.* [86] Here, one can wonder whether the Kaiser expression is still appropriated to fit the nonlinear *I-V* characteristics of individual polymer nanowires/tubes if the Coulomb interactions are strong and should be taken into account.

The Kaiser expression has been used by Long et al. [80, 90] to numerically calculate the *I-V* characteristics of individual PANI nanotube, PPY nanotubes, PEDOT nanowires, CdS nanorope, and $K_{0.27}MnO_2$ nanowire. The fitting results indicate that except at low temperatures and low bias voltages, the Kaiser generic expression can give a good description of the *I-V* characteristics of individual nanotubes/wires (Figure 7a). It indicates that the Kaiser expression (extending the Sheng model or fluctuation-induced tunneling and thermal excitation model) takes into account the microstructure feature and the conduction feature of polymer nanofibers (quasi-1D metallic conduction interrupted by small barriers). However, apparent deviation from the Kaiser expression has been evidenced in the low-temperature *I-V* curves (Figure 7b).

Particularly, we compare the values of zero-bias conductance determined from the fitting parameter (G_0) with that determined from experimental measurements (G_0, obtained from the *I-V* curve or the differential conductance). As shown in Figure 7c, the fitting parameter G_0 decreases smoothly with temperature lowering, but the experimental value G_0' sharply decreases below 80-100 K and deviates from G_0, although it becomes superposable to G_0 for temperature equal to or larger than 100 K.

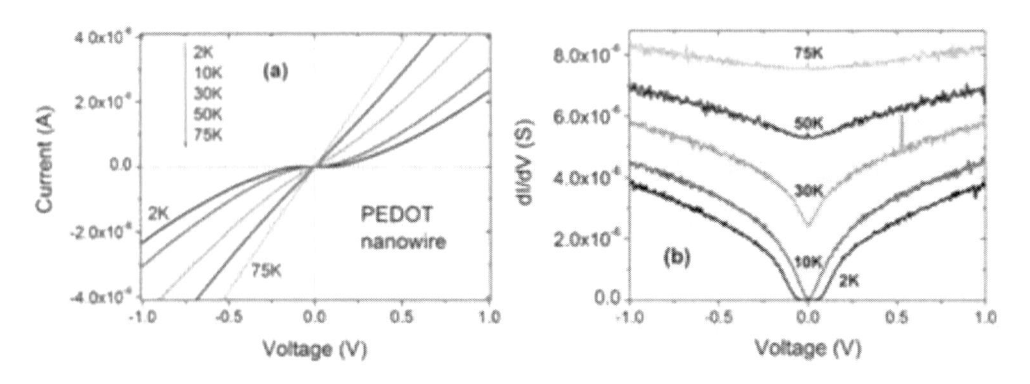

Figure 6. (a) *I-V* curves and (b) the corresponding differential conductance (d*I*/d*V*) curves of a single PEDOT nanowire at low temperatures.

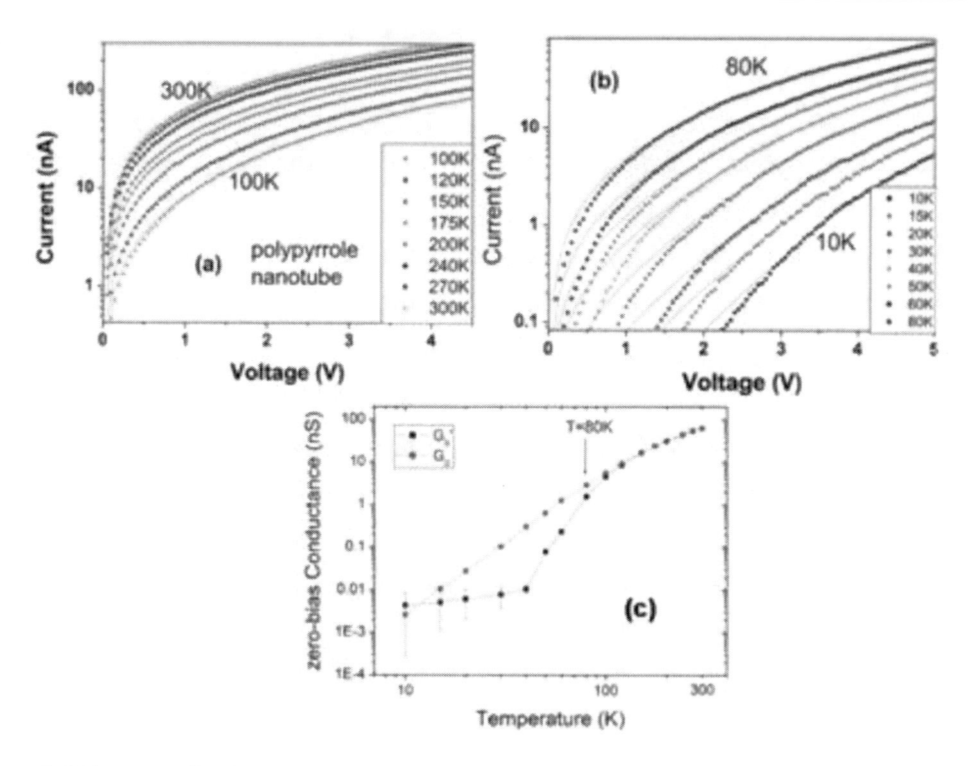

Figure 7. I-V curves of a single PPY nanotube, also showing the fits by Kaiser expression, at temperatures (a) ranging from 300 K to 100 K and (b) from 80 K to 10 K. (c) Variation of the zero-bias conductance with temperature varying, where G_0 is determined from the fitting data and G_0' is determined from the experimental data shown in (a-b) [80].

We propose that one possible reason for the deviation is that the Kaiser expression does not include the contributions from the Coulomb-gap occurring in density of states near Fermi level and/or enhanced Coulomb interactions due to nanosize effects, which become important at low temperatures and voltages. [77, 87-89]

3.4. Nanocontact Resistance

The contact resistance is often encountered when we study electronic transport in an individual CP nanotube/wire or nanofiber-based nanodevices. In this section, we discuss two kinds of nanocontact resistance: between polymer tube/fiber and metal microlead, and between two crossed polymer tubes/fibers. The nanocontact resistance between a polymer PEDOT nanowire and a Pt microlead fabricated by focused-ion-beam deposition (Figure 8a) has also been studied by Long et al. [91].

It was found that the nanocontact resistance $R_{con} \approx R_{2P}$-R_{4P} (where R_{4P} is the measured four-probe resistance and R_{2P} is the measured two-probe resistance of the same nanowire) is in the magnitude of 10 kΩ at room temperature and can reach 10 MΩ at lower temperature, which, in some case, is comparable to the intrinsic resistance of the PEDOT nanowires. On the one hand, for a semiconducting polymer nanowire in the insulating regime of the metal-insulator transition, the four-probe resistance is quite close to the two-probe resistance because the contact resistance is much smaller than the intrinsic resistance of the polymer nanowire [91].

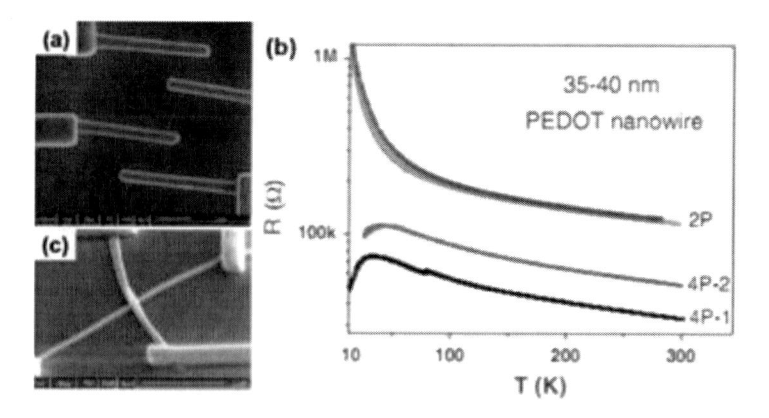

Figure 8. (a) SEM image of a single PEDOT nanowire and the attached four Pt microleads. Electrical resistance of the middle section of the nanowire could be measured by a two-probe method or a standard four-probe method. (b) Temperature dependence of four-probe (4P) and two-probe (2P) resistances for the single PEDOT nanowire [91]. (c) SEM image showing two crossed PEDOT nanowires and their attached Pt microleads [92].

On the other hand, for a nanowire that falls in the metallic regime of the metal-insulator transition (for example, the 35 nm PEDOT nanowire as shown in Figure 5b [91]), the metallic nature of the measured polymer fibers could be overshadowed in the two-probe measurement (Figure 8b) although the nanowire shows a relatively high electrical conductivity at room temperature (390-450 S cm^{-1}), because the nanocontact resistance is much larger than the intrinsic resistance of the nanowire especially at low temperatures. Nanocontact resistance has also been widely reported in carbon nanotubes and other inorganic semiconductor nanowires. So, in order to explore the intrinsic electronic transport properties of individual nanowires, especially in the case of metallic nanowires, the four-probe electrical measurement is required.

The nanocontact resistance between two crossed polymer nanotubes/wires has been studied by Long et al. [76,92]. It was found that the inter-tubular junction resistance of two crossed PANI nanotubes is very large, about 500 kΩ at room temperature, which is nearly 16 times larger than the intra-tube (intrinsic) resistance of an individual PANI nanotube (about 30 kΩ for 5-μm long nanotube) [76].

This result explains straightforwardly why an individual APNI nanotube has a much higher conductivity (30.5 S cm^{-1}) than that of a pellet of PANI nanotubes where the measured resistance is dominated by the inter-fibrillar resistance (0.03 S cm^{-1}). For crossed PEDOT nanowires (Figure 8c), the junction resistance (between the two nanowires) at room temperature can vary from 885-1383 kΩ for one sample to 370-460 MΩ for another sample, which is respectively comparable to or much larger than the intrinsic resistance of the PEDOT nanowires.

In addition, the contact resistance shows a stronger temperature dependence (R(72K)/R(300K) is about 120-141) [92]. It should be noted that the nano-junction resistance is comparable to the intrinsic resistance of polymer nanotube/wire and shows large sample-to-sample variations. Possible reasons could be attributed to the contamination of the nanotube/wire surfaces from solvent impurities or water adsorption, the variation of the junction area between the two nanotubes/wires, and the self-formation conditions of the junction.

3.5. Magnetoresistance

Since the individual polymer tubes/fibers have a much smaller size (*e.g.*, size effect induced by structural/electronic changes when synthesized in nanostructures), their magnetoresistance (MR) is found to be different from that of the bulk films and pellets. For example, Long et al. reported that the MR of single PANI nanotube and single PEDOT nanowire is positive below 10 K and increases as H^2 up to 9 T. [19, 77] Typically, a positive MR is expected for hopping conduction, because the applied magnetic field contracts the overlap of the localized state wavefunctions and it thus results in an increase of the average hopping length. It corresponds to a positive MR at sufficiently low temperatures. The weak-field *MR* with strong temperature dependence can be expressed as $\ln(R(H)/R(0)) \propto H^2 \cdot T^{-3/4}$. It is interesting found that the MR of a single nanotube/wire is much smaller than that of the nanotube/wire pellet at 9 T: $MR=[R(H)-R(0)]/R(0) < 5\%$ (2 K) for the single PANI nanotube and PEDOT nanowire, and $MR \sim 90\%$ (3 K) for the PANI nanotube pellet [19]. In addition, when the temperature increases, the MR of the single nanotube/wire becomes smaller and close to zero. No evident transition from a positive *MR* to negative one was observed. In contrast to that of single nanotube/wire, pellets of PANI and PPY nanotubes/wires show a relatively large positive *MR* at low temperatures. With temperature increasing, there is a transition from positive *MR* to small negative *MR* at about 60 K. The results indicate that the *MR* in the bulk pellet samples made of polymer nanotubes/wires is dominated by a random network of inter-fibril contacts. The small *MR* effect in individual polymer nanotube/wire has been confirmed in other samples. For example, a single Au/PANI microfiber shows a small positive *MR* (MR<4.1%) below 6 K [93]. The reason for this weak MR effect in individual polymer nanotube/wire is possibly due to the elimination of inter-nanotube/fiber contacts, small size and, relatively high conductivity of individual polymer nanotube/fiber.

4. Potential Applications of Conducting Polymer Nanotubes/Fibers

Presenting some attractive properties associated with conventional polymers, such as adjustable conductivity and ease of synthesis and processing, CPs have drawn much interest as promising materials in many fields including biomedical, microelectronics, sensors, and energy storage *etc*. In this section, we will in brief introduce some advances on applications of CPs in recent years.

4.1. Controlled Drug Release

Drug delivery devices have flourished during the last few decades and are extensively used in various kinds of treatments. And CP-based devices have been investigated to serve as electrically controlled drug delivery devices inside the body. Herein, one major challenge is to develop a drug delivery system that allows strict control of the ON/OFF state. And second, such a device must be able to deliver the drug of interest at doses that are required to obtain the therapeutic effect.

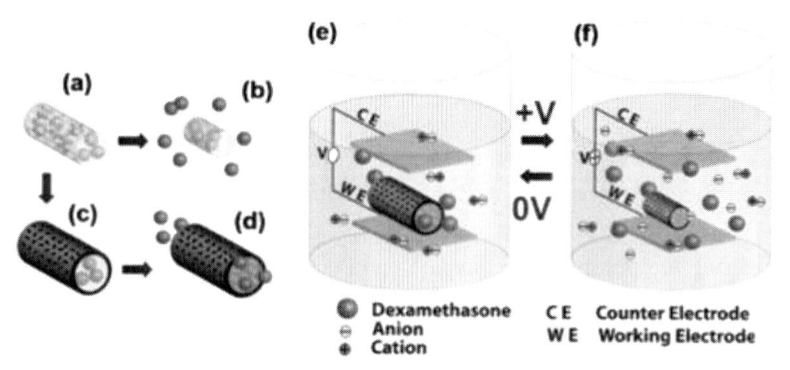

Figure 9. Schematic illustration of controlled release of dexamethasone: (a) dexamethasone-loaded electrospun biodegradable PLGA fibers; (b) hydrolytic degradation of PLGA fibers leading to release of the drug; (c) electrochemical deposition of PEDOT around the PLGA/drug composite fiber slows down the release of the drug (d); (e) PEDOT/PLGA/drug fibers in a neutral electrical condition; (f) External electrical stimulation controls the release of dexamethasone from the PEDOT nanotubes due to contraction or expansion of the PEDOT [46].

In this field, as an example, Abidian et al. [46] reported an on-demand drug realizing system based on PEDOT nanotubes. As shown in Figure 9, firstly, composite fibers of biodegradable poly(lactide-*co*-glycolide) (PLGA) and dexamethasone (drug) were fabricated using electrospinning onto a supporting electrode, and PEDOT was thereafter electrochemically grown along the surface of the template-fibers to form PEDOT nanotubes, which can slow down the release of dexamethasone. The release characteristics of dexamethasone were studied while the carrying electrode was biased at different voltages. By applying a positive voltage, electrons are injected into the chains and positive charges in the polymer chains are compensated. To maintain overall charge neutrality, counterions are expelled towards the solution and the PEDOT nanotubes contract. This shrinkage causes the drugs to come out of the ends of tubes (Figure 9f). The cumulative mass release of dexamethasone was found to increase dramatically as short voltage pulses were applied to the electrode hosting the PEDOT/PLGA/dexamethasone fibers. The result demonstrates that we can precisely release individual drugs and bioactive molecules at desired points in time by using electrical stimulation of the nanotubes.

4.2. Neural Interfaces and Guidance for Neurite Outgrowth

CPs such as PPY, PANI and PEDOT have become attractive candidates in the quest to bridge the electrode-cellular interface because they are compatible substrates for living cells.

For example, PPY doped with p-toluene sulfonate is cyto-compatible with mouse fibroblasts and neuroblastoma cells, and it can support the regrowth of regenerating axons *in vivo* [94]. Particularly, it has been shown that PPY and PEDOT nanotubes can adhere better to the surface of electrodes in comparison with their film counterparts [46-49]. The impedance of the neural microelectrodes can be significantly decreased by about two orders of magnitude and the charge-transfer capacity significantly increased about three orders of magnitude by creating PEDOT nanotubes on a gold electrode surface [46-49].

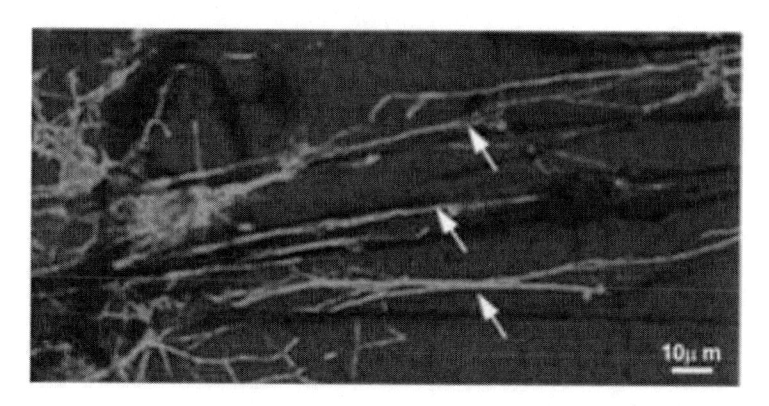

Figure 10. Fluorescence image of PC12 cells on the aligned electrospun PPY/SIBS fibers. The F-actin filaments in PC12 cell bodies and neurite outgrowths (white arrows) were labeled using Alexa Fluor488 phalloidin (green). The PPY/SIBS fibers were observed as black areas because of blocking of the background fluorescence [96].

In addition, some results demonstrate that aligned fibers have obvious advantages than random fibers in guidance of neurite outgrowth. For instance, electrical stimulation studies showed that PC12 cells, stimulation of the cells on aligned PPY/PLGA fibers resulted in longer neurites and more neurite-bearing cells than stimulation on random PPY/PLGA fibers, suggesting a combined effect of electrical stimulation and topographical guidance and the potential use of these scaffolds for neural tissue applications [95]. Similar, aligned electroactive PPY/poly(styrene-β-isobutylene-β- styrene) (SIBS) fibers are fabricated by electrospinning, and neurite outgrowth from PC12 cells could be highly orientated parallel to the aligned PPY/SIBS fibers (Figure 10). Physical interactions between the nerve cells and PPY/SIBS fibers through filopodia "sensing" were observed, which indicate a role of contact guidance as a mechanism for the observed alignment. This work highlights the capacity for electroactive PPY/SIBS fibers to support and guide nerve cell differentiation through topographic cues, which is a highly desirable characteristic in medical implants for neurological applications [96].

4.3. Field-Effect Transistors and Diodes

Field-effect transistors (FETs) based on individual CP nanofibers have been reported extensively [97-100]. For example, an ultra-short poly(3-hexylthiophene) (P3HT) field-effect transistor with effective channel length down to 5-6 nm and width ~2 nm was reported [97]. And a single nanofiber FET from electrospun P3HT exhibited a hole field-effect mobility of

0.03 cm^2/Vs in the saturation regime, and a current on/off ratio of 10^3 in the accumulation mode [98].

Recently, transistors based on aligned P3HT nanofibers were fabricated by two-fluid coaxial electrospinning technique using P3HT as core and PMMA as shell, followed by extraction of PMMA, as shown in Figure 11 [99]. It is found that the carrier mobility of the aligned fibers at a lower shell flow rate (such as 1.0 mL/h) could be dramatically improved to 1.92×10^{-1} cm^2/Vs with the on/off ratio of 4.45×10^4 possibly due to higher crystallinity and preferred orientation of P3HT with electrospun aligned nanofibers [99].

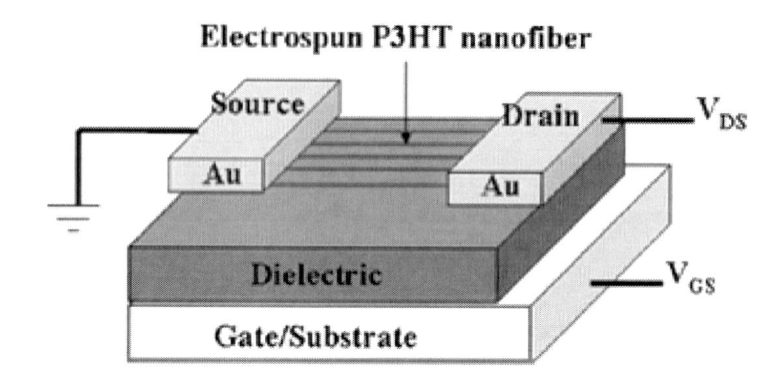

Figure 11. Schematic representation of an organic field-effect transistor based on aligned electrospun P3HT nanofibers [99].

These two values are one order of magnitude higher than those of the single P3HT fibers [98]. In addition, in the presence of a positive gate bias, electrolyte-gated transistors based on random PANI, PPY and PEDOT nanowire junction arrays demonstrated a large on/off current ratio of 978, which can vary according to the acidity of the gate medium [100].

In addition to CP nanofiber-based FETs, nanostructured CP/semiconductor (metal) diodes were also reported. As an example, Guo et al. [31] reported an organic/inorganic p-n junction nanowire consisting of PPY and CdS fabricated using an Al_2O_3 template, which displays a strong photodependent rectifying effect. It was also reported that single Au-PPY-Cd-Au nanorods exhibit diode behavior with rectifying ratio ~200 at ±0.6 V at room temperature [101]. Pinto et al. [102] reported a Schottky diode using an *n*-doped Si/SiO_2 substrate and an electrospun fully doped PANI nanofiber.

4.4. Gas Sensors

CPs have shown a promising prospect as sensor materials because of their properties to be tailored for detecting a wide range of chemical compounds, such as gases, alcohol, and so on. Normally, CP nanotube/fiber sensors have advantages of sensitivity, spatial resolution, and rapid response associated with individual nanowires along with the material advantages associated with organic conductors [103, 104].

For example, electrical conductivity of PANI can be tuned to desirable value in the range of 10^{-8} to 10^2 S/cm, and sensors fabricated from PANI nanofibers via electrospinning are faster and more reliable than conventional sensors based on thin films on detecting of water vapor, methanol, and $CHCl_3$. Especially, due to the large surface to volume ratio, uniform diameter of electrospun sensors from camphorsulfonic acid-doped PANI nanofibers also have larger responses especially for large molecule alcohols and exhibit true saturation over a given time interval in the presence of the control gas and small alcohol vapors compared to nanofiber mat sensors [105].

As shown in Figure 12a, Chen et al. [106] reported a single PANI nanofiber FET gas sensor fabricated by means of near-field electrospinning, which showed a 7% reversible resistance change to 1 ppm NH_3 with 10 V gate voltage (Figure 12b). The device with -10 V gate voltage also showed greater current decrease when it was exposed to 5, 10 and 20 ppm NH_3.

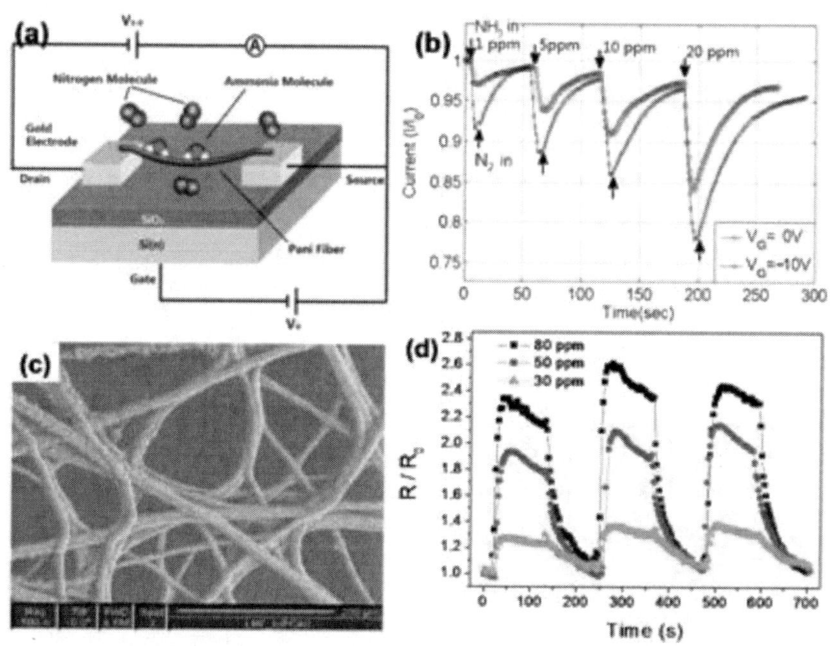

Figure 12. (a) Schematic illustration of a single PANI nanofiber FET ammonia sensor. (b) Real-time response of the single PANI fiber sensor to different concentrations of NH_3 at V_G=0V and V_G=-10V. [106] (c) SEM image of coaxial PANI/PMMA fibers covered with nano-branched PANI. (d) Rear-time resistance response of the PANI/PMMA fibers to different concentrations of NH_3 [45].

The FET characteristics of the sensor when exposed to different gas concentrations indicate that adsorption of NH_3 molecules reduces the carrier mobility in the PANI nanofiber. Namely, the effect of NH_3 molecule on PANI fiber is taking up protons from PANI, thus for forming NH_4^+ and reducing carrier concentration [106]. In addition, our group also prepared coaxial PANI/PMMA fibers by *in situ* solution deposition of nano-branched PANI on the surface of electrospun PMMA nanofibers (Figure 12c). The device showed a distinguished, reversible resistance response R/R_0 of 1.3 for 30 ppm, 2.0 for 50 ppm and 2.5 for 80 ppm NH_3 (Figure 12d).

4.5. Anticorrosion

Anticorrosion performance of PANI nanostructures on mild steel was also reported by Yang et al. [107] recently. Different 1D nanostructured PANI were synthesized in sulfuric acid solutions by direct mixed reaction and other methods. The products show an anticorrosion performance, which was studied using electrochemical measurement in 3.5% NaCl aqueous solution.

Results showed that the PANI nanofibers have uniform morphology with diameters of 60-100 nm and more excellent protective properties than conventional aggregated PANI for mild steel. This is likely due to the small diameter of PANI nanofibers that gives rise to a larger surface area of the coating film than aggregated PANI, which easily adhered to mild steel and resulted in enough reactions with PANI and mild steel, then formed compact passive layer on the surface of mild steel for anticorrosion.

4.6. Hydrogen Storage

Electrospun PANI nanofibers have also been investigated for hydrogen storage purposes due to many physisorption sites as well as chemisorption sites [63]. For PANI fibers, it is found that a reversible hydrogen storage capacity of ~3-10 wt% could be obtained at different temperature (25-125 °C) by using pressure-composition- temperature measurements. Hydrogen kinetic sorption measurements in prolonged cycles (up to 66 cycles) reveal an uptake and release of >6-10 wt% on these PANI fibers. And the hydrogen capacity and kinetics increases with increasing temperature. The higher volumetric hydrogen storage capacities in PANI fibers may be due to nanofibrillar swelling effect.

CONCLUSION

Up to date, significant progress has been made in this highly active field of nanoscale CPs. This chapter briefly summarizes and reviews three preparation methods, electrical properties, and several potential applications of conducting PANI, PPY and PEDOT nanotubes/fibers based on our results and some important contributions of other groups. Although varieties of approaches such as template-free self-assembly, hard physical template-guided synthesis, and electrospinning have been employed to synthesize and fabricate 1D conducting polymer nanostructures, facile, efficient, and large-scale synthesis of the polymer nanostructures with uniform, non-dispersed, and well-desired morphology and size, even oriented nanostructure arrays, is still desired. Several potential applications of these nanotubes/fibers in drug release, neural interfaces, field effect transistors, nano-diodes, gas sensors, anticorrosion and hydrogen storage are discussed. However, in order to fulfill their applications in nanoscale devices, there are still many challenges including the reproducibility and/or controllability of individual nanotubes/wires, stability of the doping level and improved processability of CP nanostructures. Furthermore, synthesizing conducting polymer nanostructures with highly crystalline, improved (or even metallic) electrical conductivity and/or mobility will enhance the performance of some nanodevices.

ACKNOWLEDGMENTS

This work was supported by the National Natural Science Foundation of China (Grant Nos. 11074138, 50973098, 11004114 and 50825206), the Natural Science Foundation of Shandong Province, China for Distinguished Young Scholars (Grant No.: JQ201103), the Taishan Scholars Program of Shandong Province, China and the National Key Basic Research Development Program of China (973 special preliminary study plan) (Grant No. 2012CB722705).

REFERENCES

[1] G. MacDiarmid, *Angew. Chem. Int. Edit,* 40(14), 2581-2590 (2001).
[2] J. Heeger, *Chem. Soc. Rev.*, 39(7), 2354-2371 (2011).
[3] T. A. Skotheim and J. R. Reynolds, *Handbook of Conducting Polymers*, 3rd edition. Boca Raton (2007).
[4] H. D. Tran, D. Li, and R. B. Kaner, *Adv. Mater.,* 21(14-15), 1487-1499 (2009).
[5] Y. Z. Long, M. M. Li, C. Z. Gu, M. X. Wan, J. L. Duvail, Z. W. Liu, Z. Y. Fan, *Prog. Polym. Sci.*, 36(10), 1415-1442 (2011).
[6] S. V. N. T. Kuchibhatle, A. S. Karakoti, D. Bera, S. Seal, *Prog. Mater. Sci.*, 52(5), 699-913 (2007).
[7] R. Martin, *Accounts Chem. Res.*, 28(2), 61-68 (1995).
[8] J. X. Huang, R. B. Kaner, *J. Am. Chem. Soc.*, 126(3), 851-855 (2004).
[9] M. X. Wan, *Adv. Mater.*, 20(15), 2926-2932 (2008).
[10] N. R. Chiou, A. J. Epstein, *Adv. Mater.*, 17(13), 1679-1683 (2005).
[11] J. Jang and H. Yoon, *Chem. Commun.* 720 (2003).
[12] Y. Huang, B. Dong, N. Lu, N. J. Yang, L. G. Gao, L. Tian, D. P. Qi, Q. Wu, and L. F. Chi, *Small*, 5(5),583-586 (2009).
[13] Z. M. Zhang, Z. X. Wei, M. X Wan, *Macromolecules,* 35(15), 5937-5942 (2002).
[14] M. X. Wan, Z. X. Wei, Z. M. Zhang, L. J. Zhang, K. Huang, Y. S. Yang, *Synth. Met.*, 135-136, 175-176 (2003).
[15] Z. M Zhang, Z. X. Wei, L. J. Zhang, M. X. Wan, *Acta. Mater.*, 53, 1373-1379 (2005).
[16] Z. M. Zhang ZM, M. X. Wan, Y. Wei, *Adv. Funct. Mater.*, 16, 1100-1104 (2006).
[17] Z. M. Zhang, L. Q. Wang, J. Y. Deng, M. X. Wan, *Reactive and Funct. Polym.,* 68, 1081-1087 (2008).
[18] Z. M. Zhang, J. Y. Deng, M. X. Wan, *Mater. Chem. Phys,* 115, 275-279 (2009).
[19] Y. Z. Long, Z. J. Chen, J. Y. Shen, Z. M. Zhang, L. J. Zhang, K. Huang, M. X. Wan, A. Z. Jin, C. Z. Gu, J. L. Duvail, *Nanotechnology*, 17(24), 5903-5911 (2006).
[20] L. J. Zhang, Y. Z. Long, Z. J. Chen, M. X. Wan, *Adv. Funct. Mater.,* 14(7), 693-698 (2004).
[21] Z. M. Zhang, J. Y. Deng, L. M. Yu, M. X. Wan, *Synth. Met,*158,712-716 (2008).
[22] Y. Z. Long, Z. J. Chen, Y. J. Ma, Z. Zhang, A. Z. Jin, C. Z. Gu, L. J. Zhang, Z. X. Wei, M. X. Wan, *Appl. Phys. Lett.,* 84(12), 2205-2207 (2004).
[23] M. X. Wan, *Macromol. Rapid Commun.,* 30(12), 963-975 (2009).
[24] Z. M. Zhang, M. X. Wan, Y. Wei, *Nanotechnology,* 16(12), 2827-2832 (2005).

[25] Y. Z. Long, Z. J. Chen, J. L. Duvail, Z. M. Zhang, M. X. Wan, *Physica B,* 370, 121-130 (2005).
[26] Y. Z. Long, Z. J. Chen, Z. X. Liu, Z. M. Zhang, M. X. Wan, N. L. Wang, *Chinese Phys.*, 12(4), 433-437 (2003).
[27] Z. M. Zhang, M. X. Wan, *Synth. Met.,*132, 205-212 (2003).
[28] Z. M. Zhang, J. Y. Deng, J. Y. Shen, M. X. Wan, Z. J. Chen, *Macromol. Rapid Commun.,*28, 585-590 (2007).
[29] Z. M. Zhang, Q. Li, L. M. Yu, Z. J. Cui, L. J. Zhang, G. A. Bowmaker, *Macromolecules,* 44, 4610-4615 (2011).
[30] J. L. Duvail, S. Dubois, S. Demoustier-Champagne, Y. Z. Long, L. Piraux, *Int. J. Nanotechnol.*, 5(6-8), 838-850 (2008).
[31] Y. B. Guo, Y. J. Zhang, H. B. Liu, S. W. Lai, Y. L. Li, Y. J. Li, W. P. Hu, S. Wang, C. M. Che, D. B. Zhu, *J. Phys. Chem. Lett.,* 1(1), 327-330 (2010).
[32] V. Callegari, L. Gence, S. Melinte and S. Demoustier-Champagne, *Chem. Mater.*, 21(18), 4241-4247 (2009).
[33] R. Liu, S. B. Lee, *J. Am. Chem. Soc.*, 130(10), 2942-2943 (2008).
[34] J. M. Lorcy, F. Massuyeau, P. Moreau, O. Chauvet, E. Faulques, J. Wéry, J. L. Duvail, *Nanotechnology,* 20(40), 405601 (2009).
[35] X. Y. Zhang, S. K. Manohar, *J. Am. Chem. Soc.,*126(40), 12714-12715 (2004).
[36] L. J. Pan, L. Pu, Y. Shi, S. Y. Song, Z. Xu, R. Zhang and Y. D. Zheng, *Adv. Mater.* 19(3), 461-464 (2007).
[37] Y. H. Li, J. Gong, G. H. He, Y. L. Deng, *Synth. Met.,* 161, 56-61 (2011).
[38] X. Li, S. J. Tian, Y. Ping, D. H. Kim, W. Knoll, *Langmuir*, 21, 9393-9397(2005).
[39] X. Li, M. X. Wan, X. N. Li and G. L. Zhao, *Polymer,* 50(19), 4529-34 (2009).
[40] J. H. Rong, F. Oberbeck, X. N. Wang, X. D. Li, J. Oxsher, Z. W. Niu and Q. Wang, *J. Mater. Chem.,* 19(18), 2841-2845 (2009).
[41] Z. M. Zhang, J. Sui, L. J. Zhang, M. X. Wan, Y. Wei, L. M. Yu, *Adv. Mater.,* 17, 2854-2857 (2005).
[42] Z. M. Zhang, J. Y. Deng, J. Sui, L. M. Yu, M. X. Wan, Y. Wei, *Macromol. Chem. Phys.,* 207, 763-769 (2006).
[43] L. M. Yu, Z. J. Cui, Z. M. Zhang, Q. Li, X. H. Jiang, H. Z. Zhao, *Acta. Polymerica. Sinica.,* 11, 1346-1350, (2010).
[44] S. Nair, S. Natarajan, S. H. Kim, *Macromol. Rapid Commun.,* 26, 1599-1603 (2005).
[45] C. Tang, R. Huang, Y. Z. Long, B. Sun, H. D. Zhang, C. Z. Gu, W. X. Wang, J. J. Li, *Adv. Mater. Research*, 562-564, 308-311 (2012).
[46] M. R. Abidian, D. H. Kim and D. C. Martin, *Adv. Mater.,* 18(4), 405-409 (2006).
[47] M. R. Abidian and D. C. Martin, *Biomaterials,* 29(9), 1273-1283 (2008).
[48] M. R. Abidian, K. A. Ludwig, T. C Marzullo, D. C. Martin and D. R. Kipke, *Adv. Mater.,* 21(37), 3764-3770 (2009).
[49] M. R. Abidian, J. M. Corey, D. R Kipke and D. C. Martin, *Small,* 6(3), 421-429 (2010).
[50] Z. M. Huang, Y. Z. Zhang, M. Kotaki, S. Ramakrishna, *Compos. Sci. Technol.,* 63, 2223-2253 (2003).
[51] Formhals, US patent No. 19755041(1934).
[52] D. H. Reneker, I. Chun, *Nanotechnology,* 7, 216-223 (1996).
[53] D. H. Reneker, A. L. Yarin, *Polymer,* 49(10), 2387-2425 (2008).
[54] M. M. Li, D. Y. Yang, Y. Z. Long, H. W. Ma, *Nanoscale,* 2, 218-221 (2010).

[55] B. Sun, Y. Z. Long, F. Yu, M. M. Li, H. D. Zhang, W. J. Li, T. X. Xu, *Nanoscale*, 4, 2134-2137 (2012).
[56] M. M. Li, Y. Z. Long, H. X. Yin, Z. M. Zhang, *Chin. Phys. B,* 20(4), 048101 (2011).
[57] Y. Z. Long, M. Yu, B. Sun, C. Z. Gu, Z. Y. Fan, *Chem. Soc. Rev.*, 41, 4560-4580 (2012).
[58] Greiner, J. H. Wendorff, *Adv. Polym Sci,.*219, 107-171 (2008).
[59] J. D. Schiffman, C. L. Schauer, *Polymer Rev.,* 48(2), 317-352 (2008).
[60] X. F. Lu, C. Wang, Y. Wei, *Small,* 5(21), 2349-2370 (2009).
[61] Q. Z. Yu, M. M. Shi, M. Deng, M. Wang, H. Z. Chen, *Mater. Sci. Eng. B,* 150, 70-76 (2008).
[62] N. J. Pinto, P. Carrión, J. X. Quiones, *Mater. Sci. Eng. A,*366 1-5 (2004).
[63] S. S. Srinivasan, R. Ratnadurai, M. U. Niemann, A. R. Phani, D. Y. Goswami, E. K. Stefanakos, *Int. J. Hydrogen Energy* , 35, 225-230 (2010).
[64] T. S. Kang, S. W. Lee, J. Joo, J. Y. Lee, *Synth. Met.* 153, 61-64 (2005).
[65] S. Chronakis, S. Grapenson, A. Jakob, *Polymer,* 47, 1597-1603 (2006).
[66] H. Dong , Prasad, V. Nyame, W. E. Jr Jones, *Chem. Mater.* 16(3), 371-373 (2004).
[67] Attout, S. Yunus, P. Bertrand, *Polym. Eng. Sci.,* 48, 1661-1666 (2008).
[68] Q. L. Xu, Y. Li, W. Feng, X. Y. Yuan, *Synth. Met.* 160, 88-93 (2010).
[69] H. Okuzaki, T. Takahashi, N. Miyajima, Y. Suzuki, T. Kuwabara, *Macromolecules* 39(13), 4276-4278 (2006).
[70] J. G. Park, S. H. Lee, B. Kim and Y. W. Park, *Appl. Phys. Lett.,* 81(24), 4625-4627. (2002).
[71] J. G. Park, G. T. Kim, V. Krstic, B. Kim, S. H. Lee, S. Roth, M. Burghard, Y. W. Park, *Synth. Metals,* 119, 53-56 (2001).
[72] Y. Z. Long, Z. J. Chen, N. L. Wang, Y. J. Ma, Z. Zhang, L. J. Zhang, M. X. Wan, *Appl. Phys. Lett.,* 83(9), 1863-1865 (2003).
[73] K. Ramanathan, M. A. Bangar, M. H. Yun, W. Chen, A. Mulchandani, N. V. Myung, *Nano. Lett.,* 4(7), 1237-1239 (2004).
[74] Y. Z. Long, Z. L.Chen, N. L. Wang, Z. M. Zhang, M. X. Wan, *Physica B* 325(1-4), 208-213 (2003).
[75] Y. Z. Long, Z. J. Chen, P. Zheng, N. L. Wang, Z. M. Zhang, M. X.. Wan, *J. Appl. Phys*., 93(5), 2926-2965 (2003).
[76] Y. Z. Long, L. J. Zhang, Y. J. Ma, Z. J. Chen, N. L. Wang, Z. M. Zhang, M. X. Wan, *Macromol. Rapid Commun.,* 24(16), 938-942 (2003).
[77] Y. Z. Long, L. J. Zhang, Z. Chen, K. Huang, Y. S. Yang, H. M. Xiao, M. X. Wan, A. Z. Jin, C. Z. Gu, *Phys. Rev. B,* 71(16), 165412 (2005).
[78] J. L. Duvail, Y. Z. Long, S. Cuenot, Z. J. Chen, C. Z. Gu, *Appl. Phys. Lett.* 90(10), 102114 (2007).
[79] Y. Z. Long, H. M. Xiao, Z. J. Chen, M. X. Wan, A. Z. Jin, C. Z. Gu, *Chin. Phys.* 13(11), 1918-1921 (2004).
[80] Y. Z. Long, Z. H. Yin, M. M. Li, C. Z. Gu, J. L. Duvail, A. Z. Jin, M. X. Wan, *Chin. Phys. B* , 18(6), 2514-2522 (2009).
[81] Y. Z. Long, J. L. Duvail, Z. J. Chen, A. Z. Jin, C. Z. Gu, *Chin. Phys. Lett.,* 25(9), 3474-3477 (2008).
[82] Y. Z. Long, J. L. Duvail, Z. J. Chen, A. Z. Jin, C. Z. Gu, *Polym. Adv. Technol*, 20(6), 541-544 (2009).

[83] Y. Z. Long, W. L. Wang, F. L. Bai., Z. J. Chen, A. Z. Jin, C. Z. Gu, *Chin. Phys. B* 17(4), 1389-1393 (2008).
[84] Y. Z. Long, Z. H. Yin, Z. J. Chen, A. Z. Jin, C. Z. Gu, H. T. Zhang and X. H. Chen, *Nanotechnology*, 19(21), 215708 (2008).
[85] J. H. Park, H. Y. Yu, J. G. Park, B. Kim, S. H. Lee, L. Olofsson, S. H. M. Persson, Y. W. Park, *Thin Solid Films*, 393, 129-131 (2001).
[86] B. Kaiser and Y. W. Park, *Synth. Metals*, 152(1-3), 181-184 (2005).
[87] N. Aleshin, H. J. Lee, S. H. Jhang, H. S. Kim, K. Akagi and Y. W. Park, *Phys. Rev. B* , 72(15), 153202, (2005).
[88] N. Aleshin, H. J. Lee, Y. W. Park, K. Akagi, *Phys. Rev. Lett.*, 93(19), 196601 (2004).
[89] Rahman and M. K. Sanyal, *Phys. Rev. B,* 76(4), 045110 (2007).
[90] Z. H. Yin, Y. Z. Long, C. Z. Gu, M. X. Wan, J. L. Duvail, *Nanoscale Res. Lett.* 4(1), 63-69 (2009).
[91] Y. Z. Long, J. L. Duvail, M. M. Li, C. Z. Gu, Z. Liu, S. P. Ringer, *Nanoscale Res. Lett.,* 5(1), 237-242 (2010).
[92] Y. Z. Long, J. L. Duvail, Q. T. Wang, M. M. Li, C. Z. Gu, *J. Mater. Res.,* 24(10), 3018-3022 (2009).
[93] Y. Z. Long, K. Huang, J. H. Yuan, D. X. Han, L. Niu, Z. J. Chen, C. Z. Gu, A. Z. Jin, J. L. Duvail, *Appl. Phys. Lett.,* 88(16), 162113 (2006).
[94] R. L. Williams, P. J. Doherty, *J. Mater. Sci: Mater. Med.,* 5, 429-433 (1994).
[95] J Lee JY, Bashur CA, Goldstein AS, Schmidt CE. Polypyrrole-coated electrospun PLGA nanofibers for neural tissue applications. *Biomaterials* 30, 4325-35 (2009).
[96] X. Liu, J. Chen, K. J. Gilmore, M. J. Higgins, Y. Liu, C. G. Wallace, *Journal of Biomedical Materials Research Part A,* 94A(4), 1004-1011 (2010).
[97] P. F. Qi, A, Javey, M. Rolandi, Q. Wang, E. Yenilmez, H. J. Dai, *J. Am. Chem. Soc.* 126,11774-11775 (2004).
[98] H. Q. Liu, C. H. Reccius, H. G. Craighead, *Appl. Phys. Lett.,* 87, 53106/1-3 (2005).
[99] J. Y. Chen, C. C. Kuo, C. S. Lai, W. C. Chen, H. L. Chen, *Macromolecules*, 44, 2883-2892 (2011).
[100] M. M. Alam, J. Wang, Y. Y. Guo, S. P. Lee, H. R. Tseng, *J. Phys. Chem. B*, 109, 12777-12784 (2005).
[101] S. Park, S. W. Chung, C. A. Mirkin, *J. Am. Chem. Soc.* 126, 11772-11773 (2004).
[102] N. J. Pinto, R. Gonzalez, A. T. Johnson Jr, A. G. MacDiarmid, *Appl. Phys. Lett.,* 89, 033505/1-3 (2006).
[103] J. Janata, M. Josowicz, *Nat. Mater.*, 2, 19-24 (2003).
[104] H. Liu, J. Kameoka, D. A. Czaplewski, H. G. Craighead, *Nano Lett.,* 4(4), 671-675 (2004).
[105] N. J. Pinto, I. Ramos, R. Rojas, P. C. Wang, A. T. Johnson Jr, *Sensors and Actuators B,* 129, 621-627 (2008).
[106] D. Chen, S. Lei, Y. Chen, *Sensors,* 11, 6509-6516 (2011).
[107] X. G. Yang, B. Li, H. Z. Wang, B. R. Hou, *Prog. Org. Coat.,* 69(3), 267-271 (2010).

In: Conducting Polymers
Editor: Luiz Carlos Pimentel Almeida
ISBN: 978-1-62618-119-9

Chapter 7

SYNTHESIS AND CHARACTERIZATION OF NEW COMPOSITES BASED IN NANOWIRES OF CONDUCTING POLYMERS INCLUDED IN NANOSTRUCTURED HOSTS

Marcos B. Gómez Costa, María L. Martínez, Juliana M. Juárez and Oscar A. Anunziata *
NANOTEC - Centro de Investigación en Nanociencia
Y Nanotecnología Facultad Regional Córdoba,
Universidad Tecnológica Nacional, Córdoba, Argentina

ABSTRACT

During the last few years, they have obtained great advances in the knowledge of the necessary structural and electronic requirements, so that a polymer is considered conductor (Solitons, polarons and bipolarons) in conjugated polymers. However, the challenge at the moment is in securing organic structures susceptible to provide suitable conductivity. From the structural point of view, the conductive polymers are polymeric with double bonds conjugated in their main chain, being polyheterocycles one of the groups most characteristic of this type of materials. Recently, there are many references appeared on conductive polymers with polyaniline, polyacetylene, polythiophene and polypyrrole mainly, being the method of more habitual obtaining electro polymerization of corresponding monomer. Nevertheless, the synthesis electrochemistry is slow and it is of great interest to secure synthesis effective chemicals. In this sense, polymerization with oxidizing agents is being considered like the most promising route in the obtaining of conductive structures. Here we show the synthesis and characterization of new composites using nanowires of conducting polymers included in nanostructured hosts, based on polyaniline, polypyrrole, and polyindole obtained by in-situ oxidative polymerization of corresponding pre-adsorbed monomers onto de hosts, producing hybrid composites with interesting conducting properties. In this chapter, we report the incorporation of aluminium into mesoporous materials: MCM-41, SBA-3, SBA-15 and

* E-mail address: oanunziata@scdt.frc.utn.edu.ar.

SBA-16 via the post-synthesis method, which was employed as hosts. We also discuss our results of the development of guest/host composites. The guest synthesized inside mesoporous structures, with electrical conducting or semiconducting properties, using in-situ polymerization techniques to gain understanding of the monomers adsorption mode onto the host and its influence on the final composite material. The conductive properties of the composite could be modified by varying the substituent of the aromatic ring, the host structure, the anchored sites to adsorb the monomer, and the amount of the polymers nanowires in the hosts.

1. INTRODUCTION

Mesoporous materials with a uniform mesopore structure and an extremely high specific surface area were extensively studied in the past decade [1, 2] due to their potential application as catalysts, adsorbents for large organic molecules, and host/guest chemical supporters. Typical mesoporous materials include amorphous or polycrystalline solids such as silica or transitional alumina, or modified layered materials such as pillared clays and silicates. A significant effort has been made to synthesize, within the mesoporous range, regular and well-defined pore structure materials [3-5].

A great advance in the preparation of porous materials was originated in 1992 when the Mobil Research and Development Corporation described the synthesis of mesostructured materials. These investigators reported the synthesis of a series of mesoporous silicates and are designated under the M41S family, which were synthesized by cationic surfactants of the type alkyl-trimethylammonium (structure directing agent).

The main porous solids of this family are MCM-41, MCM-48 and MCM-50 [5, 6]. MCM-41 material has a high specific surface, hexagonal symmetry and a uniform pore size distribution.

The most common mechanism in the synthesis of MCM-41 (called S^+I^-), that this is the usual mechanism of synthesis of mesoporous materials in alkaline medium, is based on electrostatic interactions between silica depolymerized negatively charged (I^-) and the cationic surfactant (S^+) that acts as a director of the structure. Is the typical mechanism of MCM-41 materials.

In acidic media, pH~2, the silicate species becomes cationic (I^+); however, cationic surfactant (S^+) can be used to synthesize mesoporous materials by the $S^+X^-I^+$ system. Here the counter ion X^- behaves as a shielding agent between S^+ and I^+. The materials synthesized through this procedure are known as “acid prepared mesostructures” (APM) or SBA [6]. Due to different precipitation conditions and charge-balance requirements, acid-derived materials (SBAs) have thicker pore walls and a framework charge different from that of base-derived mesoporous materials. For example overall framework charge of SBA-3 (acidic condition), for instance, was slightly positive, whereas the framework of MCM-41 (basic condition) was negative.

The SBA-3, SBA-15 and SBA-16 showed microporous in their structure. The SBA-15 is synthesized with the copolymer blocks as groups. The micro-pores formation is due to the partial obstruction of the PEO ($PPO_{20}PEO_{70}PPO_{20}$ -P123-) chains into the silica walls during the synthesis. This structure coexists with the mesoporous structure.

The SBA-16 synthesized under acidic conditions has a cubic structure and bimodal distribution of the pores. The SBA-16 silica mesophase with a cubic Im3m structure was prepared first by Zhao et al. [7] from the triblock copolymer surfactant Pluronic F_{127} ($EO_{106}PO_{70}EO_{106}$) and TEOS. M. Mesa et al. [8] have systematically investigated the synthesis parameters. The mesoporosity of this phase consists of two non-interpenetrating three-dimensional channel systems with spherical cavities at the intersection of the channels [9]. This structure can be expected to offer more interesting opportunities for applications involving selective host/guest interactions or diffusion of large molecules.

Many authors [10-12] have proposed the incorporation of metals (Al, Zn) via post synthesis as a promising alternative method. Considering the lack of sodium aluminate stability (aluminium source) at low pH (like SBA-3 synthesis media), the procedure used for introducing aluminium was that of postsynthesis [13].

Numerous investigations have indicated that Al-SBA-15 materials show much higher catalytic activity, compared with Al-MCM-41 [14-15]. Since pure siliceous SBA-15 materials lack acidity, active centers must be introduced into their framework (mesoporous wall). The incorporation of Al is particularly important as it gives rise to solid acid materials with acid sites associated with the presence of Al in framework positions, within the silica pore walls. Mesoporous aluminosilicates have therefore been the focus of many recent studies [11, 16, 17], because of their potential application in solid acid/base catalysis for bulky molecules activation. The traditional method of introducing Al into mesoporous silicates is the direct (mixed-gel) synthesis where, an aluminosilicate framework is formed directly from aluminate and silicate ions. Generally, the heteroatom incorporation, such as Al into microporous zeolites will introduce a charge imbalance in the framework, which is counterbalanced by protons, generating bridging hydroxyl groups (SiO_3–O_3–Al–OH, Brønsted acid sites) on these materials.

However, it is very difficult to introduce the metal ions directly into SBA's due to the easy dissociation of metal-O–Si bonds under strong acidic conditions. Thus, the post-synthesis method for the alumination of the mesoporous silicas, which are obtained under strongly acid conditions, becomes an appealing alternative [18].

Many studies have shown that aluminum can be effectively incorporated into siliceous MCM-41 and MCM-48 materials, via various post-synthesis procedures. The authors claimed that the materials produced via the post-synthesis method have superior structural integrity, acidity, and catalytic activity than those of materials having aluminum incorporated during synthesis [19-23].

The processing information at the molecular level is an intriguing and important challenge. Efforts to create electronic functions and devices based on molecules instead of bulk semiconductors are inspired by the anticipated enormous increase in computing speed and storage density.

The study of the electroactive properties of heterocyclic conducting polymers containing nitrogen atoms like polyaniline, polypyrrole, polycarbazole and their substituted derivatives [24, 1] has attracted considerable interest due to their potential industrial applications.

Among these polymers, polypyrrole displays high electrical conductivity, good environmental stability and fine for anodic electrodeposition of freestanding polypyrrole films. On the other hand, poly(para-phenylene) exhibits good thermal stability. Indole has benzene and pyrrole ring. Thus, polyindole and its derivatives may possess the properties of poly(paraphenylene) and polypyrrole in concert. However, among various aromatic-compound based conducting polymers, polyindole and its derivatives have been investigated only scarcely, although close structural similarities can be found with the polymers mentioned above [24].

Polyindole is an electroactive polymer which can be obtained either anodic or by chemical oxidation of indole, C_8NH_7. Polyindole films exhibit fairly good thermal stability, high-redox activity and stability, and slow degradation rate compared with polyaniline and polypyrrole.

This polymer and its derivatives seem to be good candidates for applications in domains including electronics, electrocatalysis and pharmacology. In its doped state, polyindole is green, with an electrical conductivity in the range of 5.10^{-4}– 8.10^{-2} S cm^{-1}, slightly depending on the nature of the counter-ion [25]. It has also been reported that polyindole films have show fairly good thermal stability, high redox activity and slow degradation rate in comparison with polypyrrole and polyaniline [26].

There are many published works studying the electrochemical properties of polyindole for use in battery electrodes [24, 27, and 28].

The encapsulation of nanosized conducting polymer filaments, such as polyaniline, into the channels of mesoporous aluminosilicate hosts (i.e. MCM-41, SBA-15, SBA-3, SBA-16) would allow using mesoporous materials in nanometer-scale for electronic devices [29-31]. Polyaniline (PANI) is one of the most studied conducting polymers, not only for being the most stable conducting polymer, but also for its electrical, optical and electrochemical properties; hence, this can be applied to a variety of industrial processes. The composites show different properties from the individual components. The performance of the synthesis and properties of new materials is constantly improved, thus the structural order of the materials can be controlled. The new mechanical, electrical, optical, photochemical, catalytic and magnetic properties form, in most cases, a synergic combination of individual components. Over the last decade, the template synthesis has been the most widely explored method of synthesis [32, 33], consisting in a host/guest reaction. The synthesis of material guest (polymer) takes place into the channels, pores or lamina of the host structure. This process requires the careful selection of host and guest to prepare the corresponding composites. The introduction of macromolecules (i.e. polymers) into host's channels confers specific properties to the resulting composites. It reveals perspectives for potential applications. The new electronic properties such as conductivity and redox behavior can be obtained with the introduction of conductor polymers. The polymer with extended π electron system showing semi-conductive properties is known as conjugated polymer. Polyaniline (PANI) can be found in a variety of forms that differ in their chemical and physical properties. The most common green protonated emeraldine shows a conductivity of ~100 S cm^{-1}, various orders of magnitude higher than those of common polymers ($<10^{-9}$ S cm^{-1}), but lower than those of typical metals ($>10^4$ S cm^{-1}) [34, 35]. The polymerization of aniline was designed to be as simple as possible. The synthesis consisted in mixing aqueous solutions of aniline hydrochloride with ammonium peroxydisulfate at a temperature of 273 K [36-39]. The efficient polymerization of aniline was achieved only in an acidic medium, where aniline exists as an anilinium cation. The electrical conductivity of these composites were studied and compared with that of other composites reported in previous works [40, 33]. In this chapter, we report the incorporation of aluminium into mesoporous materials: MCM-41, SBA-3, SBA-15 and SBA-16 via the post-synthesis method, which was employed as hosts. We also discuss our results of the development of guest/host composites. The guest synthesized inside mesoporous structure was polyaniline (PANI), polypyrrole (PPY) and polyindole (PInd). Comparative studies of the conductive properties between different Conducting Polymer/host composites were studied.

2. Synthesis and Characterization of Materials Employed as Hosts

2.1. Synthesis of Mesoporous Siliceous Materials

2.1.1. Silica Polymerization

The classical way of silica polymerization, the so called sol-gel process, is carried out in an aqueous solution as Iler [41] investigated in 1979. This process is rather uncontrolled leading to non or poorly ordered structures with broad pore size and molecular weights distributions. These porous silica are known as sol-gel materials like aerogels or xerogels. The pathways leading to ordered porous materials are very similar to the sol-gel process and will be described in the next section.

As stated before, the polymerization is carried out in an aqueous solution by adding a catalyst and a source of silica. The silica source is an important factor for the reaction conditions. Common molecular sources are alkoxysilanes like tetramethyl- and tetraethylorthosilicate or sodium silicate. Non-molecular silica source are for example already polymerized sol-gel materials which are leading to non-homogeneous solutions.

The first step of polymerization is the formation of silanol groups by hydrolysis of the alkoxide precursors, the gel, in aqueous solution:

$$\equiv Si - OR + H_2O \rightleftharpoons \equiv Si - OH + ROH$$

The actual polymerization occurs through water (oxolations) or alcohol (alcoxolations) producing condensations:

$$\equiv Si - OH + HO - Si \rightleftharpoons \equiv Si - O - Si \equiv + ROH$$
$$\equiv Si - OH + RO - Si \rightleftharpoons \equiv Si - O - Si \equiv + H_2O$$
$$\cdots \Rightarrow \text{polysilicates}$$

This polycondensation reaction leads to more or less cross linked polysilicates which are precipitating during the reaction. This dispersion is called the sol. All the reactions are strongly pH dependent. They can be catalyzed by acid or base catalyst under formation of cationic (under acidic conditions) or anionic silicate species. An acid catalyst produces weakly-crosslinked gels which easily compact under drying conditions, yielding low-porosity microporous xerogel structures.

Conditions of neutral to basic pH result in relatively micro- and mesoporous xerogels after drying. Under some conditions, base-catalyzed and two-step acid-base catalyzed gels (initial polymerization under acidic conditions and further gelation under basic conditions) exhibit hierarchical structure and complex network topology [42, 43].

2.1.2. Theory of Templating

The main concept to obtain well defined and structured polysilicates is to use a surfactant templated polymerization instead of an uncontrolled reaction. In general the Iyotropic (i.e. amphiphilic) molecules of the surfactant form a liquid crystal by aggregation in aqueous solution [1, 2]. Of cause the proper formation of the liquid crystal matrix is strongly dependent on the conditions in the solution. The structure of the liquid crystal is the so called mesostructure. Important parameters for the mesophase formation are for instance the temperature (CMT), the concentration (CMC) or the pH-value of the solution. Depending on these conditions, the structure of the mesophase can be for example ordered in spherical, cylindrical, lamellar or cubic phase or disordered. Two distinct families of phases exist. One type is based on periodic continuous minimal boundary surfaces of the phase components; the other type is micellar being based on packing of discrete micellar aggregates. Especially for the cubic phase the cubic micellar 'I' and the bicontinuous cubic phases can be well-defined. Later in this chapter some examples of materials synthesized in a micellar or bicontinuous mesophase will be given.

A detailed study of the polymorphism of lipid water systems was done by Seddon and Templer [44]. To act as a structure directing agent, the mesophase has to interact in some way with the silica precursors. There have been many different attempts to develop pathways to influence the interactions between mesophase and polycondensation reaction of the silica source. These are in particular alkaline routes (S^+I^-), acidic routes ($S^+X^-I^+$) or non-ionic (S^0I^0) and neutral (N^0I^0) routes, where: **S** is standing for the surfactant molecules, **I** for the silica source, **X**$^-$ for halides and **N** for a non ionic surfactant. The difference of the routes is mostly the kind of driving force, with can be ionic-ionic, ionic-metal in the case of the ionic routes or hydrogen-bonding interactions in the case of non-ionic routes.

The ionic routes were developed in the early 1990s by Beck et al. [1, 2] and have been intensively studied so far. They lead to monoporous materials like the MCM family. These ionic routes will not be of major interest for this work. For completeness they will just be briefly described. In contrast newer non ionic routes yield to bi-porous materials with generally larger primary mesopores and thicker walls than the counterpart's template with ionic surfactants.

The size and the range of the interactions in the solution with the silicate molecules can be furthermore varied by changing the ionic strange of the solution using inorganic salts to form halides or metals cation in alkaline or acidic medium (S^+I^-, $S^+X^-I^+$).

For the non-ionic route, co surfactants like alcohols can be added to vary the molecule interaction in solution.

2.1.3. Ionic Process

The first attempts to synthesize ordered mesoporous silica were done under ionic conditions. Under these conditions the inorganic silicates I and also the surfactant molecules S are charged depending on the pH-value of the solution.

The well known mesoporous candidates from the MCM family are typically synthesized under alkaline conditions. Under these conditions the silicates are deprotonated and form anions I^- and the surfactant molecules form cations S^+ respectively. Three different mesophases could be observed for this system leading to three structure: the hexagonal structure MCM-41, the cubic structure, MCM-48 and the lamellar structure, MCM-50. The precise reaction conditions can be found in many publications [45, 46]. The first attempts to synthesize templated mesoporous silica under acidic conditions were carried out by Huo et al. [47- 49].At these conditions the cationic surfactant molecules are surrounded by negative halides. A triple layer of surfactant, halides and the cationic silicate species ($S^+X^-I^+$) is the considerated mechanism for the mesophase formation. Mesoporous materials synthesized using these routes are for instance the SBA-3 materials.

2.1.4. Non-Ionic Process

More important for this work are the non-ionic routes templated with amphiphilic triblock copolymers. These routes are relatively new and have shown a high flexibility in tailoring the synthesis conditions and the mesostructure of the liquid crystal template. Especially the easy variation of hydrophobic/hydrophilic copolymer blocks and thus their volume fraction simply by using different chain length or co-surfactants provides a high degree of control. Commonly used block components are hydrophobic polymers like PPO, PS, PI or PE and hydrophilic polymers like PEO and PAA.

2.2. Materials with Hexagonal Symmetry

2.2.1. Si-MCM-41 and Na-Al-MCM-41

2.2.1.1. Synthesis

The Si and Al-containing-MCM-41 nanostructured material was synthesized to be applied in the preparation of composites. In this study, the material used as hosts, was obtained employing the sol– gel method.

The mesoporous silicate was synthesized by hydrolysis of tetraethylorthosilicate (TEOS) at room temperature, in an aqueous solution, using cetyltrimethylammonium bromide (CTAB) as a surfactant.

The procedure designed is described as follows: the surfactant was dissolved in de-ionized water and stirred until the solution was homogeneous and clear. After adding an adequate volume of ammonium hydroxide (25% w/w), the mixture was stirred for 5 min after which TEOS was added. The molar composition of the gel was 1 M TEOS: 1.64 M NH_4OH:0.15 M CTAB:126 M H_2O.

The reaction was stirred overnight after which the solution was filtered and washed consecutively with de-ionized water, and dried at room temperature [50].

To remove the template, the sample was heated up from ambient temperature to 723 K under N_2 flow (20 mL/min) and subsequently calcined, under air flow (10 mL/min), up to 723 K using a heating rate of 3 K/min [3].

The alumination procedure of MCM-41 was carried out as follows [13, 51]: silica MCM-41 (1 g) was stirred for 20 h at room temperature in 50 mL of water containing dissolved sodiumaluminate. The mixture was filtered, washed, dried at room temperature overnight and then calcined in air at 823 K for 5 h.

Finally, Na-AlMCM-41 sample with Si/Al = 20 was obtained, according to ICP analysis. The synthesis was effectively performed with the technique developed. The material characterization indicates the regularity of the hexagonal array of cylindrical pores in the Na-AlMCM-41.

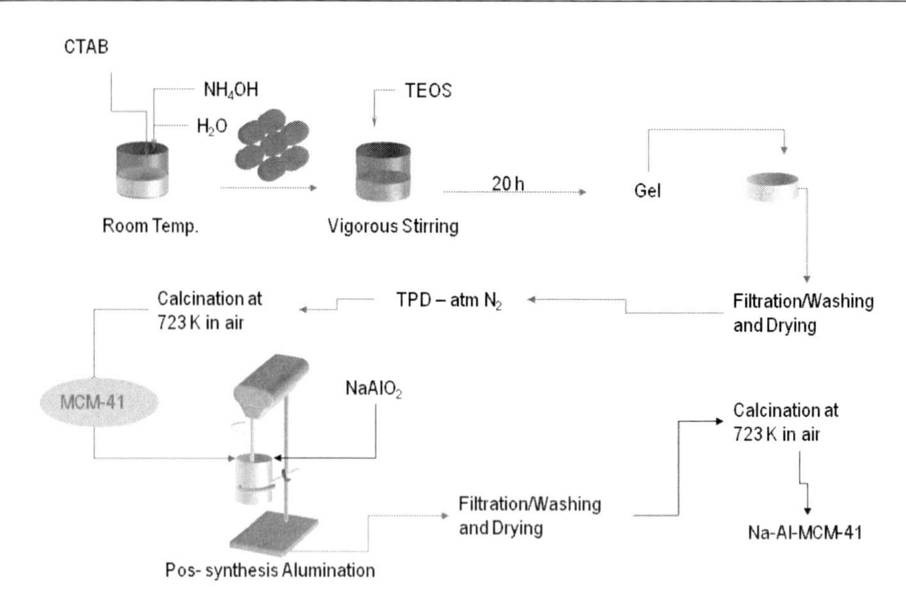

Figure 1. Na-Al-MCM-41 synthesis scheme.

2.2.1.2. Characterization

Table I shows the structural characteristic of materials. The a_O parameter is obtained from the X-ray diffraction (XRD) patterns of as made MCM-41 and Al-MCM-41 materials. The XRD patterns indicate that the hexagonally ordered structure of MCM-41 was persistent after the modification procedure (post-alumination).

2.2.2. Si-SBA-3 and Na-AlSBA-3

2.2.2.1. Synthesis

The new aluminosilicate materials, Na-AlSBA-3, were synthesized to be applied in the preparation of composites. In previous research, we presented novel results of the synthesis of new Na-AlSBA-3 material [13] since no literature was reported on aluminium-containing SBA-3 material.

In this study, the material was obtained employing the sol–gel method. The synthesis was effectively performed with the technique developed. The material characterization indicates the regularity of the hexagonal array of cylindrical pores in the Na-AlSBA-3 [13]. The mesoporous silicate was synthesized by hydrolysis of tetraethylorthosilicate (TEOS) at room temperature, in an aqueous acidic solution, using cetyltrimethylammonium bromide (CTAB) as a surfactant.

Table I. Structural characteristic of MCM-41

Sample	a_0 *(nm)	Si/Al	$Á_{BET}$,m^2/g	V(mL/g)	D **(nm)	E ***(nm)
Si-MCM-41	4.1	----	1235	0.96	2.99	1.10
Al-MCM-41	3.8	20	860	0.80	2.58	1.20

(*) $a_{o:}$ lattice parameter, $a_o=2\ d_{100}/\sqrt{3}$; V: volume of pore (**) D: pore diameter, D ≅ 4V/A; (***) E: wall thickness, E = a_o - D, (in agreement ref. [52,53]).

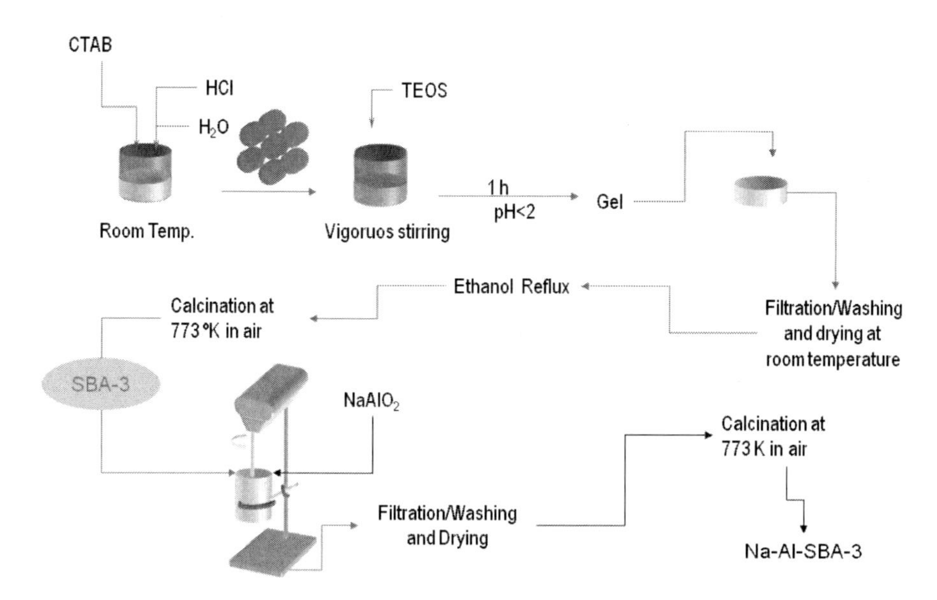

Figure 2. Na-Al-SBA-3 synthesis scheme.

The procedure designed is described as follows: the surfactant was mixed with water and HCl, 3 g of TEOS were then added and stirred to form a mixture with a molar composition of: TEOS:H_2O:HCl:CTAB = 1:130:9.2:0.12 [13]. After 45 min, a white precipitate was formed, which was then filtered, washed and dried at room temperature. The material was afterwards immersed in ethanol under reflux for 6 h in order to extract the surfactant, and calcined at 823 K in air for 6 h.

The material obtained was denoted as Si-SBA-3. The alumination procedure of SBA-3 was carried out as follows [13]: silica SBA-3 (1 g) was stirred for 20 h at room temperature in 50 ml of water containing dissolved sodiumaluminate. The mixture was filtered, washed, dried at room temperature overnight and then calcined in air at 823 K for 5 h. Finally, Al-SBA-3 sample with theoretical Si/Al = 20 was obtained.

2.2.2.2. Characterization

The presence of three Bragg angles can be distinguished in hexagonal lattice symmetry, typical of SBA-3 structure by X-ray Diffraction. Moreover, XRD patterns (no showed) indicate that the hexagonally ordered structure of SBA-3 was persistent after the modification procedure. A prominent peak, hkl =[100] as well as weaker peaks of [110] and [200] were observed in Al–SBA-3, which allowed us to corroborate, that the obtained mesoporous sample has a highly ordered pore system with a high porosity.

Table II. Structural characteristic of SBA-3

Sample	a_o *(nm)	Si/Al	$Á_{BET}$,m^2/g	V(mL/g)	D **(nm)	E ***(nm)
Si-SBA-3	3.76	----	1024	0.55	2.14	1.62
Na-Al-SBA-3	3.63	20	780	0.39	1.99	1.64

(*) a_{o}: lattice parameter, $a_o = 2\ d_{100}/\sqrt{3}$; V: volume of pore (**) D: pore diameter, $D \cong 4V/A$; (***) E: wall thickness, $E = a_o - D$, (in agreement ref. [52, 53]).

2.2.3. Si-SBA-15 and Na-AlSBA-15

2.2.3.1. Synthesis

SBA-15 was synthesized using P123 (Aldrich) as a co-polymer, 1. 5 g ($2.5x10^{-4}$ moles) of P123 was added to acid solution (48 mL of HCl 2M (pH=1), in distilled water, at 50 8C for 6 h. A homogenous solution was obtained; 7 ml (0.016 mol) of TEOS at 323K with vigorous agitation were added, by 5 min.; and then the solution was aging at room temperature for 24h. Finally, the dense gel was heated at 353 K by 48h under static conditions. The final product was filtered, washed and immersed in ethanol reflux for 6 h to extract the co-polymer dried at 373 K for 10 h. After, SBA-15 was immersed in ethanol reflux for 6 h to extract the co-polymer block. The sample was filtered and washed with distilled water and dried at 373 K and calcined at 823 K in air 6hr (1°/min), A white powder, SI-SBA-15 material was obtained. [54]

To obtain Al-SBA-15 the procedure was similar. Employing Si-SBA-15, the alumination procedure was carried out as follows [13]: silica SBA-15 (1 g) was stirred for 20 h at room temperature in a solution of different quantities of sodiumaluminate dissolved in 50 ml of water. The mixture was filtered, washed, dried at room temperature overnight and then calcined in air at 823 K for 5 h. Finally, two samples of Al-SBA-15 with Si/Al ratio equal at 50 and 32 were obtained.

2.2.3.2. Characterization

The XRD diffraction peaks of SBA-15 and Al-SBA-15, can be indexed to a hexagonal lattice with a d_{100} spacing close to 8.0nm, corresponding to a unit cell parameter a_o equal to 10 nm, based on the formula $a_o = 2d_{100}$, as was shown in Table III. After alumination, the XRD patterns indicate that all the samples retain the characteristic patterns of the hexagonal mesostructures. Textural properties of the solids were obtained from 77 K nitrogen adsorption/desorption isotherm measurements, which allowed us to calculate the specific surface areas, specific pore volumes, and mesopore size distributions, as depicted in Table III.

2.3. Materials with Cubic Symmetry

2.3.1. Si-SBA-16 and Na-AlSBA-16

2.3.1.1. Synthesis

Mesoporous silica materials with cubic Im3m structure were synthesized according to the procedure described by Kim et al. [9]. Briefly, poly (alkylene oxide)-type triblock copolymers F127 (EO106PO70EO106, MW 5 12,600) were dissolved in aqueous HCl solution. Tetraethylorthosilicate was added to the solution under continuous stirring for 15 min, at 308 K. The molar composition of the mixture was as follows: F127/TEOS/HCl/H2O = 0.004/1/4/130.

This mixture was kept under static conditions for 6 h at the same temperature. Subsequently, the mixture was placed in an oven at 323 K, over a period of 20 h; afterwards the temperature of the furnace was increased to 353 K for a 20-h aging, after that the solid product was recovered by filtration and dried at 373 K.

Table III. Physical Chemistry Characterization of SBA-15 and Al-SBA-15 hosts

Sample	Ao * (nm)	Si/Al	ÁBET,(m2/g)	V (mL/g)	D (nm) **	E (nm)***
SBA-15	10.0	----	1040	1.38	8.9	1.1
Al-SBA-15(a)	11.3	50	1020	1.32	9.1	2.2
Al-SBA-15(b)	11.7	32	960	1.26	9.3	2.4

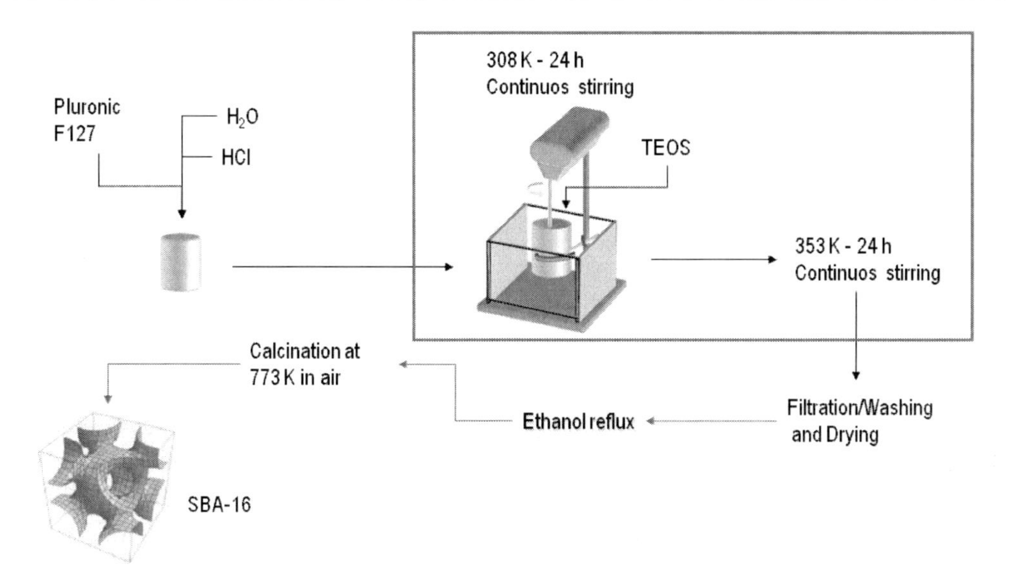

Figure 3. SBA-16 synthesis scheme.

In order to extract the surfactant, the Si-SBA-16 sample was immersed in ethanol under reflux for 6 h and then calcined at 823 K in air for 6 h. Post-synthesis alumination was employed to obtain Na-AlSBA-16, according to the procedure described previously (for Na-AlSBA-3). The Si/Al ratio of the material was 20.

2.3.1.2. Characterization

The cubic Im3m structure is a body-centered-cubic arrangement of cages with 8 apertures to the nearest neighbors as in Si-SBA-16 and Na-AlSBA-16.

The unit-cell parameter a_0 of the crystallographic structure was obtained by solving the following equation depending on the type of space group: $a_0 = dhkl.\sqrt{(Qhkl)}$, where Qhkl is: $Qhkl = h^2 + k^2 + l^2$, for a cubic space group like the cubic body centered (Im3m) SBA-16.

Table IV. Structural characteristic of SBA-16

Samples	a_0 *(nm)	Si/Al	$Á_{BET}$, m^2/g	V (mL/g)
Si-SBA-16	12.96	----	500	0.38
Al-SBA-16	12.16	20	486	0.37

3. Synthesis and Characterization of Composites

3.1. Introduction

The information processing at the molecular level is a question and a very important goal. Efforts to create electronic devices based on molecules instead conventional conductors are inspired by the enormous increase in computational speed and density data storage [1]. The MCM-4, SBA-3 and SBA-15 mesoporous molecular sieves with a channel structure packaged and a narrow distribution of pores, and SBA-16 with a structure of cavities / channels like pseudo-spheres interconnected, all offer unique opportunities for the preparation of new nanostructured composite materials.

Also, with the encapsulation and inclusion of some polymers within the cavity of these aluminosilicates, to be used as hosting, are expected new properties and applications for these materials.

In this sense the study of the adsorption of aniline, pyrrole and indole and subsequent polymerization inside host as MCM-41, SBA-3, SBA-15 and SBA-16 is very important and with great potential application in electronics (Figure 4). These materials are strongly hydrophilic, thus the study of co-adsorption of water and aniline must be considered because water molecules compete for the same active sites of both hosts.

Little information is available about the type of interaction, and adsorption sites of polar molecules on these hosting. Yariv and colleagues [55-58], have studied the adsorption of aniline over montmorillonites saturated with various cations and identified by infrared spectroscopy different types of interactions between aniline and sorption sites present in the interlayers of the material.

The composition and structure of these associations depend on the nature of the adsorption sites, depending a site is strong or weak acid or a basic site, and if sites Brønsted or Lewis acids. In the adsorption sites montmorillonites most important are hydrated [55-58]. The anilines are protonated by strong Brønsted acid sites and transformed into anilonium cations ($ArNH_3^+$). With a weak acids, anilines hydrogen bonds form bridges accepting protons of weak acids. Molecular and protonated anilines have different IR spectra.

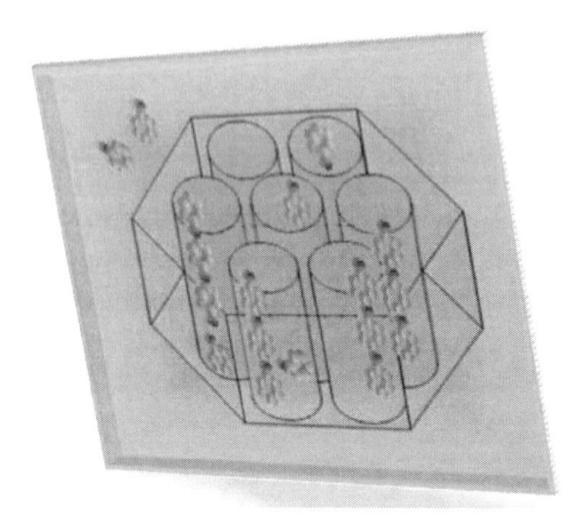

Figure 4. Scheme of aniline adsorption and consequent in-situ polymerization within the channels of the MCM-41, SBA-3 or SBA-15 [30].

Have been found two types of associations between aniline and metal cations in the interlayer of montmorillonite [55-58]: the N of the aniline molecule directly coordinated to the cation (type I), or bonds via water bridges (type II). The two types of complexes are distinguished by their IR spectra.

Type I associations are predominantly formed when the metal ions show a strong tendency to form complexes with organic bases. Bodenheimer et al. [59-62] showed that the transition elements which are among the interlayer forming stable coordination complexes with amines, which are good electron donor.

Two structures of type II were postulated for aniline derivatives, in which organic molecules can act as proton acceptor or electron donor (type A and B, respectively).

In-situ polymerization was studied by many authors on different host such as montmorillonite, SiO_2 and Al_2O_3. A composite material of Polyaniline / montmorillonite was synthesized by electrochemical methods by Yang and Chen [63] and structural changes and the effects of physical properties of these composites were studied by Lee et al. [64].

Detailed studies of a nanocomposite conductivity of polyaniline / Na^+-montmorillonite were performed by B. H. Kim et al. [65]. Choi et al. [66] synthesized polyaniline on a mesoporous silica, Si-MCM-41, to obtain a nanocomposite of polyaniline / Si-MCM and measured their physical properties using XRD analysis, transmission electron microscopy and N 2 adsorption isotherms.

Cho et al. [67] prepared a composite of a mesoporous MCM-41 sieve with conductive polyaniline and their properties measured electrorreológicas. Wu and Bein [30] synthesized and characterized PANI obtained on mesoporous aluminosilicate MCM.

Recent works of PANI / MMT and Na PANI/SBA-15 about your conductivity were published by Cho et al. [68] and Kim et al. [69], respectively.

Li et al. [70] conductive polyaniline synthesized on mesoporous silica SBA-15; XRD studies, N_2 adsorption-desorption and IR studies confirmed the existence of polyaniline in the channels of the room and measured the variation of the impedance of the composite at different levels of moisture.

In this chapter studies were carried out adsorption of aniline, pyrrole and indole in the hosts and then characterized. We are investigating the different anchoring sites in these materials and know the possible modes of adsorption of aniline, pyrrole and indole on the surface.

It is selecting the best hosting and then polymerized both aniline, pyrrole and indole adsorbed (polymerization "in situ") and make these polymers are within the channels or cavities of the hosting, in order to obtain conducting nanometric wires of polymer and new nanostructured composites with conductor or semiconductor properties.

3.2. General Procedure for Adsorbing the Monomers

All the hosts were heated at 673 K in vacuum for 1 h to eliminate mainly adsorbed water. They were then exposed to equilibrium vapors from liquid organic monomers (aniline, pyrrole and indole) during 1 h at room temperature (298 K) to obtain monomers saturated hosts according to the procedure described by Anunziata et al. [40]. In these conditions, the monomers/host saturation relation was reached. A vacuum cell with CaF_2 windows was used for FTIR studies.

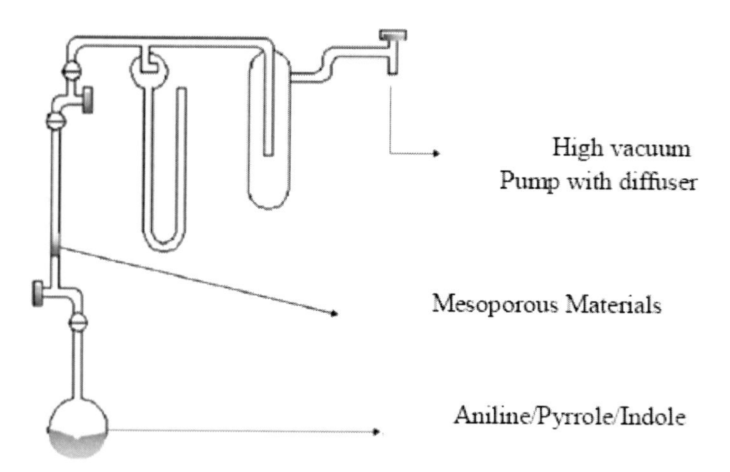

Figure 5. System used for the adsorption of various monomers onto mesoporous materials.

The spectrum was recorded after cooling the sample in the vacuum cell at room temperature. The system used for the adsorption of various monomers onto mesoporous materials (hosts) is shown in the following figure (Figure 5).

3.3. Polyaniline Composite

Aniline is a weak organic base and an aromatic compound amphiprotic, which can act as acceptor or donor of protons. Contrary to benzene, the electron density of the substituted benzene compounds is not distributed uniformly on the aromatic ring and, if the substituent is amino group, the electron density is increased especially towards the ortho-and para-. The free electron pair on the nitrogen is less able to retain a proton, so that aniline is a weaker base than the aliphatic amines. The effective load distribution calculated by the theory of molecular orbitals for aniline is N (+0.1), C in the ortho-(-0.03), C at position para-(-0.02), in the meta position C-(0) [71].

The polyaniline (PANI), figures 6 and 7, exists in a variety of forms that differ in their chemical and physical properties. Emeraldine form (conductive, green) or the protonated form of polyaniline has a conductivity in a semiconductor level of 10^0 S / cm, many orders of magnitude greater than the common polymers (insulators, $<10^{-9}$ S / cm) but smaller than the metal (typical conductors,> 10^4 S/cm). The PANI protonated (eg. Polyanilonio hydrochloride) can be converted to emeraldine base, blue, non-conductive, when it is treated with ammonium hydroxide [72].

The next section presents studies of adsorption of aniline inside the Na-Al-MCM-41, Na-Al-SBA-3, Na-Al-SBA-15 and Na-Al-SBA-16 host. It is necessary to put under consideration the co-adsorption of water as it would compete for the same active sites that aniline in all host.

In this way, also show infrared spectroscopy studies of aniline, previously adsorbed on the hosts, which was thermally dehydrated at various temperatures. Furthermore we postulate different association aniline/ host in order to determine the possible modes of adsorption of aniline on these surfaces. Subsequent, we studied the different active sites of these materials, suggesting possible modes of adsorption on surfaces. Thus, be carried out in situ aniline polymerization adsorbed on the hosting, resulting conductive or semiconductors nanometric wires polyaniline-based. There is limited information the aniline adsorption inside mesoporous aluminosilicate, used as hosting. S. Yariv and coworkers [55-57] have studied the adsorption of aniline over montmorillonites saturated with various cations, identified by infrared spectroscopy thermal different types of interactions between aniline and sorption sites present in the interlayers of material. Bodenheimer et al. [59-61], show the transition between layers of elements from stable coordination complexes with amines, which are good electron donor. Kirschhock and Fuese [62], studied the formation of charge-transfer complexes by adsorption of aniline and m-dinitrobenzene inside NaY zeolite.

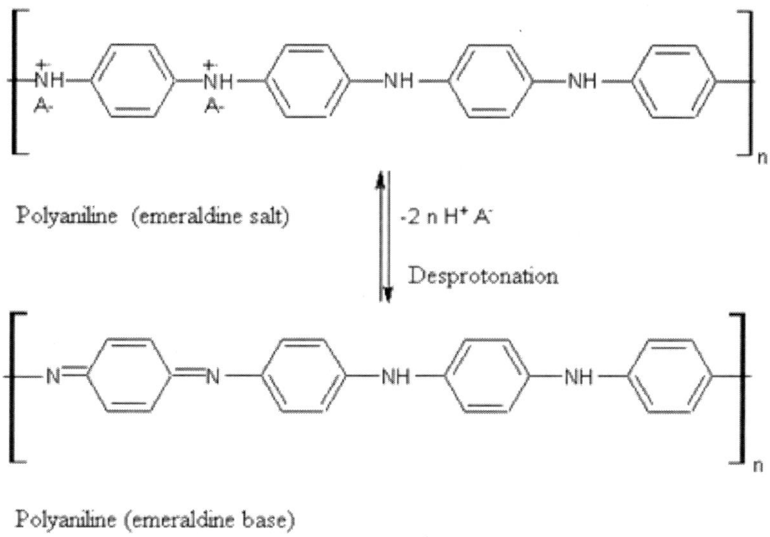

Polyaniline (emeraldine salt)

$-2\,n\,H^+A^-$

Desprotonation

Polyaniline (emeraldine base)

Figure 6. Polyaniline conductive (emeraldine salt) is deprotonated in an alkaline medium to polyaniline base. A-is an anion, i.e.: Chloride.

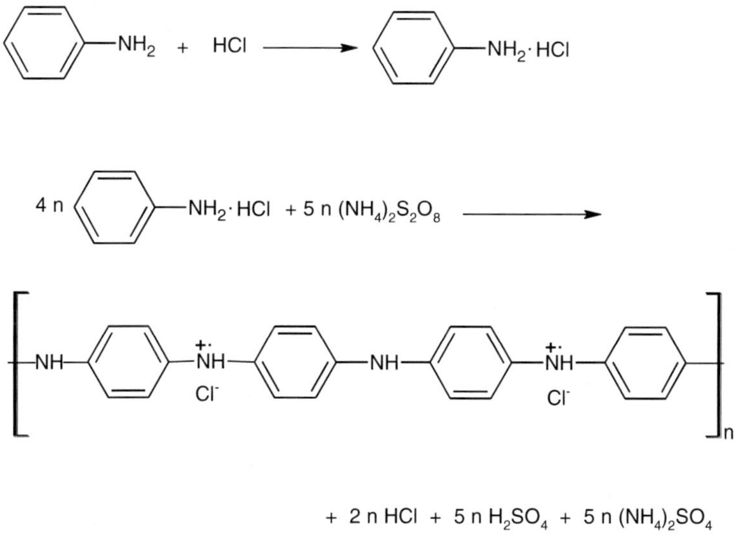

$$C_6H_5NH_2 + HCl \longrightarrow C_6H_5NH_2\cdot HCl$$

$$4\,n\,C_6H_5NH_2\cdot HCl + 5\,n\,(NH_4)_2S_2O_8 \longrightarrow$$

$$+ 2\,n\,HCl + 5\,n\,H_2SO_4 + 5\,n\,(NH_4)_2SO_4$$

Figure 7. Aniline hydrochloride preparation and its polymerization using ammonium peroxydisulfate (0.225 M), in order to obtain polyaniline (emeraldine salt).

3.3.1. Aniline Adsorption Studies

The host (Na-AlSBA-3, Na-AlSBA-16, Na-AlSBA-3 and Na-AlMCM-41) samples were dehydrated at 673 K in vacuum for 1 h. They were then exposed to equilibrium vapors from liquid aniline during 1 h at room temperature (298 K) to obtain aniline saturated hosts according to the procedure described by Anunziata et al. [40]. In these conditions, the aniline/host saturation relation was reached. The color of the samples changed from white to beige. Self-supported wafers of AN-41 (Aniline-Na-AlMCM-41), AN-15 (Aniline-Na-AlSBA-15), AN-3 (Aniline-Na-AlSBA-3) and AN-16 (Aniline-Na-AlSBA-16) were used for FTIR studies, employig a vacuum cell with CaF_2 windows. The spectrum was recorded after cooling the sample in the vacuum cell at room temperature, (Figure 8.)

When the aniline is adsorbed on the materials, the FTIR study shows one H-N-H bending and two C-C ring stretching vibrations at 1622-1625, 1604 and 1498 cm^{-1}, respectively (Figure 8).

The $N\text{-}H_2$ bending vibration overlaps with those of water and the Si-O stretching overtone of the SBA and MCM framework. Figure 8 depicts the original and deconvoluted FTIR spectra of AN-samples; curve fitting calculations were useful for determining the location of the bands.

The fitting confidence was $\chi^2 = 4 \cdot 10^{-2}$ y R^2=0.98. The deconvoluted FTIR spectra discriminate bands close to 1625 cm^{-1} corresponding to water bending and Si-O stretching overtone material. The band corresponding to H-N-H bending is placed at 1625 cm^{-1} and the C-C ring stretching vibrations are located at 1604 and 1498 cm^{-1} [33, 40]. The weak band at 1700 cm^{-1} is attributed to water physically adsorbed onto material.

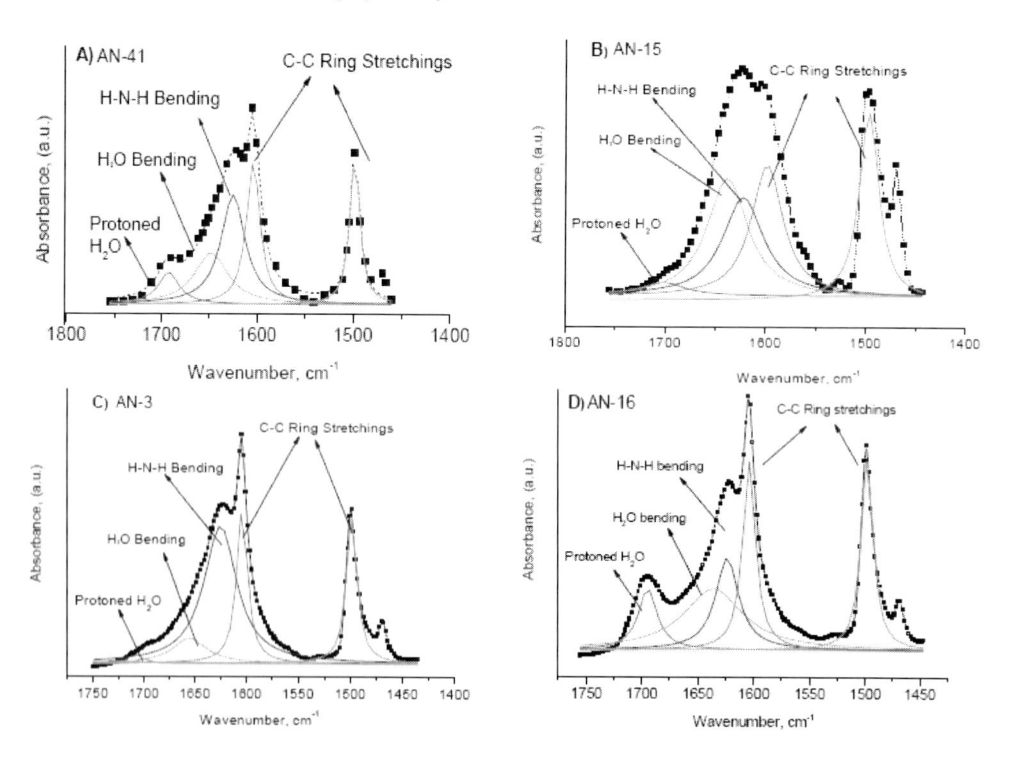

Figure 8. Deconvoluted and original FTIR spectra of: A) AN-41, B) AN-15, C) AN-3 and D) AN-16 in 1400-1800 cm^{-1} range.

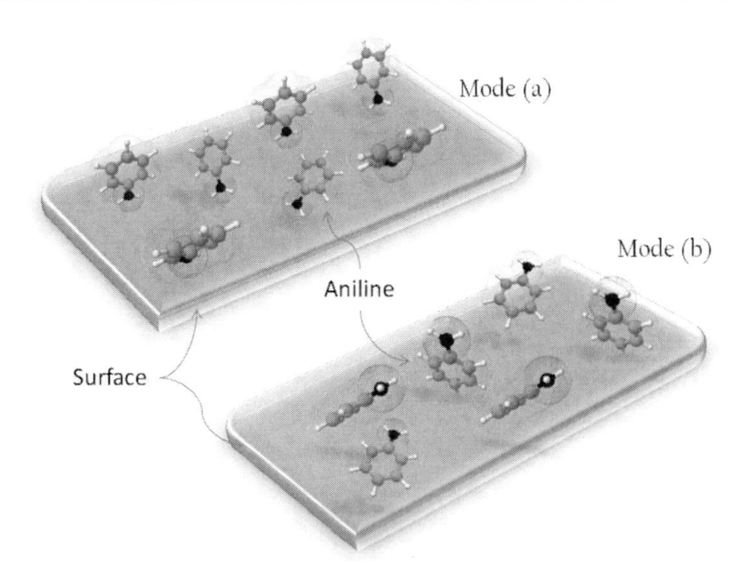

Figure 9. Modes of Aniline absorption on materials (a) aniline adsorbed through π electron system, (b) aniline adsorbed through the amino group.

We could be determined that aniline is absorbed to a great extent on sodium aluminum silicate mesoporous materials of type MCM-41, SBA-3 and SBA-15 through its π electron system remaining free amino group. While in SBA-16 interaction occurs through the amino group and π electron system. In the following figure illustrates the modes of adsorption of aniline (Figure 9).

3.3.2. Polyaniline Composites Characterization

3.3.2.1. Synthesis

The oxidative in-situ polymerization to produce polyaniline/hots composites polyaniline /Na-AlMCM-41, polyaniline /Na-AlSBA-15, polyaniline /Na-AlSBA-3 and polyaniline /Na-AlSBA-16 (PANI-41, PANI-15, PANI-3 and PANI-16, respectively) was performed. The process was carried out adding HCl (0.5 M) to 50 mg of AN-Hosts, in a beaker for 18 h, without stirring in order to obtain aniline hydrochloride. Afterwards, the system was cooled to 273 K and ammonium peroxydisulfate (0.225 M) was added. This condition was kept at least for 5 h. Finally, the material was filtered and washed with 50 ml of 0.2 M HCl and distilled water, and then dried at 333 K.

3.3.2.2. FTIR Characterization

Figure 10 shows the original and deconvoluted FTIR spectrum of PANI-41, PANI15, PANI-3 and PANI-16 in the range 1300-1775 cm^{-1}. Curve fitting calculations were useful for determining the location of the bands. The fitting confidence was $\chi^2 = 5 \cdot 10^{-5}$ and $R^2 = 0.99$. The characteristic PANI/composite bands were observed in all composites. The signal corresponding to the free amine group of the aniline near 1620-1625 cm^{-1} was not detected; evidencing that aniline is not as a free species. But in PANI-16 spectrum shows a band at a band at 1604 cm^{-1} corresponding to C-C benzene ring stretching vibration of aniline, which would prove that aniline subsists as a free species.

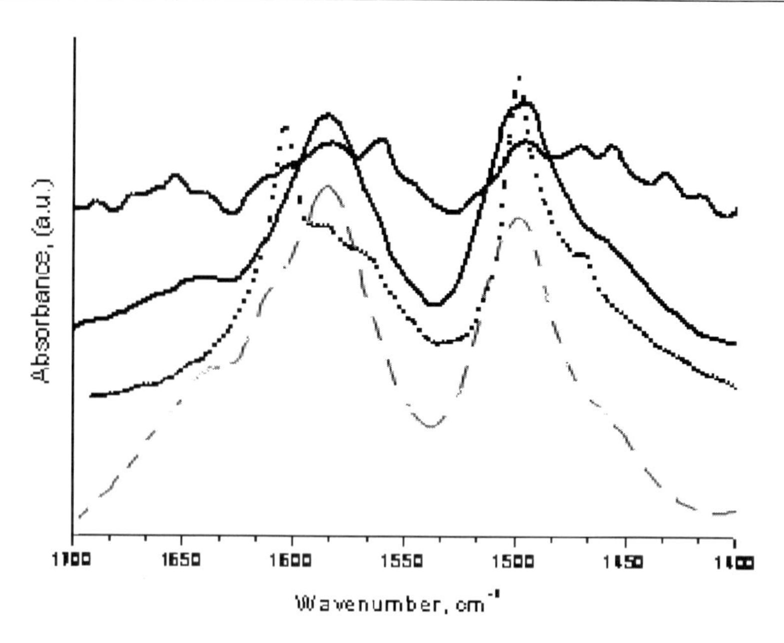

Figure 10. FTIR spectra of A: PANI-41; B: PANI-15; C: PANI-16; D: PAI-3 in the range of 1400-1800 cm^{-1}.

Table V. FTIR absorption characteristics bands of PANI, PANI-41, PANI-15, PANI-3 and PANI-16

Band Assignment	Band position, wavenumber (cm^{-1})				
	PANI	PANI-41	PANI-15	PANI-3	PANI-16
Water bending + overtone Si-O	-	1635	-	1637	-
Quinone Stretching (N=Q=N)	1584	1586	1583	1587	1585
Stretching ring (quinone)	1559	1560 (overlapped)	1560	1560 (overlapped)	1566
C-C benzene ring stretching	1490	1495	1497	1501	1499
Unidentified. Absent on pure PANI spectrum *		1373 1336	1471 1458 1433 1415	1466	1611 1488 1468 1456

* According to Ref [33].

The FTIR spectra exhibit bands associated with PANI within range quinone stretching (N=Q=N) at 1585 cm^{-1}, as well as C-C stretching of benzene ring at 1499 cm^{-1}. The assignments of the bands are found in Table V.

H. Nam et al. [73] reported a FTIR study of $(PANI)_y$ $NbMoO_6$ composites. They showed that shapes and frequencies are not identical to those of pure PANI, probably due to a difference in the orientation and protonation of the polymer chain between confined proton-rich space and in open space. In our study, the absorption bands shift from 1490 cm^{-1} for pure PANI to higher wavenumbers on the composites [3]. This shift could derive from the interaction between PANI and the host. The unidentified peaks could be produced by the interactions between the polymer and the host material. The difference in band shapes could be attributed to the crosslinking of the polymer chains within the galleries of the hosts.

3.3.2.3. XRD Studies

Further evidence is provided by X-ray diffraction studies of the polyaniline/composites, showed in Figure 11. XRD studies reported by S. Goel et al. [37] reflect variation in the molecular orientation and crystallinity of PANI nanowire and bulk PANI samples.

The bulk powder exhibits several broad peaks indicating amorphous state, and only weak crystalline peaks at 2θ ~18° and 2θ ~33° are observed. For nano-PANI samples, a peak at 2θ ~25° has been attributed to the periodicity order perpendicular to the chain direction; another important peak at ~2θ = 5.9° can be ascribed to the scattering parallel to polyaniline chains in one-dimensional structures. In our study, the diffraction pattern of composites differs significantly from that of nano PANI samples. At low angles, the samples evidence a porous structure, whereas at higher angles, the composites reveal amorphous materials (Figure 11). This pattern does not exhibit signals characteristics of PANI, since PANI is confined in the pores interacting with the host. In the Figure it is not possible to observe the polymer (as a "pure" compound) on the external surface of the crystallites in both composites.PANI-15 and PANI-41 show similar XRD patterns to their corresponding hosts. Compared with pure host samples, a noticeable diminution in the peak intensities is observed when filling pores with PANI with all hosts. This can be associated with the relatively lower contrast scattering between the pores and the walls of mesoporous silica resulting from the formation of PANI chains in the channels. This fact demonstrates that the resulting materials are composites, and that they are not formed from a combination of materials or mechanical mixture. Q. Cheng et al. [74] observed that the broad band around 2θ =23° corresponds to amorphous SiO_2, however, no appreciable peak is found at around 25°.

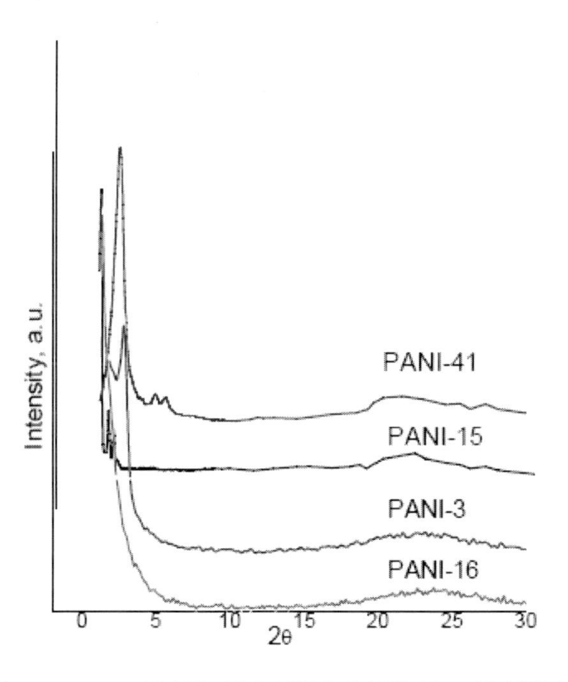

Figure 11. X-ray diffraction patterns of PANI-16, PANI-3, PANI-15 and PANI-41.

This observation implies that PANI in the nanocomposite is also amorphous, and the crystallization of its molecular chains in the pores is hindered because of the confinement effect.

3.3.2.4. BET and Thermal Studies

The surface areas of the hosts and the corresponding composites were: Na-AlSBA-16 and PANI/Na-AlSBA-16: 709 and 655 m^2/g; Na-AlSBA-3 and PANI-3: 780 and 670 m^2/g; Na-AlSBA-15 and PANI-15: 960 and 745 m^2/g; Na-AlMCM-41 and PANI-41: 860 and 748 m^2/g.

The intra-channel polymerization of aniline was proved by the difference seen in the surface area, the composites and the hosts. The reduction of surface area was 7.6, 14.1, 22.4 and 13.2 % for PANI-16, PANI-3, PANI-15 and PANI-41 composites respectively, indicating that polyaniline could be within the channels of its corresponding hosts.

3.3.2.5. TG Analysis

Figure 12 shows the TGA curve analysis for Na-AlSBA-16, PANI-16, Na-AlSBA-3, PANI-3, Na-AlSBA-15, PANI-15, Na-AlMCM-41, PANI-41 and pure PANI.

For pure PANI, the initial weight loss (303- 423 K) is caused by the loss of water from the polymer. The polymer is thermally stable up to 523 K. From this temperature on, the polymer starts to degrade slowly; yet, above 673 K, the polymer degrades rapidly.

For the Na-AlMCM-41 used as host the weight loss is about 8% around 433 K and then is stable to reach at 973 K with a total weight loss of 10%. The initial weight loss of the composite PANI-41, below 403 K, can be attributed to the loss of water and some adsorbed gases from the composite.

Thereafter, remains quite stable up to around 553 K. Above this temperature, the composite (mainly polymer) starts to lose its weight slowly, up to 923 K. Above 853 K, the composite starts to degrade very fast. The weight loss above 873 K can be attributed to the combined weight loss of the polymer and MCM. Thus, the total weight loss is about 10% for the host and for the composite is 40 %, the weight loss attributed to PANI (as a nano-guest) is about 30% w/w at 973 K.

It is interesting to see that the total weight loss for pure PANI is nearly to 50%. If we considered the weight loss of pure PANI and the guest PANI in the composite at 773 K the difference is very higher, 40 and 15% respectively. Thus, this means that as guest in the host forming the composite, the PANI becomes more stable avoid the polymer to be degrade very fast.

For the Na-AlSBA-3 used as a host, the weight loss is about 4% around 413 K and then stabilizes when reaching 973 K with a total weight loss of 8.5%. The PANI-3 initial weight loss below 403 K can be attributed to the loss of water and to some adsorbed gases from the composite. Thereafter, it remains fairly stable up to around 593 K. Above this temperature; the composite (mainly polymer) starts to lose its weight faster, up to 773 K. If we considered the weight loss of pure PANI and guest PANI in the composite at 773 K, the difference would be much wider (38% and 7.4%, respectively). Thus, this means that PANI becomes more stable as a guest in the host forming the composite, preventing the polymer from a fast degrading. The total weight loss for the host and the composite is about 8.5% and 18.5% respectively; the weight loss attributed to PANI (as a nano-guest) is about 10% w/w at 973 K. It should be noted that the total weight loss for pure PANI is near 47%.

For the Na-AlSBA-16 used as a host, the weight loss is about 4% around 408 K and then stabilizes when reaching 973 K with a total weight loss of 6%. The PANI-16 initial weight loss below 403 K can be ascribed to the loss of water and to some adsorbed gases from the composite, showing stability up to around 573 K. Above this temperature, the composite (mainly polymer) starts to lose its weight slowly, up to 973 K. The total weight loss is about 6% for the host and 9.3% for the composite; the weight loss attributed to PANI (as a nano-guest) is about 3.3% w/w.

This percentage corresponds to 0.1 g/g (weight of PANI / weight of composite) for PANI-3 and 0.03 g/g for PANI-16. In addition, the results show that PANI load is greater in PANI-3 than in PANI-16 composite. Both hosts, Na-AlSBA-3 and Na-AlSBA-16, influence the thermo-stability of the guest; however, PANI containing-Na-AlSBA-16 is more stable than PANI containing-Na-AlSBA-3. In this way, PANI containing-Na-AlSBA-15 is more stable than PANI containing-Na-AlSBA-3 but lower to PANI-16, whereas its total weight loss attributed to PANI (as a nano-guest) is about 4.7% w/w.

On the other hand, the weight loss of the polymer for the PANI/Na-AlMCM-41 composite, is around 10%. Above 873 K, Na-AlMCM-41 starts to degrade fast [33]; SBA hosts (SBA-3, SBA-15 and SBA-16), are more stable. PANI contained in SBA hosts is also more stable since, above 873 K, SBA retains its structure with PANI within its channels and the polymer becomes more stable.

In general, PANI is retained weakly for both hosts. However, PANI is more retained in Na-AlSBA-16 than in Na-AlSBA-3 host. This is due to a stronger PANI-host interaction and mostly to a diffusion effect in Na-AlSBA-16 with spherical cavities connected by small-diameter openings (channels and cavities), in contrast to Na-AlSBA-3 with uniform tubular channels and some microporosity [75].

The adsorption/desorption in the SBA-16 has been known as "pores of blockade and effect of network" [76, 77]. This is why this material displays a limitation in the adsorption and diffusion of aniline, leading to a lower amount of polymer (PANI) within SBA-16 host.

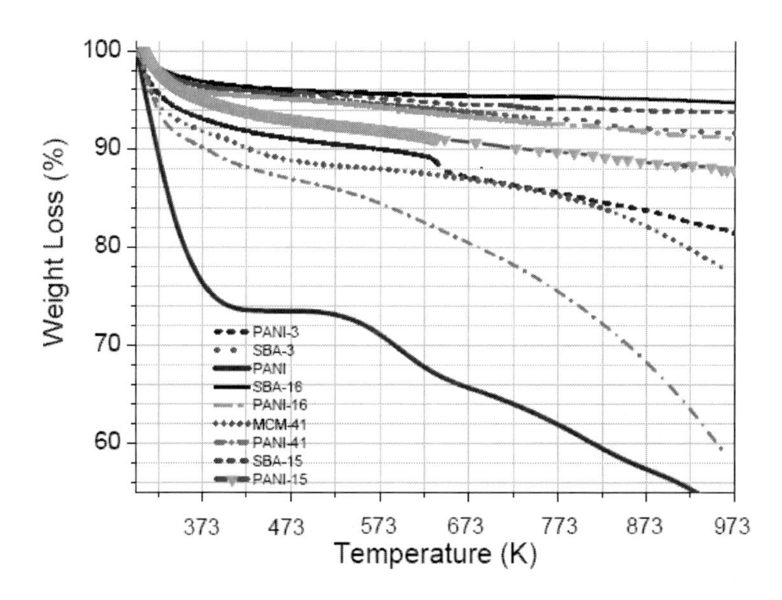

Figure 12. TGA analysis for: PANI; (SBA-3) Na-AlSBA-3, PANI-3; SBA-16) Na-AlSBA-16, PANI-16.

The higher amount of conductive emeraldine salt implies major conductivity of PANI-16 composite; yet we must consider the dependence on the structural configuration of the material, and on the type of PANI anchorage to the host and the connection of PANI nano-wires, as the case of PANI-15 which according the TG analysis shows major PANI quantity retained than SBA-16 and minor than SBA-3, but as we can see in the following section, its conductivity is minor than PANI-16, indicating that not at all PANI is as conductive emeraldine salt in the PANI-15 composite. According to TG analysis, superficial area and FTIR studies, PANI-41 and PANI-3 composite presents high formation of PANI with respect to PANI-16 while according to SEM studies, it can be stated that the higher amount of PANI does not affect superficially nor structurally the nano-structured materials (MCM-41, SBA-3, SBA-15 and SBA-16), as may be also inferred from XRD studies. Thus, we can suggest that the nano-structured hosts are preserved after aniline in-situ polymerization, in its channels/cavities and the composites were successfully prepared.

3.3.2.6. SEM Microscopy Studies of the Composites

The SEM image of Na-AlMCM-41 and PANI-41 composite particles is given in Figure 13. NaAlMCM-41 particles were less than 2 μm and they took the shape of a cubic-rectangular bar.

(a)

(b)

Figure 13. SEM images of (a) Na-Al-MCM-41 and (b) PANI-Na-AlMCM-41 composite.

(a) (b) (c)

Figure 14. SEM images of (a) PANI-3, (b) PANI-16 and (c) PANI-15.

Virtually no difference in particle surface morphology between the MCM-41 host and the PANI-41 composite material was observed, which can be served as other evidence (in agree with BET and TG results) of PANI confinement within the MCM-41 nanochannels, excluding the possibility that some nano particles or filaments of PANI is located on the outer surface.

SEM micrograph in Figure 14 shows the morphology of the grains of PANI-SBA-3, PANI-SBA-16 and PANI-SBA15.

From these studies, the well-defined external morphologies of the composites (PANI-3, PANI-16 and PANI-15), suggest that the nanomaterials employed as reservoir have a highly ordered structure, and a low level of imperfections or defects in the lattice, without any other phase on external surface, indicating that PANI is within the channels. In addition, no polymer chains are observed on the outer surface of SBA-3, SBA-16 and SBA-15 (Figure 14 a-c).

3.4. Polypyrrole Composite

Polypyrrole as one of the most promising conducting polymers has received comprehensive interests due to its excellent characteristics including easy preparation, environmental stability, high conductivity, etc. The mentioned merits confer to polypyrrole potential uses in several fields, such as sensors, actuators and electric devices [78–81]. On the other hand, the metal nanoparticles, such as silver and gold, have attracted much attention in recent years due to their interesting properties and potential applications in technological fields [82]. Silver particles have applications in catalysis, conductive inks, thick film pastes and adhesives for various electronic components, in photonics and in photography [83–86].

A conductive polymer is basically a long carbon chain having a much extended conjugation. Conjugation is understood as the alternation of single and multiple bonds. This type of structure has the fundamental property of having *p* electronic orbital extended over the entire structure. An electron in one of these levels would be highly delocalized and have high mobility, so that it would be possible to obtain electrical conduction.

The reason that a polymer with this structure not always presents conductivity, is because the electron requires a large amount of energy in order to occupy this type of orbital, since the energy separation between it and the molecular orbital employed is large. However, this energy separation can be reduced significantly by using what is known as doping [87].

This process consists in incorporating some type of structural atoms, which very strong electronegativity with respect to which possesses the carbon atoms.

Pyrrole (PY) is a heterocyclopentadiene, which contains a butadiene unit bound at both ends to a hetero-atom which has pairs of free electrons. Pyrrole representation has shown in Figure 15 wherein each ring atom, C or N, is bonded to another three via an σ bond to the atom whose formation employs three sp2 orbital, which are located in one plane and each other forming angles of 120 °. After contributing with one electron for each σ bond, each carbon atom of the ring has another free, while the nitrogen has spare two; these electrons occupy p orbital that overlap each other to generate π clouds: one above and one below the ring plane, which contain a total of six electrons, i.e. the aromatic sextet. The delocalization of π electrons stabilizes the ring; the result is an abnormally low heat of combustion, so it tends to give substitution reactions keeping the stability of the ring.

The extra pair of N, the head of the usual basicity of nitrogen compounds, is involved in the π cloud, so that is not available to be shared with acids. Accordingly, the pyrrole is a very weak base ($Kb = 2.5x10^{-14}$), compared with most amines.

For the same reason, there is a high electron density in the ring, which gives the pyrrole, a highly reactive to electrophilic substitution: are involved in reactions such as nitrosation and coupling with diazonium salts, which are characteristic of only more reactive benzene derivatives, i.e. phenols and amines. In summary, the pyrrole is best represented by a ring with an inner circle representing the aromatic sextet [88].

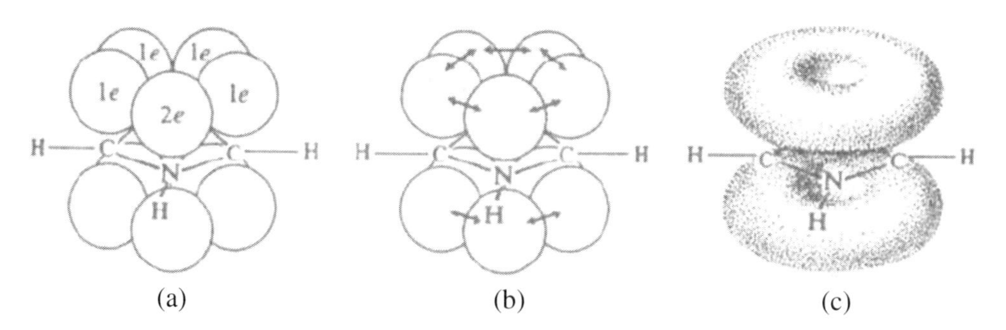

Figure 15. Molecule of the pyrrole. (a) Two electrons in the p orbital of nitrogen, one in the p orbital of each carbon. (b) Overlap of p orbitals to form π bonds. (c) Clouds above and below the ring plane, six π electrons in total, and the aromatic sextet.

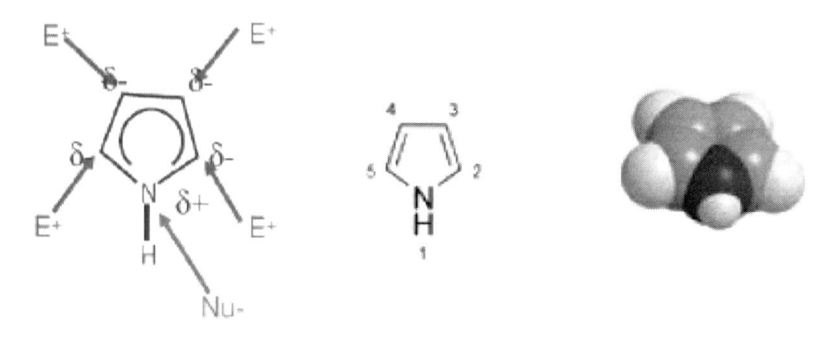

Figure 16. Scheme of electrophilic substitution of pyrrole.

By the electrophilic substitution of Pyrrole the 2nd and 5th position prevail, although the product of substitution at positions 3 and 4 is obtained in low proportion (see Figure 16).

The polypyrrole (PPY) shows a notable electrical conductivity in its oxidized form. This feature has attracted many scientists to study the electronic properties of these conjugated polymers, both doped and undoped state. In particular, the charge transport mechanism has been widely discussed. In recent years there has been increased interest in the development of nano-composites supported on mesoporous silica, due to their potential applications in catalysis and in miniaturized electronics and optical sensors [2, 54, and 89]. Numerous studies have been reported with respect to the encapsulation of guest materials such as semiconductors [90, 91] metals [92, 93] and polymers [94-98] stays within mesoporous silica. The resulting nanocomposites exhibit unique properties which differ from the materials separate. Meanwhile, there is a particular interest in materials where the conducting polymer, is confined in the channels of mesoporous materials used as host to produce new structures, even at molecular level, [30, 31, 67, 68, 99, 100].

There are only some studies in order to optimize the preparation of either polyaniline or polypyrrole, with respect to its molecular structure, molecular weight, morphology and conductivity, [101, 102]. More recently Blinova et al. [103], studied the influence of the oxidant in the synthesis of these polymers, using for this purpose ammonium persulfate in an acidic medium, determining that the oxidation of the monomers are governed by the same chemistry principles, and proposed that the polymerization reaction of the pyrrole is similar to the polymerization of aniline as shown in Figure 17.

There is not much information of the adsorption in liquid-vapor equilibrium of pyrrole on mesoporous aluminosilicates used as host. Cheng and colleagues [53, 104] have studied the adsorption of pyrrole on the channels of MCM-41 and SBA-15, and identified by infrared spectroscopy and thermal analysis, various types of interactions between pyrrole and the host.

Nakayama et al. [105], have succeeded in electrochemically polymerized polypyrrole, in presence of zeolite Y, achieving a composite film. They found that the polypyrrole included in the channels of the zeolite was in neutral form and the load must be balanced by the incorporation of cationic electrolytes.

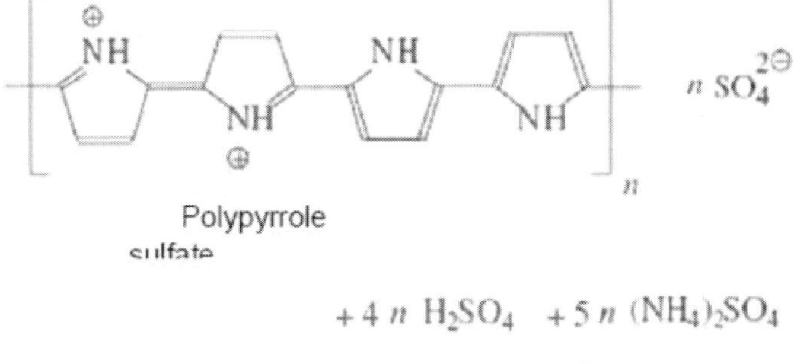

4 n NH + $5n\ (NH_4)_2S_2O_8$

HA

NH NH NH NH $n\ SO_4^{2-}$

Polypyrrole sulfate

$+ 4n\ H_2SO_4 \quad + 5n\ (NH_4)_2SO_4$

Figure 17. Pyrrole Oxidation with ammonium persulfate in acid medium.

In this section we present recent progress made on the adsorption of pyrrole and subsequent polymerization within the channels of Na-Al-SBA-3. Because they are new results, a more detailed analysis and other technical studies will be necessary to achieve complete this goal.

3.4.1. Pyrrole Adsorption Studies

SBA-3 and Na-AlSBA-3 samples, employed as hosts, were dehydrated at 673 K in a vacuum for 2 h at 2.10^{-3} mbar. Afterwards, there were exposed to equilibrium vapors from pyrrole during 24 h at 353 K to obtain pyrrole saturated hosts (Py-SBA-3 and Py-Na-AlSBA-3), according to the procedure described by Anunziata et al. [33, 40, 106]. In these conditions, the pyrrole/hosts saturation relation was reached. The FTIR spectrum of pure pyrrole is shown in Figure 18. We can see the bands assignments according to literature data. Lord and Miller [109] Publisher IR and Raman data in 1942, whereas more precise spectrum was reported by Navarro and Orza [110], employing gas-IR equipment. Table VI shows a resume of this information.

Figure 19 shows the original FTIR spectra for pure pyrrole and the adsorbed pyrrole at 353 K under vacuum over Na-Al-SBA-3 (PY/Na-Al-SBA-3). Before the pyrrole adsorption, the host was heated at 673 K for 4 hours under vacuum, desorbing the possible water molecules on the SBA-3 material. It is clear that the spectrum shown two distinctive IR zone where the pyrrole can be identified.

Table VI. Pyrrole assignment bands according to literature data

Bands assignments	Band position Wavenumber cm^{-1}
	Literature data
Stretching N-H [a, b]	3520
	3400-3405
Stretching C=C [b]	1521-1529
Stretching C-C/ C-N [c]	1473-1467
	1422-1424
Stretching C=C, C-C [c]	1396-1405
Bending CNH, Bending CCH [a,b,c]	1287-1288
Bending CCH [a,c]	1072-1074
	1049
Stretching C-C, Stretching CCH [c]	1014-1018
Bending C-H out-of-plane[a]	826-836
N-H out-of-plane [b]	736
Bending C-H out-of-plane[a]	721
	712
Ring deformation [a,c]	619-626
Bending N-H out-of-plane [b]	559
	474-479

(a) ref [111]; (b) ref [112], (c) ref. [113].

Figure 20 (A, B) depicts the original and deconvoluted FTIR spectra; curve fitting calculations were useful for determining the location of the bands. The fitting confidence was $\chi^2= 0.0011$ and $R^2= 0.9990$. Thus it can be seen vibrations of asymmetric C-N, C=C at 1440 cm^{-1}. This band appears slightly shifted to lower wavenumber respect to pure pyrrole (1470–1475 cm^{-1}), indicating that pyrrole is adsorbed trough the ring, (π electrons of (C=C).)

In the same way, is observed a band at 1420 cm^{-1} corresponding to stretching of C-N and a low signal at 1396-1400 cm^{-1}, before the deconvolution assigned to C-C vibration, additionally new band was detected at 1630-1635 cm^{-1}, corresponding to an overtone of Si-O of the host, and a deformation signal of water adsorbed.

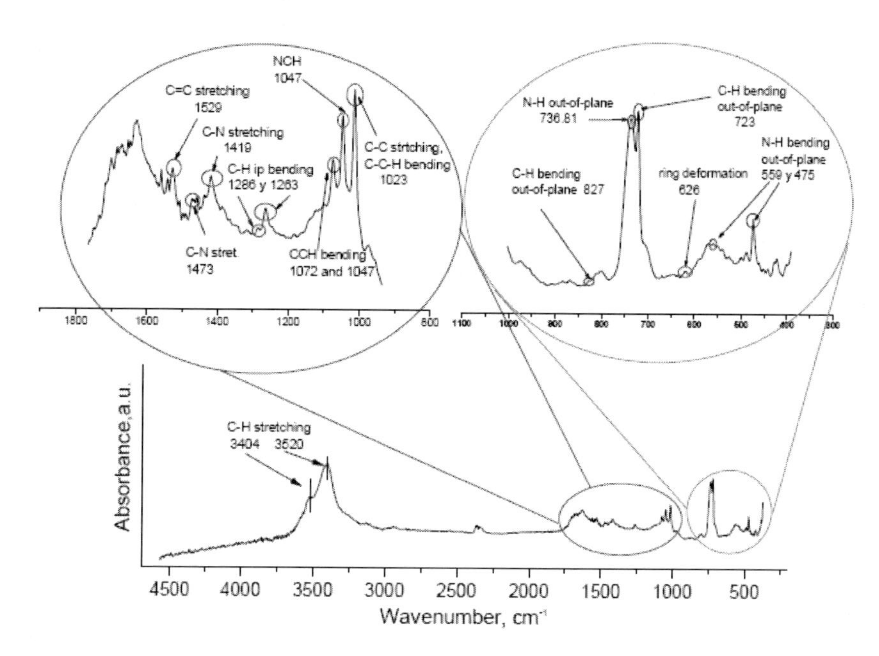

Figure 18. FTIR spectrum in the full range between 4500 - 400 cm^{-1} of the pyrrole.

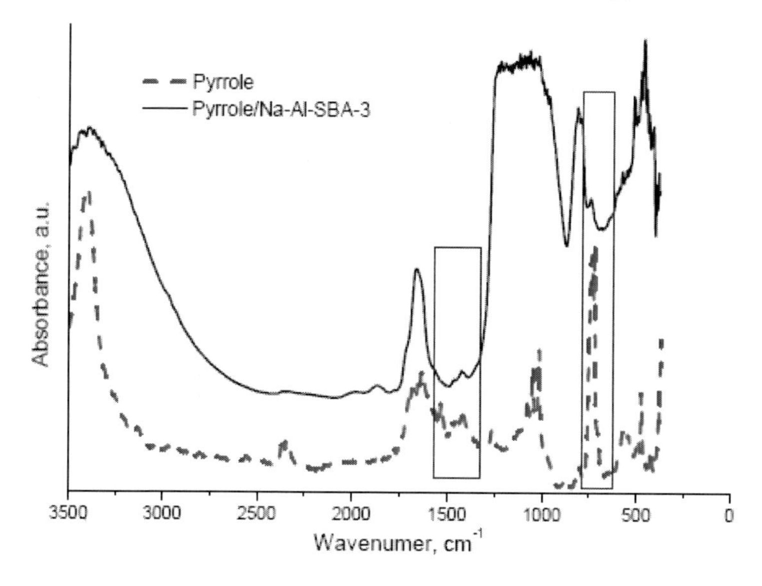

Figure 19. Spectra of pyrrole adsorbed on the material and pure pyrrole between 3500-450 cm^{-1}.

After the pyrrole adsorption, the samples were exposed to air before the FTIR analysis. As the result of this analysis, bands at 1698-1706 cm^{-1} can be assigned to re-hydratation of PY/Na-Al-SBA-3 sample.

Figure 20 B exhibits the original and deconvoluted FTIR spectra in 700-760 cm^{-1} zone. The curve fitting calculations were useful once again, for determining the better bands assignments. The fitting confidence was $\chi 2$= 0.0008 and R2= 0.9980. The bands due to a bending out-of-plane for N-H bond, at 737 cm^{-1} and C-H bond at 724 and 710 cm^{-1} are well detected.

The five member aromatic heterocyclic compounds, as pyrrole, acts as anphoteric one, interacting with basic O of the network through H bonds of NH or, as a electron donor molecule (base) to one acid cation (in our case, i.e., Na which is employed as a counter ion of tetrahedral Al of the host).

Pyrrole adsorption studies were carry out by Fostër et al. [111], over Na exchanged Faujasites, employing to pyrrole as a probe molecule to determine the acidity of these structures. They concluded that the absorption bands in the FTIR analysis of the NH group are more sensible than intermolecular interactions.

The authors found in the zone of 3500 cm^{-1}, for gaseous and liquid pyrrole, an important shifting of the absorption band to lower frequency, for the NH stretching vibrations mode. On the contrary for bending vibrations mode the shift is to higher frequency. Analyzing the FTIR spectra of pyrrole in its different phases, the NH bending out-of- plane is more evident than in the plane. In this way, the authors could determined that the NH stretching and bending of NH and CH groups outer-of-plane are a key to understand the pyrrole intermolecular interactions.

Figures 20 A and B indicate the pyrrole adsorption on the nanostructured material is carried out through its aromatic ring system. This is based on the bands assigned to stretching modes for the NH groups (737 cm^{-1}), have no shifts after adsorbed pyrrole. This would be indicative that the pyrrole is not formed via hydrogen bonds imine group with SBA-3 material used as hosting. In addition, we can observe others shift or perturbations in the bands assigned to vibration modes of the pyrrole ring toward lower wavenumbers, or lower energy, so that these links will require less energy to vibrate, suggesting that if the pyrrole molecule is adsorbed through the π-electron system.

The aromatic ring of each molecule would be destabilized energy, so that the incidence of an infrared beam, with lower frequencies is sufficient to achieve absorption energy for the bonds of these molecules. This would give an indication us that, the molecules of pyrrole would be adsorbed through aromatic ring, and the effect of the bands shifting can be interpreted like an additional interaction, between the pyrrole aromatic rings of with the Na^+ cation (Figure 21).

Uehara et al. [114], studied and characterized by means of spectroscopy, polypyrrole formed in the channels of the zeolite Y, observing that after pyrrole was adsorbed, the samples become of black color indicating the occurrence of some oxidizing polymerization. The absorption signals of polypyrrole can be differentiated of the pyrrole oligomers and monomers existing in the host. They could determine that pyrrole molecules are occluded predominantly like oligomers and or monomers in pyrrole/zeolites. Taking account the previously exposed, in the IR region of 1600-1700 cm^{-1}, two bands are observed that cannot be assigned appropriately to pyrrole; but the signals of C-C vibration bonds between rings, was located at 1671 and 1650 cm^{-1}. These signals are involved with a poor polymerization, carried out during the adsorption process. The signal at 1650 cm^{-1} was assigned to the C-C stretching between the rings of ter-pyrrole, whereas the band at 1671 cm^{-1} can be assigned to, C-C stretching between polypyrrole rings. The bands and their respective assignments can be observed in Table VII.

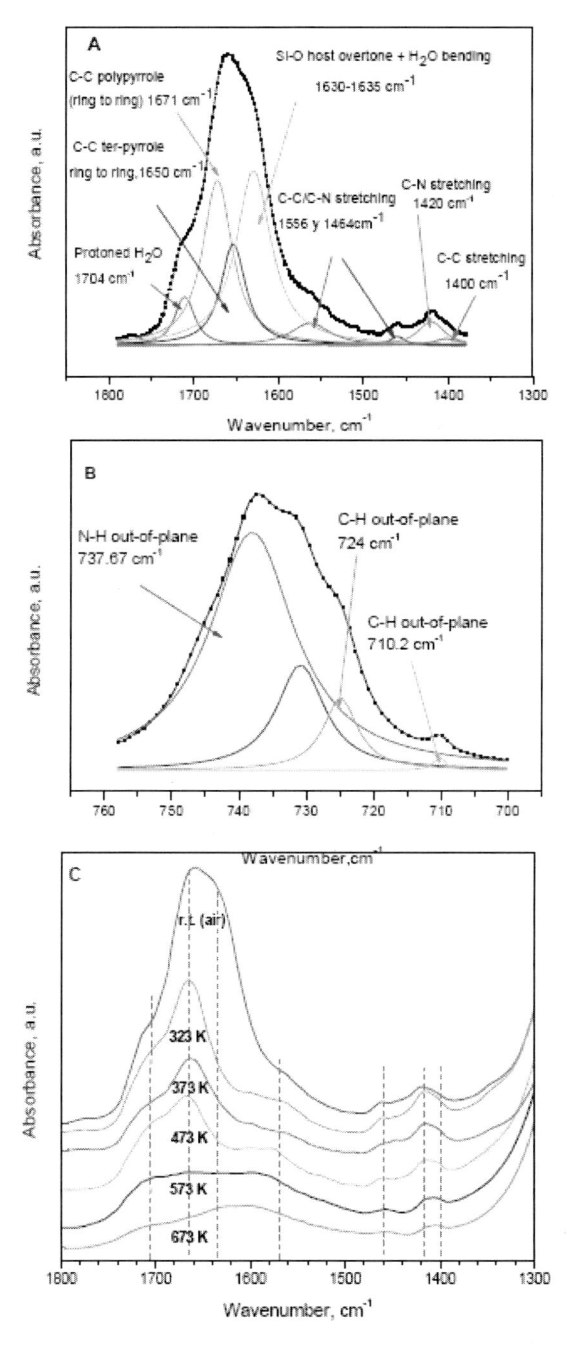

Figure 20. Pyrrole/Na-Al-SBA-3 assignments in two FTIR zone, A): 1350- 1800 cm^{-1} and B) 700-760 cm^{-1}, original spectra (line + symbol), and deconvoluted curves (colored lines); C) FTIR for Pyrrole-Na-Al-SBA-3 sample in the region of 1300-1800 cm^{-1} in air (r.t.), after heating to 323, 373, 473, 573 and 673 K under vacuum.

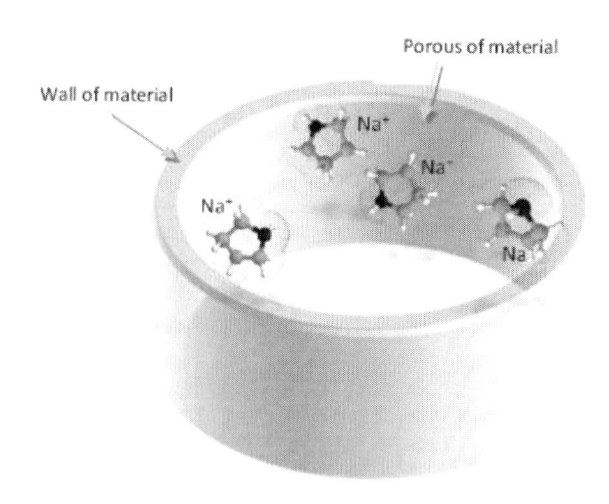

Figure 21. Scheme of pyrrole adsorbed onto Na-SBA-3 surface.

Table VII. IR absorption characteristics bands for pure Pyrrole and PY/Na-Al-SBA-3

Band Assignment	Bands position, Wavenumber, cm^{-1}		
	Literature data	Polypyrrole	PPY-Na-Al-SBA-3
Protoned water	1700	-	1700
Neutral water + overtone Si-O	1630-1635	-	1631
Over-oxidation [b]	1714	1714	1718
C-C Stretching of rings [a, b ,c]	1676	1676	1672
C-N, CNH Stretching [a, b, c]	1586	-	1585
C-C Stretching of the ring [b]	1542-1547	-	1543
C=C, C-N Asymmetric Stretching [c]	1475	1475	1471
C=C Stretching [d]	1462	1462	1462
Asymmetric C=C, C-N Stretching [c]	1456	1453	1453
Bands assigned to oligomer compound (terpyrrole)			
Asymmetric C=C, C-N Stretching [c]	1443	-	1444
Symmetric C-C Stretching [c]	1414	-	1414
Unidentified bands	-	-	1491

(a) ref [111]; (b) ref [112], (c) ref. [113].

According to the results showed in Figure 20 C, (FTIR for Pyrrole-Na-Al-SBA-3 sample in the region of 1300-1800 cm-1, after the following heat treatments: in air (a.t.), after heating to 323, 373, 473, 573 and 673 K under vacuum), it is possible to be observed that the Pyrrole/Na-Al-SBA-3 samples retain very little amount of pyrrole after heating in vacuum at 473 K, corroborating the pyrrole--Na-Al-SBA-3 interactions. As desorption temperature increases, water content of the sample and pyrrole adsorbed do not decrease in the same order. When the temperature increases, the maximum of the absorption band assigned to the overlapping of C-C bonds deformations between rings and protoned water, was shifted to higher wavenumbers and nearer of C-C deformation and protoned water signals; whereas the desorption temperature increases up to 573 K by 1 hour, the protoned water band is still present, indicating that the sample retains water after the desorption of pyrrole.

3.4.2. Polypyrrole Composite Characterization

3.4.2.1 Synthesis

The oxidative polymerization of adsorbed polypyrrole (PPY) was carried out at 298 K. One ml of ferric chloride (Cl_3Fe) 0.25M (oxidant) was added at 50 mg of materials with pyrrole adsorbed, without stirring, and left at room temperature for 24 hr. Finally, the materials were filtered, washed and dried at 323K. The samples obtained were FePPY/SBA-3 and FePPY/Al-SBA-3 composites (Table VIII). The same procedure was carried out using ammonium persulfate ($(NH_4)_2S_2O_8$) as oxidant, obtaining the samples of SPPY/SBA-3 and SPPY/AlSBA-3 composites (Table VIII).

For comparison, bulk polypyrrole was synthesized following a typical synthesis at 273K and using ammonium persulfate as oxidant [107, 108].

3.4.2.2. FTIR Characterization

The nanowires of polymers incorporated within inorganic hosts have to great interest, since it allows us to combine the characteristics of his components as a result of its interaction at molecular level.

Polypyrrole (PPY), like a promising conductive polymer, has been widely studied due to its high polarizability, high conductivity [115, 116] and electroreological properties [117-119]. This polymer exhibits different properties in comparison with the polyaniline (PANI) and other conductive polymers [120]; and so the incorporation of PPY within the channels of a nanostructured hosts, be capable of provide a material with different electrical characteristics and applications.

Although many works have been published employing polyaniline / nanostructured hosts, are few studies with polypyrrole confined within the channels of a mesoporous silica used as nanostructured hosts [120 - 122]. Polypyrrole (PPY) absorption IR spectrum, between 2000 and 450 cm^{-1} is shown in Figure 22.

As it can see several bands were assigned according to literature data [74, 104, 112, 113]. In Figure 23, the IR studies in the entire region (4000 and 500 cm^{-1}), for polypyrrole, Na-Al-SBA-3 and the composite obtained with polypyrrole/Na-Al-SBA-3 are shown. It is possible to be detected a zone where the analyses of the bands are realized, is fitted with line of color. We can to appreciate in this figure that composite displays absorption bands between 1700 and 1400 cm^{-1}, that differs from the host as well as of the host to which the adsorption of the monomer was realized (PY), previously to the polymerization.

Figure 23 shows FTIR spectra of pure polymer, aluminosilicate hosts and composites samples. The pure silica and aluminated silica show absorption bands at 1630, 1085, 964, 800 and 464 cm^{-1}. The peak at 1630 cm^{-1} is assigned to the OH bending vibrations of the adsorbed water molecules [51]. Typical asymmetric and symmetric Si-O-Si stretching vibrations are centered at 1085 and 800 cm^{-1}, respectively.

Table VIII. Polypyrrole/host samples, obtained with different oxidant agents

Oxidizing agent	SBA-3	Na-AlSBA-3
$FeCl_3$	FePPY/SBA-3	FePPY/AlSBA-3
$(NH_4)_2S_2O_8$	SPPY/SBA-3	SPPY/AlSBA-3

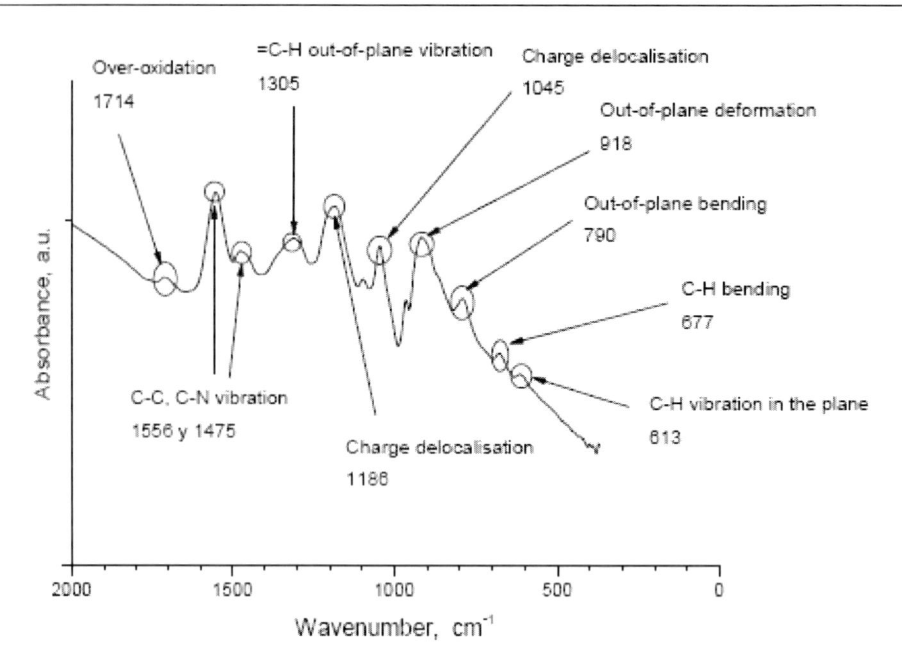

Figure 22. Polypyrrole FTIR spectrum in the range of 2000-450 cm^{-1}.

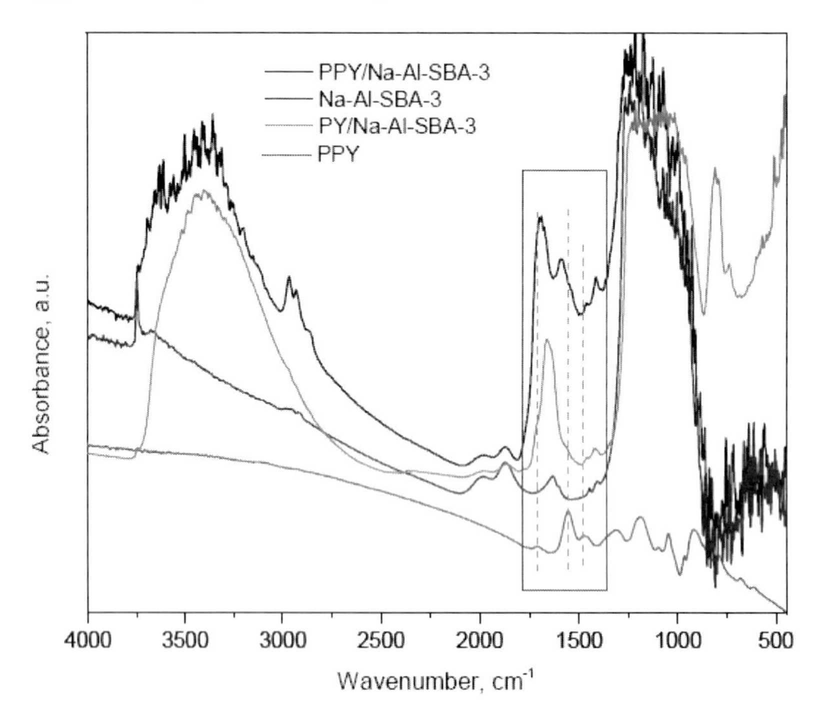

Figure 23. FTIR Spectra of PPY/Na-Al-SBA-3, PY/Na-Al-SBA-3, Na-Al-SBA-3 and pure polypyrrole between 4000 y 500 cm^{-1}.

The band at 969 cm^{-1} corresponds to Si-OH vibrations of the surface silanols, characteristic of mesoporous silica, but absent in the Na-AlSBA-3, which indicates that the surface silanol groups around 964 cm^{-1} interact with the Al species and contribute to form the Si–O–Al superficial species in the process of preparation of Na-AlSBA-3 [51].

The synthesized polypyrrole shows bands at 1556 cm^{-1} (combination of intra-ring C=C and inter-ring C–C vibration), 1475 cm^{-1} (vibrations C-C, C-N), 1305 cm^{-1} (=C-N in-plane vibration), 1186 cm^{-1} (charge dislocation), 1045 cm^{-1} (C-N), 918 cm^{-1} (C-H out-of-plane deformation), 790 C-H cm^{-1} (out-of-plane bending), 677 cm^{-1} C-H (outer bending) and 613 cm^{-1} (C-H in-plane vibration) [120, 123].

Sharp bands of PPY can be observed on the composites prepared (Figure 24). Composites show sharp polypyrrole bands around 1560, 1475, 930 and 677 cm^{-1} (Table IX). In the spectrum of FePPY/SBA-3 sample, a band at 1170 cm^{-1} (corresponding to charge dislocation band of pure PPY, 1186 cm^{-1}) can be seen overlapped with the Si-O-Si stretching band of the host. This can be attributed to the higher surface area of the SBA-3 than the aluminated sample, thus a good polymerization and a pyrrole adsorption higher than that in Al-SBA-3 were found in SBA-3.

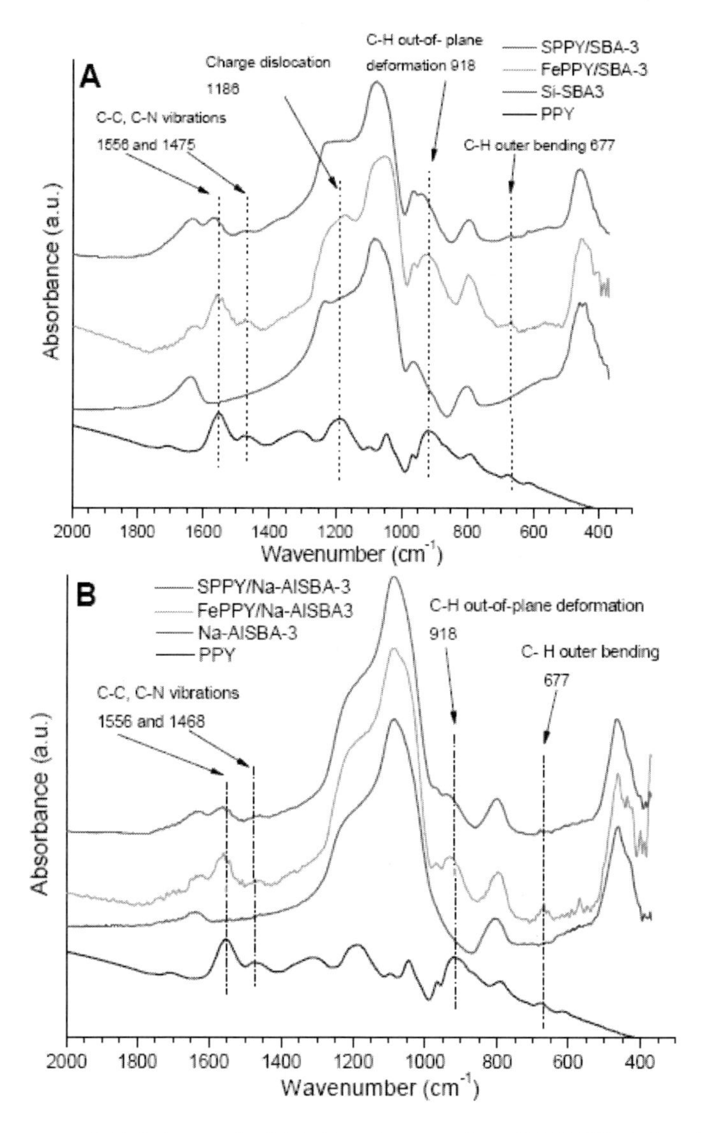

Figure 24. FTIR spectra of A) Polypyrrole, SBA-3, and composites: FePPY/SBA-3 and SPPY/SBA-3; B) Polypyrrole, Na-AlSBA-3 and composite: FePPY/Na-AlSBA-3 and PPY/Na-AlSBA-3.

Table IX. IR absorption characteristics bands for pure Polypyrrole and PPY/Na-Al-SBA-3

Band Assignment	Band position, wavenumber (cm^{-1})				
	Pure PPY	FePPY/SBA-3	SPPY/ SBA-3	FePPY/ AlSBA-3	SPPY/ AlSBA-3
H_2O bending	-	1630	1630	1630	1630
Intra-ring C=C and inter-ring C–C	1556	1558	1570	1558	1560
C-C, C-N	1475	1475	1475	1475	1475
=C-N in-plane	1305	-	-	-	-
Charge dislocation	1186	1170	-	-	-
C-N	1045	-	-	-	-
C-H out-of-plane deformation	918	926	941	929	940
Out-of-plane bending	790	-	-	-	-
Outer bending	677	677	677	677	677
C-H in-plane	613	-	-	-	-

The band at 1630cm^{-1} corresponds to water bending, typical in mesoporous materials.

The samples do not show terpyrrole bands, which would have to be found at 1444 and 1414 cm^{-1}, indicating that the pyrrole is fully polymerized [113].

The bands of the composite corresponding to the polypyrrole also show a little shift to higher wavenumbers with respect to the pure polypyrrole. This indicates that the polymer chains are shorter in the composite compared with those of the pure compound. [120]. Instead, the spectra of the SPPY/SBA-3 and SPPY/Na-AlSBA-3 show bands at 1570 and 1560 cm^{-1}, shifted to a wavenumber higher than that of the corresponding of pure PPY (1556 cm^{-1}). In general, this peak tends to shift to a lower wavenumber as the conjugated length of polymer is increased. Therefore, the PPY chain in the composite is considered to be shorter than the length of pure PPY, and shorter than the samples polymerized with $FeCl_3$ [112, 124]. The same bands in FePPY/SBA-3 and FePPY/AlSBA-3 composites are less defined than in samples polymerized with ammonium persulfate (SPPY/SBA-3 and SPPY/AlSBA-3).

3.4.2.3. Surface Area Analyses

The BET specific surface area decreases with the Al content, from 1259 m^2/g (SBA-3) to 810 m^2/g for AlSBA-3 Si/Al=10, due to the loading of the pores by guest species, as well as to the major contribution of the additional mass Al_2O_3 in the sample [51].

Table X. Surface Area and Conductivity of the hosts and composites

Sample	Surface area (BET) (m^2 g^{-1})	Area % Reduction	Conductivity (S cm^{-1})
SBA-3	1259	--	--
FePPY/SBA-3	361	71.32	$3.37.10^{-6}$
SPPY/SBA-3	393	68.78	$1.12.10^{-6}$
Na-AlSBA-3	810	--	--
FePPY/Na-AlSBA-3	289	64.32	$5.23.10^{-7}$
SPPY/Na-AlSBA-3	315	61.11	$1.28.10^{-7}$

Table X shows the area of the different composites. In all the composites, a decrease can be found in the specific area, indicating that polypyrrole could be within the channels of the corresponding SBA hosts.

3.4.2.4. TG Analysis

Figure 25 shows the TGA curve analysis for Polypyrrole; the host Na-AlSBA-3, and the composites. For pure Polypyrrole, the initial weight loss (300- 373 K) is caused by the loss of water from the polymer. The polymer is thermally stable up to 430 K. From this temperature on, the polymer starts to degrade rapidly.

The Na-Al-SBA-3 shows a constant but slight weight loss with increasing temperature. The weight loss is about 4% around 373 K and then stabilizes with a total weight loss of 8.3% when reaching 873 K. The same behavior was found for Si-SBA-3 (not shown).

In all the composites, the initial weight loss below 373 K can be attributed to the loss of water and to some gases adsorbed from the composites.

It should be noted that for FePPY/Si-SBA-3 and SPPY/Si-SBA-3, the decomposition (evacuation) of encapsulated PPY is slower than that for pure Polypyrrole. This slower decomposition of the encapsulated PPY implies diffusional constraints in the channels system [120].

If we considered the weight loss of pure PPY and guest PPY in the composite (FePPY/Si-SBA-3) at 620 K, a notable difference can be shown (30% and 18%, respectively); however, at 860 K this difference would be much wider (61.5% and 26.5%, respectively). Thus, this means that PPY becomes more stable as a guest in the host forming the composite, preventing the polymer from a fast degrading.

It is interesting to note that SBA-3-composites have more amount of PPY than AlSBA-3 composites. This fact can be seen in the difference in weight loss. At 860 K the difference of weight loss between FePPY/Si-SBA-3 and FePPY/Al-SBA-3 is approximately 12% (26% and 14%, respectively).

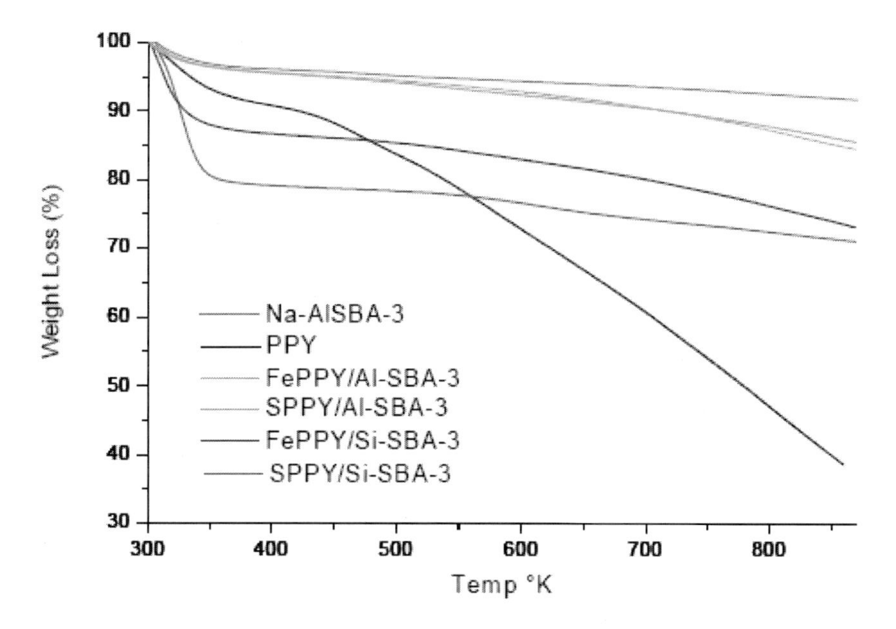

Figure 25. TGA analysis for: pure polypyrrole, Na-Al-SBA-3 (host) and the different composites.

3.5. Polyindole Composite

Indole is an aromatic heterocyclic organic compound. It has a bicyclic structure, consisting of a six-membered benzene ring fused to a five-membered nitrogen-containing pyrrole ring. Indole is a solid at room temperature. Polyindole is an electroactive polymer, which can be usually obtained after anodic oxidation of indole in various electrolytes and by chemical polymerization using oxidants like $FeCl_3$ and $CuCl_2$ [137].

Unlike the polythiophene or polypyrrole, polyindole is a polymer hardly known in the bibliography and in the few occasions the used method of synthesis is by polymerization electrochemistry. These conductive polymers, besides the mentioned difficulty in their synthesis, display the disadvantage of their low persistence of chemical agents (i.e., polyacetylene, is a very good conductor but very easily oxidized), or their difficult conformation due to their thermal instability to high temperature and/or its insolubility in conventional dissolvent, which prevents its conformed by fusion or evaporation of the dissolvent.

We can find studies dedicated to redox conducting polymers originated both, from the understanding of the theoretical concepts leading their original conduction mechanisms, and from the convenient employment of these new materials. Several works oriented to the study of pyrrole, aniline, carbazole and their derivatives were reported [126-129], however the polymerization of indole and its substituted derivatives even though all these materials, exhibit very close structural similarity, are still discussed [130-134].

H. Talbi et al. [135] studied the growing mechanisms of polyindole and the most probable structure of this polymer.

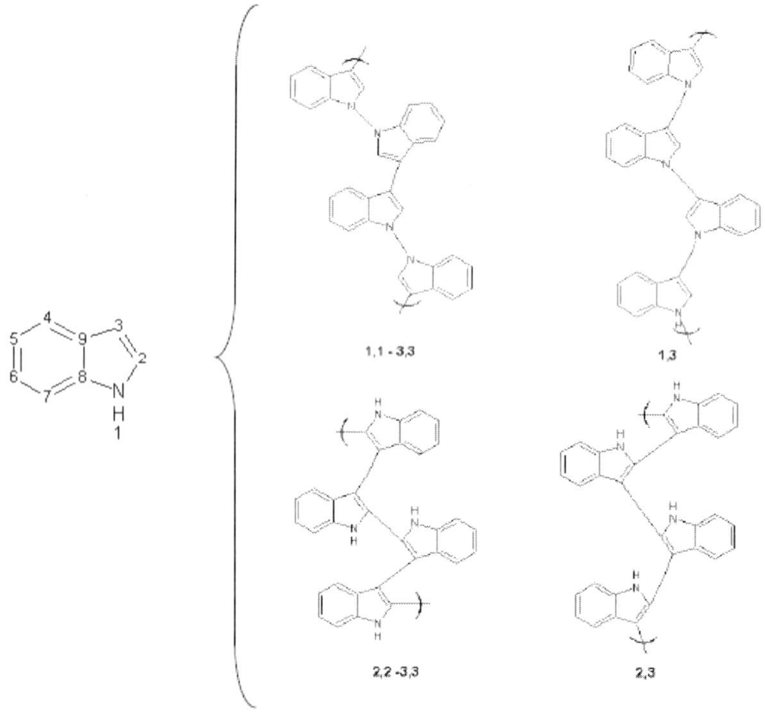

Figure 26. Representation of indole monomer and the possible structures of polyindole.

Two theoretical approaches (Hartree–Fock calculations), were used. One is based on thermodynamical control involving the most stable species obtained during the reaction. The other, more concerned with the kinetic control, is related to the study of the intrinsic reactivity of monomers and oligomers of indole. Both methods confirm that the intermonomer bonds do not involve the N atom.

The kinetic control approach allows one to predict that the C2–C3 bonds are the most likely in polyindole. Raman, FT-IR and electron energy loss (EELS) spectroscopy studies of Polyindole [136], showed the conclusion that a 2,3 structure is feasible for this polymer. The theoretical treatment of the reaction of indole polymerization estimation data, are in agreement with the experimental results, predicting the regular character of the 2,3 structure, (Figs. 26).

3.5.1. Indole Adsorption Studies

Na-Al-MCM-41 sample, employed as a host, was dehydrated at 673 K in vacuum for 1 h. After it was exposed to equilibrium vapors from indole during 24 h at 333 K to obtain indole saturated hosts according to the procedure described by Anunziata et al. [33, 40, 106].

Before the in-situ polymerization, indole adsorption on Na-AlMCM-41 was characterized by thermo-infrared spectroscopy analysis.

Figure 27 shows FTIR spectra in the region of 1300 – 1700 cm^{-1} for indole adsorbed on Na-AlMCM-41 and pure indole. When indole is adsorbed onto Na-AlMCM-41 material, Ind-AlMCM-41 shows bands at 1618, 1577, 1504, 1490, 1455, 1413, 1352 and 1334 cm^{-1} [125, 136, 137]. The band at 1615 cm^{-1} is overlap with a broad band at 1630 cm^{-1} due to the vibrations of water bending, and Si–O stretching overtone of the AlMCM-41 framework vibration [33].

The bands do not shift from their locations in the spectrum of pure indole, indicating the weak interaction between indole and the AlMCM-41 mesostructure.

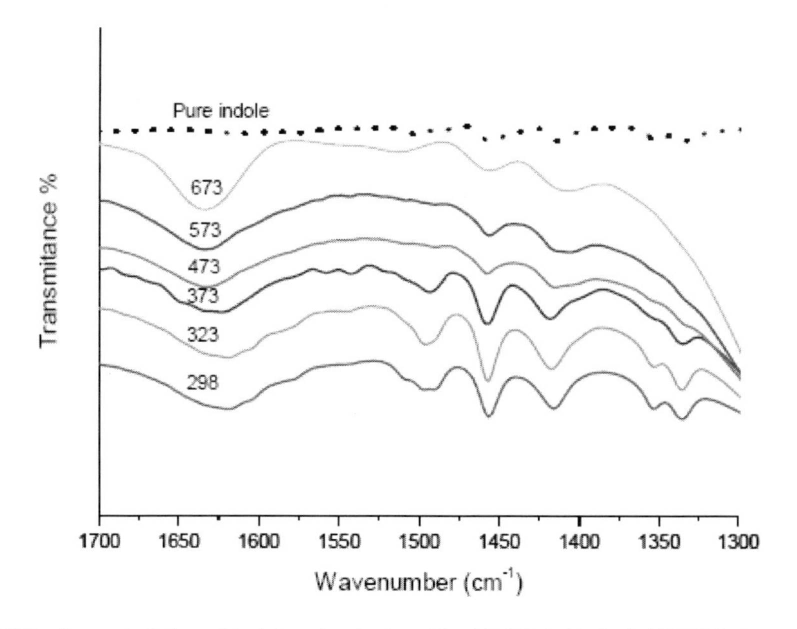

Figure 27. FTIR of pure indole and indole adsorbed on Na-AlMCM-41 (Ind-AlMCM-41) treatment in vacuum at 298, 323, 373, 473, 573 and 673 K, in the range of 1300-1700 cm^{-1}.

The sorption of the indole is expected to result mainly from the interaction of the π-electrons of the aromatic ring with electron-acceptor sites of the AlMCM-41, such as sodium cation or SiOH groups. As the sodium cation are strongly held to the Al-MCM-41 framework, due to the low capacity for electron delocalization of the MCM-41 structure, and taking into account that the structure of MCM-41 can be considered as a weak but hard base (as a function of molecular orbital theory) [40], π electrons interactions of the aromatic with structural SiOH groups of the Na-AlMCM-41 framework would be more probable. Then, indole would be held over the MCM-41 mesostructure only through such π interactions in the Ind-Al-MCM-41 sample. Table XI shows the main bands.

3.5.2. Polyindole Composite Characterization

3.5.2.1. Synthesis

Polyindole (PInd) was synthesized by chemical polymerization of indole monomer, C_8NH_7, using $FeCl_3$ as an oxidant, for characterization and fingerprint identification purposes. The oxidative polymerization of adsorbed polyindole was carried out at 298 K. Ferric chloride was used as oxidant (Cl_3Fe-0.25M), it was added to the material with indole adsorbed, without stirring, and left at room temperature for 24 hr. Finally, the materials were filtered, washed and dried at 323K. (PInd-Na-AlMCM-41)

The oxidative in situ polymerization to produce polyindole/Na-AlMCM-41 (PInd-Na-AlMCM-41) g. composite was performed similar to [106] using $FeCl_3$ as oxidant, Figure 28.

3.5.2.2. FTIR Studies

Figure 29 shows FTIR spectra of polymer and composite (PInd-Na-AlMCM-41). The bands assigned are shown in Table XI. The FTIR absorption bands of pure PInd allowed us to identify some specific PInd-Na-AlMCM-41 bands (see Table XI).

Several signals appear at 1630, 1615, 1574, 1510, 1490, 1454, 1383 and 1325 cm^{-1}, which were identified according to previous works (Table XI) [24, 125, 136, 137].

There are two main opinions on the polymerization mechanism of indole, 1,3 position or 2,3 position (Figure 26) [136]. The benzene ring is not usually involved in the polymerization process. The band at 1564 cm^{-1} could be assigned to N-H bond vibrations in the pure polymer [137] and this band shifts to a higher wavenumber in the composite. These results imply that there are still N-H bonds in the polymer backbone and in PInd-Na-AlMCM-41; however, it is more perturbed due to the polymer backbone.

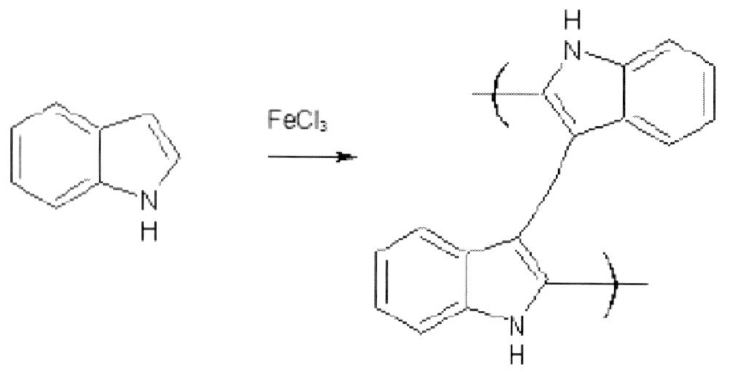

Figure 28. Indole polymerization using $FeCl_3$.

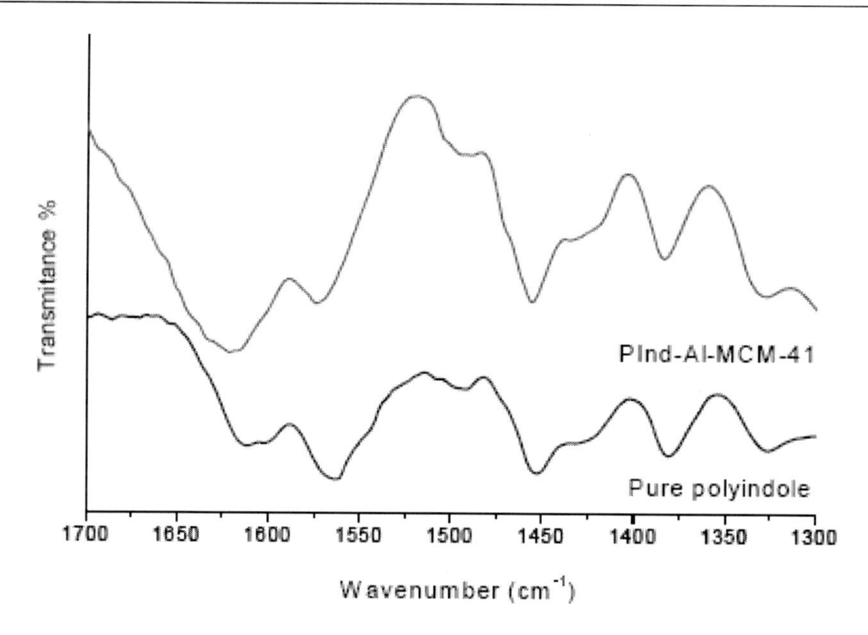

Figure 29. FTIR of pure polyindole and PInd-Na-AlMCM-41 composite in the range of 1300-1700 cm^{-1}.

Table XI. IR-band assignments of pure indole and polyindole, indole adsorbed on Al-MCM-41 (Ind-Al-MCM-41) and PInd-Na-AlMCM-41 composite

Band Assignment	Band position, wavenumber (cm^{-1}) *			
	Indole	Ind-Al-MCM-41	Polyindole	PInd-Na-AlMCM-41 composite
νNH	3415	3400	-	-
H_2O bending	1630	1630	-	1630
$\nu C_7C_8 + \nu C_5C_6 + \nu C_8C_9$	1614	1618	1612	1615
$\nu C_9C_4 + \nu C_6C_7 + \nu C_7C_8$	1577	1578	-	-
N-H Vibration	-	-	1564	1574
$\nu N_1C_2 + \delta C_2H$	1504	1508	1508	1505
$\nu C_8C_9 + \delta C_4H + \delta C_7H$	1490	1493	1493	1495
$\delta C_5H + \nu C_8N + \nu C_4C_5$	1455	1455	1452	1454
Stretch of C-N bond	-	-	1427	1429
$\delta NH + \nu C_2C_3 + \delta C_6H$	1413	1415	-	-
Stretching aromatic ring	-	-	1380	1383
$\nu C_8N + \delta C_6H + \nu C_2C_3$	1352	1354	-	-
Stretching of pyrrole ring	1336	1336	1326	1327

* [24, 136, 137].

Thus, nitrogen species may not be polymerization sites, and polymerization should mainly occur at the 2,3 position. The benzene ring is not affected in the polymerization process and 2,3 position of the pyrrole ring is responsible for the polymerization. In addition, bands at 1615, 1454 and 1380 cm-1 is induced by different stretching of aromatic ring in the polymer chain [137]. The bands observed at 1427-1429 cm^{-1} are assigned to C-N stretching.

Some PInd bands shift with respect to pure PInd, due to the interaction of PInd with the host. Adsorbed water could be observed over the material at 1630 cm−1 overlapped with 1615 cm^{-1} band.

3.5.2.3. XRD Studies

Figure 30 displays the XRD analyses of Si-MCM-41. The XRD pattern for the as-synthesized sample exhibits a strong (1 0 0) reflection peak with two small peaks (second- and third-order peaks corresponding to (1 1 0) and (2 0 0) diffraction planes), characteristic of MCM-41 material. After the template was removed by calcination, the intensity of (1 0 0) diffraction was slightly increased, indicating that proper calcinations lead to a better-defined structure of MCM-41 [13, 50, 51, 138].

Figure 30 shows that diffraction patterns of aluminated sample suggest the preservation of the MCM-41 structure after incorporation of Al by post-synthesis method.

The characteristic XRD d100 parameter is shown in Table XII.

Moreover, XRD patterns of the PInd-Na-AlMCM-41 composite (Figure 30) indicate that the hexagonal-ordered structure of MCM-41 was persistent, after the in-situ polymerization procedure, seen in Table XII. In this figure it can be noted that plane (1 0 0) in the composite, characteristic of MCM-41, shifts to higher 2θ angles. The shift in the diffraction plane indicate that the two components (polymer and host) have been successfully integrated and the structure of composite is more orderly and uniform than that in pure PInd [139].

While the composite area is significantly smaller than that for the Na-AlMCM-41 host, its characteristic structure is maintained after the PInd nanowire is within the host, in agreement with the XRD studies (Table XII).

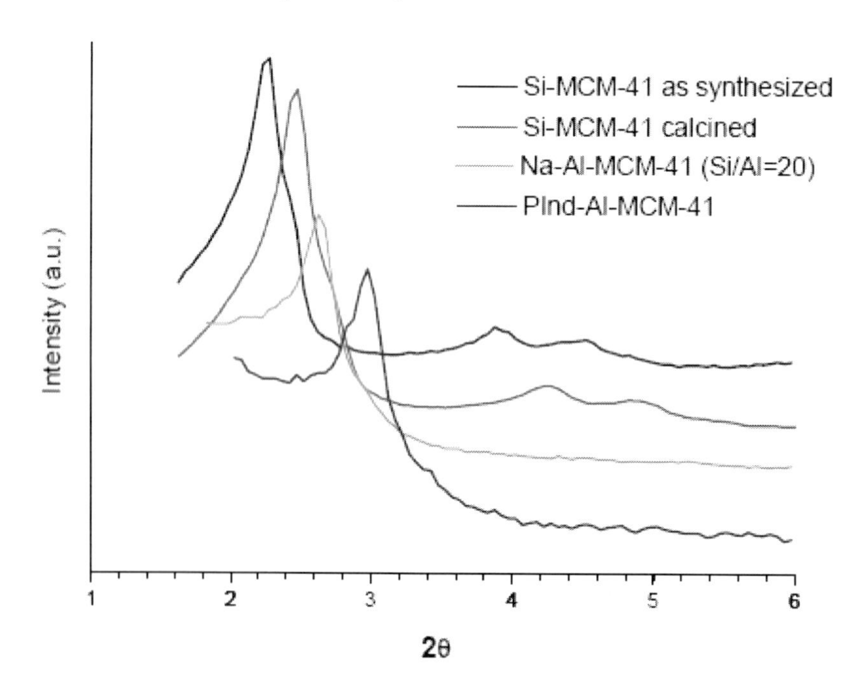

Figure 30. X-ray diffraction patterns of as synthesized and calcined Si-MCM-41, Na-AlMCM-41 and the Composite PInd-Na-AlMCM-41.

Table XII. Textural and structural properties of Si-MCM-41, Na-AlMCM-41 and PInd-Na-AlMCM-41

Sample	hkl			ao (nm)	Surface Area, m2/g
	100	110	200		
	d (nm)				
Si-MCM-41 *	39.11	22.63	19.52	4.5	---
Si-MCM-41 **	35.62	20.72	18.04	4.1	1235
Al-MCM-41	33.70	19.50	---	3.8	860
PInd-Na-AlMCM-41	30.60	---	---	3.5	332

* as synthesized; ** Calcined.

3.5.2.4. SEM and BET Studies

Figure 31 shows SEM images of the (a) host and (b) composite. To establish the morphology of the materials, the SEM studies reveal that the incorporation of PInd by in-situ polymerization were carried out in the pores of Na-AlMCM-41, having no apparent effects on the macroscopic morphology of the samples, and in which isolated clusters of PInd on the external surface of the composite are not observed.

The SEM images of the samples show aggregates of regular spherulitic-shaped particles. The particle size on both samples was approximately 300–800 nm (550 nm average) in diameter.

3.5.2.5. TG Analysis

Figure 32 shows the TGA results for Na-AlMCM-41 (a), Polyindole (b), and PInd-AlMCM-41 composite (c). For pure PInd (curve b), the initial weight loss (300 - 373 K) is caused by the loss of water from the polymer.

The polymer is thermally stable up to 430 K. From this temperature, the polymer starts to degrade slowly; but above 700 K, the polymer degrades rapidly. For the Na-AlMCM-41 used as a host (curve a), the weight loss is about 4% around 373 K and then stabilizes with a total weight loss of 8.3% w/w when reaching 873 K.

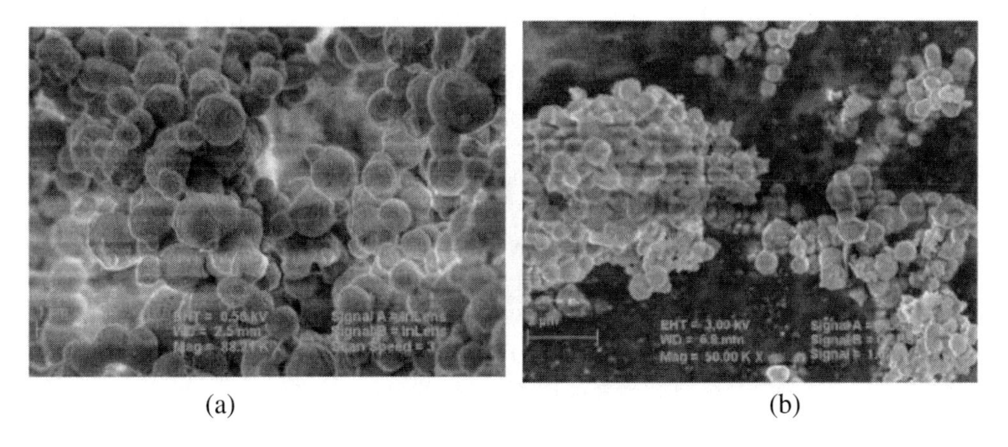

(a) (b)

Figure 31. SEM images of (a) Na-Al-MCM-41 and (b) PInd-Na-AlMCM-41 composite.

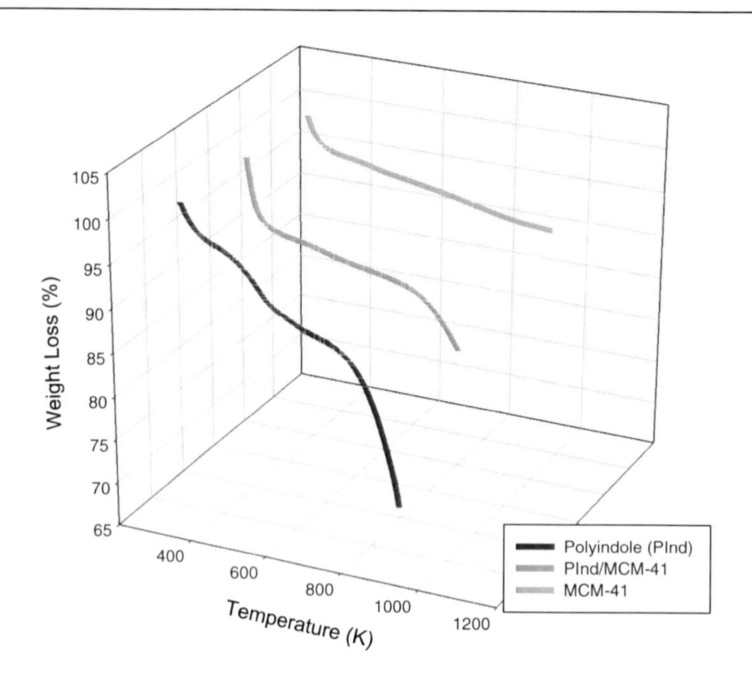

Figure 32. TGA analysis for: (a) Na-AlMCM-41; (b) Pure polyindole; (c) PInd-AlMCM-41.

The composite (curve b) initial weight lóss below 373 K can be attributed to the loss of water and to some gases adsorbed from the composite. It remains therefore fairly stable up to around 770 K. Above this temperature, the composite (the polymer mainly) starts to lose mass faster. If we considered the weight loss of pure PInd and guest PInd in the composite at 773 K, the difference would be wider (16% and 12.3%, respectively) but at 873 K this difference would be broader (17.8% and 30%, respectively).

In consequence, PInd becomes more stable as a guest in the host forming the composite, preventing the polymer from a fast degrading. The total weight loss for the host and the composite is about 8.3% and 17.8% respectively at 873 K; the weight loss attributed to PInd in the composite (as a nano-guest) is about 10% w/w at 873 K. It should be noted that the total weight loss for pure PInd is near 30%.

3.6. Comparative Studies of the Conductive Properties Between Different Conducting_Polymer/Host Composites

Conductivity properties of different Polymer/host composites are shown in Figure 33. In PANI/hosts composites PANI, the conductivity is controlled not only by the degree of chain oxidation, but also by the degree of protonation of PANI, according to the following chemical formula: $\{[(\text{–B–NH–B–NH–})_y\,(\text{–B–N=Q=N–})_{1-y}](\text{A})_x\}_n$.

In the conducting form (emeraldine salt) y is close to 0.5, B and Q are C_6H_4 rings in the benzene and quinone states, and A is a strong acid [30].

The previously mentioned decrease in the ratio indicates that, there is a higher amount of polyaniline in conductive emeraldine salt form [33, 40, 106]. Using Na-AlSBA-16 we found a larger amount of emeraldine salt with respect to other composites reported in literature (Table XII).

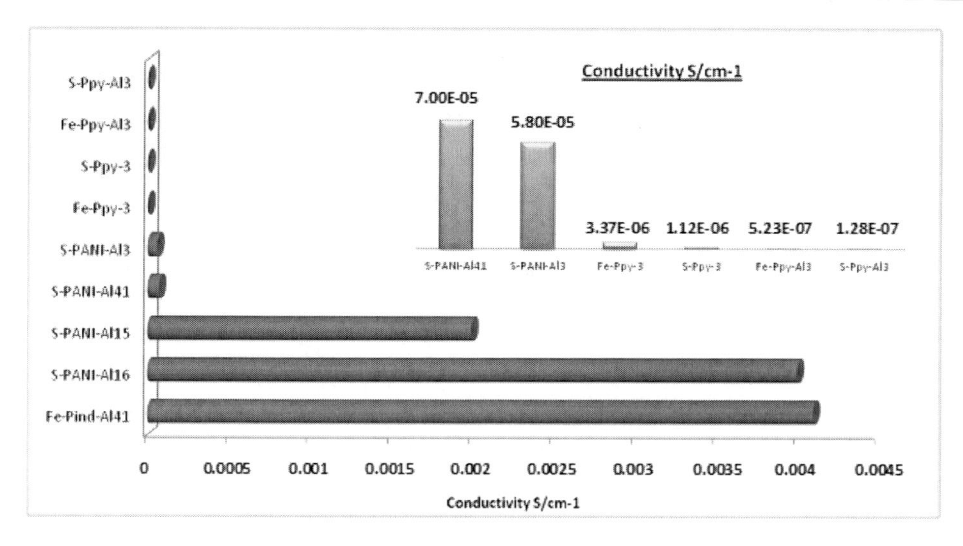

Figure 33. Conductive properties Polymer/host composites S/cm.

Table XIII. Resistivity and N=Q=N stretching / C-C stretching ratios of different PANI/hosts

Band assignments	Composites			
	PANI/ Na-Al-MCM-41	PANI/ Na-Al-SBA-15	PANI/ Na-Al-SBA-3	PANI/ Na-Al-SBA-16
Quinone / Benzene Stretching Ratio	1.2	0.75	1.27	0.56
Resistivity (Ω.cm)	14280	500	17240	250

At room temperature (298 K), the electrical conductivity of PANI-3 and PANI-16 composites is $5.8 \cdot 10^{-5}$ and $4.0 \cdot 10^{-3}$ $S \cdot cm^{-1}$ respectively. PANI-41 has an electrical conductivity of $7.0 \cdot 10^{-5}$ S/cm [33] and PANI-15 (material synthesized according to Anunziata et al., [40]) $2.0 \cdot 10^{-3}$ S/cm.

According to this information, we can deduce that most PANI is located within the channels rather than on the surface of SBA-3 and SBA-16 mesoporous materials [105]. The mesoporous materials are highly insulating. When PANI is inside the pores of meso or microporous aluminosilicates (SBA, MCM, zeolites, etc.), the conductivity of the composite can change from insulating to ionic conductors through semiconductors. Protonated polyaniline (emeraldine salt) is a conducting polymer. The mechanism of conduction in polymers, e.g., polyfuran, polyindole, polypyrrole and polyaniline, is highly complex, since such materials exhibit conductivity in a range of about ten orders of magnitude by changing their doping. To explain the electronic phenomena in these systems, the concepts of solitons, polarons and bipolarons have been used [140]. The conduction in conducting polymers (enhancing or dipping) is influenced by numerous factors: polaron length, conjugation length, overall chain length, and charge transfer to adjacent molecules [141]. In the case of PANI, its conductivity can change using different substituents bonded to the aromatic ring. Both conjugation length and redox potential are affected by their nature and substituent positions on the ring [142]. When the polymer is encapsulated in the host channels, conductivity may change. The mesoporous nanomaterial (SBA, MCM) used as a host affects the movement of the charges in the polymer. The results also show that the conductivity of the composite is lower than that of pure polyaniline. The fact that PANI-16 and PANI-15 composites present a conductivity higher than that of the other PANI/host composites arises from the larger amount of polyaniline in conductive form (emeraldine salt), described by the FTIR analysis. There is a high amount of emeraldine salt (conducting) in PANI-16 composite according to FTIR studies, and then the conductivity of this composite is higher.

According to literature data the electrical conductivity for chemically synthesized polypyrrole is 10^{-2} - 10^{-1} S cm^{-1} [103], and 10^{0} S cm^{-1} [75], polymerized with ammonium persulfate and ferric chloride respectively.

We can deduce that most PPY is located within the channels rather than on the surface of SBA-3 and AlSBA-3 mesoporous materials [74]. The mesoporous materials are highly insulating. When PPY is inside the pores of meso or microporous aluminosilicates (SBA, MCM, zeolites, etc.), the conductivity of the composite can change from insulate to ionic conductors through semiconductors. When the polymer is encapsulated in the host channels, conductivity may change. The mesoporous nanomaterial (SBA, MCM) used as a host affects the movement of the charges in the polymer. The results also show that the conductivity of the composite is lower than that of pure PPY.

SBA-3 based composites show better conductivity than Al-SBA-3 nanostructured composites, and the composites obtained by polymerization with ferric chloride show higher conductivity than the composites polymerized with ammonium persulfate.

The fact that FePPY/SBA-3 and SPPY/SBA-3 composites present higher conductivity than that of PPY/AlSBA-3 composites, arises from the larger amount of conductive PPY described by the FTIR studies. According to the FTIR spectra, the conductivity of the samples is higher in the samples with sharper polypyrrole bands; thermogravimetric analysis shows a large amount of PPY in SBA-3-composites. Generally, it is assumed that conductivity should be higher in higher degrees of crystallinity and better alignment of the chains. The PPY included in FePPY/SBA-3 is more conducting and possibly more doped.

According to literature data the electrical conductivity for chemically and electrochemically synthesized polyindole are $1 \cdot 10^{-4}$ and $5 \cdot 10^{-2}$ S cm^{-1} respectively [136].

The electrical conductivity of PInd-Na-AlMCM-41 nanocomposite at room temperature was in the range of 4.10^{-3} $S \cdot cm^{-1}$.

The composite has an electrical conductivity higher than that of the pure polymer synthesized by chemical method, approaching to pure polymer electrochemically synthesized. This behavior is attributed to the fact that the Na-AlMCM-41 host increases the movement of the electrical charges through polyindole backbone content in the mesopores. Polyindole is possibly doped by the host.

Conclusion

Polyaniline/Hosts Composites

The study of aniline adsorption by FTIR allowed us to understand better the characteristics of the active sites of the nanostructured materials employed as hosts. Diverse thermal analyses and infrared spectroscopy studies of aniline adsorbed in gas phase on mesoporous materials, Na-Al-MCM-41 type allowed us to determine that aniline interacts weakly with these structures, as much directly or through water contained in the samples. When the nanostructured material is dehydrated completely, the associations aniline-structure would be π interactions, between the aromatic ring and the structural silanol groups of the material.

When the material contains some adsorbed water molecules, before the adsorption of aniline, to the complexes π aniline-MCM-41 and aniline-SBA-15, another type of interaction would be added in which the organic base is associated to the structure through weak hydrogen bridges with residual water molecules of the material. Thus, as both Na-Al-MCM-41 as Na-Al-SBA-15 are able to retain appropriately aniline, so studies on electron acceptors sites of the material could be realized.

However to realize the polymerization in-situ of aniline adsorbed on these materials, Na-Al-SBA-15 would be more advantageous, due to it posses channels with bigger walls resisting better the high temperatures treatments. The FTIR and UV –Vis studies of PANI/Na-Al-SBA-15 suggest that is possible to obtain in situ polymerization of aniline, obtaining emeraldina variety within Na-Al-SBA-15. PANI is confined within the channels of this nanostructured material, according to the superficial area of the host and composite guest/host.

Considering the FTIR studies of PANI/SBA-3 and PANI/SBA-16, it could be observed that aniline is adsorbed in great proportion on SBA-3, through quadrupole interaction of its π electrons; with the electron acceptors sites of the host (Na^+ and SiOH), whereas the aniline adsorption on SBA-16 is also adsorbed through its amine group. The temperature programmed desorption of aniline allowed us to determine the energy of the interaction with host. The PANI FTIR spectra indicated their presence within the channels of Na-Al-SBA-3 and Na-Al-SBA-16. The aniline in situ polymerization technique to obtain PANI was effective. The FTIR analyses confirm the conductive PANI inside the channels of both materials. The spectra showed high polymerization degree using Na-Al-SBA-3. Absorption bands of free amine groups were not detected, whereas it was present in PANI/Na-Al-SBA-16, indicating that the polymerization process was incomplete. The Na-Al-SBA-16 material displays major amount of emeraldina salt (the conductive form of PANI), with respect to the other employed materials with hexagonal pore systems, probably due to the different arrange of its system channels.

The composites XRD analyses indicated that, after aniline in situ polymerization in order to obtain PANI, did not generate structural modifications, preserving their ordering to short and long range. In the diffraction patterns were not PANI signals, because it is within the channels and not on external surface like nano-clusters. Composites developed offers special characteristics to be applied in the semiconductors field, like electronics device at nanometric scale.

As it can be seen Na-Al-SBA-3 and Na-Al-MCM-41hosts, presents similarities values of area ratio respect to the corresponding composites, PANI/Na-Al-SBA-3 and PANI/Na-Al-MCM-41, being 1.25 and 1.2 respectively. This can be because of these materials display similarity in their mesoporous organization and the same structure-directing species used for their synthesis. Whereas for the mesoporous structure of Na-Al-SBA-15, with similar characteristics to the previous ones, showed inferior value, about 0.75; this could be because of the walls of mesoporous confer a greater protonation to PANI polymer, consequently the area ratio is diminished.

In the case of Na-Al-SBA-16, although its porous topology has a distribution of pseudo-spherical cavities and/or channels, this causes that the formed polymers appear mainly in their emeraldina salt appearance with high level of protonation in comparison with the previous structures. It would be possible to modify the conductive properties of the composites, varying the aromatic ring substitutes, the anchorage sites of the nanostructure where aniline would be adsorbed and the amount of PANI in the host.

The potential interest is that the polymer used to produce composite, could be altered according to the requirements of final composite. Probably the Na-Al-SBA´s hosts, could be more advantageous according to the required properties in final composite, by their mechanical properties remarkably superiors to the MCM. In this way, is very exiting to continue this study to optimize the preparations of these composites, focusing the attention in the process of preparation of PANI-hosts composites, with different conductive properties.

Polypyrrole/Hosts Composites

We have obtained a new composite, formed by polypyrrole molecular wire within the SBA-3 channels. The studies and data reported in this chapter were the first collected in this line of research. The interaction of pyrrole with Na-Al-SBA-3 mesoporous material, has been studied by FTIR, taking account that pyrrole can be used as probe molecule.

Adsorbed pyrrole can interact by two ways with the host: 1) it can link by hydrogen bond with basic Lewis sites of network oxygen of material and 2) it can donate their π electron systems to sodium basic Lewis sites, which are in the structure like counter-ion. The first interactions would be indicated by a shifting towards lower frequencies in the NH stretching vibration in the infrared, which in our case is not observed. Whereas the interaction by π electron system of pyrrole, is perceivable from the stretching of C-C and C-N was shifted towards lower frequencies. Moreover the pyrrole adsorption can be observed by the shift of C-C ring stretching. This tendency can be taken like an indication that pyrrole is adsorbed on Na-Al-SBA-3 by means of the aromatic ring.

The polymerization of monomer was realized by oxidation with ammonium persulfate; obtained polypyrrole. This reaction is governed by the same principles that aniline because of its similarity with the chemical formulation. The realized FTIR studies to the samples indicated the presence of polymer in the composite, comparing the resulting composite PPY/host with, PANI/host composite. This composite could be used in computer architecture, because it presents the particularity of being able to modify or to control its conductivity as well as its dielectric properties. One more a more detailed investigation in these points is in course.

Polyindole/Hosts Composites

Sodium aluminosilicate Na-AlMCM-41 material with longitudinal channel array was synthesized and characterized. The materials were analyzed by different methods (XRD, BET, SEM, UV-Vis and FTIR).

FTIR studies were carried out in order to analyze the nature of the interaction of indole adsorbed onto the mesoscopic material. Through infrared spectroscopy, we could observe that indole is adsorbed onto Na-AlMCM-41 in a large proportion through the π electron system with electron-acceptor sites of this material, like Na^+ cation and SiOH groups. The spectrum also shows that it is possible to obtain indole in-situ polymerization within the Na-AlMCM-41 with hexagonal pore system. The composite obtained after in-situ oxidative polymerization was analyzed by infrared spectroscopy and UV-Vis of PInd/host to corroborate the presence of polyindole. The bands that characterize this conductive polymer are present in the spectra.

The composites obtained were analyzed by X-ray diffraction, showing that after polymerization the porous structures of the materials are preserved and the PInd is found within the porous channels. The polymer is not on the surface but within the porous channels, as indicated by BET, TGA and SEM. These composites offer numerous desirable properties potentially applied in the electronic field for developing, for instance, an electronic device at a nanometric scale. The composite has an electrical conductivity higher than the pure polymer chemically synthesized close to the electrical conductivity of pure polymer electrochemically synthesized. This behavior could be ascribed to the fact that Na-AlMCM-41 host increases the movement of the electrical charges through polyindole backbone content in the mesopores. Polyindole is possibly doped by the host.

Organic molecular wires have been successfully developed, together with an inorganic Na-AlMCM-41 reservoir (hybrid composite) into the channels.

ACKNOWLEDGMENTS

OAA, MBGC, MLM CONICET Researchers; JMJ CONICET Doctoral Fellowship. Thank to CONICET Argentina, PIP N° 112-200801-00388 (2009-2012).

REFERENCES

[1] C. T. Kresge, M. E. Leonowicz, W. Roth, J. C. Vartulli, J. Beck, *Nature*, 359, 710 (1992).

[2] J. S. Beck, J. C. Vartuli, W. J. Roth, M. E. Leonowicz, C. T. Kresge, K. D. Schmitt, C. T-W. Chu, D. H. Olson, E. W. Sheppard, S. B. McCullen, J. B. Higgins, J. L. Schlenker, *J. Am. Chem. Soc.* 114, 10834 (1992).

[3] O. A. Anunziata, A. Beltramone, M. L. Martinez, L. López Belon, *J. Colloid Interf. Sci.* 315, 184 (2007).

[4] F. Chen, H-K. Luo, Y-F. Han, C. Wang, G. J. Gan, *Catal. Today*, 131, 76 (2008).

[5] Galarneau, D. D-Giscard, F. Di-Renzo, F. Fajula, *Catal. Today*, 68, 191 (2001).

[6] G. Soler-Illia, C. Sanchez, B. Lebeau, J. Patarin, *Chem. Rev.* 102, 4093 (2002).

[7] D. Zhao, Q. Huo, J. Feng, B. F. Chmelka, G. D. Stucky, *J. Am. Chem. Soc.* 120, 6024 (1998) .

[8] M. Mesa, L. Sierra, J. Patarin, J-L. Guth, *Solid State Sci.* 7, 990 (2005).

[9] T-W. Kim, R. Ryoo, M. Kruk, K. Gierszal, M. Jaroniec, S. Kamiya, O. Terasaki, *J. Phys. Chem. B*, 108, 11480 (2004).

[10] S. Zeng, J. Blanchard, M. Breysse, Y. Shi, X. Shu, H. Nie, D. Li, *Micropor. Mesopor. Mat.* 85, 297 (2005).
[11] S. Kowalak, K. Stawinski, A. Mackowiak, *Micropor. Mesopor. Mat.* 44, 283 (2001) .
[12] T. Klimova, L. Lizama, J. C. Amezcua, P. Roquero, E. Terre, J. Navarrete, J. M. Domínguez, *Catal. Today*, 98, 141 (2004).
[13] O. A. Anunziata, M. L. Martínez, M. B. G. Costa, *Mat. Lett.* 64, 545 (2010).
[14] Vinu, B. M. Devassy, S. B. Halligudi, W. Bohlmann, M. Hartmann, *Appl. Catal. A: Gen.* 281, 207 (2005).
[15] Vinu, D. P. Sawant, K. Ariga, M. Hartmann, S. B. Halligudi, *Micropor. Mesopor. Mater.* 80, 195 (2005).
[16] W. Hu, Q. Luo, Y. Su, L. Chen, Y. Yue, C. Ye, F. Deng, *Micropor. Mesopor. Mater.* 92, 22 (2006).
[17] J. M. R. Gallo, C. Bisio, L. Marchese, H. O. Pastore, *Micropor. Mesopor. Mater.* 111, 632 (2008).
[18] H. M. Kao, C. C. Ting, S. W. Chao, *J. Mol. Catal. A: Chem.* 235, 200 (2005).
[19] M. Xu, W. Wang, M. Seiler, A. Buchholz, M. Hunger, *J. Phys. Chem. B*, 106, 3202 (2002).
[20] Q. Xia, K. Hidajat, S. Kawi, *J. Catal.* 205, 318 (2002).
[21] J. M. Campelo, D. Luna, R. Luque, J. M. Marinas, A. A. Romero, J. J. Calvino, M. P. Rodríguez-Luque, *J. Catal.* 230, 327 (2005).
[22] R. Luque, J. M. Campelo, D. Luna, J. M. Marinas, A. A. Romero, *Micropor. Mesopor. Mater.* 84, 11 (2005).
[23] R. Schmidt, M. Stöcker, E. W. Hansen, D. Akporiaye, O. H. Ellestad, *Micropor. Mater.* 3, 443 (1995).
[24] Z. Cai, G. Yang, *Synthetic Met.* 160, 1902 (2010).
[25] D. Billaud, E. B. Maarouf, E. Hannecart, *Polymer* 35, 2010 (1994).
[26] P. S. Abthagir, R. Saraswathi, *Thermochim. Acta* 424, 25 (2004) .
[27] P. C. Pandey, *Sensor. Actuat. B-Chem.* 54, 210 (1999).
[28] Z. Cai, C. Hou, *J. Power Sources*, in Press, Available online 22 August 2011.
[29] K. Moller, T. Bein, *Chem. Mater.* 10, 2950 (1998).
[30] G. Wu, T. Bein, *Science*, 264, 1757 (1994) .
[31] G. Wu, T. Bein, *Science*, 266, 1013 (1994).
[32] M. Jurczyk, A. Kumar, S. Srinivasan, E. Stefanakos, *Int. J. Hydrogen Energ.* 32, 1010 (2007).
[33] O. A. Anunziata, M. B. G. Costa, R. D. Sánchez, *J. Colloid Interf. Sci.* 292, 509 (2005).
[34] O. Quadrat, J. Stejskal, *J. Ind. Eng. Chem.* 12, 352 (2006).
[35] J. Stejskal, M. Omastova, S. Fedorova, J. Prokes, M. Trchova, *Polymer*, 44,1353(2003).
[36] G. A. Rimbu, I. Stamatinb, C. L. Jackson, K. Scott, J. Optoelectron, *Adv. M.* 8, 670 (2006).
[37] S. Goel, A. Gupta, K. P. Singh, R. Mehrotra, H. C. Kandpal, *Mat. Sci. Eng. A*, 443, 71 (2007).
[38] M. Irimia-Vladu, J. W. Fergus, *Synthetic Met.* 156, 1401 (2006).
[39] J. Stejskal, I. Sapurina, *IUPAC Technical Report, Pure Appl. Chem.* 77, 815 (2005).
[40] O. A. Anunziata, M. B. G. Costa, M. L. Martínez, *Catal. Today,* 133–135, 897 (2008).
[41] R. K. Iler, *The Chemistry of Silica: Solubility, Polymerization, Colloid and Surface Properties and Biochemistry*, New York (1979).

[42] J. Brinker, *Structure of sol-gel-derived glasses. Glass Science and Technology,* 4A:169–230 (1990).
[43] C. J. Brinker, W. D. Drotning, G. W. Scherer, *In Better Ceramics Through Chemistry*, vol. 32, Materials Research Society, New York (1984).
[44] J. M. Seddon and R. H. Templer. In R. Lipowsky and E. Sackmann, eds., *Handbook of Biological Physics*, Elsevier Science B. V., 97–160 (1995).
[45] P. Behrens, A. Glaue, C. Haggenmuller, G. Schechner, *Solid State Ionics*, 101, 255 (1997).
[46] C. A. Fyfe, G. Y. Fu, *Journal of the American Chemical Society*, 117, 9709 (1995).
[47] Q. S. Huo, D. I. Margolese, U. Ciesla, P. Y. Feng, T. E. Gier, P. Sieger, R. Leon, P. M. Petroff, F. Schuth, G. D. Stucky, *Nature*, 368, 317(1994).
[48] Q. S. Huo, R. Leon, P. M. Petroff, G. D. Stucky, *Science*, 268, 1324 (1995).
[49] Q. S. Huo, D. I. Margolese, G. D. Stucky, *Chemistry of Materials*, 8, 1147 (1996).
[50] D. Kumar, K. Schumacher, C. du Fresne von Hohenesche, M. Grün, K. K. Unger, *Colloid. Surface A*, 187–188, 109 (2001).
[51] M. L. Martínez, M. B. G. Costa, G. A. Monti, O. A. Anunziata, *Micropor.Mesopor. Mater.* 144, 183 (2011).
[52] S. Laha, R. Kumar, *Micropor. and Mesopor. Mater*. 53, 163 (2002).
[53] F. Chen, A. Shen, X-J. Xu, R. Xu, F. Kooli, *Micropor. Mesopor. Mater*. 79, 85 (2005).
[54] D. Zhao, J. Feng, Q. Huo, N. Melosh, G. Fredrickson, B. Chmelka, G. Stucky, *Science*, 279, 548 (1998).
[55] S. Yariv, H. Cross, *Organo-Clay Complexes and Interactions,* Marcel Dekker, Inc., New York (2001).
[56] S. Yariv, L. Heller, N. Kaufherr, *Clays Clay Miner*. 17, 301 (1969).
[57] S. Yariv, L. Heller, Z. Sofer, Israel *J. of Chem*. 6, 741 (1968).
[58] L. Heller, S. Yariv, *Proceeding of the International Clay Conference*, 1, 741 (1969).
[59] W. Bodenheimer, B. Kirson, S. Yariv, *Israel J. Chem*. 1, 69 (1963).
[60] W. Bodenheimer, L. Keller, S. Yariv, *Israel J. Chem*. 1, 391 (1963).
[61] W. Bodenheimer, L. Keller, S. Yariv, *Clay Minerals*, 6, 167 (1966).
[62] C. Kirschhock, H. Fuese, *Microporous Materials*, 8, 19 (1997).
[63] S. Yang, K. Chen, *Synth. Met*. 135-136, 151 (2003).
[64] D. Lee, K. Char, S. Lee, Y. Park, *J. of Mater. Chem.* 13, 2942 (2003).
[65] B. Kim, J. Jung, S. Hong, J. Joo, *Macromolecules*, 35, 1419 (2002).
[66] H. Choi, M. Cho, W. Ahn, *Synth. Met*. 135-136, 711 (2003).
[67] M. Cho, H. Choi, W. Ahn, *Langmuir*, 20, 202 (2004).
[68] M. Cho, H. Choi, K. Kim, W. Ahn, Macromol. Rapid. Commun. 23 (2002) 713.
[69] J. Kim, S. Kim, H. Choi, M. Jhon, *Macromol. Rapid Commun*. 20, 450 (1999).
[70] N. Li, X. Li ,W. Geng, T. Zhang, Y. Zuo, S. Qiu, *J. Appl. Polymer Sci*. 93, 1597 (2004).
[71] O. A. Reutov, *Fundamentals of theoretical organic chemistry,* Appleton-Century-Crofts, New York (1967).
[72] J. Stejskal, R. G. Gilbert, *Pure Appl.Chem.*, (IUPAC Technical Report), 74, 857 (2002).
[73] H.-J. Nam, H. Kim, S. H. Chang, S.-G. Kang, S.-H. Byeon, *Solid State Ionics*, 120, 189 (1999).
[74] Q. Cheng, V. Pavlinek, C. Li, A. Lengalova, Y. Hea, P. Saha, *Microporous and Mesoporous Materials*, 93, 263 (2006).

[75] D. Navarrete, M. C. Criado, A. D. Villarejo, J. D. Villarejo, M. G. Molins, J. Gil, A. Salinas, J. Peña, J. P. Pariente, M. V. Regí, Liberación de fármacos en matrices biocerámicas, Real Academia Nacional de Farmacia, ISBN: 84-934430-1-8 (2006).
[76] P. Ravikovitch, A. Neimark, *Langmuir*, 18, 1550(2002).
[77] P. Ravikovitch, A. Neimark, *Langmuir*, 18, 9830 (2002).
[78] E. Smela, N. Gadegaard, *Adv. Mater.* 11, 953 (1999).
[79] G. MacDiarmid, *Angew. Chem. Int. Ed. Engl.* 40, 2581 (2001).
[80] V. Saxena, B. D. Malhotra, *Curr. Appl. Phys.* 3, 293 (2003).
[81] T. F. Otero, I. Boyano, M. T. Gortés, G. Vázquez, *Electrochim. Acta*, 49, 3719 (2004).
[82] Y. Sun, Y. Xia, *Adv. Mater.* 14, 833 (2002).
[83] Y. Wang, N. Toshima, *J. Phys. Chem., B*, 101, 5301 (1997).
[84] J. C. Lin, C. Y. Wang, *Mater. Chem. Phys.* 45, 136 (1996).
[85] R. Jin, Y. W. Cao, A. Mirkin, K. L. Kelly, G. C. Schatz, J. G. Zhang, *Science*, 294, 1901 (2001).
[86] J. R. Gould, J. R. Lenhard, A. A. Muenter, S. A. Godleski, S. Farid, *J. Am. Chem. Soc.* 122, 11934 (2000).
[87] R. Oliver, A. Muñoz, C. Ocampo, C. Alemán, E. Armelin, F. Estrany, *Chemical Physics*, 328, 299 (2000).
[88] M. Michael, C. P. Painter, *Fundamentals of Polymer Science*, Ed. Iliffe Books Ltd., 187-196 (1998).
[89] L. Bronstein, S. Polarz, B. Smarly, M. Antonictti, *Adv. Mater.* 13, 1333 (2001).
[90] X. G. Zhao, J. L. Shi, B. Hu, L. X. Zhang, Z. L. Hua, *J. Mater. Chem.* 13, 399 (2003).
[91] S. Z. Wang, D. G. Choi, S. M. Yang, *Adv. Mater.* 14, 1311 (2002).
[92] Ghosh, C. R. Patra, P. Mukherjee, M. Sastry, R. Kumar, *Micropor. Mesopor. Mater.* 58, 201 (2003).
[93] T. Asefa, R. B. Lennox, *Chem. Mater.* 17, 2481 (2005).
[94] G. S. Attard, J. C. Glyde, C. G. Goltner, *Nature*, 378, 366 (1995).
[95] S. W. Ho, T. K. Kwei, D. Vyprachticky, Y. Okamoto, *Macromolecules*, 36, 6894 (2003).
[96] D. J. Cardin, S. P. Constantine, A. Gilbert, A. K. Lay, M. Alvaro, M .S. Galletero, H. Garia, F. Marquez, *J. Am. Chem.Soc.* 123, 3141 (2001).
[97] M. Showkat, K. P. Lee, A. I. Gopalan, M. S. Kim, S. H. Choi, H. D. Kang, *Polymer*, 46, 1804 (2005).
[98] P. L. Llewellyn, U. Ciesla, H. Decher, R. Stadler, F. Schuth, K. K. Unger, *Stud. Surf. Sci. Catal.* 84, 2013 (1994).
[99] D. J. Cardin, *Adv. Mater.* 14, 553 (2002).
[100] D. Coutinho, Z. W. Yang, J. P. Ferraris, K. J. Balkus, *Micropor. Mesopor. Mater.* 81, 321 (2005).
[101] J. Rodriguez, H. Grande, T. Otero, *Handbook of organic conductive molecules and polymers*, vol. 2, Chichester: Wiley, 415–68 (1997).
[102] D. Trivedi, *Handbook of organic conductive molecules and polymers*, vol 2. Chichester: Wiley, 505–72 (1997).
[103] N. Blinova, J. Stejskal, M. Trchova, J. Prokes, M. Omastova, *European Polymer Journal*, 43, 2331 (2007).
[104] Q. Cheng, V. Pavlinek, C. Li, A. Lengalova, Y. Hea, P. Saha, *Materials Chemistry and Physics*, 98, 504 (2006).

[105] M. Nakayama, J. Yano, K. Nakaoka, K. Ogura, *Synthetic Metals*, 138, 419 (2003).
[106] M. L. Martínez, F. A. L. D'Amicis, A. R. Beltramone, M. B. G. Costa, O. A. Anunziata, *Mat. Res. Bull.* 46, 1011 (2011).
[107] N. Bloemberger, N. Laureate, *Handbook of advance Electronic and photonic Materials and Device*, Academic Press, 16, (2001).
[108] V. Shaktawat, N. Jain, M. Dixit, N. S. Saxena, K. Sharma, T. P. Sharma, *Indian Journal of Pure and Applied Physics*, 46, 427 (2008).
[109] R. Lord, F. Miller, *J. Chem. Phys.* 10, 328 (1942).
[110] R. Navarro, J. Orza, *An. Quim. Ser. A*, 79, 557 (1983).
[111] H. Föster, H. Fuese, E. Geidel, B. Hunger, H. Jobic, C. Kischhoc, O. Klepel, K. Krause. *Phys. Chem. Chem. Phys.* 1, 593 (1999).
[112] S. Lamprakopoulos, D. Yfantis, A. Yfantis, D. Schmeisser, J. Anastassopoulou, T. Theophanides, *Synthetic Metals*, 144, 229 (2004).
[113] M. Kofranek, T. Kovàr, A. Karpfen, H. Lischka, *J. Chem. Phys.* 96-6, 4464 (1992).
[114] H. Uehara, M. Miyake, M. Matsuda, M. Sato, *Journal of Materials Chemistry*, 8-9, 2133 (1998).
[115] G. Tourillon, G. Garnier, *J. Electroanal. Chem.* 135, 172 (1982).
[116] M. Biaas, A. Roy, *Appl. Polym. Sci.* 70, 2169 (1998).
[117] S. Wu, F. Zeng, J. Shen, *Polym. J.* 30, 451 (1998).
[118] J. Kim. F. Liu, H. Choi, H. Hong, J. Joo, *Polymer*, 44, 289 (2003).
[119] Y. Kim, I. Song, *J. Mater. Sci.* 37, 5051 (2002).
[120] Q. Cheng, V. Pavlinek, A. Lengalova, C. Li, Y. He, P. Saha, *Micropor. Mesopor Mat.* 93, (2006) 263-269.
[121] M. Nakayama, J. Yano, K. Nakaoka, K. Ogura, *Synth. Met.* 128, 57 (2002).
[122] J. Jang. B. Lim , J. Lee T. Hyeon, *Chem. Commun.* 1, 83 (2001).
[123] M. Kofranek, T. Kovář, H. Lischka, A. Karpfen, *Journal of Molecular Structure: Theochem*, 259, 181 (1992).
[124] Y. Furukawa, S. Tazawa, Y. Fujii, I. Harada, *Synth. Met.* 24, 329 (1988).
[125] L. Joshi, R. Prakash, *Mater. Lett.* 65, 3016 (2011).
[126] T. A. Skotkeim (Ed.), *Handbook of Conducting Polymers*, Marcel Dekker, New York (1986).
[127] G. P. Evans, in H. Gerischer, C. W. Tobias (Eds.), *Advance in Electrochemical Science and Engineering*, vol. 1, New York, 40 (1990).
[128] G. MacDiarmid, S. L. Mu, N. L. D. Somarisi, W. Mu, *Molec. Cryst. Liq. Cryst.* 121, 187 (1985).
[129] P. N. Bartlett, J. Farrington, *Bull. Electrochem.* 8, 208 (1992).
[130] R. J. Waltman, J. Bargon, *Can. J. Chem.* 64, 76 (1986).
[131] K. Jackowska, J. Bukowska, *Polish J. Chem.* 66, 1477 (1992).
[132] R. Holze, C. H. Hamann, *Tetrahedron*, 47, 737 (1991).
[133] G. Zotti, S. Zecchir, G. Schiavon, R. Seraglia, A. Berlin, A. Canavesi, *Chem. Mater.* 6, 1742 (1994).
[134] K. M. Choi, J. H. Jang, K. H. Kim, *Molec. Cryst. Liq. Cryst.* 220, 201 (1992).
[135] H. Talbi, G. Monard, M. Loos, D. Billaud, *Journal of Molecular Structure (Theochem)*, 434, 129 (1998).
[136] H. Talbi, B. Hunibert, J. Ghanbaja, D. Billaud, *Polymer*, 38, 2099 (1997).
[137] S. An, T. Abdiryim, Y. Ding, I. Nurulla, *Mater. Lett.* 62, 935 (2008).

[138] V. Meynen, P. Cool, E. F. Vansant, *Micropor. Mesopor. Mater.* 125, 170 (2009).
[139] X. Sun, J. Ren, L. Zhang, L. Chen, H. Li, R. Li, J. Ma, *Synth. Met.* 160, 2244 (2010).
[140] J. Heeger, in T. A. Skotheim (Ed.), *Handbook of Conducting Polymers*, vol. II, Dekker, New York, 729 (1986) and references therein.
[141] J. I. Kroschwitz, in J. I. Kroschwitz (Ed.), *Electrical and Electronic Properties of Polymers: A State-of-the-Art Compendium*, Wiley, New York, 1–330 (1988).
[142] M. Leclerc, G. D'Aprano, G. Zotti, *Synth. Met.* 55–57, 1527 (1993).

In: Conducting Polymers
Editor: Luiz Carlos Pimentel Almeida
ISBN: 978-1-62618-119-9

Chapter 8

FUNCTIONAL CONDUCTING POLYMERS FOR DNA SENSING

Chunhua Luo[1,2], Yiting Wang[1] and Hui Peng[1,*]

[1]Key Laboratory of Polarized Materials and Devices, Ministry of Education, East China Normal University, Shanghai, China

[2]National Laboratory for Infrared Physics, Shanghai Institute of Technical Physics of the Chinese Academy of Sciences, Shanghai, China

ABSTRACT

Conducting polymers (CPs) are polyconjugated polymers with electronic properties similar to metals, but retain the properties of conventional organic polymers. Compared to saturated polymers, CPs have different electronic structures which result in the unique properties, such as electrical conductivity, low ionization potentials and high electron affinity. For CPs in ground state (insulating or semiconducting state), π-bonds (π- π*) are partially localized that are a consequence of a phenomena called the Peierls distortion. During the doping process, the excitation across the π- π* band gap creates self-localised excitations of conjugated polymers with localized electronic states in the gap region. Such electronic structure of CPs is highly susceptible to changes in the polymeric chain environment and other perturbations of chain conformation caused by a biological recognition event, such as DNA hybridization. The changes in delocalized electronic structure or other CPs properties are manifested in altered electrical and optical properties of the CPs, and, when measured, can provide a signal for the presence of the corresponding target molecule. These advantages of CPs make them a good material for chemical sensors and biosensors. In this chapter, we provide an overview of functional CPs for DNA sensing. In the Introduction section, we briefly introduce the history of CPs and the advantages of functional CPs as sensing element for DNA detection. In the section of Electronic Structure and Synthesis of CPs, some fundamental physical principles needed to understand the properties of CPs and the methods to prepare functional CPs are addressed. In the section of Functional CPs for DNA Sensing Based on Electrochemical Methods, we discuss the DNA probe immobilization methods and transduction mechanism including amperometry, conductometry, impedimetry and

[*] E-mail address: hpeng@ee.ecnu.edu.cn

photocurrent spectroscopy, and then summarize the functional CPs reported in the literatures for DNA sensing. In the section of Functional CPs for DNA Sensing Based on Optical Methods, the detection of DNA by using cationic CPs in the solution and on solid substrate is presented. The probable future trends are presented in the last section of Summary and Future Outlook.

1. INTRODUCTION

The observation of the remarkably high electrical conductivity of a halogen-treated polyacetylene has resulted in the emergence of a new class of materials, conducting polymers (CPs), which has dramatically changed our thinking and opened up new vistas in chemistry and physics [1]. Three collaborating scientists, Alan J. Heeger, Alan G. MacDiarmid and Hideki Shirakawa who played a major role in this breakthrough were awarded the Nobel Prize in Chemistry in 2000 "for the discovery and development of electronically conducting polymers"[2-4].

CPs are polyconjugated polymers with electronic properties resembling those of metals, but retain the properties of conventional organic polymers. A number of conjugated polymers such as poly (aniline), poly (phenylenevinylene), poly (pyrrole) and poly (thiophene) have been transformed from an insulating into a highly conductive state by a "doping" process. During the doping process which may be carried out chemically or electrochemically, a polyconjugated polymer with a small conductivity, typically in the range 10^{-10} to 10^{-5} Scm^{-1}, is converted into a polymer which is in the "metallic" conducting regime (ca. 1 to 104 Scm^{-1})[3]. The doping process is reversible and the original polymer with little or no degradation of the polymer backbone can be reproduced. By controlling the doping level, the conductivity of CPs can by tuned from insulating to highly conducting. These unique properties of CPs have led to a variety of applications of these materials, such as light emitting diodes (LED) [5], electrochromic materials [6], anti-static coatings [7], solar cell [8], batteries [9], anti-corrosion coatings [10], chemical sensors and biosensors [11] and drug release systems [12-14].

DNA analysis plays an ever-increasing role in a number of areas related to human health such as diagnosis of infectious diseases, genetic mutations, drug discovery, forensics and food technology. In the early 90s, gene array technologies appeared as promising tools for the simultaneous analysis of multiple DNA sequences, in which multiple specific probe DNA fragments or oligonucleotides (ODNs) are anchored onto solid surfaces and fluorescently or radioactively tagged analyte oligonucleotides are detected [15-17]. These array technologies have had a huge impact on genomic and proteomics applications, although they have shortcomings arising from, for example, limited tagging efficiency, hazardous waste disposal and complex multi-step analysis. In order to seek faster, sensitive and label-free DNA detection, a number of novel approaches have been suggested based on optical [18-21], acoustic [22] and electrochemical [23-25] techniques. The electronic structure of CPs is highly susceptible to the changes in the polymeric chain environment and other perturbations of chain conformation caused by, for example, a biological recognition event such as DNA hybridization. The changes in delocalized electronic structure or other CPs properties are manifested in altered optical and electrical properties of the CPs, and, when measured, can provide a signal for the presence of the corresponding target molecule [26]. These advantages of CPs make them suitable materials for constructing DNA sensors. This chapter provides an overview of the applications of CPs in DNA sensing, with a special attention paid to current trends and novel applications developed recently due to the advances in nanotechnology.

2. Electronic Structure and Synthesis of CPs

Conjugated polymers consist of π-conjugated chain, where the wave functions of π-electrons of the carbon atoms overlap and which is necessary for a polymer becoming conductive. The electronic structure of conjugated polymers was described by SSH [27, 28] in terms of a quasi-one-dimensional tight-binding model in which the π electrons are coupled to distortions in the polymer backbone by the electron-phonon interaction. So the *ground state* of a conjugated polymer (insulating or semiconducting state) can be described as partially localized π-bonds (π- π*) that are a consequence of Peierls distortion which lowers the energy of the system by opening a gap at the Fermi level [29]. The *doping process* is an excitation across the π- π* band gap that creates *self-localized excitations* of conjugated polymers with localized electronic states in the gap region [29] (Figure 1). These self-localized excitations are called solitons (in degenerate ground state systems), polarons and bipolarons (in non-degenerate systems) (Figure 1). In chemical terms the doping process is an oxidation or reduction process that creates a positively or negatively charged polymer chain with oppositely charged counterions to counterbalance the charge on the polymer chain. Therefore, a polaron corresponds to a radical cation or anion and a bipolaron to a dication or a dianion.

Conducting polymers can be synthesized chemically [30] and electrochemically [31, 32], as shown in Figure 2. The reaction is usually an oxidative polymerization. If large amounts of polymer are needed, a chemical route is preferred. For example, the oxidation of pyrrole or aniline by $Fe(ClO_4)_3$ or peroxydisulfate in acid media leads to the respective conducting polymers. The polymers obtained are in an oxidized, high conductivity state containing counterions incorporated from the solution used in the preparation procedure. While, in terms of sensor applications, electrochemical polymerization is widely used because of several advantages: (i) it is performed at ambient temperatures and microelectrodes or electrodes with a large surface area can be used; (ii) the polymer film formed is confined to the electrode and its shape can thus be controlled by electrode design while the thickness can be controlled in a range from nanometers to micrometers; (iii) the properties of the CP film can be widely modulated by varying electrochemical polymerization conditions.

Electrochemical polymerization can be carried out potentiostatically, amperometrically or with potential scanning in a traditional three-electrode system and the whole process may only take a few seconds [33]. The anode can be made of a variety of materials including platinum, gold, glassy carbon, and tin or indium–tin oxide (ITO) coated glass. In the electrochemical polymerization, the choice of the solvent and electrolyte is of particular importance since both solvent and electrolyte should be stable at the oxidation potential of the monomer and provide an ionically conductive medium. For pyrrole or aniline which has a relatively low oxidation potential, the electropolymerization can be carried out in aqueous electrolytes. While, in some cases such as thiophene and benzene, organic solvents have to be used due to the limitation of high oxidation potential and low solubility of monomer in aqueous solution. Acetonitrile and propylene carbonate are widely adopted as organic solvents because of their large potential windows and high relative permittivities, which allow a good dissociation of the electrolyte and thus a good ionic conductivity. After the initial oxidation, radical cations of the monomer are formed. The next step is a dimerization reaction and then stepwise chain growth proceeds via the association of radical ions (RR route) or the association of a radical cation with a neutral monomer (RS route) [3, 34, 35]. Intensive stirring of the solution usually results in a decrease in the yield of the polymer produced, because the radical cations can react with the electrode or take part in side reactions with the nucleophilic reactants present in the solution.

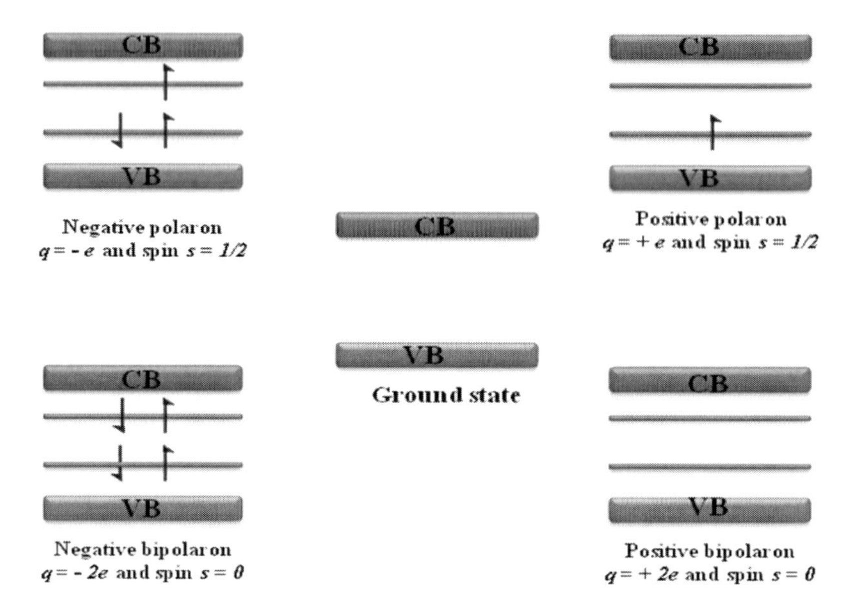

Figure 1. Band structures of self-localized excitations of a conjugated polymer with non-degenerate ground state.

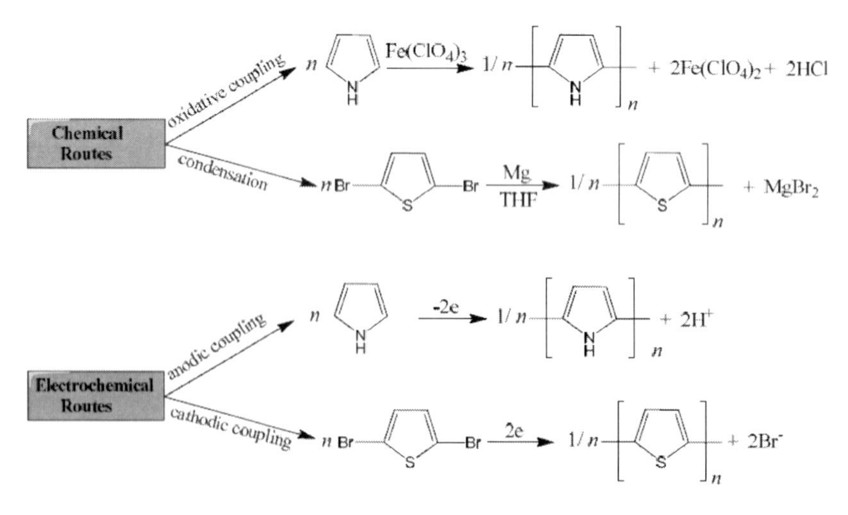

Figure 2. Electrochemical and chemical routes for the synthesis of conducting polymers.

3. Functional CPs for DNA Sensing Based on the Electrochemical Methods

Electroanalytical techniques are widely used for DNA sensing due to high sensitivity, fast response, simplicity and low cost [36-38]. As above-mentioned, the electronic structure of CPs is highly susceptible to the changes in the polymeric chain environment and other perturbations of chain conformation caused by a biological recognition event. On the other hand, CPs can be electrochemically deposited on the electrode which simplifies the sensor preparation procedures. These advantages make CPs be good materials for constructing electrochemical DNA sensors, where they act not only as substrates for immobilizing DNA probes but also as "molecular wires" to transfer the signals [38, 39].

3.1. Construction of Electrochemical DNA Sensors Based on CPs

A typical configuration of electrochemical DNA sensors based on CPs is shown in Figure 3. Single stranded DNA probes are immobilized on or within a conducting polymer layer. The target DNA is captured by base-pairing to generate a recognition signal, which is recorded through a transducer (gold, platinum, glass carbon electrode, etc.) to report this recognition event. Because the recognition event takes place at the CP/electrolyte interface and the recognition signal generated reaches the transducer through the CPs layer, the properties of the CP layer and the orientation of the immobilized DNA probes on the CP are crucial to sensor performance.

3.1.1. Immobilization of DNA Probes

The immobilization of DNA probe is critical for the overall sensor performance as it affects hybridization efficiency and transduction. Ideally, the orientation of probes should be predictable and readily accessible to the analyte DNA [40]. In general, there are three ways to immobilize DNA probes, namely electrochemical entrapment, covalent immobilization and affinity interactions.

The electrochemical entrapment method involves the electrochemical oxidation of a suitable monomer into the corresponding conducting polymer from a solution containing oligonucleotide (ODN) probes. Due to the negative charges of ODN probes, they can be entrapped into the polymer film as a dopant during the electropolymerization process.

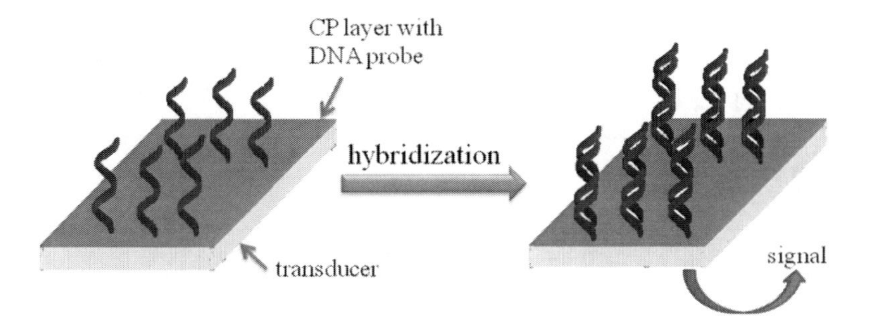

Figure 3. General electrochemical DNA sensor design based on CPs.

Wang et al. first illustrated that ODNs can act as the sole dopant during the growth of polypyrrole film while maintaining their hybridization activity [41, 42]. It was shown that relatively moderate concentrations of ODN (10^{-5} M) were sufficient to produce good films, in contrast to the 1 M NaCl needed under normal polymerization conditions, which was attributed to the multiple charges present on ODNs[42]. The incorporation of longer ODNs (20 mer) into a PPy film was irreversible, which was illustrated by the observed cation movement during oxidation/reduction cycling. The significant advantage of this immobilization method is its simplicity. Potential drawbacks include the possibility of damaging ODNs probe due to high potentials employed during polymerization, and poor target accessibility to the incorporated probe in the bulk of the resulting film [40].

Covalent immobilization can overcome the disadvantages of the electrochemical entrapment method and improve probe accessibility by target DNA [43]. Generally, ODN probes are functionalized with $-NH_2$, -COOH etc., and then are covalently attached to either a functionalized monomer or a functionalized polymer. Livache et al. developed a process that utilizes pyrrole monomer bearing an ODN to copolymerize with pyrrole allowing the immobilization of multiple probes on electrode arrays [44, 45], as shown in Figure 4. Later the same group applied a similar process to prepare an ODN array consisting of a matrix of 48 addressable 50-μm microelectrodes that was applied to detect *hepatitis C virus* in blood samples[46] which was extended to a 128 microelectrode array [47].

In another approach for covalent ODN attachment, a film is first electropolymerized from a solution containing (either exclusively or in a mixture with non-functionalized monomers) functionalized monomers, following by covalent attachment of a 5'-end modified ODN onto the functional groups in the obtained conducting polymer film. In this approach, the conducting polymer films can be prepared under conditions that are potentially incompatible with the maintenance of bioactivity of ODNs, such as organic solvents and high polymerization potentials, while the subsequent attachment of the ODN probes is performed under mild conditions that ensure ODN integrity.

Garnier et al. prepared a functionalized polypyrrole, poly(3-acetic acid pyrrole-*co-* 3-*N*-hydroxyphthalimide pyrrole), which bears an easy leaving group, *N*-hydroxyphthalimide (Figure 5a) [39, 48]. We prepared an acid functionalized polypyrrole, poly[pyrrole-*co*-4-(3-pyrrolyl) butanoic acid], where the amino-functionalized ODN was covalently attached to the polymer film by using 1-ethyl-3-(3-dimethylaminopropyl) carbodiimide (EDC) as a catalyst (Figure 5b) [38]. Thompson et al. proposed a different approach, in which a single stranded ODN probe was linked to the conducting polymer by forming a bidentate complex between Mg^{2+} and an alkyl phosphonic acid group on the polymer and the phosphate group of the ODN [49]. Recently Gautier et al. developed a process that involves electrochemical copolymerization of 3-methylthiophene and 3-(oxyalkyl)-thiophene bearing an arylsulfonamide group.

The sulfonamide terminal function of the copolymer film was electrochemically cleaved and chemically modified with N-chlorosuccinimide (NCS) which leads to sulfonyl chloride function as a prerequisite for amino-ended single-stranded DNA probe immobilization (Figure 5c) [50-52]. An advantage of this approach is the ability to electrochemically cleave immobilized ODN probes, which leads to a potential application for the addressable multiple ODN probes immobilization on an electrode array, as well as electrochemically controlled DNA delivery systems [53].

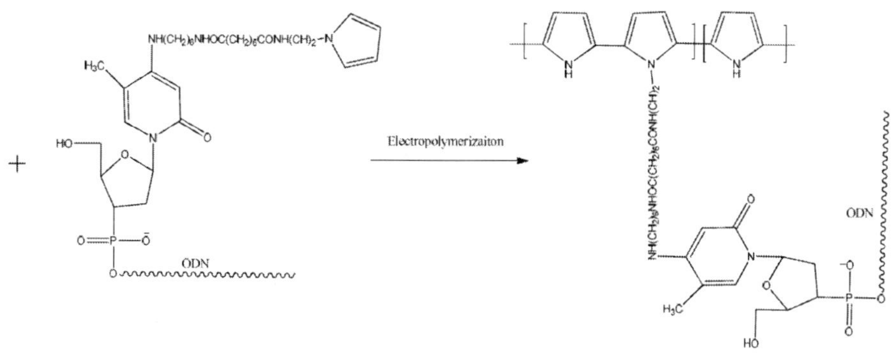

Figure 4. Schemes for covalent immobilization of ODN probes.

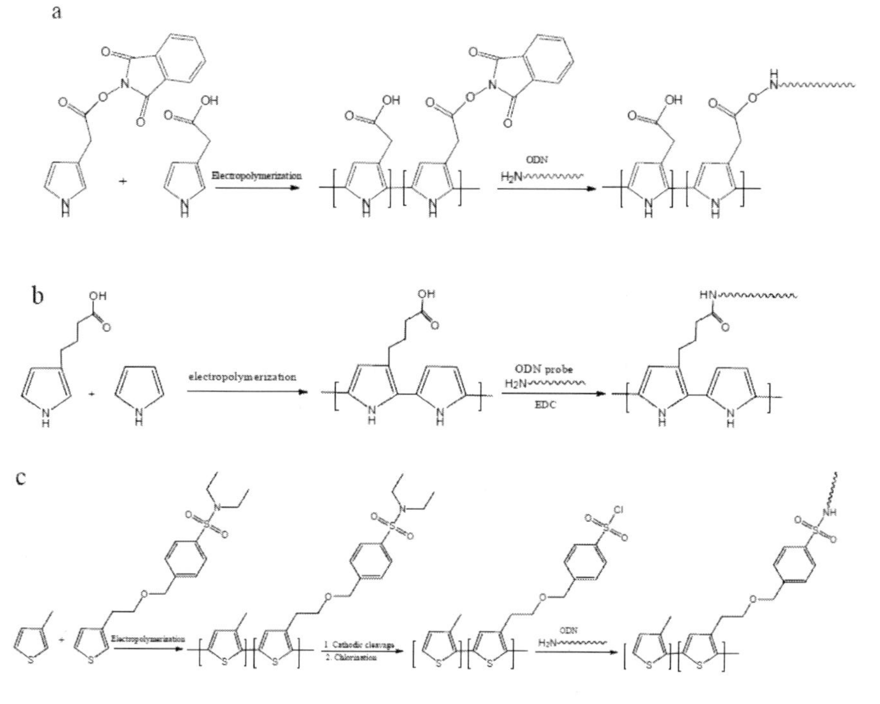

Figure 5. Schemes for covalent immobilization of ODN probes on the polymer films.

The immobilization of ODN probes on the CP film can also be achieved via affinity interaction. The most common approach involves the avidin-biotin interaction which is extremely specific and the strongest known non-covalent biological bond (association constant $K_a=10^{15}$ M). This approach can be highly versatile due to the ability to anchor different biomolecules on the same support and the sensor can be regenerated by treatment with a detergent solution that breaks the avidin/biotin bridge but does not affect the support matrix. Cosnier and Lepellec first prepared poly(pyrrole-biotin) film for grafting glucose oxidase [54]. Later Dupont-Filliard et. al exploited this principle to reversibly immobilize ODN probes [55]. Briefly a biotinylated polypyrrole film was synthesized, followed by immobilization of avidin units by biotin/avidin interaction, and exposing the film further to biotinylated ODN probes. The reversibility was achieved by a treatment in aqueous solution of sodium dodecylsulfate to cleave the avidin-biotin connection [55, 56].

3.1.2. Detection Methodologies

For electrochemical DNA sensors based on CPs, the polymer is not only used as an immobilization matrix, but also plays an active role in transduction. CPs can be reversibly doped and dedoped by using electrochemical techniques, with doping and dedoping resulting in significant changes in electrical and spectroscopic properties. These changes can be modulated by probe - analyte interactions [57, 58], and then become an analytical signal. The main detection methodologies for electrochemical DNA sensors include amperometric, conductometric, and impedimetric techniques.

3.1.2.1. Amperometric Detection and Cyclic Voltammetry

Amperometry is the most common approach in biosensors based on CPs due to the simplicity of the method and fast response. In the case of DNA sensors, the efficacy of modulating the electrical conductivity and the redox properties of the CP by hybridization plays a key role in the sensor performance. PPy/probe-modified electrodes made by using oligonucleotide probes as sole counter anions, displayed transient anodic peaks upon adding complementary target, and opposite (cathodic) signals upon spiking of the solution with non-complementary strands[41, 59]. The direction of signals (relative to the base line) was dependent on the sequence of probe dopants and of the added target oligomers. The authors suggested that such response reflected the change in the conductivity of the host PPy network caused by an increase in charge density in the case of a complementary target, or induced by electrostatic repulsion in the case of non-complementary sequences.

Cyclic voltammetry is widely used with CP based DNA sensors as the readout method, because the doping and dedoping process during potential scanning can be efficiently modulated by the hybridization event. Figure 6 illustrates voltammetric detection of DNA hybridization based on poly [pyrrole-*co*-4-(3-pyrrolyl) butanoic acid] which was electropolymerized onto a Pt electrode [38]. The cyclic voltammograms showed a decrease in the oxidation current and a positive shift in oxidation potential of the polymer film after hybridization with complementary ODN samples.

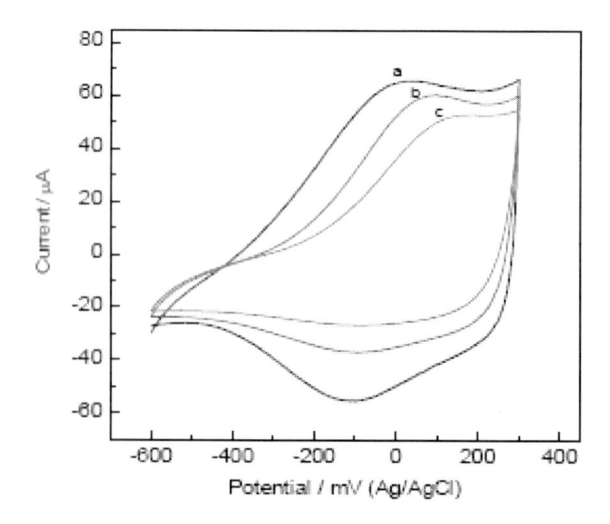

Figure 6. CV of Pt electrode modified with ODN probes grafted poly[pyrrole-*co*-4-(3-pyrrolyl) butanoic acid] film in PBS solution (a), after incubation in solution containing complementary ODN at concentrations of 3.5nM (b), and 8.8nM (c). Scan rate: 100 mV/s.

Such a change has been ascribed to the formation of bulky and rigid double stranded DNA upon hybridization, thus increasing the energy required to planarize the polymer upon oxidation. It is also found that better sensitivities are achieved by using thinner polymer films due to the larger surface to volume ratio [38, 60]. The electric properties of the conjugated back bone of CPs can even be modulated by single bases. Emge and Bäuerle functionalized bithiophene monomers with pyrimidine and triazine bases [61, 62]. After polymerization, the addition of small quantities of complementary bases strongly affected the electrochemical properties of the obtained polymer in a way similar to that described above.

3.1.2.2. Conductometric Detection

Direct conductivity measurements are widely used in CP based gas sensors [63-65]. Earlier work on indirect conductometric biosensors based on CPs was described by Kittlesen and coworkers [66]. They developed a method for the in situ determination of a polymer's conductivity as a function of an applied electrochemical potential. The experimental setup is shown in Figure 7. In this method, when the applied voltage to the microelectrodes is zero (V_d=0), sweeping the applied electrochemical potential produces a cyclic voltammogram. When V_d is made finite (0.01-0.05 V), sweeping the applied potential will produce a current between the microelectrodes, which is proportional to the polymer's conductivity. Xie et al used similar experimental setup to detect protein based on poly(3,4-ethylenedioxythiophene) nanowires functionalized with carboxylic acid side-chain functional groups which were directly grown across the electrode junction by using an electric-field-assisted method[67]. After the immobilization of a protein-binding apatmer, the dynamic range for the detection of thrombin was found to be from 1 nM to 1mM.

In another report, poly(3,4-ethylenedioxythiphene) microtubules were electrochemically synthesized in the presence of ssDNA by using polycarbonate membrane as a template, and a label-free conductometric DNA sensor was developed based on these microtubules [68]. The conductivity measurements were done at a gate potential of +0.8 V (*vs.* SCE) where the response was at its highest. The results showed that the sensor response, as well as the linear range, increased with an increase in the ssDNA length. For the sensor based on ssDNA with 20 bases, the linear range is 8.0×10^{-8} -1.0×10^{-5} g/mL^{-1} and the detection limit is 8.0×10^{-8} g/mL^{-1}.

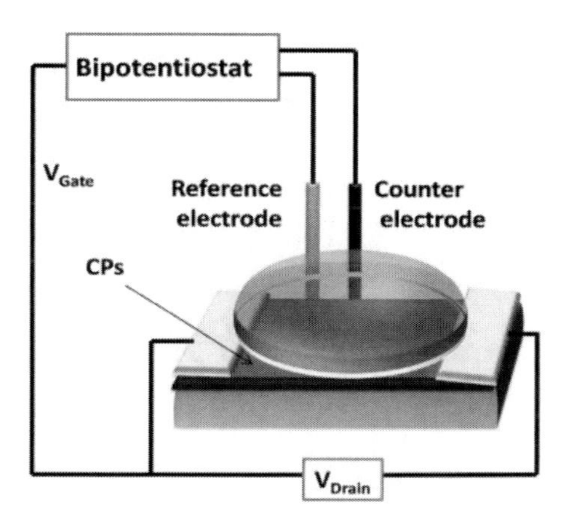

Figure 7. Schematic representation of an electrochemical setup for the determination of a polymer's conductivity in a presence of an analyte and applied electrochemical potential.

3.1.2.3. Impedimetric Detection

Electrochemical impedance spectroscopy (EIS) is a powerful tool in the study of corrosion, semiconductors, batteries and electro-organic synthesis because it provides accurate, error-free kinetic and mechanistic information. In EIS experiments, a sinusoidally varying and interrogating voltage (typically in the range of 5 to 10 mV peak-to-peak) relative to a suitable reference electrode is applied to the working electrode and the resulting current response is measured.

The real (Z_{re}) and imaginary (Z_{im}) components (or magnitude $|Z|$ and phase θ) of impedance are calculated as the ratio between the system voltage phasor and the current phasor, which are generated by a frequency response analyzer during the experiment [36]. Compared to dc techniques, EIS offers the following advantages: (i) a very small excitation amplitudes is used which causes only minimal perturbation of the electrochemical test system, reducing errors caused by the measurement techniques; (ii) data on both electrode capacitance and charge-transfer kinetics are obtained which provide valuable mechanistic information. (iii) a purely electronic model consisting of a specific combination of resistors and capacitors can be used to represent the studied electrochemical system, and then the electrochemical system can be characterized by using established AC circuit theory in terms of equivalent circuits. Figure 8 shows Randles equivalent circuit which models the electrochemical impedance of an electrode interface and is useful for interpreting many chemical systems. The Randles equivalent circuit model includes a solution resistance (R_s), a Warburg impedance (Z_ω) resulting from the diffusion of the redox probe, a double-layer capacitance (C_{dl}) and a charge transfer resistance (R_{ct}).

According to Randles equivalent circuit model, the electrode reaction is controlled by the diffusion process (mass transfer) in the low-frequency range. When $\omega \to 0$, Z_{re} and Z_{im} are expressed in following form [69]:

$$Z_{re} = R_\Omega + R_{ct} + \sigma\omega^{-1/2} \tag{1}$$

$$Z_{im} = \sigma\omega^{-1/2} + 2\sigma^2 C_{dl} \tag{2}$$

where R_Ω presents uncompensated solution resistance, C_{dl} is double-layer capacitance, R_{ct} is charge transfer, σ is the Warburg coefficient related to mass transfer and ω is angular frequency, respectively. The following relationship between Z_{re} and Z_{im} can be obtained [69]:

$$Z_{im} = Z_{re} - R_\Omega - R_{ct} + 2\sigma^2 C_{dl} \tag{3}$$

In the high-frequency range, the electrode reaction is governed by charge transfer (or kinetic step); Z_w=0 and the electrochemical impedance Z is given by the following equation [69]:

$$Z = R_\Omega + \frac{R_{ct}}{1 + R_{ct}^2 C_{dl}^2 \omega^2} - \frac{R_{ct}^2 C_{dl}\omega}{1 + R_{ct}^2 C_{dl}^2 \omega^2} j = Z_{re} + Z_{im} j \tag{4}$$

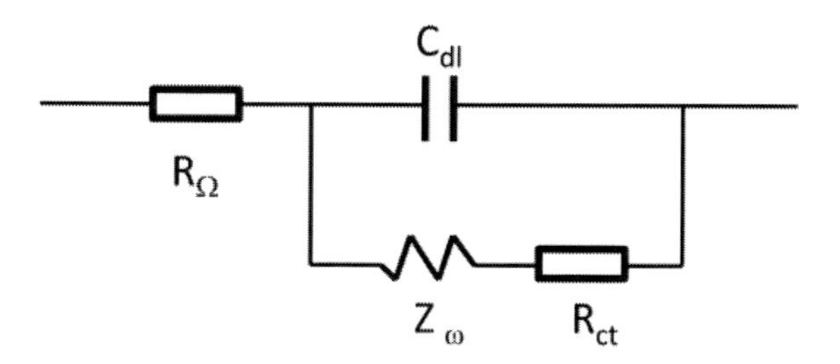

Figure 8. Equivalent circuit of a Randles cell. R_Ω presents uncompensated solution resistance, C_{dl} double-layer capacitance, R_{ct} charge transfer resistance and Z_ω Warburg impedance.

Knowing Z_{re} and Z_{im}, R_{ct} and R_Ω can be obtained from the Nyquist diagram equation:

$$(Z_{re} - R_\Omega - \frac{R_{ct}}{2})^2 + Z_{im}^2 = (\frac{R_{ct}}{2})^2 \qquad (5)$$

Furthermore, the heterogeneous standard charge-transfer rate constant (K_a^0) can be calculated [69]:

$$K_a^0 = \frac{RT}{n^2F^2AR_{ct}}\frac{1}{C_O^{*(1-\alpha)}C_R^\alpha} \qquad (6)$$

where α is the charge-transfer coefficient, A is the electrode area and C_O and C_R is the concentrations of redox couple, respectively.

The immobilization of DNA probes and hybridization event cause not only changes in the intrinsic properties of CPs film, but also changes in the various interfacial film properties, such as the capacitance and interfacial electron transfer resistance [37]. EIS can detect these changes, which can be exploited in development of truly label-less biosensing. Indeed, EIS has been successfully employed in the label-free detection of the DNA hybridization based on conducting polymers although a full mechanistic understanding of the above mentioned contributions to the overall EIS spectra change of a CPs film is still to be developed.

We have constructed label-free DNA sensors based on poly (pyrrole-co-3-pyrrolylacrylic acid) (poly(Py-*co*-PAA)) and (pyrrole-*co*-5-(3-pyrrolyl) 2,4-pentadienoic acid) (poly(Py-*co*-PPDA)) [70, 71]. The results show that there are considerable differences in the AC impedance spectra of a functionalized polypyrrole that has covalently grafted ODN probes before and after hybridization with different concentrations of complementary ODNs obtained in the presence of $Fe(CN)_6^{4-}$/ $Fe(CN)_6^{3-}$, as represented in Figure 9. The spectra were fitted based on a modified Randles equivalent circuit in which a constant phase element (CPE) was used instead of the double-layer capacitance (C_{dl}). The inclusion of a CPE instead of a plain capacitance improved the goodness of fit. In physical terms, a CPE (equation 7) reflects non-homogeneities of the surface reactions where the parameter n determines the extent of the deviation from the Randles and Ershler model:

$$CPE = A^{-1}(j\omega)^{-n} \qquad (7)$$

When $n=1$, the CPE reduces to a capacitance[36]. By using this model, the obtained EIS spectra can be well fitted. After ODN probe attachment, the values of R_{ct} increase in both materials due to the formation of negative charged ODN layer that electrostatically repels the negative redox probe, $[Fe(CN)_6]^{3-/4-}$, and inhibits interfacial electron transfer. The changes in R_{ct} value for poly(Py-*co*-PPDA) and poly(Py-*co*-PPA) films are 184 Ω and 67 Ω, respectively. After hybridization with 20.2 nM complementary ODN, the ΔR_{ct} for poly(Py-*co*-PPDA) and poly(Py-*co*-PPA) films are 173 Ω and 94 Ω, respectively.

With the advance in the nanotechnology, nanomaterials have been used to amplify the impedance change. We developed an DNA detection system, in which DNA sample fragments were entrapped in the PPy film and CdS nanoparticle labeled ODN probes were used to amplify the impedance change[72], as shown in Figure 10. The used equivalent circuit model (Figure 11) consisted of a solution resistance (R_s), an element of resistance (R_e) in parallel with capacitance (C_e) representing the Au electrode and an element of interfacial charge transfer resistance (R_{ct}) in series with a Warburg impedance (W). A constant phase element (CPE) (in parallel with W and R_{ct}) acts as a nonideal capacitor. The experimental data can be well fitted by this model and the fitting results show the increase in charge transfer resistance upon binding of complementary CdS-ODN nanoparticle probes. Besides taking the change of charge transfer resistance as the sensor response, we also suggested that the change of impedance at a fixed frequency could be used as sensor response.

Tlili et al. reported detection of DNA hybridization by using non faradic electrochemical impedance spectroscopy[73]. The DNA probes were covalently attached to the precursor copolymer, poly(3-acetic acid pyrrole, 3-N-hydroxyphthalimide pyrrole) and the impedance measurements were performed without any redox species on a DC potential of -1400 mV in order to minimize Warburg impedance and emphasize the contribution of the impedance of the PPy-DNA / electrolyte interface. At this applied potential, polypyrrole was in non-doped and semiconducting state; therefore no parasite electrochemical reaction occurred during measurement.

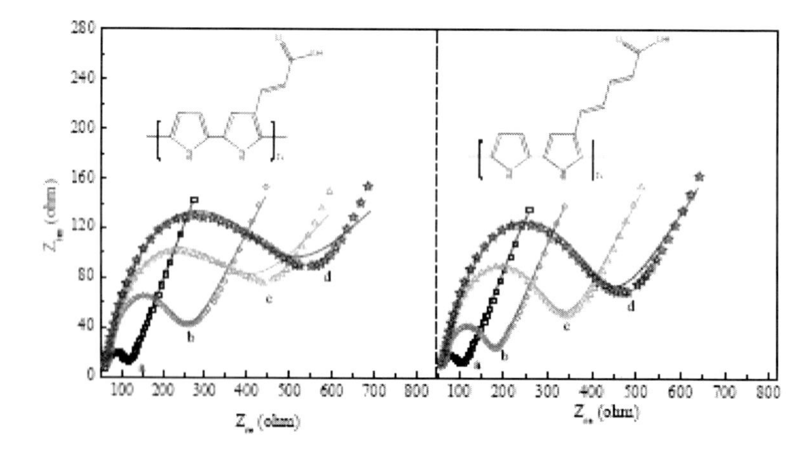

Figure 9. Nyquist plots ($-Z_{im}$ *vs.* Z_{re}) for electrochemical impedance measurements based on electrodes coated with poly(Py-*co*-PAA) (A) and poly(Py-*co*-PPDA) (B) in PBS solution (pH 7.4) containing 5.0 mM $Fe(CN)_6^{3-}$ / $Fe(CN)_6^{4-}$. (a) Before immobilization of the ODN probes; (b) after immobilization of ODN probes; after hybridization with (c) 20.2 nM and (d) 200.2 nM complementary ODNs. In all cases, the experimental data are shown as symbols and the fitting curves obtained with the equivalent circuit model as solid lines.

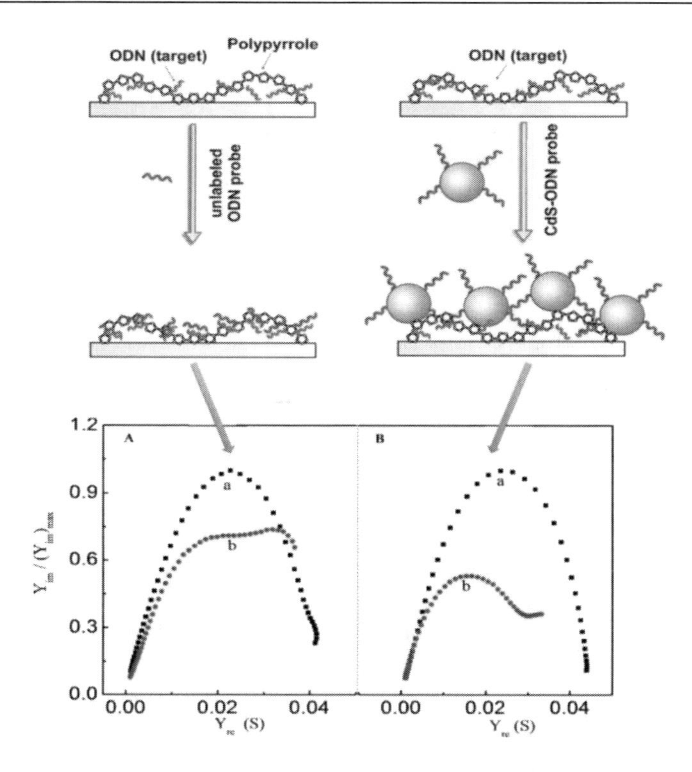

Figure 10. Amplification of impedometric response with CdS nanoparticles. Normalised admittance spectra of the gold electrodes with deposited PPy / target ODN film in PBS solution containing 5.0 mM $Fe(CN)_6^{3-}$ / $Fe(CN)_6^{4-}$ before (curve a) and after (curve b) incubation with 3.6 μM ODN probe (A) or CdS-ODN probe (B).

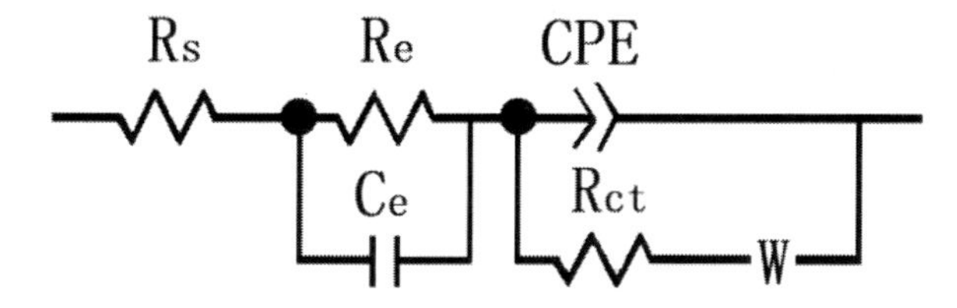

Figure 11. Scheme of the equivalent electrical circuit. R_s is solution resistance; R_{ct} charge transfer resistance, W is Warburg impedance and CPE is constant phase element. The gold electrode is presented by a resistance R_e in parallel with capacitance C_e.

The obtained impedance spectra were fitted by using the modified Randles equivalent circuit with a CPE in order to reflect the non-homogeneity of the layer. While in this case, the charge transfer resistance decreased upon grafting of DNA probe and increased upon hybridization with complementary DNA. The authors explained these based on the effect of the negative charge of *ss*-DNA and *ds*-DNA and of their conformational structures. In the case of grafting of *ss*-DNA onto functionalized polypyrrole which could be considered as p-type semiconductor under the experimental conditions, the grafting of negatively charged *ss*-DNA leads to increase in the majority carrier density and then a decrease of resistance of space charge region. On the other hand, the *ss*-DNA is in a flexible random structure and can penetrate into the polymer pores to increase the ionic concentration in the polymer film. After hybridization, the *ds*-DNA is in the helix formation, resulting in the significant stiffness of the functionalized polypyrrole which leads to a decrease in intrinsic conjugation of the polymer backbone and causes the increase in the charge transfer resistance.

The decrease in the impedance caused by the hybridization is wildly observed in polythiophene systems. A terthiophene polymer bearing a carboxyl group was polymerized on a glassy carbon electrode and used for the preparation of DNA sensor [74]. A reduction in total impedance which was recorded at an open circuit potential without any redox species was obtained after hybridization, and the highest differences in admittance (=1/|impedance|) were observed at ~1 kHz. The mechanism for the decrease in the impedance was not investigated in detail and the authors attributed the decrease in impedance to the higher conductivity of double stranded ODN compared with *ss*-ODN. We synthesized a terthiophene bearing an unsaturated side chain (3-((2':2'',5'':2'''-terthiophene)-3''-yl) acrylic acid, TAA) and used its polymer for DNA hybridization detection [75]. The impedance spectra were measured without redox probe and at 800 mV under which the polymer is in oxidized state. The hybridization caused a decrease in the impedance as shown in Figure 12, similar to the results of the reference [74]. The mechanism were investigated by using electrochemical quartz crystal microbalance (EQCM) and we found that the dopants play an important role in the impedance change [76]. When poly(TAA) film dopped with $CF_3SO_3^-$ which provided the best sensor response was cycled by potential scan in acetonitrile solution containing only 0.05 M of tetrabutylammonium trifluoromethanesulfonate, the frequency decreased during polymer oxidation process, as shown in Figure 13A. This illustrated that the anions, $CF_3SO_3^-$, from the solution moved into the polymer film, resulting in the increase of mass on the electrode surface. During the reduction process, the frequency increased, indicating that the anions moved out of the film. However, when poly(TAA) was circled in PBS buffer solution, the frequency increased during the oxidation process and decreased during the reduction process, as shown in Figure 13B, which indicated that in aqueous PBS solution the dominant ion movement in poly(TAA) film was cation movement. For the DNA sensor based on poly(TAA) film, the hybridization caused an increase in the negative charge due to the formation of duplex DNA and facilitated the cation movement during the doping process which resulted in the increase in the conductance of polymer film. In the cases of polypyrrole systmes, the dopant are usually Cl^{-1} or ClO_4^- which is small ion and in aqueous solution the dominant ion movement is anion movement [77, 78].

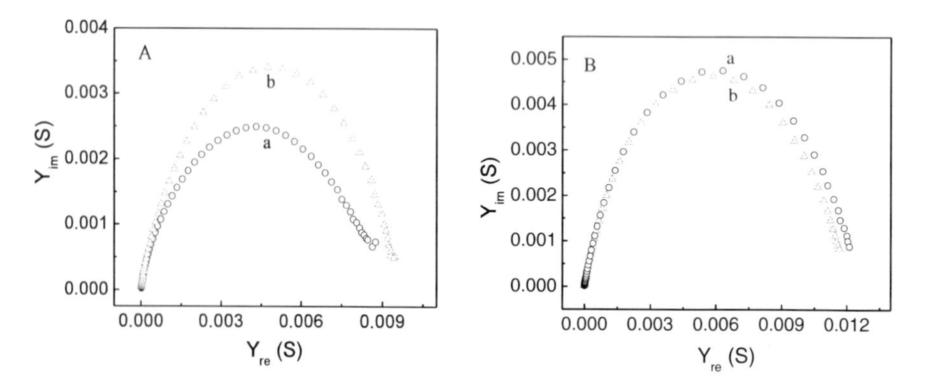

Figure 12. Admittance spectra of: PTAA films doped with TBAFMS (A) and $LiClO_4$ (B). (a): After immobilization of the ODN probes and (b) after incubation with 4.03 μM complementary ODN.

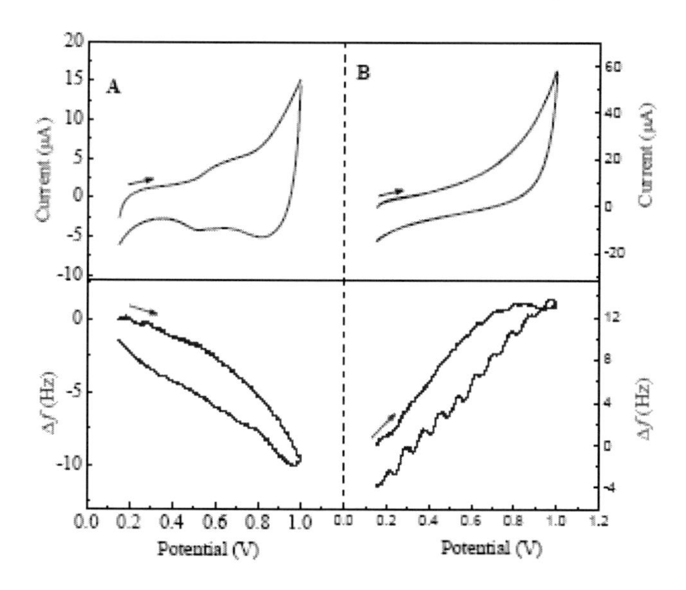

Figure 13. Cyclic voltammograms of poly(TAA) film and corresponding frequency changes in an acetonitrile solution containing 0.05 M TBAFMS (A) and PBS buffer solution (B). Scan rate: 100 mV/s.

The increase in the negative charge after hybridization results in the enhanced repellence to the anion exchange at the interface of film and electrolyte which causes an increase in the charge transfer resistance.

Gautier et al. prepared functionalized polythiophene matrix for a label-free DNA hybridization detection by copolymerizing an arylsulfonamide modified thiophene with 3-methylthiophene[52, 79]. The impedance spectra were recorded without any reodx species at +1.1 V. The hybridization caused the reduction in the impedance which was ascribed to the formation of the double helix structure liberating the surface from the random coil conformation of the ss-DNA and restoring a partial anionic exchange at the interface between film and electrolyte. Later, the same group compared non-Faradaic impedance spectra with Faradaic impedance spectra of the same system to reveal the different changes in impedance modulus[80]. Generally, Faradaic impedance measurement discusses the kinetic of electron transfer processes and non-Faradaic impedance is related to the alterations of capacitance and molecular layer organization [36]. The authors found that the impedance features obtained by Faradaic approach are dramatically different from those obtained by non-Faradaic approach. For non-Faradaic impedance, the hybridization caused a decrease in semicircle diameter in Nyquist diagram, while an increase of semicircle diameter was observed in Faradaic measurements due to the electrostatic repulsion between the negative charge of the redox probe and negatively charged DNA. Furthermore, the length of target DNA also affected the impedance change. ss-DNA target with 675 bases, that is much longer than the probe (42 bases), caused an increase of impedance modulus, which was diametrically opposite to the change in non-Faradaic impedance observed upon hybridization with 37 bases DNA target. The authors ascribed this difference to the different organization of the DNA modified layer after hybridization. When the probe and the target have the same length, the hybridization resulted in an opening of the interface to the access of mobile ions in solution. When the target was much longer than the probe, the double helix was extended to the solution by a flexible single-stranded DNA sequence, which prevented the access of the anions and caused an increase of impedance.

3.1.2.4. Photocurrent Detection

Photoelectrochemical techniques have been employed increasingly for investigating thin photoconducting films and corrosion layers on metals and alloys [81, 82], because of the simplicity of the experimental setup, the possibility of monitoring in-situ surface changes with time, and the ability to investigate very thin films. Due to these advantages, photoelectrochemical techniques have also been used for DNA sensing [83-85] and DNA damage detection [86, 87].

The initial work on photocurrent spectroscopy as a transduction tool for a direct detection of DNA hybridization based on CPs has been done by Lassalle et al. [88-90]. The photocurrent response under white light illumination was analyzed for a polypyrrole copolymer with one monomer unit modified by grafted ODN which was exposed to a blank buffer solution, and buffer solutions containing either complementary or non-complementary ODNs. When exposed to the blank buffer and after interaction with noncomplementary ODN, the pyrrole-ODN showed similar photocurrent response, while, the photocurrent was much lower following hybridization. The reason for the decrease of photocurrent is not clear. It may originate from different physiochemical behavior of the polymer film caused by the formation of DNA double helix. The band gap energies for a direct electronic transition in these three cases were estimated, with values of 2.9, 2.85, and 3.1 eV, for the polymer, the polymer after interaction with noncomplementary ODN and the hybridized film, respectively. These values were higher than that of polypyrrole (e.g. 2.2 to 2.6 eV), indicating less photosensitive film due to the presence of oligonucleotides. The photocurrent evolution during hybridization revealed that the kinetics of the process could be followed. However, these sensing films were not completely characterized in terms of sensitivity, reproducibility and the calibration curve.

3.2. Functional CPs for Contructing Electrochemical DNA Sensors

CPs used for constructing electrochemical DNA mainly involve polypyrrole, polythiophene, polyanline and their derivatives. Among them, polypyrrole and its derivatives are extensively used in design of biosensor due to their good biocompatibility and polymerization at neutral pH [91]. A DNA sensor based on polypyrrole for rapid detection of *Escherichia Coli* was prepared by using 25 base pair oligonucleotide probe as dopants[92]. A DNA concentration of 1 μg / μL was detected by cyclic voltammetry and the analysis was finished in 15 min. A pulsed amperometric detection of target DNA in PCR-amplified amplicons with platinum electrodes modified by single-stranded DNA (20 bases) entrapped within polypyrrole has also been reported [93]. The detection time was 30–35 min and the sensor response to complementary target was higher than that to noncomplementary by a factor of at least 6-8. Komarova et al. prepared ODN-doped PPy sensor films to detect the pathogen *variola major* by means of chronoamperometry [94]. It was established that thinner films with smaller dopants and a higher concentration of dopant ions produced stronger amperometric signals. Blocking of the film surface with fragmented half thymus DNA resulted in complete disappearance of the non-specific signal when ultra-thin (Langmuir-

Blodgett) films were tested, while the specific signal from complementary ODN remained unaffected. Additionally, lowering the potentials during the hybridization reduced the non-specific signal. Under optimal conditions a detection limit of 1.6 fmol of target ODN in 0.1 ml (16 pM) was achieved. For this type of DNA sensors based on polypyrrole, the obvious advantages are simplicity and fast detection, however, the steric hindrance and poor accessibility of the entrapped probe in the film to the analyte result in poor hybridization efficiency and greatly reduce the sensitivity and selectivity.

In order to overcome these disadvantages, variety of funtionalized pyrrole monomers have been developed for covalently immobilizing DNA probe and constructing electrochemical DNA sensors, as shown in Figure 14. A pyrrole monomer bearing an ODN (Figure 14, monomer 1) was synthesized and copolymerized with pyrrole to realize addressable multiple probes immobilization by Livache et al. [44, 45]. The hybridization was detected by photocurrent spectroscopy [90]. A biotinylated pyrrole monomer (Figure 14, monomer 2) was also prepared by the same group [55]. The obtained biotinylated polypyrrole film could be used for reversible biotinylated ODN probes immobilization via biotin/avidin interaction. Thompson et al. prepared pyrrole monomer modified with a phosphonic acid group (Figure 14, monomer 3) [49]. The DNA probe was immobilized onto the obtained polymer film with the help of magnesium cations which served as a bridge between the phosphonic acid group of the grafted polymer and the phosphate group of the oligonucleotide probe. This type of linkage makes the oligonucleotide offset from the surface of the polymer, giving it some freedom of movement and easing the effect of steric hindrances on the hybridization event. The hybridization was detected by cyclic voltammetry, while the sensitivity was not investigated.

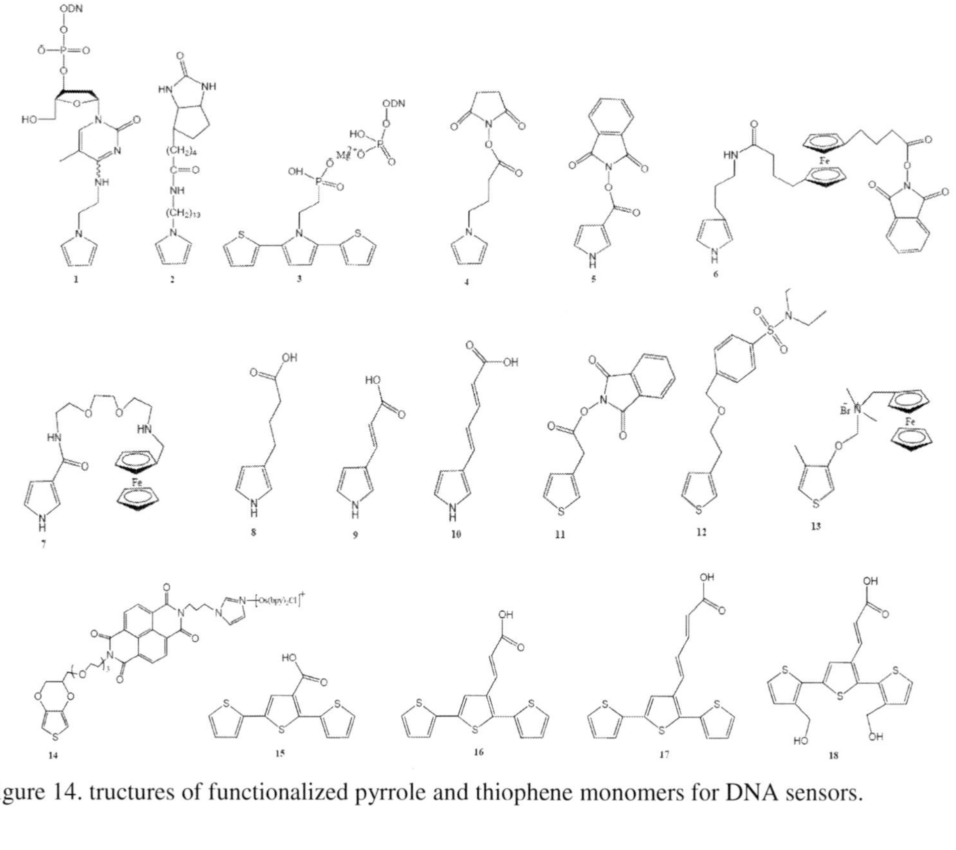

Figure 14. tructures of functionalized pyrrole and thiophene monomers for DNA sensors.

A pyrrole monomer bearing an easy leaving group, pyrrole-NHS (Figure 14, monomer 4), was prepared through the reaction of 4-pyrrolyl butanoic acid with N-hydroxysuccinimide (NHS) by Ionescu et al. [95]. Its homopolymer was used to covalently anchor an amino-21-mer oligonucleotide probe for detecting a short cDNA sequence from *West Nile Virus* (WNV) by amperometric method[95]. After incubation with a target model of the WNV cDNA, the modified electrode was further incubated in a complementary biotinylated 15-mer WNV cDNA solution followed by the specific attachment of a biotinylated glucose oxidase via an avidin bridge. The hybridization event was then monitored at 0.6 V vs Ag/AgCl by amperometric detection of H_2O_2, generated by the enzyme marker in the presence of glucose. Since the product of the enzyme-catalyzed reaction was detected, the sensitivity of the sensor was related to the permeability of the poly (pyrrole-NHS) film. With the aim of increasing the permeability, the polymer film was overoxidized until it lost the conductivity. A relatively short hybridization period (2 h) allows the convenient quantification of the WNV DNA target in the range 10^{-10}–10^{-15} g ml^{-1} and the detection limit is 1 fg ml^{-1}.

Because 3-substitution of pyrrole is more favorable than *N*-substitution with regard to maintaining the high intrinsic conductivity of the polymer, Garnier et al. prepared a functionalized pyrrole monomer bearing another easy leaving group, *N*-hydroxyphthalimide (Figure 14, monomer 5) [39, 48]. An amino-substituted ODN was grafted onto the precursor copolymer, poly(3-acetic acid pyrrole-*co*- 3-*N*- hydroxyphthalimide pyrrole), by a direct chemical substitution of the easy leaving group, *N*-hydroxyphthalimide. The detection limit was found to be 2 nM. A pyrrole monomer bearing a relatively long butanoic acid side chain (Figure 14 monomer 8) prepared by us, was expected to position the ODN probe away from the copolymer backbone and allow easy hybridization with the complementary ODN sequences[38]. In both cases, the hybridization was detected by cyclic voltammetry and the sensitivities are related to the redox properties of polymers. In order to increase the sensitivity of DNA detection, a ferrocenyl-functionalized pyrrole (Figure 14, monomer 6) was prepared [96] and ODN probes were subsequently grafted onto the obtained ferrocenyl-functionalized polymer film. In this arrangement the ferrocenyl group on the polymer is used as a probe for DNA detection to produce narrow and reversible redox signature due to its high sensitivity to changes in the electronic and steric environment. Upon hybridization, a shift of the oxidation wave of ferrocenyl groups to more positive potentials and a decrease in oxidation current were observed. The results were explained by the decrease of the permeability of the polymer film to dopant ions and changes in the polymer backbone conformation. The estimated detection limit was 2 pM of the target ODN molecule. However, this strategy would be unsuitable for a multiprobe addressing on a chip, because the electropolymerization was carried out first in an organic solvent, followed by the grafting of ONA probes in an aqueous medium. Bouchet et al developed a technique allowing one-step electro-addressing of probes on a microarray to acheive sensitive and label-less multidetection of DNA targets in solution[97]. Firstly, a pyrrole-ferrocene derivative (Figure 14, monomer 7) that could be electropolymerized in an aqueous medium was synthesized. The electrochemical copolymerization of this monomer with pyrrole monomers bearing different sequences of ODN and 2-(1H-pyrrol-3-yl) ethanol was then carried out by using a miniaturized graphite electrode network. The hybridization resulted in the changes of the ferrocene oxidation currents. A good selectivity between Human Immunodeficiency Virus and Hepatitis B Virus targets was achieved and the detection limit was up to 100 pM.

The effect of the 'linker' group (functionalized side chain that links the polymer backbone and the bioprobe) on the resulting sensor properties was investigated [70, 71]. Pyrrole monomers with conjugated side chains which provide binding sites for DNA probes were synthesized (Figure 14, monomer 9 and 10). The motivation to use a conducting polymer with conjugated carbon side chains is based on the idea that extension of the main chain conjugation into the side chain may improve polymer susceptibility to the changes caused by DNA hybridization. Figure 9 shows the sensors responses based on the copolymers of these two functionalized monomers (3-pyrrolylacrylic acid (PAA) and 5-(3-pyrrolyl) 2,4-pentadienoic acid (PPDA)) with pyrrole, illustrating longer side chain improves the sensor response. It is also confirmed that copolymers with conjugated side-chain functionalization have superior properties for use in biosensor applications compared to those with saturated side chain. The resulting sensor had good selectivity and a detection limit of 0.5 nM.

Functionalized polythiophenes are another type of CPs used for DNA sensing. Generally, most thiophenes with functional groups suitable for immobilization of biomolecules (such as $-NH_2$ and -COOH) have been difficult to electropolymerize, because such functional groups exhibit substantial nucleophilicity and attack the radical cation intermediates formed during electropolymerization, hence inhibiting the polymerization process [98]. In order to overcome this disadvantage, different ways have been suggested. Cha et al. synthesized a thiophene monomer whose carboxyl group is protected by an easy leaving group (Figure 14, monomer 11)[99]. After polymerization, the amino-modified ODN probes were covalently attached to the polymer film by taking place of the leaving group. The hybridization was detected by cyclic voltammetry and the discrimination of single base mismatch was achieved. Gautier et al. developed 3-(oxyalkyl)-thiophene bearing an arylsulfonamide group (Figure 14, monomer 12) [50-52]. The sulfonamide terminal functions of the obtained copolymer film can electrochemically cleaved and chemically modified with N-chlorosuccinimide (NCS), resulting in sulfonyl chloride functions which can be used for direct immobilization of amino-ended single-stranded DNA probe. The feasibility of constructing a label free DNA sensor based on its copolymer has been illustrated by using EIS[52, 79]. A ferrocene-functionalized cationic polythiophene (Figure 14, monomer 13) was prepared by Le Floch et al and was used as an indicator for hybridization detection by using square wave voltammetry[100]. Neutral peptide nucleic acid (PNA) was used as a capture probe. After hybridization with complementary target, the resulted negatively charged duplex interacted with the ferrence-functionalized cationic polythiophene through an attractive electrostatic interaction. The detection limit was 0.5 nM. The discriminations of one and two mismatched DNA targets were achieved. Tansil et al. synthesized naphthalene diimide grafted with ethylenedioxythiophene monomer (Figure 14, monomer 14) as an efficient DNA intercalator [67]. The Os^{2+}/Os^{3+} redox centers provided current signals for DNA detection. But the direct electrochemical detection of this electroactive intercalator was not sensitive enough and the detection limit only reached around 100 pM and a very low current output (~10 nA) even with a more sensitive square wave voltammetric method. In order to amplify the electrochemical signal output, the intercalated monomers were electropolymerized to form PEDOT with a greater electroactivity, then further copolymerized with EDOT–OH. After this amplification procedure, this signal was improved 100 times more, and detection of 20 pM target was satisfactorily achieved.

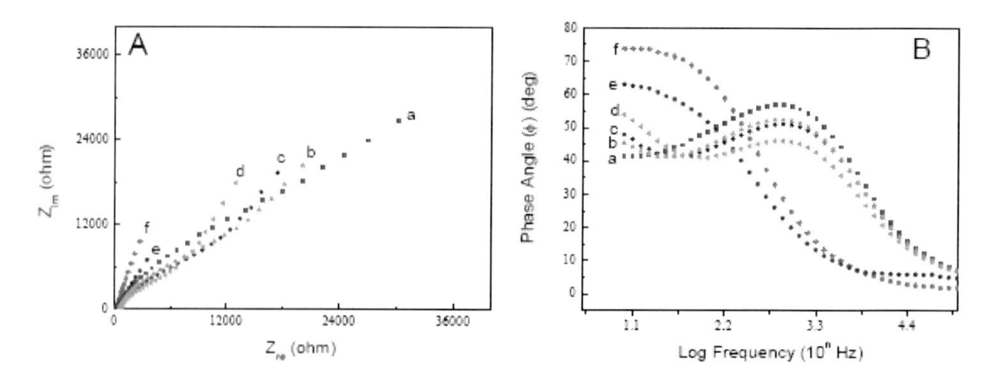

Figure 15. (A) Nyquist plot of a PTAA film showing Z_{im} and Z_{re} (a) after 434 μM probe attachment but before hybridization; (b) after 0.136 nM com; (c) after 13.6 nM com; (d) after 1.36 μM com where complementary targetwas applied in order of increasing concentration and where impedance was measured at a 200 mV constant bias potential in a tris–EDTA solution without the $Fe(CN)_6^{3-/4-}$ redox probe. The Nyquist plot of a PTAA film showing Z_{im} and Z_{re} (e) after probe attachment but before hybridization; (f) after 1.36 μM com where impedance was measured at a 900mV constant bias potential in a tris–EDTAsolution without the $Fe(CN)_6^{3-/4-}$ redox probe. (B) Phase Angle Bode plot of the PTAA film shown in (A).

In another approach, a terthiophene monomer bearing a carboxyl group (Figure 14, monomer 15) was synthesized, which can be easily electropolymerized on a glassy carbon electrode to form a homopolymer and solved the electropolymerization problem of thiophene monomer functionalized with carboxyl group [74]. A DNA sensor based on this polymer showed a good specificity. The sensor responses for one and two based mismatch DNA targets were only 14.3 % of that for the complementary target. This polymer was further used as a matrix for hydrazine-catalyzed ultrasensitive detection of DNA and proteins [101]. The detection limit for the DNA was estimated to be 450 aM (that corresponds to 2700 DNA molecules in 10 μL). We synthesized terthiophene monomers bearing a conjugated side chain (Figure 14, monomer 16 and 17) and investigated their DNA sensing properties as active substrates [75, 102, 103]. The electropolymerization conditions and the oxidation state of the polymer were found to affect their sensing properties. For example, the DNA sensing properties of poly(3-((2':2'',5'':2'''-terthiophene)-3''-yl)acrylic acid) films (PTAA, monomer 16) were evaluated using electrochemical impedance spectroscopy in their reduced and oxidized states in dilute tris–EDTA buffer[102], as shown in Figure 15. The general form of the impedance diagrams (Figure 15A) for the film in the 'reduced' state was of a transmission line-type element in the mid-range of frequency that curved downwards to the real axis at high frequency and upwards to a limiting capacitive element at low frequency. Hybridization of complementary ODN with the surface-bound probe had the effect of decreasing the impedance range over which the transmission line element was observed: the transition to the capacitive element occurred at lower impedance. This transition decreased in impedance with increasing hybridization of complementary ODN. With the film in the 'oxidized' state, only the first part of the transmission line element was observed curving towards the real axis. The phase angle Bode plot in Figure 15B (curve a–d) shows that the most obvious changes in the phase angle for the 'reduced' state of the film occurred at around 1000 Hz where the peak maximum decreased with increasing concentration of ODN target. The change in the phase angle, ΔΦ, is linearly dependant on the logarithm of ODN target concentration. For the 'oxidised' state of the film (Figure 15B, spectra e and f), hybridization caused an increase in Φ peak value at about 10 Hz. The response of the film in the 'oxidised' state to the same concentration of target DNA was higher than that in the 'reduced' state.

The solvent-induced collapse of the microstructure of redox-active polythiophene derivatives which could affect the DNA sensing properties was investigated by using poly (5-(2':2'', 5'':2'''0-terthiophene)-3''-yl)] (2E, 4E) penta-2,4-dienoic acid) (PTPDA, the monomer 17 in Figure 14) [104]. The PTPDA film was prepared by using tetrabutylammonium bistrifluoromethanesulfonimidate ($TBA(CF_3)_2(SO_3)_2$) or tetraethylammonium tetrafluoroborate ($TEABF_4$) as the electrolyte in CH_2Cl_2. The obtained PTPDA films were characterized in different solutions by cyclic voltammetry, as shown in Figure 16. When the films was cycled in monomer-free CH2Cl2 solvent containing the corresponding electrolyte, a sharper oxidation and reduction peak occurred for PTPDA film made with $TBA(CF_3)_2(SO_2)$ (Figure 16, b), while the PTPDA film made with TEABF4 showed a less distinct polymer oxidation peak (Figure 16, a). The sharper oxidation and reduction peak of PTPDA film made with $TBA(CF_3)_2(SO_2)_2$ indicated the formation of more uniform polymer chains and a more reversible redox process likely facilitated by a faster exchange of ions in and out of the film. When these two films were potentially cycled in aqueous PBS buffer, the distinct peaks that were observable upon cycling in the organic solvent vanished and the current measured diminished progressively with cycling in the aqueous solution (Figure 16, c and d). This significant flattening of demonstrated that the redox process in both films was substantially hindered and there was little cation or anion movement that would otherwise accompany the film oxidation–reduction cycle. The CV behaviour of both films in PBS buffer indicated that their microstructure has collapsed with the ejection of dopants (and CH_2Cl_2) that supported the reversible redox activity observed in monomer-free solution. When potentially cycled in aqueous PBS buffer containing $Fe(CN)_6^{3-/4-}$ redox probe, these two films showed the different behaviors. With the large hydrophobic $(CF_3)2(SO_2)_2^-$ anion as the dopant, reversible behaviour of the couple was observed, around the expected redox potential (Figure 16, f). The current however, progressively decreased with cycling. With the smaller hydrophilic dopant BF4 , the cathodic wave for the redox couple was not observed. Again, the current decreased with cycling. The collapse of the film microstructure due to the ejection of dopants was confirmed by scanning electron microscopy (SEM: Figure 17). The surface morphology of the freshly prepared films ($PTPDA/TBA(CF_3)_2(SO_2)_2$, Figure 17A; PTPDA/TEABF4, Figure 17C) showed on the 100 nm scale a very rough, sharply peaked surface indented with many pores. This morphology was retained for the $PTPDA/TBA(CF_3)_2(SO_2)_2$ film after 20 h soaking in PBS (Figure 17B). However, for the PTPDA/TEABF4 film, soaking in PBS resulted in the surface morphology smearing out and the porosity filling in (Figure 17D). These images confirm the microstructure collapse of the PTPDA/TEABF4 film on soaking in aqueous buffer. The effects on the PTPDA/ $TBA(CF_3)_2(SO_2)_2$ film were either on a size scale not observable by SEM or were more marked in the interior of the film, below the outer surface. The collapse of the film microstructure significantly affects its DNA sensing properties. After 20 h soaking in PBS, the ODN probes were grafted to the films. Only the $PTPDA/TBA(CF_3)_2(SO_2)_2$ film responded to the target ODN. This result also indicated that the use of the hydrophobic $(CF_3)_2(SO_2)_2^-$ anion led to a more favorable surface configuration for detection of surface-bound ODN than the use of the hydrophilic BF_4^- anion.

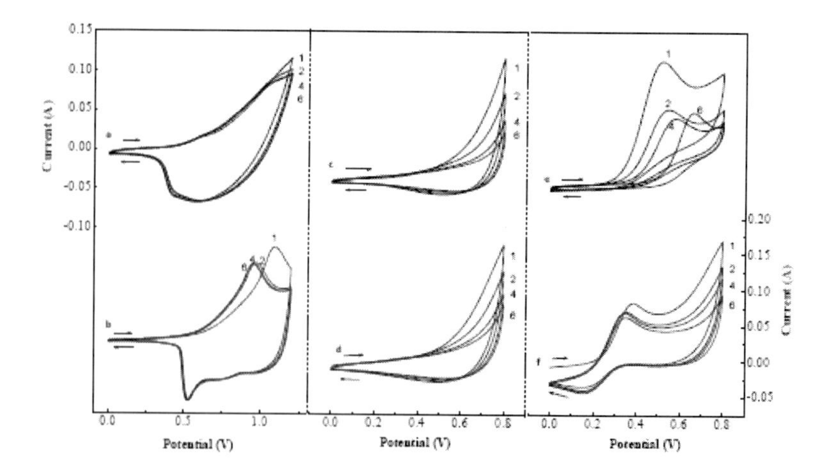

Figure 16. Cyclic voltammograms of PTPDA films in monomer-free CH_2Cl_2 containing 50 mM (a) $TBA(CF_3)_2(SO_2)_2$ or (b) $TEABF_4$, PBS buffer (c and d), and PBS buffer containing 5 mM of the $Fe(CN)_6^{3-/4-}$ redox couple (e and f). The films were made in a monomer solution containing 3 mM TPDA and 50 mM $TEABF_4$ (a, c, e) or $TBA(CF_3)_2(SO_2)_2$ (b, d, f) in CH_2Cl_2.

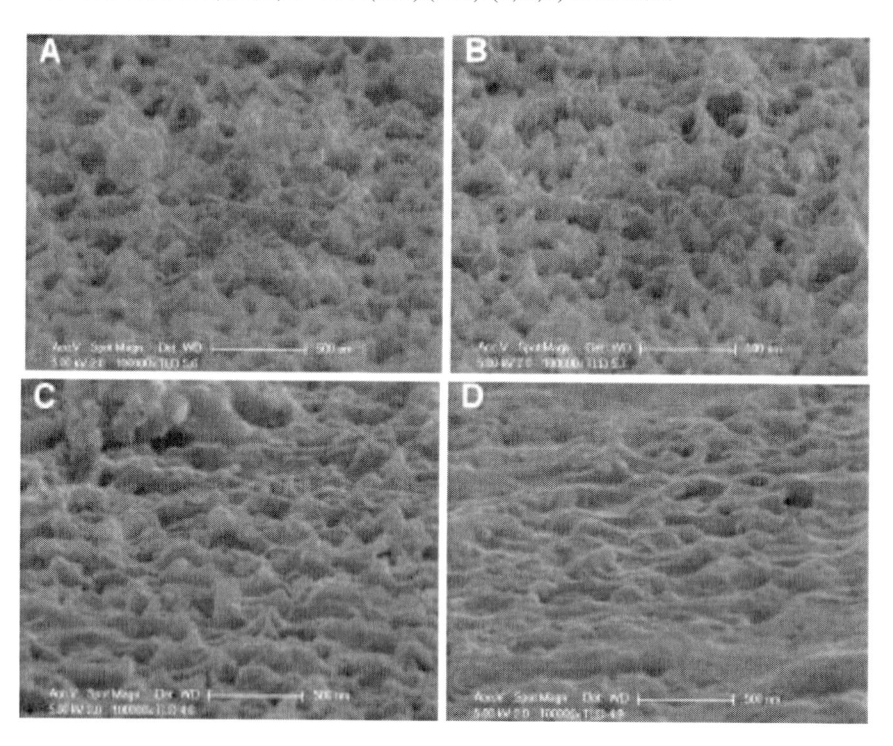

Figure 17. SEM images of dried PTPDA film made with 50 mM (A and B) $TBA(CF_3)_2(SO_2)_2$ or with (C and D) $TEABF_4$ in CH_2Cl_2. Films soaked for 20 h in PBS buffer prior to SEM are shown in B and D.

In order to address the problem of the collapse of film microstructure, a terthiophene monomer having a hydrophilic side-chain was synthesized, 3-[3',3'''-bis(hydroxymethyl)-2':2'',5'':2'''-terthiophene)-3''- yl](E) acrylic acid, (HTAA) (Figure 14, monomer 18). The methylhydroxy substituent in HTAA, absent in TAA and TPDA, was intended to endow its polymer PHTAA with greater electrochemical stability in aqueous solution arising from its greater hydrophilicity.

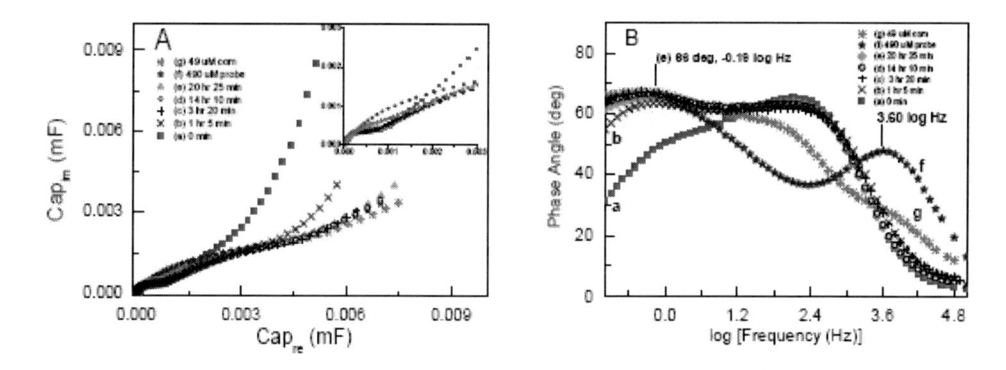

Figure 18. (A) Complex capacitance plot of PHTAA film made with $TBA(CF_3)_2(SO_2)_2$ over 20 h in PBS buffer containing 5 mM $Fe(CN)_6^{3-/4-}$. (B) Phase angle plot of the PHTAA films. After the 20 h, the films were immersed in 490 μM probe ODN solution (curve f) and then incubated in a 49 μM complementary target ODN solution (curve g). The inset in A is the magnification about the origin.

Figure 18 represents the capacitance diagrams (Cole–Cole plots) and correspond phase angle plots of PHTAA film doped with $(CF_3)_2(SO_2)_2^-$ exposed to PBS solution containing 5 mM $Fe(CN)_6{}^{3-/4-}$. The capacitance diagrams showed a semi-circular arc at high frequency (the low capacitance section of the diagram) that did not significantly change over 20 h of immersion whereas some change occurred at low frequencies (Figure 18A), which indicated much better resistance of PHTAA film to the collapse of microstructure in aqueous solution than that of PTPDA[104]. The advantages of PHTAA polymer in mitigating the microstructure collapse effects were investigated by grafting ODN probes and incubating into 49 μM target ODN solution. After the ODN probe grafting and hybridization, there were significant changes in the high-frequency part of the diagram for the film–ODN complex (Figure 18A, g and f). These variations were much clearer in impedance phase angle plots (Figure 18B, g and f). Compared with less hydrophilic PTAA and PTPDA[103, 104], the more hydrophilic PHTAA gave a detection of surface hybridisation which was a significant improvement over the signal observed for less hydrophilic polythiophenes[103, 104].

Compared to polypyrrole and polythiophene, polyaniline is less used in DNA sensor designs, due to its polymerization commonly being carried out in acidic environment which is not suitable for biomolecules. However, polyaniline undergoes two redox processes, and is mechanically resilient and environmentally stabile, which make it attractive material for the DNA sensing applications. Gu et al reported the immobilization of DNA on a thin layer of polyaniline/poly (acrylic acid) (PANI/PAA) composite polymer film that was electrodeposited on a boron-doped diamond (BDD) thin films [105]. The carboxylic acid residues in the polymer film act as the binding sites for DNA attachment, whilst the conductive polymer matrix enhanced the electron-transfer between DNA and the diamond surface. The direct oxidation of guanine and adenine in the double helix DNA were used to detect the hybridization event. The advantage of such design is minimal non-specific DNA adsorption, which was confirmed by the fluorescence microscopy and cyclic voltammetry measurements. This PANI/PAA modified BOD electrode was further characterized by electrochemical impedance spectroscopy [106]. The impedance measurement showed changes in the impedance modulus as well as electron-transfer resistance upon the probe DNA immobilization and after the hybridization with target DNA. A good selectivity between the complementary DNA targets and the one-base mismatch targets was demonstrated, as well as the sensor reusability. Davis et al. investigated the differences between polyethylenimine, polyaniline and polydiaminobenzene modified screen-printed carbon electrodes containing single-stranded DNA by AC impedance approach [107]. Complementary DNA hybridization gave rise to a lowering of the capacitance of the electrode/polymer film in solution. Arora et al reported an ultrasensitive DNA hybridization biosensor based on polyaniline electrochemically deposited onto Pt disc electrode [108]. Activated avidin was covalently attached to PANI film, followed by the immobilization of 5'-biotin end-labeled oligonucleotide probe via biotin and avidin interaction. The hybridization was detected by using both direct electrochemical oxidation of guanine and redox electroactive indicator methylene blue (MB). Compared to a direct electrochemical oxidation of guanine, hybridization detection using MB resulted in the enhanced detection limit by about 100 times and reached 2 fM. This sensor has been utilized for direct detection of *Escherichia coli* by immobilizing a 5'-biotin-labeled *E. coli* probe using a differential pulse voltammetry technique in the presence of methylene blue as a DNA hybridization indicator[109]. The detection limits for complementary target probe, *E. coli* genomic DNA and *E. coli* were 0.009 ng/μL, 0.01 ng/μL and 11 *E. coli* cells/mL without PCR amplification and it could be used 5-7 times at temperatures of 30-45 °C.

Gao et al proposed a novel approach for DNA detection by using polyaniline as hybridization indicator[110]. A mixed monolayer of thiolated peptide nucleic acid (PNA) capture probe and 4-aminothiophenol was first self-assembled on a gold electrode. The interaction of PNA with the sample DNA formed a heteroduplex, followed by further hybridization with Horseradish peroxidase (HRP) labeled oligonucleotide detection probes, which resulted in the immobilizing HRP on the biosensor surface. A mixture of aniline and H_2O_2 in a phosphate buffer (pH4.0) was applied to the biosensor. HRP catalyzed the oxidative polymerization of aniline and polyaniline produced were exclusively deposited on the hybridized nucleic acid molecules. The DNA was quantified by the electroactivity of the deposited PANI by using square wave voltammetry. This DNA sensor has shown a good sensitivity with a detection limit of 1.0 fM and good selectivity for one-, two- and three-base mismatched sequences, which resulted from the dual dependence of the signal-amplification process on template and catalyst that greatly lowered non-hybridization-related background noise. This biosensor was further used to detect three mRNAs, His4, RCA1, and GAPDH, and the results obtained were in good agreement with those of Northern blot analysis on the same samples.

3.4. Nanostructured CPs for Contructing Electrochemical DNA Sensors

The advances in nanotechnology have opened new ways for DNA sensors design. The sensitivity and other attributes of sensor can be improved by using nanomaterials which have superior physical and chemical properties over their bulk counterpart because of the effects such as the quantum confinement, mini size effect, surface effect and macro-quantum tunnel effect. Until now, conducting polymer nanomaterials with different morphologies such as nanoparticle, nanowire and nanotube have been prepared by chemical or electrochemical methods (see a review paper [111]). These nanostructured materials have been used to constructing the electrochemical DNA sensors. Ramanathan et al. [112] reported the preparation of biologically functionalized polypyrrole nanowires by the electropolymerization from an aqueous solution of pyrrole monomer in the presence of a model biomolecule (avidin- or streptavidin-conjugated ZnSe/CdSe quantum dots) and within a 100 or 200 nm wide and 3 μm long channels between gold electrodes. After addition of biotin –DNA, the avidin- and streptavidin-polypyrrole nanowires generated a rapid change in resistance to as low as 1 nM of biotin-DNA. The authors suggested the possibility of single-molecule detection by adjusting the nanowire's conductivity to a value closer to the lower end of the semiconductor. Cai et al prepared polypyrrole and multi-walled carbon nanotubes functionalized with carboxylic group (MWNTs-COOH) nanocomposite for indicator free DNA hybridization detection [113]. MWNTs-COOH were first modified to the glassy carbon electrode and ODN probes were doped within the electropolymerized PPy films by serving as the sole counter anion during the growth of the film. After hybridization, a decrease in impedance was observed due to the higher conductivity of double stranded ODNs compared with ssODN. The process of optimization of hybridization conditions revealed that the sensitivity of the sensor increased dramatically with an increase in the amount of multi-walled carbon nanotubes (MWNT) used. The response (difference of logarithmic impedance values) obtained was five times larger when the optimum amount of MWNT was used (as compared to the sensor without MWNTs). The estimated detection limit of this simple method was 10 nM. The detection limit of this sensor was further improved to 0.05 nM by the formation of metallized double-stranded DNA [114].

The template or template-free preparations of nanostructred polyaniline have been widely investigated [111]. Chang et al prepared highly organized PANI nanotube array on a graphite electrode by using a thin nanoporous aluminium layer as a template and constructed an ultrasensitive electrochemical nucleic acid biosensor based on this array [115]. The carboxyl group ended oligonucleotide probes were covalently immobilized on these PANI nanotubes. The differential pulse voltammetry was used to detect hybridization by employing daunorubicin as the indicator. This sensor showed good sensitivity with a detection limit of 1.0 fM. The authors contributed the advantages of the sensor to a collective effect due to the large effective surface area of the nanotube array which enhanced conductivity and faster kinetics for the analyte than in the case of disordered structures. Liang et al prepared large arrays of oriented polyaniline nanowires by controlled nucleation and growth during a stepwise electrochemical deposition process, without the use of porous membrane supports [116]. A DNA sensor was fabricated based on such nanowires on the glassy carbon electrode [117]. Oligonucleotide probes with phosphate groups at the 5' end were covalently attached to the PANI nanowires. The hybridization events were detected by differential pulse voltammetry (DPV) measurement using methylene blue (MB) as an indicator and the detection limit of 1.0×10^{-12} mol L^{-1} was acheived. We demonstrated a template-free method to prepare polyaniline nanotubes from a solution containing poly(methyl vinyl ether alt-maleic acid) (PMVEA) and using ammonium persulfate as the oxidant [118]. The obtained nanotubes (Figure 19A) were deposited on a glassy carbon electrode, followed by grafting of amino-modified ODN probes onto the residual carboxylic acid functionalities on the nanotubes. The potential pulse amperometry technique was used to obtain a direct and fast electrochemical readout for hybridization (Figure 19 B). The estimated detection limit was 3.4×10^{-10} mol L^{-1}.

Table 1 summarized the characteristics of electrochemical DNA sensors based on different functionalized CPs.

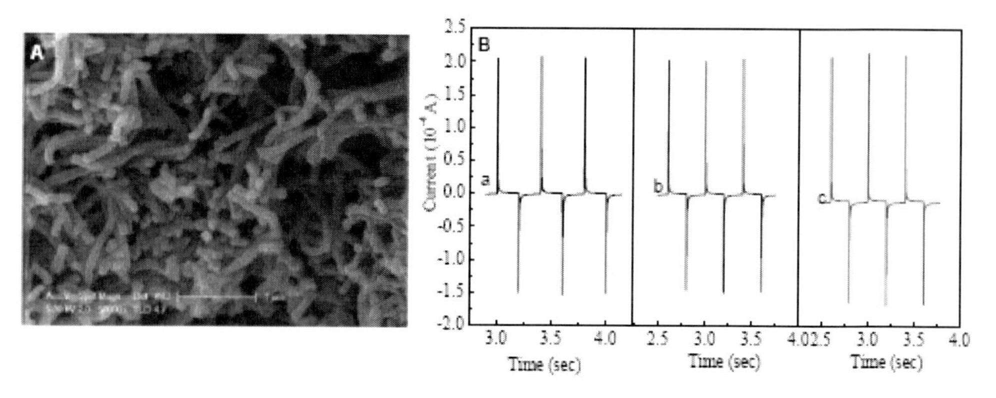

Figure 19. (A) SEM images of PANI nanotubes. (B) Potential pulse amperometry response of the PANI/PMVEA nanotube electrode with grafted ODNs (a), after incubation in a solution of noncomplementary ODNs (4.03 μM) (b) and after incubation in a solution containing complementary ODNs (4.03 μM) (c).

4. Functional CPs for DNA Sensing Based on the Optical Methods

Similar to their electric properties, the optical properties of conducting polymers are alos closely related to their electronic ground state. Therefore their optical properties can be straightforwardly varied by chemical substitution on the basic monomer unit which changes the electronic structures. On the other hand, minor perturbations in the conjugation of the polymer induced by, for example, disordering of the chain backbone, will lead to a change in optical properties such as adsorption and fluorescence , which can be used to develop novel detection methodologies to transduce physical and chemical information into a measurable signal [129-131].

4.1. Cationic CPs for Optical DNA Detection in Solution

In general, cationic CPs are used for the optical DNA detection in solution because they are able to interact with DNA via electrostatic forces. A series of cationic water-soluble polyfluorene, polythiophene and polyarylene derivatives were synthesized and used for the DNA detection in solution. The signal transduction mechanisms are mainly fluorescence resonance energy transfer (FRET) and analyte-induced conformational changes of CPs.

For FRET systems, the CP and the acceptor are chosen to meet the requirement of the spectroscopic overlap between the emission of the CP and the absorption of the acceptor according to Förster theory [132]. When excited, excitons migrate along the backbone of the CP and the exciton energy can be trapped by a bound acceptor molecule, which results in a quenched fluorescence signal of the CP and an amplified fluorescence signal of the acceptor [133, 134]. The Bazan and Heeger research groups pioneered investigations using cationic water-soluble CPs to detect specific DNA sequences via FRET [135].

Table 1. Characteristics of electrochemical DNA sensors based on CPs

Matrix	Immobilization Method	Detection Method	Detection Limit	Selectivity	Ref.
Polypyrrole	ODN entrapment	Amperometric	-	-	[41]
Polypyrrole	-	Amperometric flow cell	$6.1x10^{-16}$ in 20 μL (gives ~30 pmol dm^{-3})	-	[59]
Polypyrrole	ODN entrapment	Potential pulse amperometry	0.37 ng mL^{-1}	-	[93]
pTPTC3-PO3H2 / PPY	magnesium cations serving as a linker	amperometric	18-mer:0.16 fmol 27-mer: 3.5 fmol, respectively.		[119]
poly[pyrrole-COOH, pyrrole-ODN_{probe}]	Covalent attachment of ODN probe	Amperometry	1 pM		[120]
poly(pyrrole-NHS)	Covalent attachment of ODN probe	Amperometric	1 fg mL^{-1}	-	[95]
Polypyrrole	ODN entrapment	Chronoamperometric	16 pmol dm^{-3} (100 pg ml^{-1})	-	[94]
Polypyrrole	ODN entrapment	Cyclic voltammetry			[92]
Poly(3-pyrrolyl actetic acid)	Covalent attachment of ODN probe	Cyclic voltammetry (oxidation current)	~ 2 nmol dm^{-3}	-	[39]
Ferrocenyl groups-grafted polypyrrole	Covalent attachment of ODN probe	Cyclic voltammetry (oxidation current)	2 pmol dm^{-3}	-	[96]
poly[pyrrole-*co*-4-(3-pyrrolyl) butanoic acid]	Covalent attachment of ODN probe	Cyclic voltammetry (oxidation current)	~ 1-2 nmol dm^{-3}	-	[38]
polypyrroles bearing ferrocene and oligonucleotide	Covalent attachment of ODN probe	Cyclic voltammetry	100 pM		[97]
MWNT–COOH / Polypyrrole	-	Differential pulse voltammetry	2.3×10^{-14} mol/L	three-base mismatch was obviously distinguished	[121]
Polypyrrole/MWNT	ODN entrapment	differential pulse voltammetry	8.5×10^{-11} mol/L		[122]
Polypyrrole on MWCNT	ODN entrapment in PPy	AC impedance	10 nmol dm^{-3}	Mismatch: One base: 51 % Two base: 18% Three base: 8 % of fully complementary response	[113]
poly [3-acetic acid pyrrole, 3-*N*-hydroxyphthalimide pyrrole)]	Covalent attachment of ODN probe	AC impedance	0.2 nM		[73]
Polypyrrole and CdS-labeled probes	Target ODN entrapped in PPy film	AC impedance (change in charge transfer resistance)	~1 nmol dm^{-3}	one-, two-, and six-base mismatches: 62.6 %, 55.2 % and 18.7 % of the signal from complementary ODNs	[72]
Poly (pyrrole-*co*- 3-pyrrolylacrylic acid)	Covalent attachment of ODN probe	AC impedance	0.98 nM		[71]

Table 1. (Continued)

Matrix	Immobilization Method	Detection Method	Detection Limit	Selectivity	Ref.
Polypyrrole/MWNTs-COOH	ODN entrapment	AC impedance	5.0×10^{-11} mol/L	Mismatch: One base:35.7 % Three base:-5.8 % of fully complementary response	[114]
Poly(pyrrole-co-5-(3-pyrrolyl) 2,4-pentadienoic acid)	Covalent attachment of ODN probe	AC impedance	0.5 nM	Mismatch: One base:27 % Three base:38 % of fully complementary response	[70]
PPy/MWNTs-COOH	Covalent attachment of ODN probe	AC impedance	5.0 pM		[123]
Au-Ag/Polypyrrole nanocomposite	Covalent attachment of ODN probe	AC impedance	5.0×10^{-10} mol/L	Three base mismatch: 8 % of fully complementary response	[124]
Polypyrrole nanowire	avidin-biotin link	Conductometry			[112]
ODN-grafted polypyrrole (using phosphoramidite blocks)	Covalent attachment of ODN probe	a. indirect fluorescence b. microgravimetric c. SPRi	a. 1-10 pmol dm^{-3} b. 10 nmol dm^{-3} c. 250 nmol dm^{-3}	-	[125]
Polypyrrole	ODN entrapment	electrogenerated chemiluminescence	0.1 pM		[126]
poly(3,4-ethylenedioxythiophene)	ODN entrapment	Conductimetry	8.0×10^{-8} g/L		[68]
poly(thiophen-3-yl-acetic acid 1,3-dioxo-1,3-dihydro-isoindol-2-yl ester)	Covalent attachment of ODN probe	Cyclic voltammetry	1 nmoL	TG mismatch: 29.3% TC mismatch 24.3%	[99]
Ferrocene functionalized polythiophene		Square Wave voltammetry	0.5 nM	One and tow- base mismatches can be distinguished	[100]
poly-5, 2':5', 2''-terthiophene -3'-carboxylic acid	Covalent attachment of ODN probe	Differential pulse voltammetry	450 aM		[101]
Poly-3-((2':2'',5'':2'''-terthiophene)-3''-yl) acrylic acid	Covalent attachment of ODN probe	AC impedance			[75]
Poly(3'-carboxyl-5,2';5',2-terthiophene)	Covalent attachment of ODN probe	AC impedance		Mismatch: One base: 14.5%; two base: 14.5 % of fully complementary response	[74]
poly(3-methylthiophene-co-M1) M1: 3-(oxyalkyl)-thiophene bearing an arylsulfonamide group	Covalent attachment of ODN probe	AC impedance			[52, 79, 80]
Polaniline/ graphite oxide	Physical adsorption	Square wave voltammetry	29.34 ug/mL	-	[127]

Matrix	Immobilization Method	Detection Method	Detection Limit	Selectivity	Ref.
Polyaniline	Self-assembly monolayer of ODN probes	Square wave voltametry	1.0 fM	Mismatch: One base:30 %; two base:15 % ; three base: ~0% of fully complementary response	[110]
polyaniline/poly (acrylic acid) / boron-doped diamond	Covalent attachment of ODN probe	Cyclic voltammetry			[105]
Polyaniline nanotube	Covalent attachment of ODN probe	Differential pulse voltammetry	1.0 fM		[115]
polyaniline	avidin-biotin link	Differential pulse voltammetry	0.0054 attogram/ul		[108]
Polyaniline nanowire	Covalent attachment of ODN probe	Differential pulse voltammetry	1 pM	Mismatch: three base: ~0% of fully complementary response	[117]
Avidin-modified polyaniline	avidin-biotin link	Differential pulse voltammetry	0.009 ng/uL		[109]
polyaniline/poly (acrylic acid) / boron-doped diamond	Covalent attachment of ODN probe	AC impedance	20 nM		[106]
Polyaniline/Gold Nanoparticle	Covalent attachment of ODN probe	AC impedance		One base mismatch can be distinguished	[128]
Polyaniline nanotube	Covalent attachment of ODN probe	potential pulse amperometry	0.34 nM		[118]

A solution initially contains a cationic conjugated polymer, poly((9,9-bis(6'-*N*,*N*,*N*-trimethyl ammonium)-hexyl)-fluorene phenylene) iodide (CP **1** in Figure 20) and a peptide nucleic acid (PNA) sequence labeled with a chromophore dye (fluorescein, C*). PNA is a neutral DNA analogue and therefore there is no electrostatic interaction between the charged polymer and the PNA. In the absence of DNA, there were no electrostatic interactions between CP **1** and PNA-C*, and the distance between CP **1** (donor) and C* (acceptor) was too far for efficient FRET to occur.

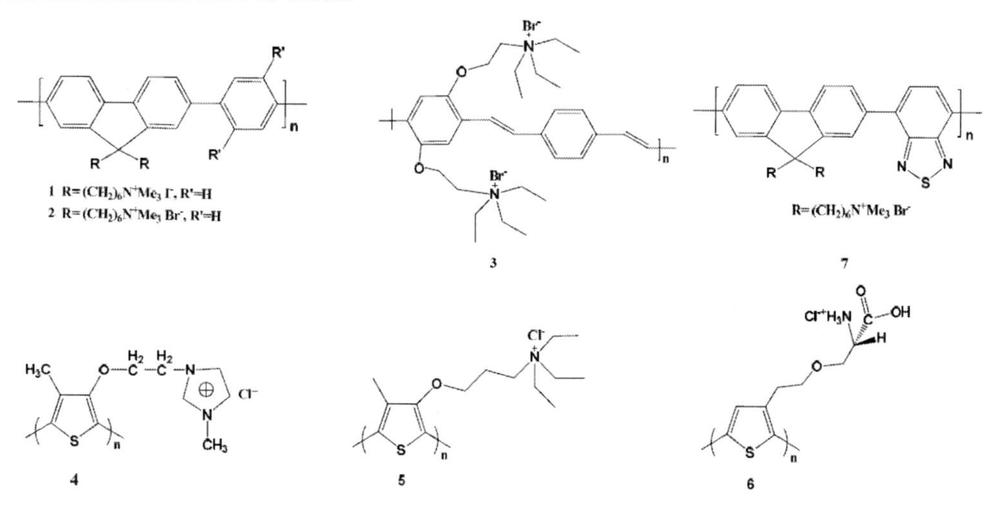

Figure 20. Molecular structures of cationic CPs.

When the complementary ssODN is added to the solution, the hybridization results in the formation of a charged complex of ODN/PNA-C*. The electrostatic interactions bring the charged ODN/PNA-C* complex within the distance required for FRET between CP **1** (donor) and C* (acceptor) and the FRET takes place. In this case conjugated polymer excitation afforded fluorescein emission more than 25 times higher than that obtained by exciting the dye. This procedure permitted detection of 10 pM of target ODN with a standard fluorometer. Further investigation of the energy transfer process between the above cationic conjugated polymer and fluorescein using ultra-fast pump-dump-emission spectroscopy has shown that both electrostatic and hydrophobic interactions contribute to the formation of the cationic conjugated polymer/PNA-C*/ODN complexes [136]. Additionally, these groups developed a strategy for detection of SNPs with fluorescein-labelled PNA (PNA-FL), CP **2** and single-strand-specific S1 nuclease enzyme[137]. Long ssDNA target sequences were digested by the S1 nuclease, leaving intact only those regions that hybridize to the PNA probe. Any PNA/DNA mismatches resulted in complete DNA digestion. Therefore, FRET from CP **2** to fluorescein takes place only for the perfect PNA-FL/DNA complex. The use of S1 enzyme allows for unambiguous SNP identification and also lowers the assay's background signal by reducing size discrepancies between the target DNA and CP.

Since PNA is more expensive than DNA, the same scheme has been adapted for use of C*-labeled ODN as an optical probe instead of PNA [138]. The idea is based on the premise that the electrostatic interaction between dsODN and the cationic conjugated polymer would be stronger than that of ssODN and the conjugated polymer due to the higher charge density of the dsODN. Non-complementary ODN would interfere with the FRET process between ssODN-C* and the conjugated polymer. Indeed, it was observed that a three-fold increase in fluorescence intensity from fluorescein (C*) relative to that of a noncomplementary strand when complementary ssDNA target was included in the solution. The energy transfer from CP **1** to the chromophore reached the maximum when the ratio of polymer chains to DNA strands was approximately 1:1.

We synthesized poly({2,5-bis[2-(*N,N*-diethylamino)-1-oxapropyl]-*p*-phenylenevinylene} -*alt*-*p*- phenylenevinylene) dibromide ($PPVNEt_2Br_2$, CP **3** in Figure 20) and applied it to the DNA detection based on FRET, as pictured in Figure 21. In the presence of Cy3 labeled ODN in SSC buffer (pH7.9), the fluorescence of Cy3 can be clearly detected when exited at 400nm (there is no emission of Cy3 upon excitation at 400 nm), as shown in Figure 22. This illustrated efficient FRET from $PPVNEt_2Br_2$ to Cy3. 5 nm red shift in the Cy3 emission peak in the presence of the polymer was observed due to an increase in polarity in the vicinity of Cy3 by the interaction with the cationic polymer[129]. After hybridization with complementary ODN, the FRET ratios progressively decreased with increased concentration of complementary ODN. One likely reason is that the polymer chain, which is in conformation of a flexible random coil, strongly interacts with single strand ODN through the electrostatic and hydrophobic interactions that bring them in intimate contact. On the formation of stiffer ODN duplex structure the distance between polymer and dye increases resulting in decrease of FRET efficiency.

The DNA detection based on the conformational changes of cationic CPs was pioneered by The Leclerc group[139]. A cationic polythiophene derivative, poly(1H-imidazolium-1-methyl-3-{2-[(4-methyl-3-thienyl)-oxy]ethyl}chloride (CP **4** in Figure 20) was synthesized. The aqueous solution of the cationic polymer appeared yellow with a UV-Vis absorption around 392 nm, attributed to a random-coil conformation. Upon addition of 1.0 equiv. (per monomer unit of a polymer) of 20-base ssDNA to CP **4**, the color of the solution changed from yellow to red in about 5 min due to the formation of a so-called "duplex" between CP **4** and the ssDNA probe driven by electrostatic attraction. The formation of "duplex" lead to a highly planar and conjugated polymer conformation with a peak absorption at long wavelength (λ_{ab} = 520 nm). When the complementary ODN was added, the mixture becomes yellow again (absorption maxima at 421 nm) presumably due to formation of a more stable "triplex" complex between the cationic polymer and dsODN. In addition to UV-Vis spectroscopy, fluorometric detection of ODN hybridization was also illustrated. The complexation of the cationic polymer with ssODN led to a decrease in fluorescence of this polythiophene derivative, while hybridization with complementary ODN resulted in a fivefold increase in the fluorescence intensity at an emission maximum of 530 nm. The spectroscopic data suggested that this is due to the helical structure of the polythiophene backbone that wraps with greater affinity to dsODN than to ssODN, combined with better solubility of the nonstoichometric triplex structure which limits interchain quenching. After the optimization of detection procedure, the specific identification of 20 and 50-mer targets was reported to be at the zeptomolar (zM) level [140].

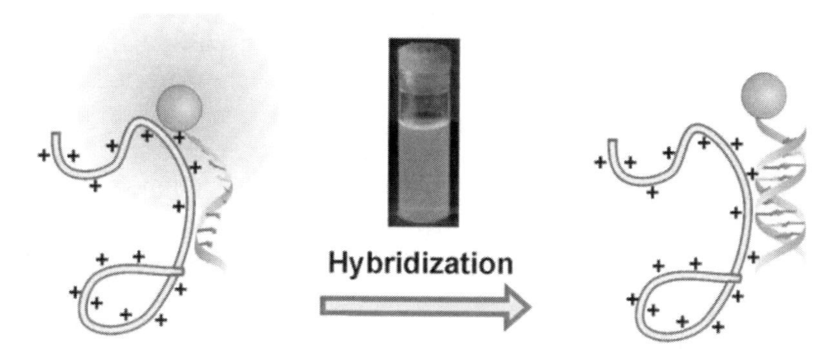

Figure 21. Scheme of CP 3 applied to the DNA detection based on FRET.

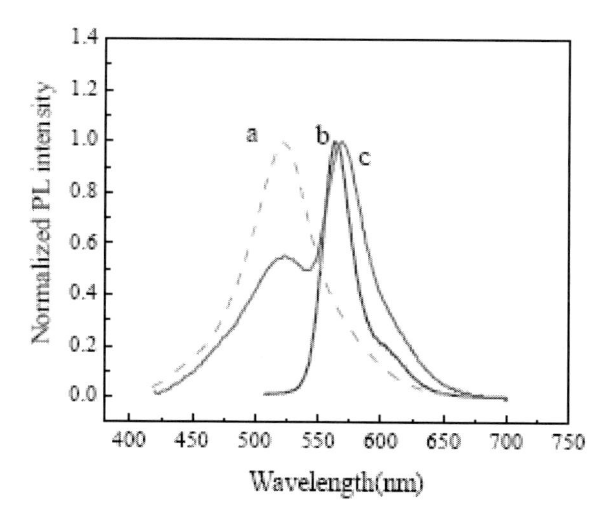

Figure 22. Normalized emission spectra of (a) $PVNEt_2Br_2$ upon excitation at 400 nm, (b) ODN-Cy3 (5'-Cy3-TCGGCATCAATACTCATC-3') in SSC buffer upon excitation at 488 nm and (c) $PPVNEt_2^+$ / ODN-Cy3 (0.6 μM)) duplex in SSC buffer upon excitation at 400 nm.

Another polythiophene derivative, poly[3-(3′-N,N,N-triethylamino-1′-propyloxy)-4-methyl-2,5-thiophene hydrochloride] (PMNT, CP **5** in Figure 20), was used to monitor the cleavage of ssDNA by a single-strand-specific S1 nuclease and hydroxyl radicals[141]. PMNT exists in a random-coil conformation and exhibits a relatively short absorption wavelength of 394 nm. When ssDNA was added to the solution of PMNT, a "duplex" was formed through electrostatic interactions in which PMNT takes on a highly conjugated and planar conformation and exhibits a relatively long absorption wavelength with a dramatic colour change from yellow to pink-red. However, if ssDNA is hydrolysed into small fragments by the S1 nuclease or hydroxyl radical, the "duplex" cannot form. In this case, PMNT remains in a random-coil conformation and exhibits a relatively short absorption wavelength with a colour change to yellow. The most important characteristic of this method is direct visualization of the DNA cleavage by the "naked-eye", which makes it more convenient than other methods that rely on instrumentation. This polymer was further used in a a colorimetric strategy for sensing pH-driven conformational conversion of DNA i-motif structure [135, 142].

Nilsson reported a similar optical response by using a zwitterionic polythiophene derivative CP **6** [143]. Introduction of a single-stranded oligonucleotide in this zwitterionic polythiophene derivative solution will induce a decrease of the intensity and a red-shift of the fluorescence of the polymer. The authors attributed these changes to aggregation-associated interchain interactions induced by electrostatic and hydrogen-bonding interactions between CP **6** and the oligonucleotide and also to polymer backbone planarisation caused by disruptions of internal interactions between the amino and carboxyl groups on the polymer side chains. On addition of a complementary oligonucleotide, the intensity of the emitted light was increased and blue-shifted which was attributed to interaction between hybridized dsDNA and CP **6**, resulting in a less elongated conformation of CP **6**. The detection limit of this method reached ~10^{-11} moles.

4.2. Optical DNA Detection on Solid Substrates

Strong electrostatic interactions between cationic conjugated polymer and charged ODN sequences, as discussed above, has also been utilized to develop a DNA detection assay on a solid surface [144]. Amino-functionalized PNA-C* (Cy5) probes were covalently immobilized to a functionalized glass slide followed by a hybridization step with charged complementary ODN sequences. Subsequent incubation with luminescent cationic polymer solution, poly[(9,9'-bis(6'-*N,N,N*-trimethylammonium)-hexyl)-fluorene–*co*–alt-4,7-(2,1,3-benzothiadiazole) dibromide] (PFBT, CP **7** in Figure 20), led to electrostatic complex formation between the negatively charged PNA-C*/ODN duplex and the cationic polymer. Excitation of the polymer in a triplex PNA-C*/ODN/PFBT complex resulted in FRET from the polymer to the reporter dye Cy5 with more than an order of magnitude amplification of the Cy5 emission intensity, confirming the benefits of the light harvesting properties of the conjugated polymer. In the same report it was also demonstrated that the luminescent properties of the PFBT can be used for label-free DNA detection upon the formation of PNA/ODN/PFBT complexes.

Leclerc and colleagues reported a unlabelled PNA-based DNA sensor on a solid substrates by using poly(1H-imidazolium, 1-methyl-3-[2-[(4-methyl-3-thienyl)oxy]ethyl]-, chloride) (CP **4** in Figure 20)[145]. After hybridization, the polymer strongly interacts with the negatively-charged backbone of the complementary oligonucleotides bound to PNA probes, allowing transduction of hybridization into a fluorescence signal. However, the microarray exhibited a lower sensitivity than that achieved in homogeneous media. In order to improve the sensitivity, the same group introduced a dye-labelled probe and the fluorescence chain reaction to microarray-based multiplex detection[146]. Self-assembled molecular structures composed of Cy3-tagged oligonucleotide probes and a cationic poly(1H-imidazolium, 1-methyl-3-[2-[(4-methyl-3-thienyl) oxy]ethyl]-bromide) were formed in this assay. The rapid and efficient energy transfer between the polymeric transducer (λab = 408 nm, λem = 530 nm) and Cy3 (λ_{ab} = 540 nm, λ_{em} = 570 nm) within the molecular aggregates lead to a massive intrinsic amplification of the fluorescence signal. About three hundred DNA molecules in a volume of 0.4 μL could be detected, which was greater by a factor of ~1500 than that obtained with unlabelled ssDNA/polythiophene complexes.

Other optical techniques, such as surface plasmon resonance (SPR) and surface plasmon resonance imaging (SPRi), have also been used in detection of DNA hybridization [147, 148]. Spatially address detection 'spots' on the SPRi substrate were prepared by a one-step electrocopolymerization of pyrrole and pyrrole grafted with ODNs using a pipette tip as an electrochemical cell. The detection limit of the sensor was in the range of 15 nM for a 15-mer ODN target. The advantage of the SPRi technique lay in the possibility of generating real-time kinetic data for the multi-parallel hybridization processes. Inganäs et al. also demonstrated the use of a SPR technique in DNA detection [149]. The zwitterionic conjugated polymer CP **6** was spin-coated onto a gold plated chip, and annealed. Subsequently the film was swollen with a buffer solution containing an ODN which was 'imprinted' onto the polymer film using electrostatic/ hydrogen bonding interactions. The incubation of the sensory film in a solution of complementary ODN brings about hybridization and a distinct change in SPR signal.

5. Summary and Future Outlook

This review outlines the advances in applications of CPs to DNA sensing. The unique electronic structures of CPs make them be good materials for e DNA detection. Numerous functionalized CPs have been synthesized and used in preparation of DNA sensors with considerable sensitivity and selectivity. In most related research work, stability and activity of bioactive moieties immobilized within CPs have often been of a concern, while less concern has been shown for the stability of the polymer itself. To develop CP-based gene sensors to be used in field and other applications both a shelf life of the sensor and the polymer properties such as chemical and electrochemical stability during storage and measurements should be addressed. In the electrochemical sensors, a challenge that is not often addressed is adhesion between the polymer layer and an underlying electrode. Generally, CPs films are electrodeposited on a physical substrate and the resulted film is adherent to it mainly due to hydrophobic interaction. It has been known that reduction and oxidation processes of a conducting polymer are companied with the migration of ions into and out of film, resulting in dynamic changes of the dimensions of the polymer that would generate interfacial shear stress, ultimately resulting in delamination of the polymer film from the substrate. New approaches to covalent binding of CP to the substrates may prevent these undesirable effects. The advancement of microfabrication and nanotechnology in recent years has significantly been speeding up the evolution of biosensor design. Microfabrication technology can produce geometrically well-defined, highly reproducible structures and surface areas. The combinations of CPs with such devices are expected to produce DNA detection devices with high sensitivity and ability for multi-analyte determinations.

It is clear that DNA biosensors based on water-soluble CPs have undergone rapid development in recent years. Major challenges in this field include intractable self-quenching due to aggregation in water, relative poor selectivity due to static electronic interaction and noticeable photobleaching. To circumvent these limitations, novel designs for polymer chemical structures are desired to optimize the backbone structure and pendant groups.

ACKNOWLEDGMENTS

The authors are grateful to the financial support of Shanghai Pujiang Program (Grant No. 11PJ1403000), NCET and PCSIRT.

REFERENCES

[1] H. Shirakawa, E. J. Louis, A. G. MacDiarmid, C. K. Chiang, A. J. Heeger, *J. Chem. Soc., Chem. Commun.*, 578 (1977).
[2] A. J. Heeger, *Angew. Chem. Int. Ed. 40*, 2591 (2001).
[3] A. G. MacDiarmid, *Angew. Chem. Int. Ed. 40*, 2581 (2001).
[4] H. Shirakawa, *Angew. Chem. Int. Ed. 40*, 2575 (2001).
[5] P. K. H. Ho, J.-S. Kim, J. H. Burroughes, H. Becker, F. Y. L. Sam, T. M. Brown, F. Cacialli, R. H. Friend, *Nature 404*, 481 (2000).
[6] B. Sankaran, J. R. Reynolds, *Macromolecules 30*, 2582 (1997).
[7] M. A. Khan, S. P. Armes, *Adv. Mater. 12*, 671 (2000).
[8] G. G. Wallace, C. O. Too, D. L. Officer, P. C. Dastoor, *MRS Bulletin 30*, 46 (2005).
[9] L. W. Shacklette, J. E. Toth, N. S. Murthy, R. H. Baughman, *J. Electrochem. Soc. 132*, 1529 (1985).
[10] D. E. Tallman, G. Spinks, A. Dominis, G. G. Wallace, *J. Solid State Electrochem. 6*, 73 (2002).
[11] L. Dai, P. Soundarrajan, T. Kim, *Pure Appl. Chem. 74*, 1753 (2002).
[12] S. Geetha, C. R. K. Rao, M. Vijayan, D. C. Trivedi, *Anal. Chim. Acta 568*, 119 (2006).
[13] H. Xu, C. Wang, C. Wang, J. Zoval, M. Madou, *Biosens. Bioelectron. 21*, 2094 (2006).
[14] M. R. Abidian, D.-H. Kim, D. C. Martin, *Adv. Mater. 18*, 405 (2006).
[15] A. C. Pease, D. Solas, E. J. Sullivan, M. T. Cronin, C. P. Holmes, S. P. Fodor, *Proc. Natl. Acad. Sci. USA 91*, 5022 (1994).
[16] S. P. Fodor, R. P. Rava, X. C. Huang, A. C. Pease, C. P. Holmes, C. L. Adams, *Nature 364*, 555 (1993).
[17] S. P. Fodor, J. L. Read, M. C. Pirrung, L. Stryer, A. T. Lu, D. Solas, *Science 251*, 767 (1991).

[18] C. Peter, M. Meusel, F. Grawe, A. Katerkamp, K. Cammann, T. Boerchers, *Fresenius J. Anal. Chem. 371*, 120 (2001).
[19] J. Liu, S. Tian, L. Tiefenauer, P. E. Nielsen, W. Knoll, *Anal. Chem. 77*, 2756 (2005).
[20] H. Peng, C. Soeller, J. Travas-Sejdic, *Chem. Commun.*, 3735 (2006).
[21] H. Peng, L. Zhang, T. H. M. Kjaellman, C. Soeller, J. Travas-Sejdic, *J. Am. Chem. Soc. 129*, 3048 (2007).
[22] Y. Okahata, K. Niikura, *Denki Kagaku oyobi Kogyo Butsuri Kagaku 66*, 7 (1998).
[23] J. Wang, *Electroanalytical Methods for Biological Materials*, 27 (2002).
[24] L. Alfonta, A. K. Singh, I. Willner, *Anal. Chem. 73*, 91 (2001).
[25] E. Katz, Y. Weizmann, I. Willner, *J. Am. Chem. Soc. 127*, 9191 (2005).
[26] D. T. McQuade, A. E. Pullen, T. M. Swager, *Chem. Rev. 100*, 2537 (2000).
[27] W. P. Su, J. R. Schrieffer, A. J. Heeger, *Phys. Rev. Lett. 42*, 1698 (1979).
[28] W. P. Su, J. R. Schrieffer, A. J. Heeger, *Phys. Rev. B 22*, 2099 (1980).
[29] A. J. Heeger, *Reviews of Modern Physics 73*, 681 (2001).
[30] R. D. McCullough, *Adv. Mater. 10*, 93 (1998).
[31] J. Heinze, *Synth. Met. 43*, 2805 (1991).
[32] G. G. Wallace, G. M. Spinks, L. A. P. Kane-Maguire, *Conductive Electroactive Polymers: Intelligent Materials Systems, Second Edition*, 2002.
[33] G. G. Wallace, M. Smyth, H. Zhao, *TrAC, Trends Anal. Chem. 18*, 245 (1999).
[34] S. Sadki, P. Schottland, N. Brodie, G. Sabouraud, *Chem. Soc. Rev. 29*, 283 (2000).
[35] G. Inzelt, M. Pineri, J. W. Schultze, M. A. Vorotyntsev, *Electrochim. Acta 45*, 2403 (2000).
[36] E. Katz, I. Willner, *Electroanalysis 15*, 913 (2003).
[37] J. Wang, *Chem. Eur. J. 5*, 1681 (1999).
[38] H. Peng, C. Soeller, N. Vigar, A. Kilmartin Paul, B. Cannell Mark, A. Bowmaker Graham, P. Cooney Ralph, J. Travas-Sejdic, *Biosens. Bioelectron. 20*, 1821 (2005).
[39] F. Garnier, H. Korri-Youssoufi, P. Srivastava, B. Mandrand, T. Delair, *Synth. Met. 100*, 89 (1999).
[40] S. Cosnier, *Anal Bioanal Chem 377*, 507 (2003).
[41] J. Wang, M. Jiang, A. Fortes, B. Mukherjee, *Anal. Chim. Acta 402*, 7 (1999).
[42] J. Wang, M. Jiang, *Langmuir 16*, 2269 (2000).
[43] A. I. Minett, J. N. Barisci, G. G. Wallace, *React. Funct. Polym. 53*, 217 (2002).
[44] T. Livache, A. Roget, E. Dejean, C. Barthet, G. Bidan, R. Teoule, *Nucleic Acids Res. 22*, 2915 (1994).
[45] N. Lassalle, A. Roget, T. Livache, P. Mailley, E. Vieil, *Talanta 55*, 993 (2001).
[46] T. Livache, B. Fouque, A. Roget, J. Marchand, G. Bidan, R. Teoule, G. Mathis, *Anal. Biochem. 255*, 188 (1998).
[47] T. Livache, H. Bazin, P. Caillat, A. Roget, *Biosens. Bioelectron. 13*, 629 (1998).
[48] P. Godillot, H. Korri-Youssoufi, P. Srivastava, A. El Kassmi, F. Garnier, *Synth. Met. 83*, 117 (1996).
[49] L. A. Thompson, J. Kowalik, M. Josowicz, J. Janata, *J. Am. Chem. Soc. 125*, 324 (2003).
[50] J.-F. Pilard, C. Cougnon, J. Rault-Berthelot, A. Berthelot, C. Hubert, K. Tran, *J. Electroanal. Chem. 568*, 195 (2004).
[51] S. Dubey, B. Fabre, G. Marchand, J. F. Pilard, J. Simonet, *J. Electroanal. Chem. 477*, 121 (1999).

[52] C. Gautier, C. Cougnon, J.-F. Pilard, N. Casse, *J. Electroanal. Chem. 587*, 276 (2006).
[53] C. Gautier, C. Cougnon, J.-F. Pilard, N. Casse, B. Chenais, *Anal. Chem. 79*,7920(2007).
[54] S. Cosnier, A. Lepellec, *Electrochim. Acta 44*, 1833 (1999).
[55] A. Dupont-Filliard, A. Roget, T. Livache, M. Billon, *Anal. Chim. Acta 449*, 45 (2001).
[56] G. Bidan, M. Billon, M. L. Calvo-Munoz, A. Dupont-Fillard, *Mol. Cryst. Liq. Cryst. 418*, 255 (2004).
[57] Z. Shen, P. E. Burrows, V. Bulovic, S. R. Forrest, M. E. Thompson, *Science 276*, 2009 (1997).
[58] D. G. Lidzey, D. C. Bradley, S. F. Alvarado, P. F. Seidler, *Nature 386*, 135 (1997).
[59] J. Wang, M. Jiang, B. Mukherjee, *Anal. Chem. 71*, 4095 (1999).
[60] H. Korri-Youssoufi, A. Yassar, *Biomacromolecules 2*, 58 (2001).
[61] A. Emge, P. Bauerle, *Synth. Met. 102*, 1370 (1999).
[62] P. Baeuerle, A. Emge, *Adv. Mater. 10*, 324 (1998).
[63] H. Bai, G. Shi, *Sensors 7*, 267 (2007).
[64] A. G. MacDiarmid, *Synth. Met. 84*, 27 (1997).
[65] A. G. MacDiarmid, A. J. Epstein, *Synth. Met. 69*, 85 (1995).
[66] G. P. Kittlesen, H. S. White, M. S. Wrighton, *J. Am. Chem. Soc. 106*, 7389 (1984).
[67] H. Xie, S.-C. Luo, H.-h. Yu, *Small 5*, 2611 (2009).
[68] K. Krishnamoorthy, R. S. Gokhale, A. Q. Contractor, A. Kumar, *Chem. Commun.*, 820 (2004).
[69] A. J. Bard, L. R. Faulkner, *Electrochemical Methods: Fundamentals and Applications*, 1980.
[70] H. Peng, C. Soeller, J. Travas-Sejdic, *Macromolecules 40*, 909 (2007).
[71] H. Peng, C. Soeller, N. A. Vigar, V. Caprio, J. Travas-Sejdic, *Biosens. Bioelectron. 22*, 1868 (2007).
[72] H. Peng, C. Soeller, M. B. Cannell, G. A. Bowmaker, R. P. Cooney, J. Travas-Sejdic, *Biosens. Bioelectron. 21*, 1727 (2006).
[73] C. Tlili, H. Korri-Youssoufi, L. Ponsonnet, C. Martelet, N. J. Jaffrezic-Renault, *Talanta 68*, 131 (2005).
[74] T.-Y. Lee, Y.-B. Shim, *Anal. Chem. 73*, 5629 (2001).
[75] H. Peng, L. Zhang, J. Spires, C. Soeller, J. Travas-Sejdic, *Polymer 48*, 3413 (2007).
[76] H. Peng, L. Zhang, J. Spires, C. Soeller, J. Travas-Sejdic, *Proceedings of IEEE- sensors 2007.*
[77] C. Weidlich, K. M. Mangold, K. Juettner, *Electrochim. Acta 50*, 1547 (2005).
[78] G. Inzelt, V. Kertesz, A.-S. Nyback, *J. Solid State Electrochem. 3*, 251 (1999).
[79] C. Gautier, C. Cougnon, J.-F. Pilard, N. Casse, B. Chenais, M. Laulier, *Biosens. Bioelectron. 22*, 2025 (2007).
[80] C. Gautier, C. Esnault, C. Cougnon, J.-F. Pilard, N. Casse, B. Chenais, *J. Electroanal. Chem. 610*, 227 (2007).
[81] K. Leitner, J. W. Schultze, U. Stimming, *J. Electrochem. Soc. 133*, 1561 (1986).
[82] F. Di Quarto, S. Piazza, M. Santamaria, C. Sunseri, *Handb. Thin Film Mater. 2*, 373 (2002).
[83] J. Liu, L. de la Garza, L. Zhang, N. M. Dimitrijevic, X. Zuo, D. M. Tiede, T. Rajh, *Chem. Phys. 339*, 154 (2007).
[84] S. Liu, C. Li, J. Cheng, Y. Zhou, *Anal. Chem. 78*, 4722 (2006).
[85] Q. Li, G. Luo, J. Feng, D. Cai, Q. Ouyang, *Analyst 125*, 1908 (2000).

[86] M. Liang, S. Jia, S. Zhu, L.-H. Guo, *Environ. Sci. Technol. 42*, 635 (2008).
[87] M. Liang, L.-H. Guo, *Environ. Sci. Technol. 41*, 658 (2007).
[88] N. Lassalle, P. Mailley, E. Vieil, T. Livache, A. Roget, J. P. Correia, L. M. Abrantes, *J. Electroanal. Chem. 509*, 48 (2001).
[89] N. Lassalle, E. Vieil, J. P. Correia, L. M. Abrantes, *Synth. Met. 119*, 407 (2001).
[90] N. Lassalle, E. Vieil, J. P. Correia, L. M. Abrantes, *Biosens. Bioelectron. 16*,295(2001).
[91] A. Ramanavicius, A. Ramanaviciene, A. Malinauskas, *Electrochim. Acta 51*, 6025 (2006).
[92] M. I. Rodriguez, E. C. Alocilja, *IEEE Sensors Journal 5*, 733 (2005).
[93] A. Ramanaviciene, A. Ramanavicius, *Anal. Bioanal. Chem. 379*, 287 (2004).
[94] E. Komarova, M. Aldissi, A. Bogomolova, *Biosens. Bioelectron. 21*, 182 (2005).
[95] R. E. Ionescu, S. Herrmann, S. Cosnier, R. S. Marks, *Electrochem. Commun. 8*, 1741 (2006).
[96] H. Korri-Youssoufi, B. Makrouf, *Anal. Chim. Acta 469*, 85 (2002).
[97] A. Bouchet, C. Chaix, C. A. Marquette, L. J. Blum, B. Mandrand, *Biosens. Bioelectron. 23*, 735 (2007).
[98] G. Li, G. Kossmehl, H.-P. Welzel, G. Engelmann, W.-D. Hunnius, W. Plieth, H. S. Zhu, *Macromol. Chem. Phys. 199*, 525 (1998).
[99] J. Cha, J. I. Han, Y. Choi, D. S. Yoon, K. W. Oh, G. Lim, *Biosens. Bioelectron. 18*, 1241 (2003).
[100] F. Le Floch, H.-A. Ho, P. Harding-Lepage, M. Bedard, R. Neagu-Plesu, M. Leclerc, *Adv. Mater. 17*, 1251 (2005).
[101] M. J. A. Shiddiky, M. A. Rahman, Y.-B. Shim, *Anal. Chem. 79*, 6886 (2007).
[102] B. Spires John, H. Peng, E. Williams David, E. Wright Bryon, C. Soeller, J. Travas-Sejdic, *Biosens Bioelectron 24*, 934 (2008).
[103] J. B. Spires, H. Peng, D. E. Williams, C. Soeller, J. Travas-Sejdic, *Electrochim. Acta 55*, 3061 (2010).
[104] J. B. Spires, H. Peng, D. Williams, J. Travas-Sejdic, *J. Electroanal. Chem. 658*,1(2011).
[105] H. Gu, X. Su, K. P. Loh, *Chem. Phys. Lett. 388*, 483 (2004).
[106] H. Gu, X. d. Su, K. P. Loh, *J. Phys. Chem. B 109*, 13611 (2005).
[107] F. Davis, A. V. Nabok, S. P. J. Higson, *Biosens. Bioelectron. 20*, 1531 (2005).
[108] K. Arora, N. Prabhakar, S. Chand, B. D. Malhotra, *Biosens. Bioelectron. 23*, 613 (2007).
[109] K. Arora, N. Prabhakar, S. Chand, B. D. Malhotra, *Anal. Chem. 79*, 6152 (2007).
[110] Z. Gao, S. Rafea, L. H. Lim, *Adv. Mater. 19*, 602 (2007).
[111] J. Jang, *Adv. Polym. Sci. 199*, 189 (2006).
[112] K. Ramanathan, M. A. Bangar, M. Yun, W. Chen, N. V. Myung, A. Mulchandani, *J. Am. Chem. Soc. 127*, 496 (2005).
[113] H. Cai, Y. Xu, P.-g. He, Y.-z. Fang, *Electroanalysis 15*, 1864 (2003).
[114] Y. Xu, Y. Jiang, H. Cai, P.-G. He, Y.-Z. Fang, *Anal. Chim. Acta 516*, 19 (2004).
[115] H. Chang, Y. Yuan, N. Shi, Y. Guan, *Anal. Chem. 79*, 5111 (2007).
[116] L. Liang, J. Liu, C. F. Windisch, Jr., G. J. Exarhos, Y. Lin, *Angew. Chem. Int. Ed. 41*, 3665 (2002).
[117] N. Zhu, Z. Chang, P. He, Y. Fang, *Electrochim. Acta 51*, 3758 (2006).
[118] L. Zhang, H. Peng, P. A. Kilmartin, C. Soeller, J. Travas-Sejdic, *Electroanalysis 19*, 870 (2007).

[119] C. d. S. Riccardi, H. Yamanaka, M. Josowicz, J. Kowalik, B. Mizaikoff, C. Kranz, *Anal. Chem. 78*, 1139 (2006).
[120] F. Garnier, B. Bouabdallaoui, P. Srivastava, B. Mandrand, C. Chaix, *Sens. Actuators. B: Chem. B123*, 13 (2007).
[121] G. Cheng, J. Zhao, Y. Tu, P. He, Y. Fang, *Anal. Chim. Acta 533*, 11 (2005).
[122] H. Qi, X. Li, P. Chen, C. Zhang, *Talanta 72*, 1030 (2007).
[123] Y. Xu, X. Ye, L. Yang, P. He, Y. Fang, *Electroanalysis 18*, 1471 (2006).
[124] Y. Fu, R. Yuan, Y. Chai, L. Zhou, Y. Zhang, *Anal. Lett. 39*, 467 (2006).
[125] T. Livache, E. Maillart, N. Lassalle, P. Mailley, B. Corso, P. Guedon, A. Roget, Y. Levy, *J. Pharm. Biomed. Anal. 32*, 687 (2003).
[126] Z. Chang, J. Zhou, K. Zhao, N. Zhu, P. He, Y. Fang, *Electrochim. Acta 52*, 575 (2006).
[127] J. Wu, Y. Zou, X. Li, H. Liu, G. Shen, R. Yu, *Sens. Actuators. B: Chem.B104*,43(2005).
[128] S. Tian, J. Liu, T. Zhu, W. Knoll, *Chem. Mater. 16*, 4103 (2004).
[129] M. Leclerc, *Sensors Update 8*, 21 (2001).
[130] M. Leclerc, *Adv. Mater. 11*, 1491 (1999).
[131] M. Bera-Aberem, H.-A. Ho, M. Leclerc, *Tetrahedron 60*, 11169 (2004).
[132] T. Förster, *Annalen der Physik 437*, 55 (1948).
[133] B. Liu, G. C. Bazan, *Chem. Mater. 16*, 4467 (2004).
[134] F. Feng, F. He, L. An, S. Wang, Y. Li, D. Zhu, *Adv. Mater. 20*, 2959 (2008).
[135] B. S. Gaylord, A. J. Heeger, G. C. Bazan, *Proc. Natl. Acad. Sci. USA 99*, 10954 (2002).
[136] Q.-H. Xu, S. Gaylord Brent, S. Wang, C. Bazan Guillermo, D. Moses, J. Heeger Alan, *Proc. Natl. Acad. Sci. USA 101*, 11634 (2004).
[137] B. S. Gaylord, M. R. Massie, S. C. Feinstein, G. C. Bazan, *Proc. Natl. Acad. Sci. USA 102*, 34 (2005).
[138] B. S. Gaylord, A. J. Heeger, G. C. Bazan, *J. Am. Chem. Soc. 125*, 896 (2003).
[139] H.-A. Ho, M. Boissinot, M. G. Bergeron, G. Corbeil, K. Doré, D. Boudreau, M. Leclerc, *Angew. Chem. Int. Ed. 41*, 1548 (2002).
[140] K. Dore, S. Dubus, H.-A. Ho, I. Levesque, M. Brunette, G. Corbeil, M. Boissinot, G. Boivin, M. G. Bergeron, D. Boudreau, M. Leclerc, *J. Am. Chem. Soc. 126*, 4240 (2004).
[141] Y. Tang, F. Feng, F. He, S. Wang, Y. Li, D. Zhu, *J. Am. Chem. Soc. 128*, 14972 (2006).
[142] X. Liu, Y. Tang, L. Wang, J. Zhang, S. Song, C. Fan, S. Wang, *Adv. Mater. 19*, 1471 (2007).
[143] K. P. R. Nilsson, O. Inganas, *Nat Mater 2*, 419 (2003).
[144] B. Liu, G. C. Bazan, *Proc. Natl. Acad. Sci. USA 102*, 589 (2005).
[145] F. Raymond, H.-A. Ho, R. Peytavi, L. Bissonnette, M. Boissinot, F. Picard, M. Leclerc, M. Bergeron, *BMC Biotechnol. 5*, 10 (2005).
[146] A. Najari, H. A. Ho, J.-F. Gravel, P. Nobert, D. Boudreau, M. Leclerc, *Anal. Chem. 78*, 7896 (2006).
[147] P. Guedon, T. Livache, F. Martin, F. Lesbre, A. Roget, G. Bidan, Y. Levy, *Anal. Chem. 72*, 6003 (2000).
[148] S. Szunerits, N. Knorr, R. Calemczuk, T. Livache, *Langmuir 20*, 9236 (2004).
[149] P. Bjoerk, N.-K. Persson, K. Peter, R. Nilsson, P. Asberg, O. Inganaes, *Biosens. Bioelectron. 20*, 1764 (2005).

In: Conducting Polymers
Editor: Luiz Carlos Pimentel Almeida

ISBN: 978-1-62618-119-9

Chapter 9

CONDUCTING POLYMERS IN LITHIUM BATTERIES

David Lepage, Christian Kuss and Steen B. Schougaard*
Departement of chemistry, University of Quebec at Montreal,
Montreal, Canada

ABSTRACT

Conducting polymers have a number of intrinsic properties that make them uniquely suited as components in lithium batteries. As such, many new studies that combine the subject of lithium batteries and conducting polymers have been published since the last complete review of Novak et al. [1]. Initially, conducting polymers were considered for the cathode materials as they are electrochemically active, stable, have high electronic conductivity in the doped state and have great porosity for the electrolyte. The intrinsic specific capacity is however quite poor and an extensive body of work has therefore been conducted to modify the monomer, the polymerization method, as well as, adding an electroactive species to the polymer chain or as counter-ions. Nevertheless, ceramic electroactive particles, combined with carbon and a non-conductive polymer binder still serve as the cathode in most commercial batteries. Alternative composite systems are being developed, and some of the more creative approaches include using conducting polymers both as conducting coatings and as the conducting matrix that provide ionic and electronic transport between the current collector and the electroactive particle. In this chapter we review the latest developments (since 1997) of conducting polymers as the redox active charge storing unit, as well as, the ionic and electronic conducting matrix used in lithium batteries electrode.

1. INTRODUCTION

To achieve sustainable energy production, fossil fuels will need to be replaced, most likely by a combination of technologies. Biomass, wind, hydroelectric, geothermal and solar are currently being developed to this end [2]. Part of the solution will likely be temporary and mobile storage of electricity by the use of battery systems. Particularly the replacement of the

* E-mail address: schougaard.steen@uqam.ca.

combustion engine vehicle by electric alternatives is currently under development. To date many battery technologies have been proposed, with lithium batteries being one of the most promising candidates [3]. Hence, this chapter will focus on the potential role of conducting polymers in lithium batteries.

The history of the lithium battery begins in the 1970s, when researchers became interested in lithium redox chemistry for energy storage. To this end, many compounds and many ideas were proposed, but none achieved major impact on the battery market [4-7]. The breakthrough for the commercialization of lithium batteries came as late as 1991, when Sony launched the first commercial rechargeable lithium battery product [8]. In the following years, but particularly during the first decade of the new millennium, lithium batteries became increasingly established in small portable devices, like music players, laptops and cell phones [9]. To date, a major obstacle for the application of lithium batteries in electric or hybrid electric vehicles is their cost.

Presently, an optimistic price for lithium batteries is around 700 to 1250 U.S $ per kilowatt-hour, while a price of around 250 U.S $ would be necessary to compete with fossil fuel engines [10,11]. Also of concern is their energy density. Lithium batteries offer around 130 watt-hours per kilogram (Wh/kg) (achieving around 150 km with an economically sensibly sized car battery), while for driving long distances (more than 500km) without recharge about 1500 Wh/kg are necessary [11].

To give an overview, this chapter will start with the basic concepts of lithium batteries to set the stage for the unique utility of the polymer. The main focus will however be on the polymer materials themselves, their characterization, their use in batteries as charge storage electrodes and their use as additives to improve current battery systems.

2. Concepts in Lithium Batteries

Compared to other battery technologies, lithium batteries provide the highest energy by volume (Watt x hour / liter) and by weight (Watt x hour/ kilogram). From an environmental perspective, the rechargeability and the unheard-of longevity of lithium batteries are the key point for their large scale use [12, 13]. All these properties are based on specific characteristics of the lithium element: it has the most electropositive potential of all metals (-3.04 V *vs.* standard hydrogen electrode) and is one of the lightest elements (molecular weight of 6.94 g/mol). Also very important are the large resources of lithium, which makes lithium extraction very economical [11]. *E.g.* the 10,000-square-kilometer Salar de Uyuni salt deposit in Bolivia presents the biggest lithium resource that is currently exploited [14]. Moreover, large scale lithium battery producers are implementing new recycling procedures to recover lithium and other battery components [15-17].

Batteries in general consist of two electrodes that are connected ionically through an electrolyte system, with one of the electrodes serving as electron donor (anode) and the other as electron acceptor (cathode). These electrodes are physically separated by a permeable membrane and electrically connected by an external circuit (Figure 1). The functions of the electrodes invert when the battery switches from charge to discharge or *vice versa*, *i.e.* the anode becomes the cathode and the cathode becomes the anode. We will adhere throughout this chapter to a common simplification, naming electrodes only by their function during discharge.

Every battery part anode, cathode, binder (which binds active electrode particles to a current collector) and electrolyte has already been subjected to well researched optimization [18]. Initially, a metallic sheet of lithium was used as anode, but problems with security and durability arose when dendrite growth on the surface of the lithium occurred. A successful workaround used special ion conducting polymer separators, [19, 20] however, in general lithium insertion compounds are used (host 2 in Figure 1). These include graphite [21], $Li_4Ti_5O_{12}$ [22], silicon [23], tin oxide [24] and others [25]. Normally, the electrolyte is chosen according to the anode. Several combinations of electrolytes (solid [26] and nonaqueous [27]) have been investigated, however a mixtures of carbonates (ethylene, diethyl and dimethyl) with the lithium hexafluorophosphate ($LiPF_6$) salt as ionic charge carrier seems to be the best compromise for the moment [28].

Insertion compounds are also used as active cathode materials (host 1 in Figure 1). A variety of compounds have been synthesized for example $LiCoO_2$ [29], $LiFePO_4$ [30], $LiMn_2O_4$ [31] and conducting polymers [1]. In general, a distinction is made between lithium batteries in which both electrodes are based on insertion materials and batteries that have metallic lithium as one of the electrodes, the former being termed lithium-ion while the latter is termed simply lithium battery. We do not make this distinction here, as measurements in most of the cited work are made relative to metallic lithium in "half" cell setup, while the final "full" cell application would most likely entail an insertion type counter electrode.

During the electrochemical cycling of insertion compounds, lithium needs to pass through the bulk material. This is a slow process, *e.g.* diffusion coefficients of Li are ~10^{-14} cm^2/s in $LiFePO_4$ [32], ~10^{-9} cm^2/s in $LiCoO_2$ [33] and ~10^{-8} to ~10^{-10} cm^2/s in $LiMn_2O_4$ [34]. To bypass this problem, these materials are synthesized as nano-sized powders which decrease the diffusion path. This, however, impedes electronic contact between the nanoparticles and adds to the already low electronic conductivity of ceramic insertion compounds, *e.g.* $LiFePO_4$ ~10^{-9} S/cm [35], $LiCoO_2$ ~10^{-4} S/cm [36] and $LiMn_2O_4$ ~10^{-6} S/cm [37].

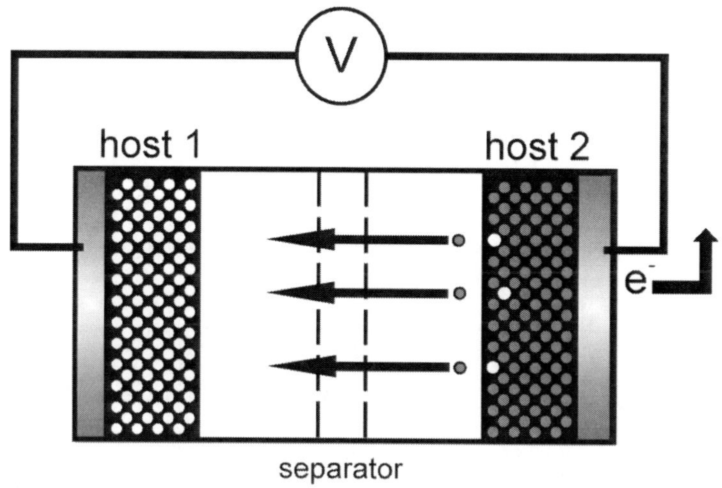

Figure 1. Representation of a lithium battery in operation. Grey circles represent lithium, host 1 is the cathode and host 2 the anode. Arrows indicate the transport of lithium ions through the electrolyte, as well as, the electrons (e^-) through the external circuit.

2.1. Advantages of Conducting Polymers for Lithium Batteries

Of key importance to the use of conducting polymer in lithium batteries is the combination of charge storage capabilities with good electronic and ionic conductivities.

The following list summarizes the intrinsic benefits of conductive polymers, focusing on its use as a cathode material or as part of a composite, supporting the insertion compound.

Cathode: Conducting polymer as charge storage material

1. Electronic conductivity: The transfer of electrons between the bulk of the charge storage material to the current collector of the lithium battery should be fast. High electronic conductivities of conducting polymers provide for a high rate of charge and/or discharge of the lithium battery.
2. Charge storage: The active electrode material has the crucial task to accept, store and release charges. Conducting polymers can host charges by reduction and oxidation of the polymer network, which goes along with ion insertion (doping) or deinsertion (undoping). The battery is considered fully charged when the conducting polymer is completely doped.
3. Ionic conductivity: While electrons are exchanged with the active material, ionic components need to balance the exchanged charges. Conducting polymers are commonly permeable to the electrolyte. Through swelling they form an open gel-like structure, which allows easy access to the electrolyte. Furthermore, the transfer of lithium ions into the bulk polymer gives access to storage units throughout the polymer and not only on the surface.

Composite: Conducting polymer and insertion compound

1. Electronic conductivity: Common insertion compounds are ceramics of intrinsically poor electronic conductivity. Hence, conducting additives, such as conducting polymers, are necessary for use of these materials in lithium battery electrodes.
2. Binder: Many ceramic materials are micro- or nanopowders. As such, polymeric binders are used to bind the particles to the current collector. Conducting polymers can at the same time bind the particles and provide electronic transport.
3. Protective layer: Some insertion compounds react with the electrolyte. This may be countered by adding a protective layer of conducting polymer.

Other reasons for implementation of conducting polymers in lithium batteries will be discussed throughout the chapter.

2.2. Storing Charges in Conducting Polymer for Lithium-Ion Batteries

A conducting polymers ability to store charges can be estimated using theoretical considerations. For simple conducting polymers, the storage of one electron requires between one and four monomer units [38-44]. Theoretical capacities based on this charge storage ability are reported in Table 1. Importantly, as in most of the literature, the dopant agent is not taken into consideration, since it is considered part of the electrolyte.

Table 1. Theoretical capacity of some conducting polymer

Polymer	Abbr.	Mw (g/mol)	Theoretical Capacity (mAh/g)
Polypyrrole	Ppy	65	100-400
Polythiophene	PTh	82	80-319
Poly(3,4-ethylenedioxythiophene)	Pedot	142	37-148
Polyaniline	Pani	93	72-288

Material capacities reported in the literature are often difficult to compare, the reason being the complex dynamics of charge storage in batteries. Reporting the amount of inactive materials and other relevant information like electrode thickness and densities is crucial and often lacking. For example the thickness and the density of the cathode may change the electrochemical response substantially [45, 46]. Moreover, in lithium batteries, the factors limiting the capacity at high charge/discharge rates are difficult to identify [47]. For this reason, we report the capacity value at low rates of discharge. When possible the cited value is obtained at a discharge rate of C/10, where the battery is completely discharged in ten hours at constant current. A 1C rate represents a complete constant current discharge in one hour.

2.3. Transport of Li^+ in Conducting Polymer

As mentioned above, facile ion conduction is a key aspect for the application of conducting polymers in lithium batteries, especially in high rate applications. *E.g.* when the conducting polymer is used as a cathode material, ions move through the conducting polymer to reach their storage point. Similarly, when the conducting polymer is used as a conducting coating, the lithium passes through the polymer (Figure 2). A measure frequently used to compare ionic transport in battery materials is the diffusion coefficient (D).

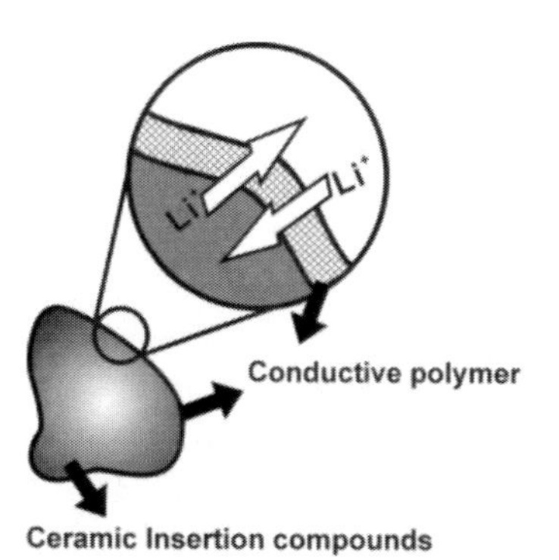

Figure 2. Schematization of the surface of a composite insertion compound with a conducting polymer coating.

We report in Table 2 some select examples of the lithium ion diffusion coefficient (D_{Li}) in conducting polymers.

Table 2. Diffusion coefficient of lithium (D_{Li}) in conducting polymers

Polymer / dopant	D_{Li} (cm^2/s)	Ref.
Ppy/polyvinylphosphate	$4.8x10^{-10}$	[48]
Pedot/PSS	~$3x10^{-7}$	[49]
Ppy/ClO_4^-	~10^{-9}	[50]

PSS: poly (styrenesulfonate).

Diffusion of lithium ions does not always fully describe the limiting kinetics of the ionic charge exchange during cycling. It has previously been shown by cyclic voltammetry (CV), electrochemical impedance spectroscopy (EIS) and electrochemical quartz crystal microbalance (EQCM) that some negative charge carriers like ClO_4^- are mobile, while others are practically immobilized in Ppy [51]. Importantly, the diffusion of small anions may be orders of magnitude faster than the diffusion of lithium in the polymer, *e.g.* $D_{ClO_4^-}$ in Ppy is approximately 5 x 10^{-7} cm^2/s [52]. This is of significance, when using conducting polymers as charge storage materials, as contrary to the standard solid state lithium insertion compounds, it may be the *anion* that is displaced during cycling.

2.4. Characterization of Conducting Polymers for Lithium Batteries

Conventional materials analysis methods are used extensively to characterize conducting polymers for lithium batteries. For example, imaging techniques like scanning electron microscopy (SEM) and transmission electron microscopy (TEM) are used for micro-structure determination, while crystal and molecular structure is determined by X-ray diffraction (XRD) and spectroscopic methods (NMR, IR, Raman *etc.*) that can be employed both *in* and *ex-situ*. A common problem encountered when developing conducting polymer/solid state insertion compound composites is the need to determine the polymer contents. One often employed technique is thermogravemetric analysis in oxidative atmosphere. The drawback of this technique is that the insertion compound may be oxidized during analysis and that the maximum temperature may be insufficient for complete combustion. Alternatively, C/H/N combustion elemental analysis may be employed. This technique is not affected by the oxidation of the insertion compound, and incomplete combustion is less of a concern due to the elevated temperatures (1700-1800 °C) reached during the analysis [53, 54].

The electrochemical performance of new materials is normally tested using research sized coin cell batteries. The first step in the process is the fabrication of the electrodes [46]. When the sample is in powder form, these are prepared by mixing the active material with conducting additives (carbon) and binder (Polyvinylidene fluoride) before it is deposited on the current collector. Metallic lithium is frequently used as the negative electrode, as it, under most practical experimental conditions, will act as a pseudo reference electrode. The electrolyte, the battery casing and the separator are chosen based on their inertness relative to the system being studied. Similarly, battery assembly takes place in an argon atmosphere glove box to avoid reaction of electrolyte and lithium metal with oxygen and moisture.

The most basic electrochemical testing is accomplished using chronopotentiometric measurement, which yields the initial charge storage capacity incl. the energy density, the capacity fade and columbic efficiency (recuperated charge divided by charge stored).

Importantly, these figures-of-merit are dependent on the applied current density, the potential limits for current reversal, temperature and the details of the electrochemical cell, including the density, thickness and microstrucutre of the electrode. Therefore, as a minimum, charge and discharge curves at several rates should be presented. Other often employed electrochemical techniques include cyclic voltametry (CV), current/potential interrupt techniques (GITT, PITT) and electrochemical impedance spectroscopy (EIS).

EIS operates by applying an alternating voltage to the system, while scanning a range of modulation frequencies. Concurrently, the current is measured as amplitude and shift relative to the modulation. By splitting the current into a sum of an in-phase and an out-of-phase current, two resistances can be determined. For increased ease of interpretation, the overall resistance (now called impedance) is given as a complex number, with the real part being the in-phase resistance and the complex part being the out-of-phase resistance. Obtaining reproducible EIS spectra is generally straight forward, yet EIS interpretation is often complex. However, in favorable cases it is possible to determine parameters, such as electrolyte conductivity, interface capacitance and diffusion coefficients using EIS [55].

In the domain of lithium batteries, EIS has been used to investigate aging [56,57], to determine the diffusion coefficient of lithium in insertion compounds [58-60], to study commercial [61] and high power lithium batteries [62]. Specifically for conducting polymers, the variation of the impedance during electropolymerization [63] and doping/undoping effects on the electrochemical response [64] has been studied.

Given the importance of the doping level on the charge storing ability and on the electrical conductivity it is surprising that techniques other than the electrochemical ones are not more commonly used to probe this quantity. *E.g* X-ray photoelectron spectroscopy may be used to yield information on the oxidation states of the constituent elements [65,66], while attenuated total reflection Fourier transform infrared (ATR-FTIR) spectroscopy offers *in situ* data on the appearance and disappearance of vibrations characteristic to doping [67,68].

The list of factors that affect the suitability of a given material in lithium batteries is generally quite extensive, which leads to a desire to extrapolate data from sources other than the battery literature. For example, the polymer should be chemically resistant to the electrolyte, also at elevated temperature or large overpotentials. Poly (alkylenedioxythiophene)s like Pedot are considered stable conducting polymers at high temperatures (188°C) in benzonitrile [69], and even during ageing in acidic or alkaline media, Pedot exhibits better resistance than polypyrrol [70]. In general, Pedot is a good choice of conducting polymer for a lithium battery application. However, when examining the polymer at 100°C in air it ages quickly [71]. The latter is clearly not of any concern in a lithium battery, but emphasizes the importance of examining the conducting polymer in the unique environment that exists in the lithium battery.

3. CONDUCTING POLYMER AS CATHODE

Novak's et al. review from 1997 addressed the large variety of monomers that could be used in battery applications [1]. In this chapter, we further focus on recent conducting polymer preparation methods relevant for application in lithium batteries. There are numerous polymerization reactions, but two techniques are most often reported: chemical oxidation, and electrochemical oxidation (Potentiostatic, potentiodynamic and galvanostatic). By electrochemistry it is relatively easy to change the reaction conditions. For example Otero and coworkers studied the capacity *vs.* lithium of three polymers (Ppy, PTh and poly-3-methylthiophene) while modifying electrolyte concentrations, the deposition potential and temperature of the electrosynthesis [72]. Dalmolin and coworkers used Pani electrodeposited onto reticulated vitreous carbon to increase the macroscopic surface area, obtaining a capacity of ~60 mAh/g [73]. Also copolymers of pyrrole and thiophene [74] or 3,4-ethylenedioxythiophene and thiophene [75] have been produced by electropolymerization and been used in functional lithium batteries.

While electropolymerization achieves good results, chemical oxidation is often simpler and more economical to employ. For example thiophene and 3,4-ethylenedioxythiophene have been polymerized using $FeCl_3$ [76,77]. The main variable during chemical oxidation is the choice of counter ion. For example battery performance differences have been observed for PF_6^- and BF_4^- doped polyaniline, whereby Pani:PF_6^- exhibited better discharge capacities in part due to a better electronic conductivity [78,79]. One of the few examples where oxidative polymerization is not used, is the work of Patel and co-workers who developed a block polymer consisting of poly(3-hexylthiophene) for electronic and poly(ethylene oxide) for ionic conductivity [80]. This work illustrates very elegantly the design flexibility inherent to polymer chemistry, which is difficult if not impossible to duplicate in the solid state ceramic materials used in batteries today.

3.1. Strategies to Increase the Capacity of Conducting Polymers

The intrinsically low capacity of the conducting polymers compared to that of insertion compounds can be bypassed by adding a redox active species to the polymer. Following this strategy, Park and co-workers bound chemically a ferrocene group as electroactive species to the pyrrol monomer. The respective copolymer pyrrol-ferrocene:pyrrol exhibited a capacity of 65 mAh/g at a rate of C/5 compared to 20 mAh/g for pure polypyrrol [81]. Another possibility is to electrostatically attach the redox active couple, *i.e.* the dopant of the conducting polymer acts as a redox active couple. For example, the polymer poly(2,5-dihydroxy-1,4-benzoquinone-3,6-methylene) was used to dope Pedot in a lithium battery to obtain a capacity of 129 mAh/g at a discharge rate of C/20 [82]. Zhou and coworkers doped the conducting polymers Ppy, Pth and Pedot with $Fe(CN)_6^{4-}$ for an enhancement of the capacity to 145, 125 and 115 mAh/g respectively [83]. Polypyrrole doped by the phosphomolybdate anion $[PMo_{12}O_{40}]^{3-}$ has also been proposed but the high molecular weight of the anion makes the composite less attractive [84].

3.2. Macroscopic Structures of the Conducting Polymer

The macroscopic configuration of the polymer plays a role in its properties as a charge storage electrode. *E.g.* the polymer polyaniline has been produced in nanotubes and nanofibres to increase the polymer surface and thereby the capacitive charge storage [85]. Furthermore, several materials presented in this section form films, among others Ppy. These films can be used as flexible, bendable and mechanically robust cathodes for lithium batteries

[86, 87] avoiding the use of a current collector. Wang et al. measured the mechanical strength of Ppy films by stretching the films with 30% tensile strain 2000 times without breakage [88]. A complete flexible battery was proposed with Ppy fibres as cathode and a film of carbone nanotubes as anode. The capacity of the battery was ~20 mAh/g after the 10th cycle. [89]. Combined, this recent development emphasises the importance of the macroscopic aspects of optimizing the conducting polymer structure in the lithium battery.

4. CONDUCTING POLYMERS AS ADDITVES

Conducting polymer nanocomposites are already used in many domains to stabilize colloidal suspensions as well as to improve physical, mechanical and electrical proprieties [90]. The same general properties are used in lithium batteries.

4.1. $LiFePO_4$ and Conducting Polymers

One of the most promising cathode materials for lithium batteries is *olivine* $LiFePO_4$, because it is environmentally friendly and inexpensive [30]. Another advantage of the $LiFePO_4$ system is its excellent stability, over 30 000 stable charge/discharge cycles have been shown, at a charge-discharge rate of 15C/5C [91]. The main drawbacks of $LiFePO_4$ are, however, its poor lithium diffusion in the bulk particle [59,60,92-94] and its low electronic conductivity [30,95]. These are normally overcome by reducing the size of the particle and by adding a conducting carbon coating [96-98]. Conducting polymers are used as precursors for these coatings, as replacements for carbon coatings and/or to provide structural integrity in a composite conducting polymer/$LiFePO_4$ electrode.

4.1.1. Conducting Polymer Coatings on $LiFePO_4$

The purpose of coating $LiFePO_4$ is to decrease the resistance between the active particle and the current collector by ensuring the entire surface of the particle is in good electronic contact with other particles (Figure 3).

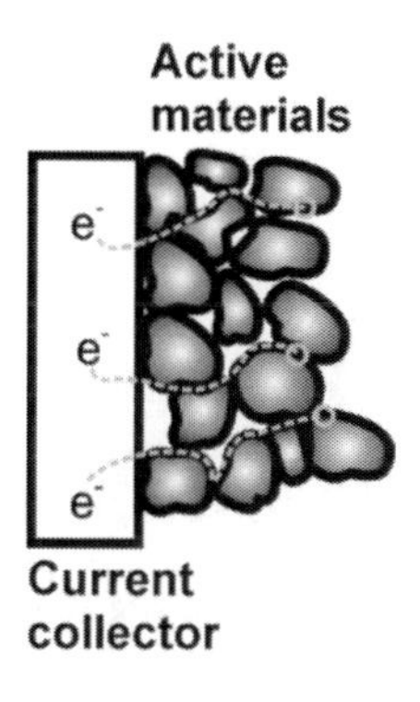

Figure 3. Schematization of a cathode a lithium battery.

To produce these coatings, several polymerization methods have been developed. One of the two most common methods is Electro-Deposition (E.D.) which is accomplished by applying a potential to a suspension of particles, electrolyte and monomer. Potentiostatic, potentiodynamic and galvanostatic techniques are commonly used for E.D. The second common polymerization method is driven by a chemical oxidant ($FeCl_3$, $(NH_4)_2S_2O_8$, Fe(III)tosylate. *etc.*) which is added to a suspension of particles and monomer. After particle recovery, the product is washed and dried before use in the battery. A modification of the chemical oxidation uses the intrinsic oxidative power of pre-oxidized $LiFePO_4$ for polymerization, achieving excellent coatings on $LiFePO_4$ [99].

Tables 3 to 7 show a compilation of different composites of conducting polymers and $LiFePO_4$. We report the capacity of the material at slow discharge rate, normally C/10, and the method of polymerization.

Table 3. Composite C-$LiFePO_4$ and Ppy formed by electrodeposition

Capacity (mAh/g)	Technique	Comment	Pol. (%wt)	C (%wt)	Ref.
~135	C.V.	Deposited on stainless-steel mesh	16	0	[100]
~160	C.V.	Deposited on stainless-steel mesh	16	0	[101]
122 to 154	C.V.	Deposited on stainless-steel mesh Capacity varied depending the electro-polymerization	20 to 30	0	[102]
132	C.V.	Deposited on stainless-steel mesh Discharge at C/5	N/S	0	[81]

N/S: not specified.
C.V.: Cyclic Voltammetry.

Table 4. Composite C-$LiFePO_4$ and Ppy formed by chemical oxidant

Capacity (mAh/g)	Oxidant	Comment	Pol. (%wt)	C (%wt)	Ref.
~130	$FeCl_3$	Doping agent *p*-toluenesulfonate	10	10	[103]
~140	$(NH_4)_2S_2O_8$	Doping agent *p*-toluenesulfonate	3	20	[101]
~148	$(NH_4)_2S_2O_8$	Doping agent *p*-toluenesulfonate	7	20	[101]
~120	$(NH_4)_2S_2O_8$	Doping agent *p*-toluenesulfonate	15	20	[101]
145	Fe(III)tosylate	--	4	15	[104]
155	Fe(III)tosylate	Cycled at 55 °C	4	15	[104]
148	$FeCl_3$	Special doping with polyethyleneglycole Discharge at C/5	N/S	0	[105]
156	$FeCl_3$	Special doping with polyethyleneglycole No carbon coating	N/S	0	[106]
147	$FeCl_3$	Discharge at C/5	21	5	[107]
153	$FeCl_3$	Special doping with polyethyleneglycole Discharge at C/5	>21	5	[107]
164	$FeCl_3$	The amount of carbon is not specified Discharge at C/5 No carbon coating	10	N/S	[91]

N/S: not specified.

Table 5. Composite C-$LiFePO_4$ and PTh

Capacity (mAh/g)	Oxidant	Comment	Pol. (%wt)	C (%wt)	Ref.
~132	$FeCl_3$	Discharge at C/5	5.9	20	[108]
~157	$FeCl_3$	Discharge at C/5	10.6	20	[108]
~150	$FeCl_3$	Discharge at C/5	12.9	20	[108]

Table 6. Composite C-$LiFePO_4$ and Pedot

Capacity (mAh/g)	Oxidant	Comment	Pol. (%wt)	C (%wt)	Ref.
190	$FeCl_3$	The amount of carbon is not specified Discharge at C/5 No carbon coating Doping agent *p*-toluenesulfonate	20	N/S	[91]
175	$FeCl_3$	The amount of carbon is not specified Discharge at C/5 No carbon coating Doping agent *p*-toluenesulfonate	10	N/S	[91]
166	$(NH_4)_2S_2O_8$	$LiFePO_4$ 25 nm No carbon coating	7	20	[109]
163	Li_xFePO_4	Special polymerization with the Li_xFePO_4 No carbon coating	7.1	0	[99]

N/S: not specified.

Table 7. Composite of C-$LiFePO_4$ and Pani

Capacity (mAh/g)	Oxidant	Comment	Pol. (%wt)	C (%wt)	Ref.
~156	$(NH_4)_2S_2O_8$		3	20	[101]
~158	$(NH_4)_2S_2O_8$		7	20	[101]
~142	$(NH_4)_2S_2O_8$		13	20	[101]
120	$FeCl_3$	15 %wt polymer composite show similar result	10	20	[110]
140	$FeCl_3$	20 %wt polymer composite show similar result	25	20	[110]
~157	$(NH_4)_2S_2O_8$		3.4	20	[32]
~162	$(NH_4)_2S_2O_8$		5.7	20	[32]
~162	$(NH_4)_2S_2O_8$		7.3	20	[32]
~160	$(NH_4)_2S_2O_8$		13.8	20	[32]
~165	N/A	Pani is used to restricts particle growth Discharge at 0.6C	N/A	12	[111]

N/A: not applicable.

Care must be taken when comparing these capacities, as they are calculated from battery capacities, where other parts of the cathode might store charges as well. To add to the confusion, the capacity may be calculated relative to the active material mass or relative to the mass of the composite. Astonishingly, capacities that are higher than the theoretical capacity of $LiFePO_4$ (170mAh/g) have been reported, leading to the conclusion that an adequate description of the system is missing. We also provide the quantity of polymer (Pol.) and the amount of carbon added (C) where available to improve comparability.

4.1.2. Conducting Polymer as Carbon Precursor for Coating $LiFePO_4$

Wang and co-workers used Pani during the synthesis of $LiFePO_4$ to restrict particle growth. The polymer was subsequently decomposed thermally to a conducting coating of carbon[111].

Similarly, Ppy has been used in conjunction with a high temperature treatment to regenerate air-aged $LiFePO_4$ [112].

4.1.3. $LiFePO_4$-Conducting Polymer Composites as a Free Standing Film

The utilization of conducting polymers allows for film-type macrostructures, which consist of a matrix of conducting polymer and $LiFePO_4$ (Figure 4). The free standing film electrodes are attractive because they hold promise of decreasing the amount of inactive electrode material (binders, conductive carbon additives, current collector). Moreover they can be used directly in a battery without further modification. As shown in Figure 4, the insertion compound resides inside the conducting polymer and is in the ideal case evenly distributed. An example of this macrostructure was made by Wang and co-workers, who produced a free standing film of Ppy and $LiFePO_4$ [113].

Javier et al. went further by using the block co-polymer poly(3-hexylthiophene)-b-poly(ethylene oxide) and Lithium bis(trifluoromethane sulfone)-imide (LiTFSI) to bind the cathode particles together and ensure electronic conduction, while at the same time replacing the liquid electrolyte with the ethylene oxide section of the polymer for lithium ion conduction [114].

In conclusion, various composites of conducting polymer and $LiFePO_4$ are now available. While comparison between the performances of these materials is difficult, the cited reports bear witness of great variety of polymerization techniques and monomer choice, as well as, a number of composites with promising properties. More importantly, there is now sufficient evidence that the conducting polymers can replace carbon coating and/or binder in lithium batteries that industry is beginning to examine this field [115,116].

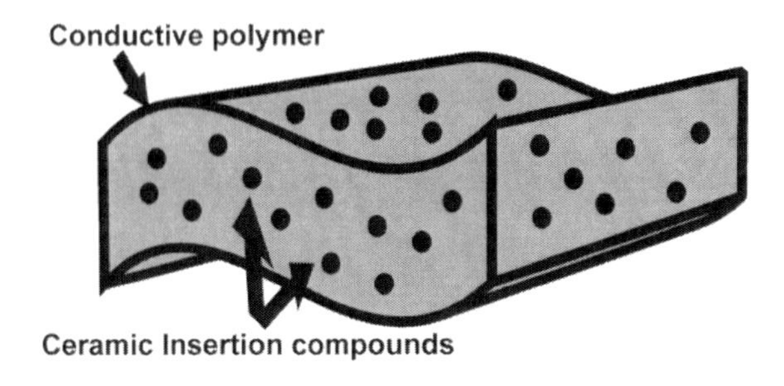

Figure 4. Schematization of a free standing film.

4.2. Manganese Oxide and Conducting Polymer

Arguably the most popular cathode based on manganese for lithium batteries is $LiMn_2O_4$ [31], because it is cheap and considered easy to produce. The two major concerns with this product is manganese dissolution into the electrolyte and the high lithium insertion voltage (~4.15 vs. Li/Li^+), which damages the electrolyte [117]. Several coatings have been proposed to protect $LiMn_2O_4$ for example borate compounds or lithium salts. The latter is obtained by mixing acetylacetone with $LiMn_2O_4$, followed by thermal decomposition [118]. (See Yi et al. for a recent review of $LiMn_2O_4$ coatings [119]). Alternatively, conducting polymers have been proposed as a protective layer. Data for different coatings of conducting polymers on $LiMn_2O_4$ is collected in Table 8. Liu and co-workers produced a composite of nanoparticulate MnO_2 with Pedot nanowires by reaction of $KMnO_4$ with Pedot, increasing the capacity of the pristine polymer by a factor of 4 at a mole ratio MnO_2:Pedot of 1.4:1 [125].

The origin of the difficulty encountered in decreasing manganese dissolution and improving the cyclability of $LiMn_2O_4$ with the conducting polymer seems unclear. One possible explanation may be a problem between the reaction environment and the chemical stability $LiMn_2O_4$ during the coating. This problem was studied by Du Pasquier and co-workers [126], who observed the formation of MnO_2 due to the acidification of the reaction environment during the polymerization. Careful choice of experimental conditions is therefore necessary to avoid degradation reactions during coating.

4.3. Vanadium Oxide and Conducting Polymer

Vanadium oxide has been extensively studied as cathode for lithium batteries [127]. The layered V_2O_5 [128] is particularly interesting for composites with conducting polymers because the polymer can be inserted between the layers of vanadium. The purpose of combining conducting polymer with V_2O_5 is to increase reversibility and lithium capacity in the reaction $V_2O_5 + xLi^+ + xe^- \rightarrow Li_xV_2O_5$. It was found that the space between the layers of $V_2O_5 \cdot xH_2O$ was sufficiently large (~14 Å) to include conducting polymers [129] (Figure 5).

Table 8. Conducting polymer/ $LiMn_2O_4$ composites

Capacity (mAh/g)	Oxidant	Comment	Pol. (%wt)	C (%wt)	Ref.
128.5	$LiMn_2O_4$/ $HClO_4$	Ppy/ $LiMn_2O_4$	6.8	10	[120]
~125	$LiMn_2O_4$/ $HClO_4$	Discharge at 0.37C Ppy/ $LiMn_2O_4$	15	0	[121]
~120	$Li_{1,03}Mn_{1,97}O_4$/ $HClO_4$	Pedot /$Li_{1,03}Mn_{1,97}O_4$ Discharge at C/5	20.3	0	[122]
~115	$Li_{1,02}Mn_2O_{4.05}$/ $HClO_4$	Pedot /$Li_{1,02}Mn_2O_{4.05}$/ Discharge at C	22.3	0	[123]
78	$Li_{1,01}Mn_{1.97}O_4$/ $HClO_4$	Pedot /$Li_{1,01}Mn_{1.97}O_4$ Discharge at C/5 Other rates in the article	21	0	[124]

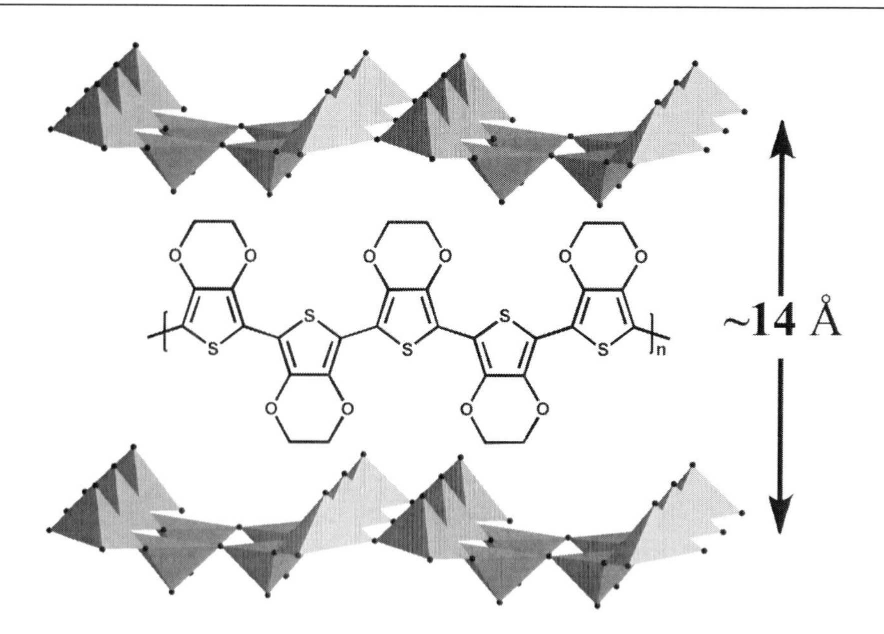

Figure 5. Schematization of a composite V_2O_5/Pedot (image modified from Murugan [130]).

A layered V_2O_5/Pedot composite was presented by Murugan using microwave assisted synthesis to include one to two layers of Pedot between each layer of V_2O_5 [130].

Leroux and coworkers synthesized a nanocomposite with Pani chains between layers of V_2O_5, which exhibited a capacity of 165 mAh/g after reoxidation of vanadium with oxygen [131], and a composite of V_2O_5 with Ppy and Pani [132]. Going forward from these initial results, the influence of synthesis conditions [133] and the doping [134] in the composite were studied.

Different conducting polymers, Pani, PTh and Ppy/V_2O_5 composites were produced by Goward and coworkers showing that $(Pani)_{0.44}$ /V_2O_5 exhibits the best properties [135]. The same group of researchers tried different compositions of the composite Ppy/V_2O_5, finding that the increase of polypyrrole contents reduces electronic conductivity, but increases the specific area, which assists the insertion of lithium into V_2O_5 [136]. The composite Ppy/V_2O_5 was further modified by using pyridinesulfonic acid (PSA) as additive during the polymerization. This favors the polymerization of Ppy inside the V_2O_5 structure. The resulting material exhibits a capacity of 130 mAh/g after 30 cycles, compared to 90 mAh/g without PSA [137].

Nanobeam V_2O_5 have also been used to increase contact between particles and consequently increase electronic conductivity of the electrode. Kim and coworkers used this strategy in conjunction with a coating of Ppy to produce a composite with a capacity of 176 mAh/g after 40 cycles [138]. Finally, Song and coworkers combined these two technologies, synthesizing a composite of nanobeam Pedot/V_2O_5 with PSA additive. This composite exhibited a capacity of 186 mAh/g at 10 C and a fair stability (approximately 9% degradation) over 150 charge/discharge cycles at a rate of 10 C [139].

In conclusion, composites of V_2O_5 with conducting polymers are interesting hybrid compounds, because of their large initial capacities. More over, recent improvements in the use of conducting polymers with targeted design of nanoparticles are a promising step towards reducing capacity fading during high power cycling.

4.4. Other Insertion Compound/Conducting Polymer Composites

There are a large number of redox active lithium ion insertion compounds other than the ones mentioned above [140]. Some of these have been linked with conducting polymers (Table 9). $LiCoO_2$ is already in widespread commercial use and conducting polymers can only contribute little to its further success. Common for other composites in table 9 is that they have not received widespread attention in the literature. Depending on the compound, this is due to high toxicity, low stability, or high cost, as well as, other considerations.

Table 9. Composite insertion compound and conducting polymer

Capacity (mAh/g)	Oxidant	Comment	Pol. (%wt)	C (%wt)	Ref.
~272	E.D.	Pedot/$LiCoO_2$	N/S	5	[141]
136	E.D.	Poly(3-methylthiophene)/$LiCoO_2$	N/S	2.5	[142]
215	$Co_{0.2}CrO_x$/ $HClO_4$	Ppy/$Co_{0.2}CrO_x$ (7,15 and 21% wt Pol. was also reported)	11	10	[143]
~300	$Ag_2V_4O_{11}$/ $HClO_4$	Ppy/ $Ag_2V_4O_{11}$ irreversible	7	5	[144]
~100	$FeCl_3$	Pedot/MoS_2	N/S	25	[145]

N/S: not specified.
E.D.: Electrodeposition.

4.5. Conducting Polymer and Lithium Sulfur Batteries

The search has been intense for materials that allow a large amount of lithium to be stored per redox center in order to increase the capacity. Importantly, the host must also be light, stable and reversible. Particularly interesting are therefore sulfur batteries. The discharge mode offers an interesting chain of reactions for a total of 2 e^- exchange per sulfur atom [146]:

$$S_8 \rightarrow Li_2S_8 \rightarrow Li_2S_6 \rightarrow Li_2S_4 \rightarrow Li_2S_2 \rightarrow Li_2S$$

Unfortunately, side reactions create polysulphides (Li_xS_y) which are soluble in common electrolytes and will react with the anode [147]. Moreover, the different forms of Li-S compounds that are created during the process have very low electronic conductivity. To counter this, conducting polymers may chemically bind redox active sulfur groups or can simply be mixed with the sulfur.

Wang and coworkers used a suspension of sulfur in a pyrrol solution and added the oxidant to trap the sulfur in the conducting polymer. The capacity of this mixture was 1280 mAh/g for the first cycle and decreased to 630 mAh/g after 20 cycles [148]. By using a surfactant, Sun and co-workers created a nano-wire network of polypyrrol, which was able to contain sulfur by simple mixing of both products. The capacity of the first discharge was 1222 mAh/g decreasing to 570 mAh/g after 20 cycles [149].

To chemically bind the sulfur, it is possible to use a monomer containing sulfur like 2,5-dimercapto-1,3,4-thiadiazole (DMcT). The bonds between monomer units can be formed and broken in a largely reversible redox reaction. During the polymerization (charge) reaction S–S bonds are formed and they are reduced (broken) to thiols during the depolymerization (discharge). A capacity of 184 mAh/g was obtained by mixing Pani and DMcT. The conducting polymer was added to assist the formation of poly-DMcT by interacting with the monomer at the molecular level [150]. The same interaction was proposed for the mixture of poly(*o*-toluidine) and 2,5-dimercapto-1,3,4-thiadiazole [151]. To improve the electrochemical reversibility, Yu and coworkers added organomonothiol to control the extent of polymerization of DMcT. The mixture PAni/DMcT/ 2-Mercaptobenzothiazole in a ratio of 5/3/2 gives good columbic efficiency, but a capacity of only ~40 mAh/g after 50 cycles [152].

Another strategy is the utilization of the monomer bis(2-aminophenyloxy)disulfide, a polymer of which contains S-S bonds in a side chain. Both doping/undoping of the polymer and breakage/formation of the disulfide bonds, contributes to the battery capacity. A first cycle discharge of 230 mAh/g was achieved this way, which deteriorated to ~60 mAh/g after 7 cycles [153]. The last strategy involves the modification of a Ppy backbone by adding two sulfur functions in position 3 and 4 of pyrrol and subsequent addition of sublimed sulfur. The capacity reached 515 mAh/g at the first cycle and degraded only little to 452 mAh/g after 20 cycles [154].

In conclusion, the low cyclability of lithium batteries based on elemental sulfur can be increased using sulfur in a composite material with conducting polymers. However, a large performance gap remains between the cyclability of ceramic insertion compounds and sulfur batteries.

4.6. Conducting Polymers and Insertion Compounds for Negative Lithium Battery Electrodes

While lithium metal is the simplest anode for lithium batteries some problems exist with dendrite growth and electrolyte stability. These can in large be overcome using insertion compounds. Since the first use of graphite anodes in commercial lithium batteries [8] many other types of carbon electrodes were investigated like carbon nanotubes (CNTs) and carbon nanofibers (CNFs). As with other electrode materials, lithium insertion into the carbon needs to be reversible, the carbon needs to have good electronic and ionic conductivity and exhibit good mechanical strength. To improve these properties, composite materials with conducting polymers have been developed. In an early study by Veeraraghavan et al., a graphite/PPy composite was fabricated to change the solid electrolyte interface (SEI) [155]. This interface forms due to partial electrolyte decomposition and is observed as a nonreversible capacity loss and changes in the kinetics of lithium ion exchange [18]. Just like ceramic powder insertion materials, CNTs need a matrix to be used in a lithium battery anode. As such, the composite CNT/Pedot with a stable capacity around 250 mAh/g has been developed by Chen et al.[156]. A different path was followed by Ji and co-workers, who used polyacrylonitrile/polypyrrole bicomponent nanofibers as a precursor for carbon with the aim of obtaining a material with high porosity and structural integrity [157]. After thermal carbonization, capacities 20% higher than the theoretical capacity of graphite were observed at the 0.1C rate, possibly due to capacitive effects of the high surface area material.

To avoid the problems related to the SEI layer formation, other insertion materials have also been developed for lithium battery anodes. Silicon [23] and tin oxide [24] are very promising anode materials, due to their high specific capacities (*e.g.* $Li_{4,4}Si$ corresponds to 4200 mAh/g). However, side reactions and large volume changes during charge and discharge result in a loss of electronic connection with the conducting agent (normally carbon) [158]. For this reason conducting polymers have been used as the binding polymer matrix. In fact, standard conducting polymer (Ppy [159,160] and Pedot [161]) decrease the initial irreversible capacity loss of silicon anodes. With these results as the back-drop, Liu and coworkers synthesized Poly(9,9-dioctylfluorene-co-fluorenone-co-methylbenzoic ester)(PFFOMB)/Si composites, which contain polymer functionalities that cater to the adhesive and the electronic properties of the polymer. The composite yielded an impressive 1400mAh/g after 650 cycles [162].

Tin-based amorphous composite oxides (TCO) contain Sn(II)-O as the active center for lithium insertion, which accepts eight lithium ions per formula unit [163]. More over, SnO_2 compounds are very promising anode materials but, like silicon, exhibit a strong variation of volume during lithium insertion, entailing a rapid decay of electrode performance. Again, conducting polymers have been proposed to maintain cohesion in the electrode. He and coworkers synthesized SnO_2/PAni and obtained a capacity of 657 mAh/g with only 0.092% capacity loss after 80 cycles [164,165]. For the composite SnO_2/Ppy, Yuan and coworkers proposed 18 %wt Ppy, [166], whereas the best reported results were obtained with 21 %wt (40 % of the initial capacity (1055mAh/g) after 20 cycles (430mAh/g)) [167]. Conducting polymer composites with TiO_2 (anatase) as anode material were considered to increase lithium diffusivity. Ppy and Pedot increased the diffusion coefficient of lithium to $8x10^{-7}$ and $1.4x10^{-6}$ cm^2/s respectively [168]. The effects of polypyrrole on the performance of nickel oxide has also been studied. After 30 cycles, the capacities of NiO and NiO/Ppy composites were 119 and 436 mAh/g, respectively [169].

CONCLUSION

The research carried out on conducting polymers for lithium batteries over the past 15 years shows the unique possibilities inherent to the tailorability of the conducting polymers. Arguably, the most rapidly expanding subsection of this research field has been composite structures with nanoparticles of solid state lithium insertion compounds. The technology readiness level (TRL) in terms of long term cyclability and fabrication know-how suggests practical application of these composites in the near future. The development of conducting polymer composites for application in lithium batteries is evolving rapidly. Especially the use of conducting polymers to mitigate large volume change anode processes, such as lithium insertion into silicon, should find widespread interest.

Alternatively, the use of molecular charge storage units offers design flexibility and opportunities that are second to none. These storage units can be covalently anchored to the backbone or included as the charge compensating dopant of the oxidized polymer. As such, the resulting structure immobilizes the charge storage molecule and ensures facile ionic and electronic transport during battery cycling.

Combined, conducting polymers in new battery architectures and chemistries is an exciting and vibrant research field. Further, this field is poised to expand rapidly, as we continue our search for new materials that can store energy from intermittent "green" sources and bring about a fossil fuel free future.

ACKNOWLEDGMENTS

The authors acknowledge partial financial supports from the Natural Sciences and Engineering Research Council of Canada (NSERC) Grant no. CRD 385812-09. The authors also gratefully acknowledge Prof. Xueliang (Andy) Sun for his helpful discussions and proof reading of this manuscript.

REFERENCES

[1] P. Novak, K. Muller, K. S. V. Santhanam, O. Haas, *Chem. Rev.*, 97, 207-282 (1997).
[2] A. Cho, *Science*, 329, 786-787 (2010).
[3] M. Armand, J. M. Tarascon, *Nature*, 451, 652-657 (2008).
[4] G. Feuillade, P. J. Perche, *Appl. Electrochem.*, 5, 63-69 (1975).
[5] M. S. Whittingham, *Science*, 192, 1126-1127 (1976).
[6] D. W. Murphy, P. A. Christian, *Science*, 205, 651-656 (1979).
[7] G. C. Farrington, J. L. Briant, *Science*, 204, 1371-1379 (1979).
[8] T. Nagaura, K. Tozawa, *Prog. Batteries Sol. Cells,* 9, 209 (1990).
[9] B. Scrosati, J. Garche, *J. Power Sources*, 195, 2419-2430 (2010).
[10] B. Dunn, H. Kamath, J. M. Tarascon, *Science*, 334, 928-935 (2011).
[11] R. F. Service, *Science*, 332, 1495 (2011).
[12] J. M. Tarascon, M. Armand, *Nature*, 414, 359-367 (2001).
[13] K. Zaghib, M. Dontigny, A. Guerfi, P. Charest, I. Rodrigues, A. Mauger, C. M. Julien, *J. Power Sources*, 196, 3949-3954 (2011).
[14] J. Friedman-Rudovsky, *Science*, 334, 896-897.
[15] J. Xu, H. R. Thomas, R. W. Francis, K. R. Lum, J. Wang, B. Liang, *J. Power Sources*, 177, 512-527 (2008).
[16] M. J. Lain, *J. Power Sources*, 97–98, 736-738 (2001).
[17] M. Contestabile, S. Panero, B. Scrosati, *J. Power Sources*, 92, 65-69 (2001).
[18] R. Marom, S. F. Amalraj, N. Leifer, D. Jacob, D. Aurbach, *J. Mater. Chem.*, 21, 9938-9954 (2011).
[19] J.Y. Song, Y.Y. Wang, C.C. Wan, *J. Power Sources*, 77,183-197 (1999).
[20] A. Manuel Stephan, *Eur. Polym. J.*, 42, 21-42 (2006).
[21] M. Mohri, N. Yanagisawa, Y. Tajima, H. Tanaka, T. Mitate, S. Nakajima, M. Yoshida, Y.Yoshimoto, T. Suzuki, H. Wada, *J. Power Sources*, 26, 545-551 (1989).
[22] T. Ohzuku, A. Ueda, N. Yamamoto, *J. Electrochem. Soc.*, 142, 1431-1435 (1995).
[23] S. C. Lai, *J. Electrochem. Soc.*, 123, 1196-1197 (1976).
[24] M. Winter, J. O. Besenhard, *Electrochim. Acta*, 45, 31-50 (1999).
[25] Z. J. Wei-Jun, *J. Power Sources*, 196, 13-24 (2011).

[26] N. Kamaya, K. Homma, Y. Yamakawa, M. Hirayama, R. Kanno, M. Yonemura, T. Kamiyama, Y. Kato, S. Hama, K. Kawamoto, A. Mitsui, *Nat. Mater.*, 10, 682-686 (2011).

[27] K. Xu, *Chem. Rev.*, 104, 4303-4418 (2004).

[28] R. Fong, U. von Sacken, J. R. Dahn, *J. Electrochem. Soc.*, 137, 2009-2013 (1990).

[29] K. Mizushima, P. C. Jones, P. J. Wiseman, J. B. Goodenough, *Mater. Res. Bull.*, 15, 783-789 (1980).

[30] A. K. Padhi, K. S. Nanjundaswamy, J. B. Goodenough, *J. Electrochem. Soc.*, 144, 1188-1194 (1997).

[31] M. M. Thackeray, W. I. F. David, P. G. Bruce, J. B. Goodenough, *Mater. Res. Bull.*, 18, 461-472 (1983).

[32] W. M. Chen, L. Qie, L. X. Yuan, S. A. Xia, X. L. Hu, W. X. Zhang, Y. H. Huang, *Electrochim. Acta*, 56, 2689-2695 (2011).

[33] C. Y. Yao, T. H. Kao, C. H. Cheng, J. M. Chen, W. M. Hurng, *J. Power Sources*, 54, 491-493 (1995).

[34] Y. H. Rho, K. Dokko, K. Kanamura, *J. Power Sources*, 157, 471-476 (2006).

[35] R. Amin, P. Balaya, J. Maier, *Electrochem. Solid-State Lett.*, 10, A13-A16 (2007).

[36] S. Levasseur, M. Ménétrier, C. Delmas, *Chem. Mater.*, 14, 3584-3590 (2002).

[37] J. Molenda, W. Kucza, *Solid State Ionics*, 117, 41-46 (1999).

[38] J. R. Heinze, B. A. Frontana-Uribe, S. Ludwigs, *Chem. Rev.*, 110, 4724-4771 (2010).

[39] T. F. Otero, M. Bengoechea, *Langmuir*, 15, 1323-1327 (1999).

[40] K. E. Aasmundtveit, E. J. Samuelsen, L. A. A. Pettersson, O. Inganäs, T. Johansson, R. Feidenhans'l, *Synth. Met.*, 101, 561-564 (1999).

[41] U. Barsch, F. Beck, *Electrochim. Acta*, 41, 1761-1771 (1996).

[42] M. J. Swann, G. Brooke, D. Bloor, J. Maher, *Synth. Met.*, 55, 281-286 (1993).

[43] D. Ofer, R. M. Crooks, M. S. Wrighton, *J. Am. Chem. Soc.*, 112, 7869-7879 (1990).

[44] R. J. Waltman, J. Bargon, A. F. Diaz, *J. Physi. Chem.*, 87, 1459-1463 (1983).

[45] D. Y. W. Yu, K. Donoue, T, Inoue, M. Fujimoto, S. Fujitani, *J. Electrochem. Soc.*, 153, A835-A839 (2006).

[46] T. Marks, S. Trussler, A. J. Smith, D. Xiong, J. R. Dahn, *J. Electrochem. Soc.*, 158, A51-A57 (2011).

[47] M. Park, X. Zhang, M. Chung, G. B. Less, A. M. Sastry, *J. Power Sources*, 195, 7904-7929 (2010).

[48] J. M. Davey, S. F. Ralph, C. O. Too, G. G. Wallace, *Synth. Met.*, 99, 191-199 (1999).

[49] A. Lisowska-Oleksiak, K. Kazubowska, A. Kupniewska, *J. Electroanal. Chem.*, 501, 54-61 (2001).

[50] M. J. Ariza,T. F. Otero, *Colloids Surface A*, 270-271, 226-231 (2005).

[51] P. M. Dziewoński, M. Grzeszczuk, *J. Phys. Chem. B*, 114, 7158-7171 (2010).

[52] C. Deslouis, T. El Moustafid, M. M. Musiani, B. Tribollet, *Electrochim. Acta*, 41, 1343-1349 (1996).

[53] E. Pella, *Elemental organic analysis—Part 1: Historical developments*, 112-116 (1990).

[54] E. Pella, *Elemental organic analysis—Part 2: State of the art*, 28-32 (1990).

[55] J. R. Macdonald, W. B. Johnson, *Impedance Spectroscopy*, 2 ed., John Wiley and Sons Inc. (2005).

[56] S. Brown, N. Mellgren, M. Vynnycky, G. Lindbergh, *J. Electrochem. Soc.*, 155, A320-A338 (2008).

[57] N. Mellgren, S. Brown, M. Vynnycky, G. Lindbergh, *J. Electrochem. Soc.*, 155, A304-A319 (2008).
[58] Y. R. Zhu, Y. Xie, R. S. Zhu, J. Shu, L. J. Jiang, H. B. Qiao, T. F. Yi, *Ionics*, 17, 437-441 (2011).
[59] K. Tang, X. Yu, J. Sun, H. Li, X. Huang, *Electrochim. Acta*, 56, 4869-4875 (2011).
[60] P. P. Prosini, M. Lisi, D. Zane, M. Pasquali, *Solid State Ionics*, 148, 45-51 (2002).
[61] T. Osaka, T. Momma, D. Mukoyama, H. Nara, *J. Power Sources*, 205, 483-486 (2012).
[62] D. Andre, M. Meiler, K. Steiner, C. Wimmer, T. Soczka-Guth, D. U. Sauer, *J. Power Sources*, 196, 5334-5341 (2011).
[63] G. S. Popkirov, E. Barsoukov, R. N. Schindler, *J. Electroanal. Chem.*, 425, 209-216 (1997).
[64] P. Ferloni, M. Mastragostino, L. Meneghello, *Electrochim. Acta*, 41, 27-33 (1996).
[65] K. Z. Xing, M. Fahlman, X. W. Chen, O. Inganäs, W. R. Salaneck, *Synth. Met.*, 89, 161-165 (1997).
[66] G. Greczynski, T. Kugler, W. R. Salaneck, *Thin Solid Films*, 354, 129-135 (1999).
[67] H. Neugebauer, *J. Electroanal. Chem.*, 563, 153-159 (2004).
[68] C. Kvarnström, H. Neugebauer, A. Ivaska, N. S. Sariciftci, *J. Mol. Struct.*, 521, 271-277 (2000).
[69] G. Heywang, F. Jonas, *Adv. Mater.*, 4, 116-118 (1992).
[70] H. Yamato, M. Ohwa, W. Wernet, *J. Electroanal. Chem.*, 397, 163-170 (1995).
[71] P. Rannou, M. Nechtschein, *Synth. Met.*, 101, 474-474 (1999).
[72] T. F. Otero, I. Cantero, *J. Power Sources*, 81-82, 838-841 (1999).
[73] C. Dalmolin, S. R. Biaggio, R. C. Rocha-Filho, N. Bocchi, *Electrochim. Acta*, 55, 227-233 (2009).
[74] M. I. Sanchez De Pinto, H. T. Mishima, B. A. López De Mishima, *J. Appl. Electrochem.*, 27, 831-838 (1997).
[75] C.-C. Chang, L.-J. Her, J.-L. Hong, *Electrochim. Acta*, 50, 4461-4468 (2005).
[76] L. Liu, F. Tian, X. Wang, Z. Yang, M. Zhou, *React. Funct. Polym.*, 72, 45-49 (2012).
[77] L. Zhan, Z. Song, J. Zhang, J. Tang, H. Zhan, Y. Zhou, C. Zhan, *Electrochim. Acta*, 53, 8319-8323 (2008).
[78] K. S. Ryu, K. M. Kim, S. G. Kang, J. Joo, S. H. Chang, *J. Power Sources*, 88, 197-201 (2000).
[79] K. S. Ryu, K. M. Kim, S. G. Kang, G. J. Lee, J. Joo, S. H. Chang, *Synth. Met.*, 110, 213-217 (2000).
[80] S. N. Patel, A. E. Javier, G. M. Stone, S. A. Mullin, N. P. Balsara *ACS Nano*, 6, 1589-1600 (2012).
[81] K.-S. Park, S. B. Schougaard, J. B. Goodenough *Adv. Mater.*, 19, 848-851 (2007).
[82] N. Oyama, T. Sarukawa, Y. Mochizuki, T. Shimomura, S. Yamaguchi, *J. Power Sources*, 189, 230-239 (2009).
[83] M. Zhou, J. Qian, X. Ai, H. Yang, *Adv. Mater.*, 4913-4917 (2011).
[84] P. Gómez-Romero, M. Lira-Cantú, *Adv. Mater*,. 9, 144-147 (1997).
[85] F. Cheng, W. Tang, C. Li, J. Chen, H. Liu, P. Shen, S. Dou, *Chem. Eur. J.*, 12, 3082-3088 (2006).
[86] J. Z. Wang, S. L. Chou, H. Liu, G. X. Wang, C. Zhong, S. Yen Chew, H. Kun Liu, *Mater. Lett.*, 63, 2352-2354 (2009).

[87] I. Sultana, M. M. Rahman, S. Li, J. Wang, C. Wang, G. G. Wallace, H. K. Liu, *Electrochim. Acta*, , 60, 201-205 (2012).
[88] C. Wang, W. Zheng, Z. Yue, C. O. Too, G. G. Wallace, *Adv. Mater.*, 23, 3580-3584 (2011).
[89] J. Wang, C. Y. Wang, C. O. Too, G. G. Wallace, *J. Power Sources*, 161, 1458-1462 (2006).
[90] R. Gangopadhyay, A. De, *Chem. Mater.*, 12, 608-622 (2000).
[91] H. C. Dinh, S. I. Mho, I. H. Yeo, *Electroanalysis*, 23, 2079-2086 (2011).
[92] E. Markevich, M. D. Levi, D. Aurbach, *J. Electroanal. Chem.*, 580, 231-237 (2005).
[93] Y. Zhu, C. Wang, *J. Phys. Chem. C*, 114, 2830-2841 (2010).
[94] G. K. P. Dathar, D. Sheppard, K. J. Stevenson, G. Henkelman, *Chem. Mater.*, 23, 4032-4037 (2011).
[95] S.-Y. Chung, J. T. Bloking, Y.-M. Chiang, *Nat. Mater.*, 1, 123-128 (2002).
[96] M. Gaberscek, R. Dominko, J. Jamnik, *Electrochem. Commun.*, 9, 2778-2783 (2007).
[97] N. Ravet, J. B. Goodenough, S. Besner, M. Simoneau, P.Hovington, M. Armand, *The Electrochemical Society and The Electrochemical Society of Japan,* 1999, *Meeting Abstracts 99–2*, Oct 17 – 22.
[98] Z. Chen, J. R. Dahn, *J. Electrochem. Soc.*, 149, A1184-A1189 (2002).
[99] D. Lepage, C. Michot, G. Liang, M. Gauthier, S. B. Schougaard, *Angew. Chem. Int. Ed.*, 50, 6884-6887 (2011).
[100] Y.-H. Huang, K.-S. Park, J. B. Goodenough, *J. Electrochem. Soc.*, 153, A2282-A2286 (2006).
[101] Y.-H. Huang, J. B. Goodenough, *Chem. Mater.*, 20, 7237-7241 (2008).
[102] I. Boyano, J. A. Blazquez, I. de Meatza, M. Bengoechea, O. Miguel, H. Grande, Y. Huang, J. B. Goodenough, *J. Power Sources*, 195, 5351-5359 (2010).
[103] G. X. Wang, L. Yang, Y. Chen, J. Z. Wang, S. Bewlay, H. K. Liu, *Electrochim. Acta*, 50, 4649-4654 (2005).
[104] Y. Yang, X.-Z. Liao, Z.-F.Ma, B.-F.Wang, L. He, Y.-S. He, *Electrochem. Commun.*, 11, 1277-1280 (2009).
[105] A. Fedorková, R. Oriňáková, A. Oriňák, I. Talian, A. Heile, H. D. Wiemhöfer, D. Kaniansky, H. F. Arlinghaus, *J. Power Sources*, 195, 3907-3912 (2010).
[106] A. Fedorková, R. Oriňáková, A. Oriňák, H. D. Wiemhöfer, D. Kaniansky, M. Winter, *J. Solid State Electrochem.*, 14, 2173-2178 (2010).
[107] A. Fedorková, A. Nacher-Alejos, P. Gómez-Romero, R. Oriňáková, D. Kaniansky, *Electrochim. Acta*, 55, 943-947 (2010).
[108] Y.-M. Bai, P. Qiu, Z.-L Wen, S.-C. Han, *J. Alloys Compd.*, 508, 1-4 (2010).
[109] A. Vadivel Murugan, T. Muraliganth, A. Manthiram, *Electrochem. Commun.*, 10, 903-906 (2008).
[110] G. Lei, X. Yi, L. Wang, Z. Li, J. Zhou, *Polym. Adv. Technol.*, 20, 576-580 (2009).
[111] Y. W. Yonggang Wang, E. Hosono, K. Wang, H. Zhou, *Angew. Chem. Int. Ed.* 47, 7461-7465 (2008).
[112] X. Xia, Z. Wang, L. Chen, *Electrochem. Commun.*, 10, 1442-1444 (2008).
[113] J.-Z. Wang, S.-L. Chou, J. Chen, S.-Y. Chew, G.-X. Wang, K. Konstantinov, J. Wu, S.-X. Dou, H. K. Liu, *Electrochem. Commun.*, 10, 1781-1784 (2008).
[114] A. E. Javier, S. N. Patel, D. T. Hallinan, V. Srinivasan, N. P. Balsara, *Angew. Chem. Int. Ed.*, 50, 9848-9851 (2011).

[115] J. B. Goodenough, K.-S. Park, S. B. Schougaard, *United Stated Patent* No. 11/447510 (2006).
[116] S. B. Schougaard, M. Gauthier, C. Kuss, D. Lepage, G. Liang, C. Michot, *United Stated Patent* No. 87343-87334 (2009).
[117] G. Amatucci, J. M. Tarascon, *J. Electrochem. Soc.*, 149, K31-K46 (2002).
[118] G. G. Amatucci, A. Blyr, C. Sigala, P. Alfonse, J. M. Tarascon, *Solid State Ionics*, 104, 13-25 (1997).
[119] T. F. Yi, Y. R. Zhu, X. D. Zhu, J. Shu, C. B. Yue, A. N. Zhou, *Ionics*, 15, 779-784 (2009).
[120] A. H. Gemeay, H. Nishiyama, S. Kuwabata, H. Yoneyama, *J. Electrochem. Soc.*, 142, 4190-4195 (1995).
[121] S. Kuwabata, S. Masui, H. Yoneyama, *Electrochim. Acta*, 44, 4593-4600 (1999).
[122] C. Arbizzani, M. Mastragostino, M. Rossi, *Electrochem. Commun.*, 4, 545-549 (2002).
[123] C. Arbizzani, A. Balducci, M. Mastragostino, M. Rossi, F. Soavi, *J. Electroanal. Chem.*, 553, 125-133 (2003).
[124] C. Arbizzani, A. Balducci, M. Mastragostino, M. Rossi, F. Soavi, *J. Power Sources*, 119-121, 695-700 (2003).
[125] R. Liu, J. Duay, S. B. Lee, *ACS Nano*, 4, 4299-4307 (2010).
[126] A. Du Pasquier, F. Orsini, A. S. Gozdz, J. M. Tarascon, *J. Power Sources*, 81-82, 607-611 (1999).
[127] M. S. Whittingham, *Chem. Rev.*, 104, 4271-4302 (2004).
[128] B. Araki, C. Mailhe, N. Baffier, J. Livage, J. Vedel, *Solid State Ionics, 9–10, Part 1*, 439-444 (1983).
[129] M. G. Kanatzidis, C. G. Wu, H. O. Marcy, C. R. Kannewurf, *J. Am. Chem. Soc.*, 111, 4139-4141 (1989).
[130] A. V. Murugan, *Electrochim. Acta*, 50, 4627-4636 (2005).
[131] F. Leroux, B. E. Koene, L. F. Nazar, *J. Electrochem. Soc.*, 143, L181-L183 (1996).
[132] F. Leroux, G. Goward, W. P. Power, L. F. Nazar, *J. Electrochem. Soc.*, 144, 3886-3895 (1997).
[133] M. Lira-Cantu, P. Gomez-Romero, *J. Electrochem. Soc.*, 146, 2029-2033 (1999).
[134] F. Huguenin, M. T. d. P. Gambardella, R. M. Torresi, S. I. C. de Torresi, D. A. Buttry, *J. Electrochem. Soc.*, 147, 2437-2444 (2000).
[135] G. R. Goward, F. Leroux, L. F. Nazar, *Electrochim. Acta*, 43, 1307-1313 (1998).
[136] H. P. Wong, B. C. Dave, F. Leroux, J. Harreld, B. Dunn, L. F. Nazar, *J. Mater. Chem.*, 8, 1019-1027 (1998).
[137] I. Boyano, M. Bengoechea, I. de Meatza, O. Miguel, I. Cantero, E. Ochoteco, J. Rodríguez, M. Lira-Cantú, P. Gómez-Romero, *J. Power Sources*, 166, 471-477 (2007).
[138] Y. Kim, Q.-T. Ta, H.-C. Dinh, P. K. Aum, I.-H. Yeo, W. I. Cho, S.-i. Mho, *J. Electrochem. Soc.*, 158, A133-A138 (2011.
[139] H. M. Song, D. Y. Yoo, S. K. Hong, J. S. Kim, W. I. Cho, S. I. Mho, *Electroanalysis*, 23, 2094-2102 (2011).
[140] B. L. Ellis, K. T. Lee, L. F. Nazar, *Chem. Mater.*, 22, 691-714 (2010).
[141] L. J. Her, J. L. Hong, C. C. Chang, *J. Power Sources*, 161, 1247-1253 (2006).
[142] C. C. Chang, L. J. Her, J. L. Hong, W. L. Ho, S. J Liu, *J. New Mater. Electrochem. Syst.*, 11, 49-54 (2008).

[143] R. P. Ramasamy, B. Veeraraghavan, B. Haran, B. N. Popov, *J. Power Sources*, 124, 197-203 (2003).
[144] J. W. Lee, B. N. Popov, *J. Power Sources*, 161, 565-572 (2006).
[145] A. V. Murugan, M. Quintin, M.-H Delville, G. Campet, C. S. Gopinath, K. Vijayamohanan, *J. Power Sources*, 156, 615-619 (2006).
[146] B. Scrosati, J. Hassoun, Y. K. Sun, *Energy Environ. Sci.*, 4, 3287-3295 (2011).
[147] X. Ji, L. F. Nazar, *J. Mater. Chem.*, 20, 9821-9826 (2010).
[148] J. Wang, J. Chen, K. Konstantinov, L. Zhao, S. H. Ng, G. X. Wang, Z. P. Guo, H. K. Liu, *Electrochim. Acta*, 51, 4634-4638 (2006).
[149] M. Sun, S. Zhang, T. Jiang, L. Zhang, J. Yu, *Electrochem. Commun.*,10, 1819-1822 (2008).
[150] N. Oyama, T. Tatsuma, T. Sato, T. Sotomura, *Nature*, 373, 598-600 (1995).
[151] L. Yu, X. Wang, J. Li, X Jing, F. Wang, *J. Electrochem. Soc.*, 146, 1712-1716 (1999).
[152] L. Yu, X. Wang, J. Li, X Jing, F. Wang, *J. Electrochem. Soc.*, 146, 3230-3233 (1999).
[153] Y. Z. Su, W. Dong, J. H. Zhang, J. H. Song, Y. H. Zhang, K. C. Gong, *Polymer*, 48, 165-173 (2007).
[154] Z. Shi chao, Z. Lan, Y. Jinhua, *J. Power Sources*, 196, 10263-10266 (2011).
[155] B. Veeraraghavan, J. Paul, B. Haran, B. Popov, *J. Power Sources*, 109, 377-387 (2002).
[156] J. Chen, Y. Liu, A. I. Minett, C. Lynam, J. Wang, G. G. Wallace, *Chem. Mater.*, 19, 3595-3597 (2007).
[157] L. Ji, Y. Yao, O. Toprakci, Z. Lin, Y. Liang, Q. Shi, A. J. Medford, C. R. Millns, X. Zhang, *J. Power Sources*, 195, 2050-2056 (2010).
[158] C. S. Wang, G. T. Wu, X. B. Zhang, Z. F. Qi, W. Z. Li, *J. Electrochem. Soc.*, 145, 2751-2758 (1998).
[159] Z. P. Guo, J. Z. Wang, H. K. Liu, S. X. Dou, *J. Power Sources*, 146, 448-451 (2005).
[160] L. Sun, Y. Shi, L. Chu, Y. Wang, L. Zhang, J. Liu, *J. Appl. Polym. Sci.*, 123, 3270-3274 (2012).
[161] L. Yue, S. Wang, X. Zhao, L. Zhang, *J. Mater. Chem.*, 22, 1094-1099 (2012).
[162] G. Liu, S. Xun, N. Vukmirovic, X. Song, P. Olalde-Velasco, H. Zheng, V. S. Battaglia,;Wang, L.;Yang, W. *Adv. Mater.* 2011, *23*, 4679-4683.
[163] Y. Idota, T. Kubota, A. Matsufuji, Y. Maekawa, T. Miyasaka, *Science*, 276, 1395-1397 (1997).
[164] Z. Q. He, W. P. Liu, L. Z. Xiong, H. Shu, X. M. Wu, S. Chen, K. L. Huang, *Chin. J. Inorg. Chem.*, 23, 813-816 (2007).
[165] Z. Q. He, L. Z. Xiong, W. P. Liu, X. M. Wu, S. Chen, K. L. Huang, *J. Cent. South Univ. Technol.*, 15, 214-217 (2008).
[166] L. Yuan, J. Wang, S. Y. Chew, J. Chen, Z. P. Guo, L. Zhao, K. Konstantinov, H. K. Liu, *J. Power Sources*, 174, 1183-1187 (2007).
[167] L. Cui, J. Shen, F. Cheng, Z. Tao, J. Chen, *J. Power Sources*, 196, 2195-2201 (2011).
[168] P. M. Dziewonski, M. Grzeszczuk, *Electrochim. Acta*, 55, 3336-3347 (2010).
[169] N. H. Idris, J. Wang, S. Chou, C. Zhong, M. M. Rahman, H. Liu, *J. Mater. Res.*, 26, 860-866 (2011).

In: Conducting Polymers
Editor: Luiz Carlos Pimentel Almeida

ISBN: 978-1-62618-119-9

Chapter 10

CONDUCTING ORGANIC POLYMERS IN HYBRID, ORGANIC AND DYE SENSITIZED SOLAR CELLS

Youhai Yu[*], Gerardo Teran-Escobar[†], Jose M. Caicedo, Irene Gonzalez-Valls and Monica Lira-Cantu

Centre d'Investigacio en Nanociencia i Nanotecnologia (CIN2, CSIC-ICN),
Laboratory of Nanostructured Materials for Photovoltaic Energy,
Escola Tecnica Superior d Enginyeria (ETSE), Bellaterra, Barcelona, Spain

ABSTRACT

Modern research on conductive polymers was initiated back in 1977 with the publication "*Electrical conductivity in doped polyacetylene*" by a group in the University of Pennsylvania. [1] By the year 2000, the seminal contribution of Professors Heeger, MacDiarmid and Shirakawa was awarded with the Nobel Prize of Chemistry. Research in the early 80s was focused on the improvement of the electrical conductivity of conducting organic polymers (COP), with the main goal of replacing electrical metallic wiring with these inexpensive and light-weight polymeric materials. It was until the late 80's and early 90s when the discovery of novel optical properties of these conducting polymers envisaged new possible applications in optoelectronic devices, including both light-emitting diodes and solar cells. The discovery of the photovoltaic (PV) effect on organic molecules was first observed in early 1960 with the observation that many common dyes present semiconducting properties. These dyes where among the first organic materials to exhibit PV properties. [2] The experience obtained by the latter, together with the discovery of the electrical conductivity and optical properties of COPs [3] permitted, in the 1980s, the development of the first photovoltaic device fabricated completely with organic polymers. [4, 5] Since then, the research on the PV properties of COPs has been growing steadily. One of the most investigated polymer in PV is poly(p-phenylenevinylene) (PPV) and its derivatives. [6-9] At the beginning of 2000, the most important polymer devices, where assembled using MEH PPV; Sariciftci et al. reports studies on the photophysics of mixtures of conjugated polymers with fullerenes [10]; in

[*] E-mail address: youhai.yu@cin2.es.
[†] E-mail address: gerardo.teran@cin2.es.

this moment we start to see bilayer of donor-acceptor (D/A) devices. Consequently, interpenetrating phase separated D/A network composites i.e bulk hererojunction (BH), would appear to be ideal for photovoltaic materials. [11] In the last years, we have seen a constantly growth of new organic materials, been PPV and PS derivates the most representatives, and that shows higher efficiencies. [12-16] The synthesis of new organic materials follows well defined direct guidelines, according to the knowledge and results acquires from intense research. Overcome actual frontiers on intrinsic limitations like exciton diffusion length (10 nm), large exciton binding energies (0.5 V typically), and electron mobility (10^{-9} cm^2/Vs) are the challenge of the today materials; also the optical absorption characteristics in the UV, visible and near infrared as well as its electromagnetic absorption characteristics, are the several properties need to be consider for a photopolymer to be used in solar-energy conversion.

In this work, we present a brief overview of the contribution of conducting organic polymers in last-generation Solar Cells, also known as Excitonic solar cells: hybrid solar cells (HSC), organic solar cells (OSC) and dye sensitized solar cells (DSSC). An introduction of our most recent research work on each of these types of solar cells is also included.

1. Dye Sensitized Solar Cells (DSSC)

The history of dye sensitized solar cells (DSSC) started in 1972 with a chlorophyll sensitized zinc oxide (ZnO) electrode. For the first time, photons were converted into an electric current by charge injection of excited dye molecules into a wide band gap semiconductor. [17] But the efficiency of these devices was poor. The research on dye-sensitized solar cells gained considerable impulse, when Grätzel and co-workers greatly improved the interfacial area between the organic donor and inorganic acceptor by using nanoporous TiO_2 electrodes, with a roughness factor of 1000 times dramatically increased the light harvesting efficiency and solar cells of 7 % efficiency were obtained in 1991. [18] This triggered a boom in research activities and today cells of over 12 % are state of the art. [19, 20] In a DSSC, an organic dye adsorbed at the surface of an inorganic wide-band gap semiconductor is used for absorption of light and injection of the photoexcited electron into the conduction band of the semiconductor. The nanoporous TiO_2 ensures a dramatic enlargement of the contact area between the dye and semiconductor; while the positive charges are transported by the liquid electrolyte. In DSSC, the initial photoexcitation occurs in the light absorbing dye as shown in figure 1. Nanoporous semiconductors such as TiO_2 not only act as support for dye sensitizer but also function as electron acceptor and electronic conductor. Subsequent injection of electrons from the photo-excited dye(D^+) into the conduction band of semiconductors results in the flow of current travelling across the nanocrystalline TiO_2(nc-TiO_2) film to the charge collecting electrode and then to the external circuit. Sustained conversion of light energy is facilitated by regeneration of the reduced dye sensitizer either via a reversible redox couple (O/R), which is usually I^{3-}/I^- or via the electron donation from a p-type semiconductor. The photoanode, made of a nanoporous dye-sensitized n-type semiconductor, receives electrons from the photo-excited dye sensitizer which is thereby oxidized to D^+. The neutral dye sensitizer (D) can be regenerated by the oxidation reaction (R->O) of the redox species dissolved in the electrolyte. The mediator R will then be regenerated by reduction at the cathode (O->R) by the electrons circulated through the external circuit. A high-performance DSSC requires the counter electrode to be highly catalytic and high conductive.

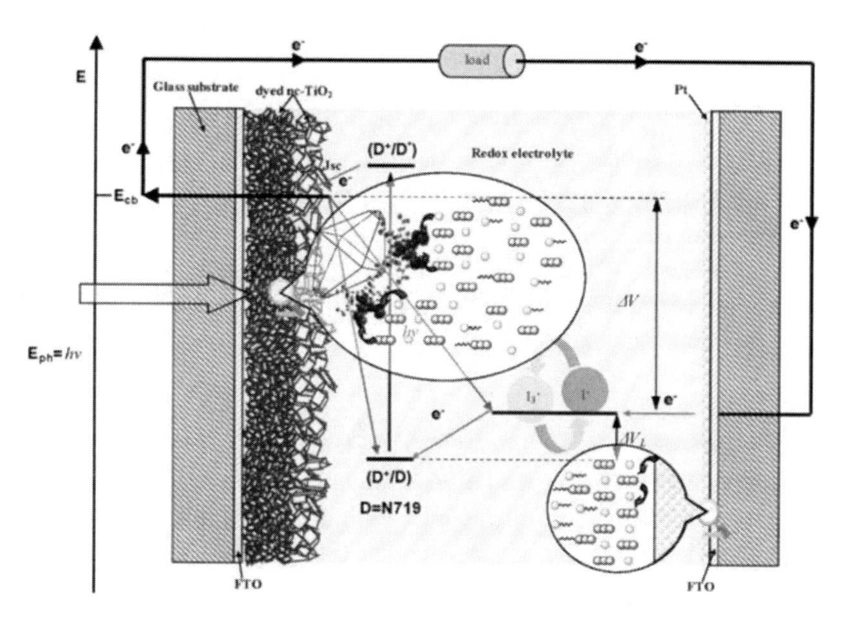

Figure 1. A schematic representation of a Dye Sensitized Solar Cell (DSSC) and its working mechanism.

Hence, platinum (Pt), which is a good catalyst for the reduction in the redox species, such as triiodide/iodide, is usually used as the counter electrode of the DSSC. [21] However, the noble Pt remarkably increases the cost of the DSSC. In addition, the best Pt counter electrode of the DSSC is produced by a high-temperature hydrolysis process. The high-temperature hydrolysis process brings a problem to deposit Pt on a flexible plastic substrate, which is needed for the flexible DSSCs. So many groups have investigated other materials to replace Pt as the counter electrodes of the DSSCs, such as carbon black, conductive polymers, and conducting polymer/ carbon composites. [22-39]

1.1. Counter Electrode with Conducting Polymer

Conducting polymers are suitable materials for counter electrodes. Thus Saito et al. used poly(3,4-ethylenedioxythiophene)(PEDOT) doped with p-toluenesulfonate (PEDOT-TsO) or polystyrenesulfonate (PEDOT-PSS) as catalysts for Counter electrodes of DSSCs. [23, 40] PEDOT-TsO showed better performance than PEDOT-PSS. Cyclic voltammetry experiments confirmed that PEDOT-TsO shows a similar electro-catalytic effect for I^{3-}reduction to that of Pt nanoparticles. The V_{oc} and FF increase with the thickness of the PEDOT-TsO. An efficiency of 4.6% was reached with this material, the PV performance parameters being J_{sc} =11.2 mA/cm^2, V_{oc} = 670 mV and FF = 61%. An Australia group use vapor phase polymerization method to deposit conducting conjugated polymer composite on substrate and use it as counter electrode for DSSS. [24] With poly[3,4-ethylenedioxythiophene: para-toluenesulfonate] (PEDOT:PTS) counter electrode in a dye sensitized solar cell, they obtained maximum J_{sc} of 9.7 mA/cm^2, V_{oc} of 759 mV, fill factor of 0.71 with a power conversion efficiency of 5.25% for glass based wet type dye sensitized solar cell, under illumination of 100 mW/cm^2. Recently another research group in Australia first report a facile synthesis of new PEDOT-on-ITO-PEN plastic counter electrodes for DSSC by electrodeposition of PEDOT film onto conducting plastic (ITO-PEN). [25] The new PEDOT-on-ITO-PEN plastic electrodes performed equivalently or better than platinised FTO glass or the commercially available Pt/Ti alloy on ITO-PEN and it's much cheaper than the later one. Other kinds of conducting polymer such as polypyrrole(PPy) also can be used to fabricate Pt free counter electrode. [26] PPy nanoparticle was synthesized and coated on a conducting FTO glass to construct PPy counter electrode used in DSSC. The PPy electrode has smaller charge-transfer resistance and higher electrocatalytic activity for the I^2/I^- redox reaction than that Pt electrode does. Overall energy conversion efficiency of the DSSC with the PPy counter electrode reaches 7.66%, which is higher (11%) than that of the DSSC with Pt counter electrode. In order to further reduce the cost of DSSC, Lee et al. prepared a flexible composite film on a plastic substrate graphene with coated by PEDOT. The film is used as counter electrode in dye-sensitized solar cells (DSSCs) instead of platinum/transparent conducting oxides (TCOs). The performance of the TCO and Pt free counter electrode rivals that of convenient DSSCs. [27] Polyaniline (PANI) is also a very good conducting polymer which can be used as a substitute for platinum to construct the counter electrode in DSSCs. Many groups have prepared PANI based counter electrode with microporous [28] or nanostructure [29]. They also studied the in situ chemical polymerization method [30, 31] or electrospinning method [32] to construct a low production cost PANI counter electrode of DSSC.

1.2. Counter Electrode with Conducting Polymer–Carbon Composites

Conducting materials such as polypyrrole and polyaniline can be used along with carbon black as a composite catalyst for the counter electrode. Ikeda et al. used a polyaniline-loaded carbon black (PACB) as a counter electrode. [33] A viscous mixture of PACB with the ionic liquid 1, 3-diethyleneoxide-derivatized imidazoliumiodide (EOI) was formed and this paste was applied between the photoelectrode and bare FTO-glass as the counter electrode. The best concentration of the PACB is 10% in the PACB-EOI mixture, yielding 3.48% conversion efficiency (J_{sc} = 12.8 mA/cm^2, V_{oc} = 580 mV, FF = 47%). The significant improvement of Ikeda's methods is result in the dispersion of the carbon black in the polymer material, which is related to an increase in the interface between these materials, and the addition of the EOI, which can penetrate into the pores of TiO_2 for transferring the charge. Luiz and coworker have prepared the polyaniline doped with Keggin-type POMs and used it as counter electrode for DSSC. Devices with J_{sc} of 2 mA/cm^2 and V_{oc} higher than 0.60 V were obtained. But the efficiency is still quite low (0.29%). [34]

Recently, a flexible composite electrode, which is composed of conducting polyaniline (PANI) as electroactive material and flexible graphite (FG) as conducting substrate, has been fabricated by in situ chemical polymerization to substitute for the expensive Pt counter electrode used in DSSCs. A device with the composite counter electrode shows an overall conversion efficiency of 7.36%, which is comparable to 7.45% of that with Pt electrode under the same test condition. Facile charge-transfer and low sheet resistance of the composite electrode are suggested to be responsible for high performance of the DSSC using such counter electrode. [35] Other kinds of Nanographite/polyaniline (NG/PANI) composite films were developed by Lin's group. The nanographite/aniline (NG/ANI) particle was firstly synthesized by a reflux method and served as the monomer for the electro-polymerization of the NG/PANI composite films. The DSSC employing the NG/PANI (20 mC/cm^2)) counter electrode exhibited a higher short-circuit current density (J_{SC}) but lower fill factor (FF), and gave a comparable power-conversion efficiency of 7.07%, as compared to that of a DSSC containing a sputtered Pt CE (eta = 7.19%). [36] Polyaniline nanofibers can also be deposited on graphitized polyimide carbon films to develop a flexible counter electrode for DSSC. The flexible counter electrode exhibited very low charge transfer resistance and series resistance due to the high electrocatalytic activity of the polyaniline nanofibers and the high conductivity of the flexible graphitized polyimide film. In combination with a dye-sensitized TiO_2 photoelectrode and electrolyte, the photovoltaic device with the polyaniline counter electrode shows an energy conversion efficiency of 6.85% under 1 sun illumination. [37] In 2008, dye-sensitized solar cells with a MWCNT/PEDOT:PSS film as the counter electrode were demonstrated by Ouyang et al. [38] The counter electrodes were fabricated by a low-cost solution processing at room temperature. The devices exhibit high photovoltaic performance, which is close to the DSSCs with Pt as the counter electrode. The photovoltaic performance is significantly higher than that of the devices with a MWCNT/PSSA as the counter electrode. This difference is due to the different composite structures.

Table 1. Performances of DSSCs applying different types of conducting polymer counter electrodes

Substrate	Catalyst on CE	Area (cm^2)	Light irradiation (mW/cm^2)	Jsc (mA/cm^2)	Voc (mV)	FF(%)	η (%)	Ref
Ito glass	PEDOT-TsO	0.35	100	11.2	670	61	4.6	[23]
ITO/PEN	PEDOT:PTS	0.25	100	9.7	795	71	5.25	[24]
ITO/PEN	PEDOT	0.25	100	14.1	787	73	8.0	[25]
FTO glass	PPy		100	15.01	740	69	7.66	[26]
FTO glass	PANI		100	14.6	714	69	7.15	[28]
FTO glass	PANI		100	12.5	723	55	4.95	[29]
FTO glass	PANI	0.15	100	10.7	810	64	5.6	[30]
FTO glass	PANI		100	8.39	668	47	2.64	[31]
FTO glass	PANI/CSA–PLA	0.16	100	12.01	700	63	5.3	[32]
FTO glass	Carbon black and polyaniline	0.24	100	12.8	580	47	3.48	[33]
FTO glass	Carbon black and PANI-POMs	0.25	100	2.43	700	17	0.29	[34]
Flexible graphite (FG)	PANI	0.15	100	15.41	657	727	7.36	[35]
ITO glass	NG/PANI	0.16	100	20.14	710	49	7.07	[36]
Graphitized poly-imide carbon film	PANI	0.5	100	12.4	834	66	6.85	[37]
FTO glass	MWCNT/PEDOT:PSS	0.25	100	15.5	660	63	6.5	[38]
FTO glass	PEDOT/MWCNT	0.25	100	17.0	720	66	8.08	[39]

In MWCNT/PSSA, the MWCNT is wrapped by the non-conjugated PSSA, which lowers the charge transfer from the redox species to the counter electrode, while in MWCNT/PEDOT:PSS, the MWCNT is wrapped by the conductive PEDOT, which does not affect the charge transfer. Later a Taiwan research group use chemically polymerized poly(3,4-ethylenedioxythiophene) (PEDOT) and MWCNT to prepare Pt free counter electrode in a DSSC and achieved best cell performance of 8.08% with short-circuit current density, open-circuit voltage and fill factor of 17.00 mA/cm^2, 720 mV and 0.66, respectively. The performance of the DSSCs containing the PEDOT coated electrode was compared with sputtered-Pt electrode. The high photocurrents were attributed to the large effective surface area of the electrode material resulting in good catalytic properties for I^{3-} reduction. [39]

1.3. Conducting Polymer as Hole Transport Materials

Because of the encapsulation problem posed by the use of liquid in the conventional wet-type DSSC, much work is being done to make a solid state DSSC. [41, 42] The use of solvent free electrolytes in the DSSC is supposed to offer very stable performance for the device. To construct a solid-state DSSC, a solid p-type conductor should be chosen to replace the liquid electrolyte. Conducting polymer materials have been used in nc-TiO_2 based DSSC to transport hole carriers from the dye cation radical to the counter electrode instead of using the I^{3-}/I^- redox species. Solid state DSSCs using conductive polymers such as polythiophene derivatives and polyaniline (PANI) derivatives were attempted to construct using a spin-coating technique and related dip-coating methods. At first, poly(3-butylthiophene) [43] and poly(octylthiophene) [44] were examined using some organic dyes with no carboxyl groups or N3 dye as sensitizer, giving very poor results. Employment of poly(4-undecy-2,2'-bithiophene) [45] and poly(3-undecy-2,2'-bithiophene) [46] also resulted in giving poor performance with fill factor (ff=~40%) and short circuit current density (J_{sc}=~70 μA/cm^2) under comparable conditions. With regards to polyaniline (PANI), the spin-coating technique using a PANI dispersion with high conductivity or electrochemical deposition of polyaniline gave very poor performance. [47] Nanostructures PANI also have been applied as HTM in DSSCs. In order to suppress the overgrowth and aggregation of PANI nanofibers, high gravity was introduced to control the synthesis of PANI nanofibers polymerization. By application of these PANI nanofibers as hole conductor media in quasi-solid-state DSSC, it was found that the electrochemical behaviors are strongly dependent on the high gravity that is applied for the synthesis of the polyaniline nanofibers. [48] Kitamura et al. use porous structure of the carbon increases the interface between the polypyrrole and counter electrode to improve the charge exchange. [49]. A carbon counter electrode can improve the fill factor and the efficiency will then increase. A modest 0.62% efficiency was reported for 10 mW/cm^2 irradiance (J_{sc} = 0.104 mA/cm^2, V_{oc} =716 mV, and FF = 78%). Somani et al. applied conducting polyaniline in solid state solar cells sensitized with methylene blue in 2004. [50] This solid state DSSC was fabricated using conducting polyaniline coated electrodes sandwiched with a solid polymer electrolyte, poly(vinyl alcohol) with phosphoric acid. It exhibits good photoresponse to visible light because of the presence of illumination enhances the electrochemical reaction (doping of polyaniline by migration of anions).

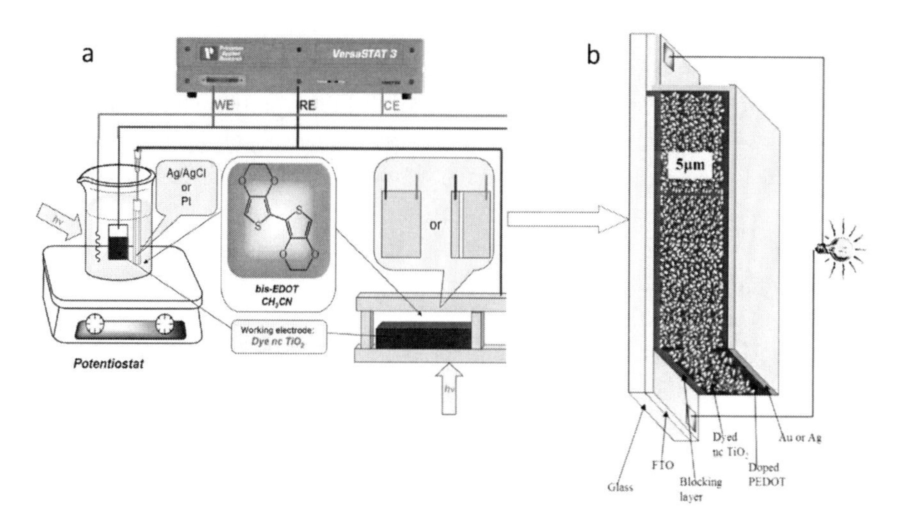

Figure 2. a: The scheme of four kinds of the photo-electrochemical polymerization (PEP) of bis-EDOT. b: A schematic representation of a ss-DSC applying PEDOT as HTM.

Recently, Lee and his coworkers present the fabrication of high efficient quasi-solid state DSSCs with the composite electrolytes containing polyaniline-loaded carbon black (PACB), an efficiency of 5.81% was obtained, under AM1.5 illumination. [51]

Recent attempts using region-regular poly(3-hexylthiophene) (P3HT) for hole-transport layers of N3 or N-719 sensitized DSSC led to slightly improved performances(η=~0.8%). [52-54] Yanagida's group recently succeeded in fabricating P3HT-based DSSCs with η=2.70% under illumination with AM1.5 solar light irradiation. [55] Employment of very thin nc-TiO_2 layers (thickness ~ 400 nm) prepared using the spin-coating technique, oleophilic dye molecules such as HRS-1, and ionic liquid containing t-BP and LiTFSI as additives led to the success. Almost the same result has been reported using the organic dye D102 by Ramakrishna et al. [56] Among the conducting polymer, the poly(3,4-ethylenedioxythiophene):poly(styrenesulfonate) (PEDOT:PSS) is well-known to have a respectable wide range of conductivity(10^{-3}-500 S/cm), high level of transparency, and chemical and thermal stability with safe handling. PEDOT/PSS is now successfully applied to organic light emitting diodes and organic thin film solar cells as charge-transporting layers. Aqueous PEDOT/PSS dispersions have secondary and tertiary structures as nanosize particles ranging from 10 to 90 nm, and the conductivity seems to depend on the particle size and structures. The nanometer-sized pores of nc-TiO_2 layers (~30nm) could not be filled deep inside by simple methods that normally rely on capillary force or gravity. Although micrometer-thick porous nc-TiO_2 electrodes that work as electron acceptors with respectable electron-transporting ability are the most important characteristics in DSSCs, such nanoporous structures are the most difficult obstacle for conducting polymer to infiltrate into the nanospace of nc-TiO_2 layers of DSSCs. In order to get perfect charge separation between dye molecules and polymer phases, conducting polymer should be organized in the pores as hole conductors that nicely contact with dye molecules adsorbed on the TiO_2 surface. However, deposition in nanoporous materials cannot be easily achieved by traditional methods such as evaporation or spin coating. A sophisticated route is to in situ synthesize HTM polymers within the nanopore of the dyed nc-TiO_2 electrode. Yanagida's' group reported for the first time the construction of polypyrrole-based DSSC with in situ photo-electrochemical polymerization (PEP) method. [57] They found that the photovoltaic performance can be improved by introduction of lithium perchlorate ($LiClO_4$) in device layers. Further, the cell characteristics were slightly improved (J_{sc}=0.07 mA/cm^2, V_{oc}=630 mV under reduced light intensity (22 mW/cm^2) when N3 dye was replaced by oleophilic dye, $Ru(dcb)_2(pmp)_2$ (pmp=3-(pyrrole-1-ylmethyl)-pyridine) as a sensitizer. [57] The in situ PEP can be successfully applied to effective pore filling of PEDOT in porous nc-TiO_2 voids by employing the more oxidizable bis-ethylenedioxythiophene (bis-EDOT, Eox=0.5 V vs Ag/Ag^+) instead of EDOT (Eox=1.0 V vs Ag/Ag^+). They also investigate the influence of doped anions on PEDOT as hole conductors for solid-state DSSCs. [58, 59] But the PEP method they used is complicated and not easy to control, so the author modified the PEP method and have developed a new thin layer photo-electrolytic cell based on the modification of our previous reported method. [60] Our work aims for the synthesis of the polymers by different electrochemical techniques: constant-voltage [58, 60, 61] and constant-current [62] with the possibility to simplify the methodology and be able to improve reproducibility of the technique.

The initial results shown here (Figure 3 and Table 2) support that we have successfully developed a new thin layer photo-electrolytic cell based on the modification of in-situ photo-electrochemically polymerization technique which allow us to synthesize the polymers by different electrochemical techniques: constant-voltage and constant-current. We employed four kinds of PEP method to fabricate PEDOT based solid state DSSCs. We use PEDOT as the first examples of the conducting organic polymer. By applying Ru complex dye Z907 as the sensitizer, the good overall conversion efficiency of around 2% (η) at AM 1.5 illumination (100 mW/cm^2) for all SSDSSCs were reached. From the SEM images we can see not all the pores of the TiO_2 nano-paticles were filling with PEDOT, we believe that we can improve the SSDSCs performance with the optimization of PEP condition. [63] Further efforts such as change other kinds of dye or polymer, preparing more porous TiO_2 films, and optimizing the PEP condition are in progress.

A new direction in the field of DSSC, is the incorporation of a polymer or organic semiconductor that combines the functions of light-absorption and charge (hole) transport in a single material, and also is able to replace the dye and hole transporting layer.

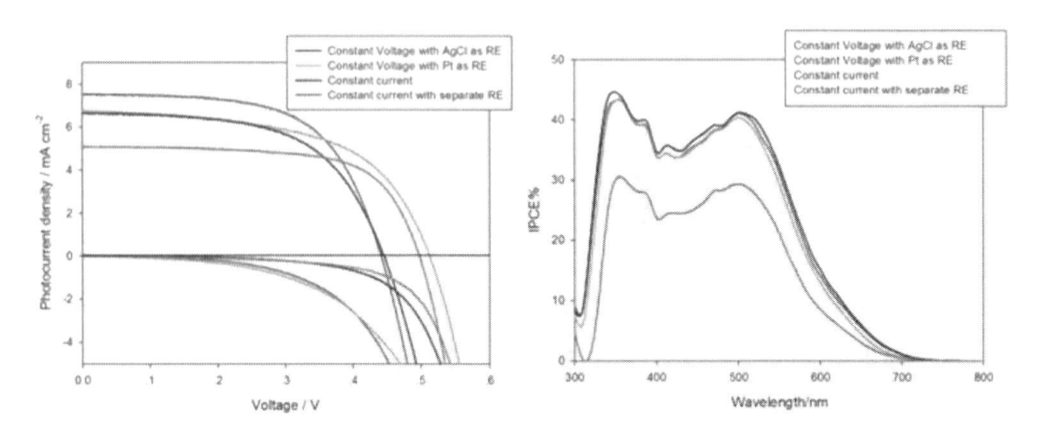

Figure 3. Photocurrent-voltage characteristics (left) and ICPE spectra (right) for ss-DSCs applying the Z907 dye applying PEDOT as HTM.

Table 2. Performance of ss-DSCs with different PEP methods

Method	V_{oc}(V)	J_{sc}(mA)	ff	η(%)
Constant Voltage and Ag/AgCl as RE	0.444	7.52	60.8	2.03
Constant Voltage and Pt as RE	0.510	6.75	56.9	1.96
Constant Current	0.447	6.64	57.9	1.72
Constant Current and separate RE	0.497	5.08	66.9	1.69

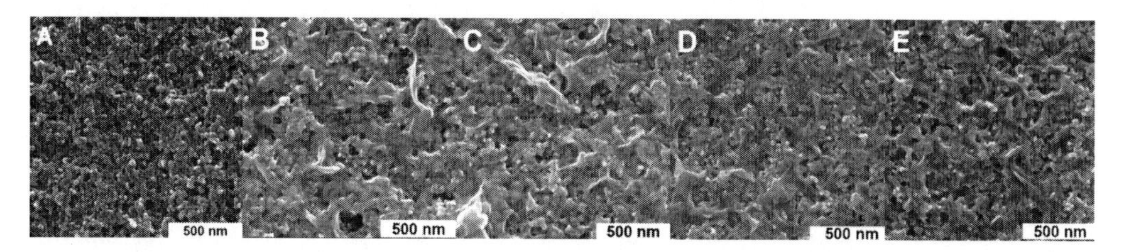

Figure 4. SEM image of top view of Z907 dye-sensitized TiO_2 photoelectrode before (A) and after (B-E) polymerization to form a PEDOT layer. B) Constant Voltage and Ag/AgCl as RE. C) Constant Voltage and Pt as RE. D) Constant Current. E) Constant Current and separate RE.

Figure 5. SEM images of cross section and SEM EDX images. A: TiO_2 after dye absorption. (Red line in EDAX images) B: Constant Voltage and Ag/AgCl as RE. (Blue line) C: Constant Voltage and Pt as RE. (Cyan line) D: Constant Current. (Green line) E:Constant Current and separate RE. (Brown line).

The excitation energies and valence band offsets of optical π-conjugated semiconducting polymers allow electron transfer to the conduction band of the inorganic semiconductor when the polymer is excited across the $\pi - \pi^*$ absorption band. By the short diffusion length of excitons (5-20 nm) is necessary to create a large area of interface between the two materials. The advantageous concept of creating a large interface on a nanoscopic scale of donor and acceptor materials has also been utilized in photovoltaic cells based on a blend of a conjugated polymer with a fullerene derivative and a blend of two conjugated polymers with different electron affinities. [11, 41] In principle, the conjugated conducting polymer can be widely used in both wet type or solid, quasi-solid state DSSC as counter electrode or hole transport materials. This kind of materials can reduce the cost of the DSSC and leads it to a successful replacement for existing technologies in "low density" applications.

2. Hybrid Solar Cells (HSC)

Hybrid solar cells (HSC) are photovoltaic devices where the active layer is made by the combination of organic and inorganic materials. Essentially, the generation of the exciton takes place at the interlayer between both materials where the electron-hole pair splits. Hybrid materials take advantage of the synergy behind the properties of the organic and the inorganic counterparts. For example, it is possible to tune the optical band gap of the inorganic semiconductors by several different means, like the modification of the absorption coefficients by tuning the size of nanoparticles. The availability of preparing thin organic/inorganic hybrid materials with various semiconducting polymers has opened the door to the construction of a new class of devices, where both components are photoactive. Moreover, polymers and nanoparticles can be processed from solution at room temperature, enabling the manufacture of large area, flexible, and lightweight devices.

Some inorganic semiconductors, manufactured as processable nanoparticles colloids like TiO_2 or ZnO, can be tailored at the nanoscale by varying the size of the nanoparticles. The latter modifies the band gap and the absorption/emission spectra. [64] One effective strategy for hybrid solar cell fabrication is the use of blends of nanocrystals with semiconducting polymers as bulk heterojunction (BHJ). [65-68] The excitons created upon photoexcitation are separated into free charge carriers very efficiently at interfaces between the organic and the inorganic semiconductor. The solubility of the n-type and p-type components in the same solvent is an important issue. Organic semiconductors are commonly dissolved in organic solvents, whereas the inorganic semiconducting nanoparticles are commonly dispersed in aqueous solvents. By using different strategies, like ligand-exchange, the nanoparticles can be made soluble in common organic solvents to finally fabricate a polymer/oxide blend as bulk heterojunction with optimal interfacial interaction.

The utilization of semiconductor nanoparticles (TiO_2, ZnO, CdS, etc) embedded into semiconducting polymer blends are promising for several reasons [66]:

1) Inorganic semiconductor nanoparticles can have high absorption coefficients and higher photoconductivity as compared to many organic semiconductor materials.
2) The n- or p- type character of the nanocrystals can be varied by synthetic routes.
3) Band gap of inorganic nanoparticles as a function of nanoparticles size. If the nanoparticles become smaller than the size of the excitation in the bulk semiconductor ($\approx$10 nm), the electronic structure of such small particles is more like those of giant molecules than an extended solid. The electronic and optical properties of such small particles depend not only on the material of which they are composed, but also on their size. [69-74]

A general rule of thumb for donor polymers in single junction hybrid solar cells is to reduce its band gap to an optimal 1.5-1.6 eV and keep its LUMO level above the inorganic acceptor conduction band. Thus, the application of oxides like, TiO_2, with large band gap (3-3.3 eV) is a perfect combination for the fabrication of HSC. Moreover, metal oxides offer high physical and chemical stability, and thus have been widely studied as a material for polymer photovoltaic conversion. [75-77] For a conjugated polymer/titanium dioxide HSC, two main approaches have been developed: filling thin films of nanostructured TiO_2 with conjugated polymers to produce photovoltaic cells [78, 79] or randomly mixing the polymer and the TiO_2 nanocrystal to fabricate polymer hybrid solar cells. [80, 81] Among semiconductor materials, semiconductor oxides have been applied in HSCs with great success due to the possibility to overcome polymer charge-transfer limitations by the high electron-injection properties observed from the oxide. According to this concept, different approaches to create hybrid polymer solar cells have been explored using inorganic semiconductors like

TiO_2, ZnO, CdSe, CdS, Nb_2O_5, etc.). [78, 82-87] However, among those inorganic semiconductor, the metal oxides, TiO_2 and ZnO are at the core of intense research efforts concerning photovoltaic energy conversion (see Figure 6) because of their special qualities such as easy of fabrication, non-toxicity and relatively low cost. Thus the intense research around these materials, we can find a long list of reviews, [88-90] including metal oxide/polymer solar cells. [91-94]

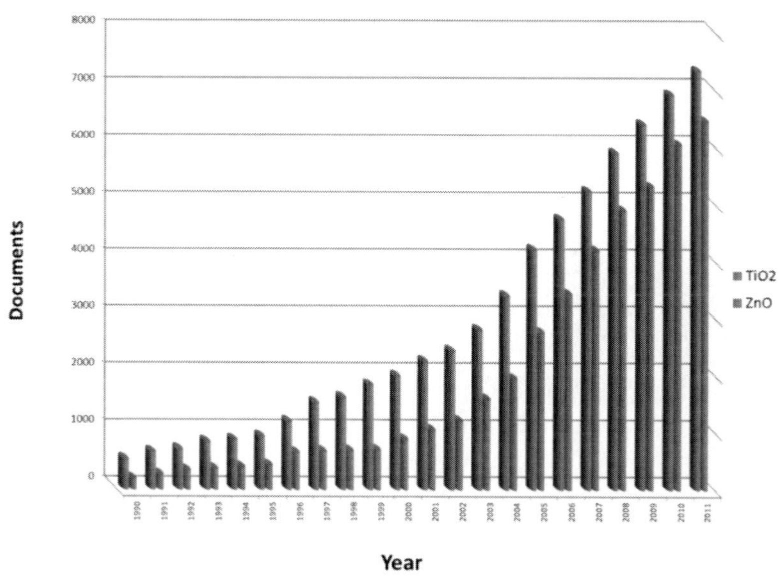

(Source: Scopus).

Figure 6. Evolution research documents related to TiO_2 and ZnO the last 20 years.

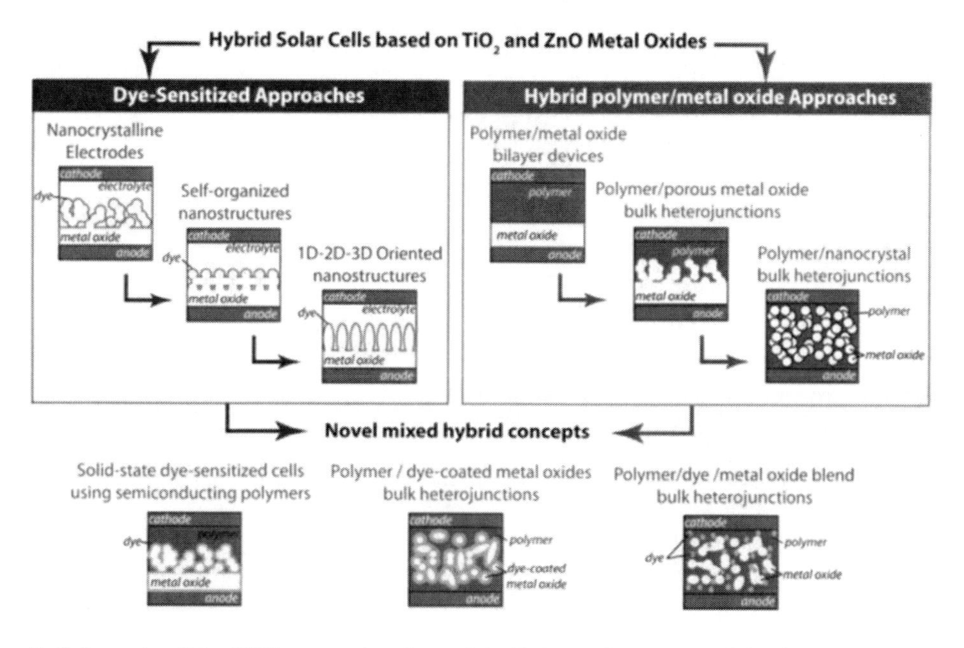

Figure 7. Schematic of the HSC approaches (copyright *Polymer International*, Society of Chemical Industry [107]).

Table 3. Example of some hybrid polymer/semiconductor oxide solar cells

Structure	Jsc (mA/cm^2)	Voc (V)	FF (%)	H (%)	Ref
Bilayer Structures					
ITO/TiO_2/MEH-PPV/Hg	0.32	0.92	52	0.15	[120]
FTO/TiO_2/P3HT/Ag	0.85	0.60	67	0.34	[121]
FTO/TiO_2/P3HT/Ag	1.22	0.72	51	0.45	[79]
Nanostructured structures					
FTO/nTiO_2/P3HT/Ag	1.22	0.72	51	0.45	[79]
ITO/TiO_2/MEH-PPV/Au	3.25	0.86	28	0.71	[122]
ITO/nZnO/P3HT/Au	2.18	0.36	44	0.35	[123]
Nanocrystal Structures					
ITO/PEDOT:PSS/TiO_2(nrods):P3HT/Al	2.62	0.69	63	1.14	[124]
ITO/PEDOT:PSS/TiO_2(nrods):N3:P3HT/Al	4.40	0.78	65	2.20	[125]
ITO/PEDOT:PSS/TiO_2-(oligo-3HT-(Br)COOH/Al	2.82	0.65	64	1.19	[126]
ITO/PEDOT:PSS/ZnO(ncrystal):MDMO-PPV/Al	2.40	0.81	59	1.60	[96]
ITO/PDEOT:PSS/P3HT:ZnO/Li/Al	3.50	0.83	50	1.40	[127]
ITO/PEDOT:PSS/ZnO:P3HT/Al	5.20	0.75	52	2.0	[95]

The combination of polymers and TiO_2 or ZnO nanoparticles where the polymers acts like electron donor and hole conducting material, is a well researched system [79, 93, 95-100]. The efficiency of such devices, have a lot of dependence on the charge transfer complexes, and the rate of these transfer processes are closely connected to the energy level alignment between the materials at the interface. So there is a general need to understand the basic nature of the materials and interfaces at a molecular level. Therefore, extensive research work has been performed to measure the properties of the interfaces between metals and organic materials. An interface dipole may be formed between the materials if there´s is a charge transfer from the molecule to the substrate. [101-106]

In a basic configuration, the oxide semiconductor works like electron transport material (ETM) and the organic semiconductor as the hole transport material (HTM), where their power conversion efficiencies (PCE) have reached more than 2% (See table 3). [94, 95, 108-117] The interaction between semiconductor oxides and conjugated polymer can be brought about by the formation of bi-layers or as blends of oxide and polymer, in the latter case resembling the well-known bulk heterojunction solar cell.

2.1. Low Band Gap Polymers as a Donor Material

The band gap (Eg) of a polymer, is the difference between the HOMO (highest occupied molecular orbital) and the LUMO (lowest unoccupied molecular orbital), and the common units reported are electronvolt (Ev). It can be measured by techniques such as UV-vis spectroscopy and cyclic voltammetry. A polymer with low band gap character must show less than 2 eV. Reducing the band gap at which organic semiconductors absorb light permits to enhance the power conversion efficiency [128-130]. Few examples of low band gap polymers have been described in the literature, (Table 4). There are different factors, which affect the band gap that should be taken into account when designing new polymers with low band gap, like intra-chain charge transfer, substituent's effect, π-conjugation length, etc. [131]

Table 4. Examples of some conjugated polymers applied in solar cells

Polymer	HOMO (eV)	LUMO (eV)	Eg(eV)	Ref.
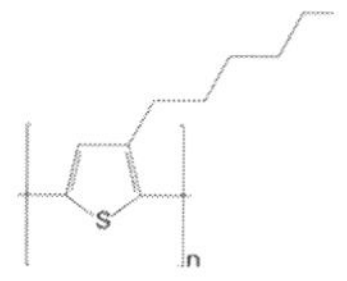 P3HT	-5.2	-3.2	2.0	[151]
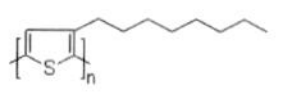 P3OT	-5.25	-3.55	1.8	[152]
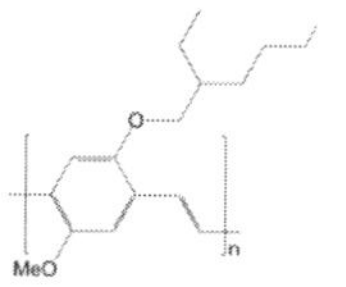 MEH PPV	-5.3	-2.9	2.4	[153]
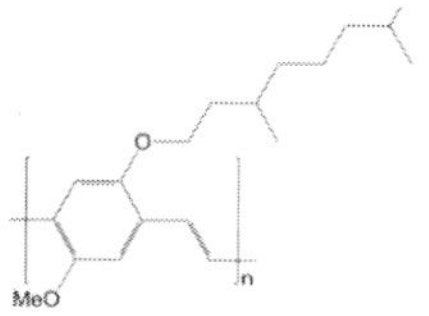 MDMO PPV	-5.3	-3.0	2.3	[96]
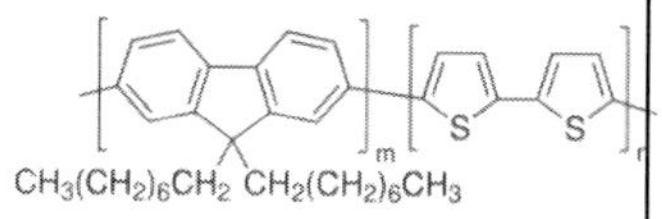 $CH_3(CH_2)_6CH_2$ $CH_2(CH_2)_6CH_3$ F8T2	-5.5	-3.1	2.4	[154]
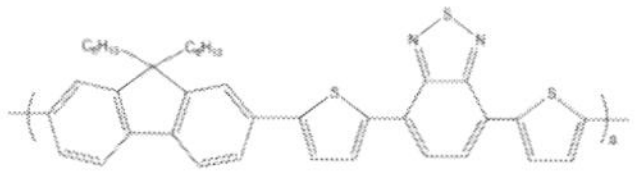 PFDTBT	-5.5	-3.6	1.9	[155]

Table 4. (Continued)

Polymer	HOMO (eV)	LUMO (eV)	Eg(eV)	Ref.
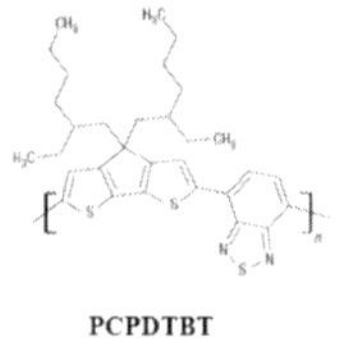 PCPDTBT	-4.9	-3.5	1.4	[156]
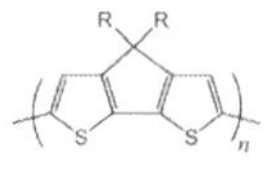 PDOCPDT	-5.15	-3.35	1.8	[157]
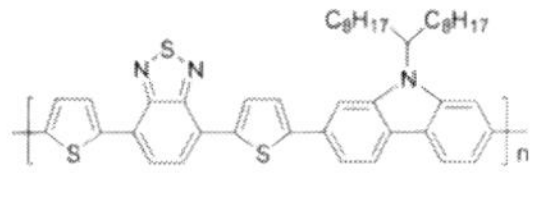 PCDTBT	-5.5	-3.6	1.9	[148]
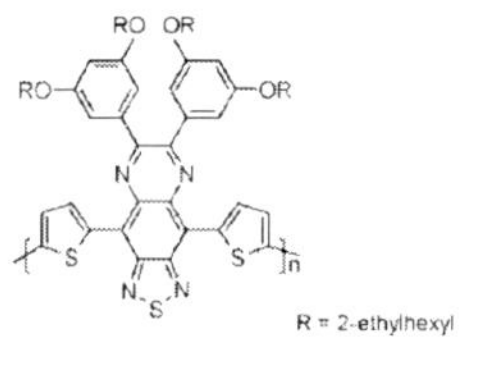 PBTTQ (R = 2-ethylhexyl)	-4.7	-3.75	0.94	[145]
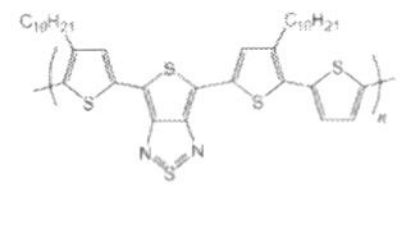 PDDTT	-4.71	3.59	1.12	[158]
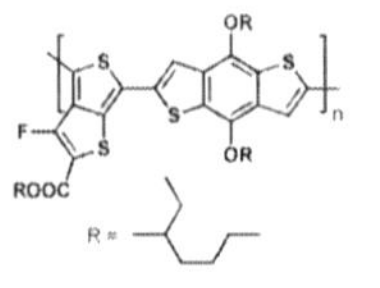 PTB7	-5.15	-3.31	1.84	[159]

Polymer	HOMO (eV)	LUMO (eV)	Eg(eV)	Ref.
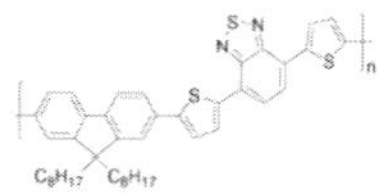 APFO-3	-5.84	-3.53	2.31	[160]
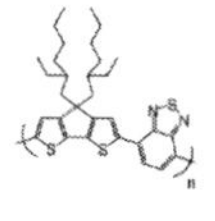 PCPDTBT	-5.30	-3.57	1.73	[161]

There are different ways to decrease the band gap of a polymer, and the effect of the different changes, can affect the morphological, mechanical and physical properties of the material.

- The fused ring system: applied in a copolymer system as an electron acceptor unit coupled with an electron donor unit. The stability is attributed to the quinoid structure formed in the resonance forms of the polymer. [131, 132]
- The substituents on the donor and acceptor units can affect the band gap. The energy level of the HOMO of the donor can be increased by attaching electron donating groups (EDG), such as thiophene and pyrrole. Similarly, the energy level of the LUMO of the acceptor is lowered, when electron withdrawing groups (EWG), such as nitrile, thiadiazole and pyrazine, are attached. This will result in improved donor and acceptor units, and hence, the band gap of the polymer is decreased. [133]
- The side groups also have another effect. It has been shown that the electronic band structure can be tuned without tuning the band gap by addition of EWG and EDG and this can be of great importance when the energy level alignment in a OPV device is taken into account. [134-137]
- In the donor/acceptor copolymer the intra-chain charge transfer is shown by the electron affinity; thus the electron affinity is higher at the acceptor unit compared with the donor unit. [138-141]
- The intermolecular interactions also affect the band gap; i.e. the polymer P3HT orders in solid phase causes a red shift in the whole absorption spectrum, and as a result, a lower band gap is achieved. [142]
- The π-conjugation length is of great importance since a torsion in the polymer back bone causes a decrease in the conjugation length and the band gap increases, thus a high π-conjugation length results in a low band gap polymer. [143]

The HOMO/LUMO energy levels need to be optimized to achieve an efficient charge transfer. Specifically, the LUMO levels of the donor polymer need to be near to the acceptor LUMO or conduction band (CB) edge to minimize energy loss during electron transfer. Yet, it has to be higher than the exciton binding energy for the exciton dissociation to take place.

On the other hand, the HOMO level of the donor polymer should be as low as possible to maximize the open circuit voltage (V_{oc}). Additionally, the polymers need to have a high degree of planarity and be able to self assemble into an organized structure with enhanced packing and charge transport via treatment such as thermal and solvent annealing. Other important aspects to consider, like high hole mobility in the donor polymer (up to 0.2 cm^2/Vs in P3HT) [144], or a band gap between 1.5-1.6 eV should be optimal. Recent researches on minimizing the band gap to values in the range of 1 eV, [145-147] have shown poor cell efficiencies. The authors attributed these low efficiencies to a low hole mobility, and low LUMO offset. The latter indicates that the design of polymers with ultralow band gap needs to be combined with high carrier mobility and optimized energy levels.

More recently, a low band gap polymer, Poly[[9-(1-octylnonyl)-9H-carbazole-2,7-diyl]-2,5-thiophenediyl-2,1,3-benzothiadiazole-4,7-diyl-2,5-thiophenediyl] (PCDTBT) has been published. Besides its low band gap of 1.7 eV it also shows a low HOMO level, making it an ideal choice for OPVs. [148] Other new class of fluorine-based polymers has been obtained [149, 150], with clearly advantages like:

1) The incorporation of fluorine thieno [3,4-b]-thiophene results in a lower HOMO level and increased V_{oc}
2) The polymer backbone quinoidal structure can be stabilized and the planarity is improved by the thieno [3,4-b]-thiophene moiety, leading to a high hole mobility,
3) The polymer forms an effective morphology with fullerene derivatives

2.2. Hybrid Solar Cells Using TiO_2 and ZnO Metal/Oxide Polymer Blends

As mentioned above, TiO_2 and ZnO are two of the most promising materials to be applied in HSCs due to their low cost, ease of fabrication and environmentally friendly nature. Polymer solar cells based on various nanoscale morphologies of nanocrystal metal oxides, ranging from well dispersed nanoparticles or nanorods to well-connected nanoporous or aligned nanorod structures grown directly on substrates, have been reported, and are good options to enhance the performance of the solar cells. Until now, the best polymer/metal oxide hybrid solar cell is based on the BHJ architecture where blends of dispersed nanocrystals and conducting polymers are applied as the active material. However, due to the strong incompatibility between inorganic nanocrystals and polymers the precarious control of the BHJ morphology of the two intermixed components becomes more challenging than with the conventional polymer/fullerene hybrid solar cells. [162]

2.3. Interface Oxide Semiconductor/Polymer

Excitons in organic films can typically travel less than 20 nm before recombining, [120, 163] the electron acceptor must be intermixed on the nanometer length scale with the organic semiconductor in order to obtain high charge separation yield. The adsorption of molecules on oxides surfaces is a key process in the fabrication of electronic devices, catalysis, photoelectrolysis, corrosion, sensor development, etc. These oxides adopt a vast number of structural geometries and at an electronic level they can exhibit metallic character or behave as semiconductors or insulators. [164] On highly ionic oxides, the binding on an adsorbate is frequently due to pure electrostatic interactions with the oxide substrates. [165, 166] The presence of O vacancies induces electronic states within the oxide band gap that make possible bonding interactions with the orbitals of adsorbates. For example, in the case of MgO (100), the electronic charge is highly localized on the O vacancy sites [167], thus, a large fraction of this charge can be transferred to the adsorbates present on its surface [168]. In a much less ionic oxide like TiO_2, the degree of charge localization on the O vacancy sites is less pronounced and part of the excess electronic charge is distributed on the neighboring cations. [169, 170] The effect of these O_{vac} in semiconductor oxides and the type of semiconductor polymer deposited on its surface is directly related to the type of metal atoms within the polymer and has high influence on the final photovoltaic properties of the device. The predominant interaction between a semiconductor oxide, like the TiO_2, and a polymer, like the MEH-PPV, can be realized through electrostatic forces, Van der Waals or hydrogen bridges, etc. In the case of P3HT, where a S-atom is a main constituent of the polymer backbone, the interaction could be moderately different. In general, the interaction of sulfur with oxide surfaces has been subject of many research works, and a few studies can be find in the literature analyzing in detail the bonding of sulfur to well defined oxide surfaces. [171] Sulfur adsorption on TiO_2 surfaces analyzed by tunneling microscopy (STM) [172] indicate that, at room temperature, the S-atom can be adsorbed on the titanium rows of TiO_2 [173] to form TiS_x.

The interaction oxide semiconductor-sulfur becomes important, since the research of polymers with sulfur compound in the structure (see Table 4) are of high interest due to their better adhesion. The strength, by consequence, depends mostly on the interface chemistry and the atomic scale morphology. The adhesion of organic-inorganic compounds in this boundary interlayer, can be a result of a combination of different interactions, all of them with different strong, like covalent, electrostatic and dispersive. At the end, the contribution of each interaction, will affect the final properties of the solar cell [79, 174]. Therefore, efficient dissociation of excitons demands optimized nanoscale morphology where the interconnected TiO_2 nanoparticles form a uniformly distributed network in the blend that minimizes the effective exciton-interface distance. Among many others, a critical issue concerns the adhesion of its organic and inorganic components, critically affecting the resulting mechanical, thermal and optoelectronic properties. When considering a polymer/oxide hybrid, the covalent bonds are not expected to be the mayor contribution to adhesion since; in general, the polymer does not form covalent bonds with the inorganic material; however electrostatic interactions occur between the ions of the surface and the partially charged atoms in the polymers due to the iconicity of the metal oxide. [175] Actually, we know that intense electrostatic interactions occur between the ions of the surface of the semiconductor oxide and the partially charged atoms in the polymers due to the iconicity of the metal oxide. This is the case of poly(3-hexilthiophene) (P3HT), for which large atomic partial charges are found [11, 80, 83, 175]; accordingly, comparatively strong adhesion between P3HT and TiO_2 is expected. Besides the inorganic substrate, it is also crucial to consider the morphology of the polymer; the chains can be distorted as a result of the interactions with the oxide surface. A reasonable expectation is that the adhesion is a result of the formation of the larger number of electrostatic interactions, minimizing the effect caused by the distortions. By consequence, strong link of the polymer to the inorganic substrate is necessary to give rise to an efficient photoconversion. One way to improve the energy exchange between oxide semiconductor and polymer is to enhance the interaction between both; According to these, recent studies, have demonstrated the important role that oxygen vacancies play in the adsorption process: the presence of O vacancies induces electronic states within the oxide band gap that make possible bonding interactions with the orbital's of the adsorbates (Figure 8). [165] In the case of TiO_2 the electronic charge in the O vacancies site it´s no pronounced, and redistributed to the neighboring cations [164] Thus, is well known that in the chemisorptions the electrostatic forces are the prominent interaction between the polymer and the titania, however, the covalent bond can it be present, and this kind of interaction, even in short quantity, can affect directly the performance of a device. We can understand this kind of interaction, like a way to share electrons from the polymer donor to the oxide acceptor, using the covalent bond like a bridge; could it be seen like a push-pull composite that can enhance the performance.

In a recent work from our laboratory, we have assembled a bilayer hybrid solar cell, with the structure FTO/TiO_2-P3HT/Ag with the objective of research the interaction between oxide-polymer materials. To elucidate this interaction we applied the XPS analysis in depth profiling mode. While XPS is a surface sensitive technique, a depth profile of the sample in terms of XPS quantities can be obtained by combining a sequence of ion gun etch cycles interleaved with XPS measurements from the current surface. Figure 9 shows the depth profiling starting from the surface of the sample, in this case P3HT. To elucidate the interaction between TiO_2-P3HT, we can separate the results in 2 different regions. For the sulfur 2p region (Figure 9A) at the surface, a signal at 164 eV, that corresponds to an organic-sulfur binding energy, specifically, for the unbound thiol species. [176] At the 3rd strip off, we can see a right shift towards 161-163eV. The amorphous titanium oxysulfide, where S atoms bond to Ti, exhibit a S2p peak around 160-163 eV. [177] Our results are in good agreement with the work reported by Hebenstreit et al. that shows the S2p at 162 eV due to the replacement of some S atoms by O atoms on the TiO_2 (110) surface. [178]

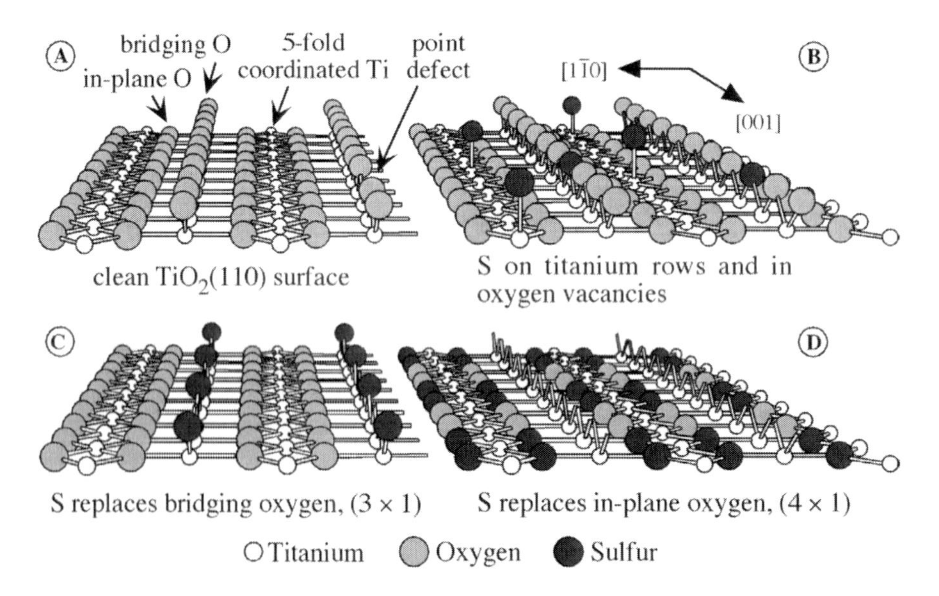

Figure 8. Atomic models of (A)the bulk TiO2 (110) Surface; (B) S Adsorbed at RT; (C) S Adsorbed at 300 oC; (D) S adsorbed at 400 oC. Copyright, Surface Science, Elsevier [172].

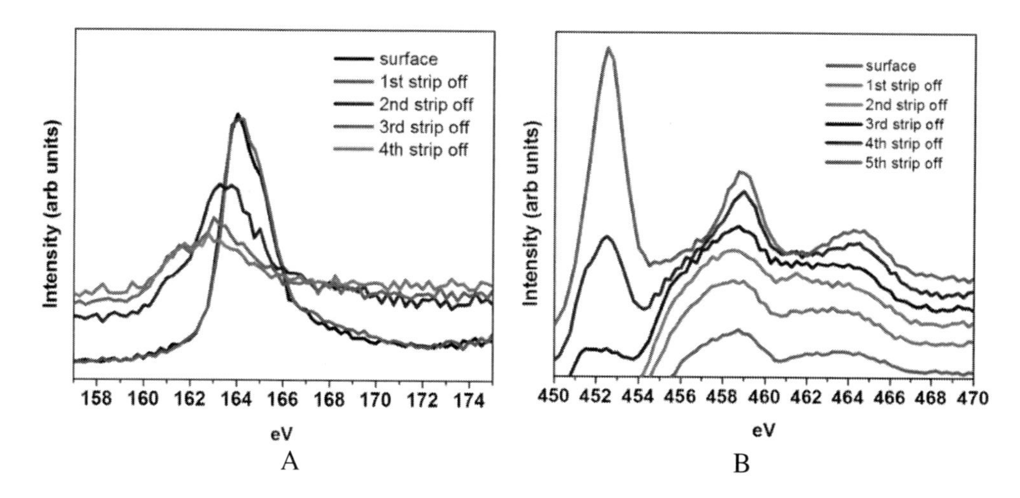

Figure 9. Depth profiling XPS scan; A) Sulfur 2p region, B) Ti 2p peak exposed to the P3HT.

These previous studies indicate the formation of Ti-S bonds in TiO_2. Supporting this theory, the region of Ti 2p (Figure 9B), show us the different valence states of Ti starting from the surface of the P3HT. The titanium peak of the surface covered with P3HT, clearly exhibits a Ti^{3+} shoulder (≈457 eV) [178] that decrease at increasing depth. A peak at less energy (≈453 eV) then appears, that corresponds to the metallic Ti [179]. The exchange of O atoms per S atoms of the P3HT, only takes place at the surface of TiO_2, but can affect the resistance of the circuit, regarding to the formal bond that is formed between booth materials. The effect that have this interaction, in a complete device assembled, is directing related to the performance of the device, affecting the resistance series, that have an effect over the V_{oc} and the J_{sc}; while this kind of interactions can be controlled, the performance could it be optimized, increasing the electronic properties.

3. Organic Solar Cells (OSC)

In general, we have talked about organic solar cells, and their variants like DSSC, or HSC, but in strict terminus, a solar cell is defined by the material(s) where the absorption and separation of electrons is done, namely active area. In DSSC and HSC, the active area is composed by a blend of organic-inorganic materials, commonly a semiconductor oxide and a dye (organometallic structure) in DSSC, or a polymer in HSC. In a pure OSC, the blend is made by organic materials, where surely we´re going to find inorganic materials in the structure, like electron conductor layer, hole conductor layer, or semitransparent electrode (commonly ITO or FTO).

Nowadays, a common structure of OSC, is composed by 5 different layers (figure 10); this design, came as a result of intense research in the last decade. In general, a nanoscale interpenetrating network is obtained by blending a donor (polymer) and an acceptor, where actually that place is occupied by fullerene structures. Normally, is commonly used the P3HT polymer as donor material; however a lot of new materials with well performance have been designed (Table 2). Most commonly, charge extraction is achieved using ITO (indium thin oxide), which has a high work function. In a regular structure, the ITO (anode role) acts like´s a hole extracting material through a hole transport layer (commonly poly (3,4-ethylenedioxythiophene) poly(styrene sulfonate):PEDOT:PSS), and the electrons are extracted from the low work function metal (like Aluminum). In an inverted structure, the ITO (cathode role) extracts the electrons trough an electron transport layer (commonly a semiconductor oxide like TiO_2 or ZnO) and the holes are extracted from the metal electrode (commonly silver). Actually, the inverted structure has received more attention, as a result of the high oxidation that can suffer a metal of low work function [180]; this affect directly the stability of the inverted device under ambient conditions. [181]

During the last decade, significant progress has been made in the development of new low band gap electron donor polymers (Table 4) with absorption spectra extending to larger wavelengths of the solar spectrum. Their HOMO and LUMO levels have been engineered for a better spectra match than those of P3HT when blending with PCBM for a higher V_{oc}. The V_{oc} is directly linked to the respective energy levels of the donor and acceptor materials, so that fine tuning these parameters will be linked to material paring and tuning of the absorption and may influence the exciton dissociation energy. [182]

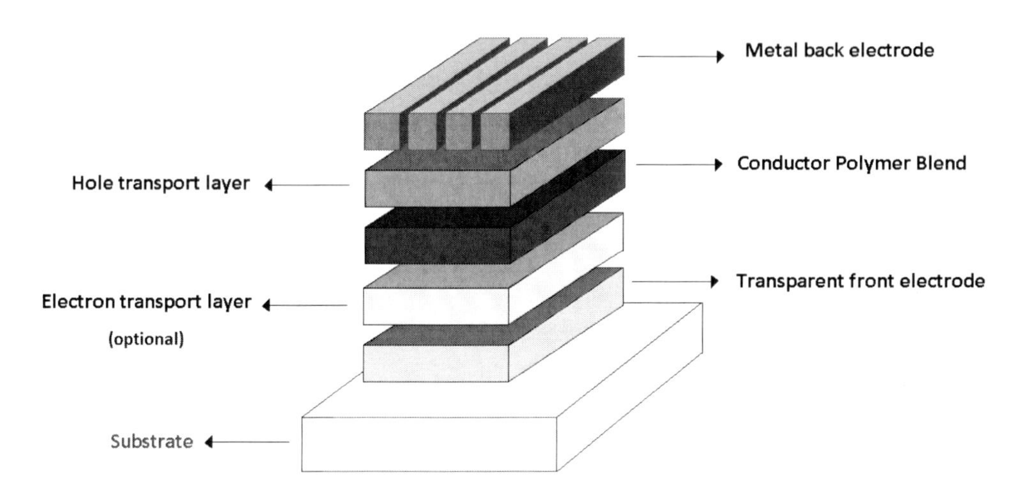

Figure 10. Schematic representation of an Organic Solar Cell showing each of the layers that constitutes a complete device.

According to Scharber´s rule, the decrease in the optical gap should be done by lowering the energy of the donor material´s LUMO and if possible maintaining or reducing as much as possible the HOMO to promote high values of V_{oc} [183]; however, precautions must be taken, so that the HOMO is maintained at a relatively lower energy level than the oxidation threshold of oxygen (around -5.3 eV) to obtain stable materials. [184]

3.1. Acceptor Materials

The Fullerene C_{60} have unique chemical and physical properties, because it's highly efficient electron acceptor character and electronic absorption throughout the UV-vis spectral region; these special properties are directly related to the very high symmetry of the C_{60} molecule, in which 60 equivalent carbon atoms are at the vertices of a regular truncated icosahedrons: each carbon site on C_{60} is trigonally bounded to other carbon atoms. Absorbs strongly in the UV range and much more weakly in the visible region for the symmetry reasons. [185] For construction of optoelectronic devices, the charge separation originating from photoinduced electron-transfer is of utmost importance. To optimize this process the contact between the fullerene end the electron-donor material must be very close. The main process of the OSC to convert sunlight into electricity is as follows: The light absorbing material with a band gap in the visible region (that can be excited for photons with energy situated in the visible region) absorbs photons that excite the electron from the ground state to the excited state, and bound electron-hole pairs (excitons) are created. The excitons diffuse to the donor acceptor interface where excitons dissociate into free charge carriers after overcoming the binding energies. The free charge carriers transport to the respective electrodes under the internal electric fields, resulting in the generation of photocurrent.

Comparatively with the research on donor low band gap polymers, fewer studies have been dedicated to acceptor optimization, being the best advance with the PC70BM, due to its asymmetrical structure. [159, 186, 187] Also, some variations have been achieved to the C60, by tuning to a high LUMO level of the molecule, achieving higher V_{oc} [188]; which means that an optimization could it be possible tuning the band gap of the acceptor.

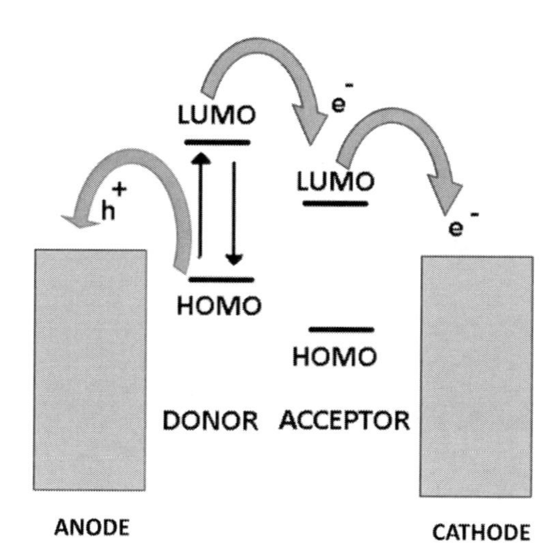

Figure 11. Diagram of electron and hole movement in a photovoltaic device.

3.2. Buffer Layers (Hole´S Conductor Material (HCM)

For their practical use, the long-term performance stability of OSC, as well as their high PCE, is necessary. In contrast to considerable efforts to improve the PCE of OSC's, research in the stability of OSC's has been undertaken only during the last few years. The degradation mechanism of OSC's and key technologies to produce long-term stability are still being investigated. For polymer-based OSC's, poly(3,4-ethylenedioxithiophene): poly(styrene sulfonate) (PEDOT:PSS) is conventionally introduced as a buffer layer to obtain high PCE (Figure 12). In our experience, the use of PEDOT: PSS can change dramatically the efficiency of a device, (increasing 40% to 50% the global PCE, and around 60% the fill factor). Poly(3,4-ethylenedioxythiophene) or PEDOT is an electrochemically stable conjugated polymer which can be oxidized (doped) to a state of high electrical conductivity, while maintaining moderate transparency. Oxidized PEDOT with poly(4-styrenesulfonate) or PPS, is used in thin films as hole transport layer in proto-type polymer based, light emitting devices and OSC. The PEDOT:PSS work function is on the order of 5.0 to 5.3 eV and leads to a built-in voltage of o.7 to 1.0 V, depending which metal is combined.

However, the PEDOT PSS has a low work function, so that the hole injection could be a limiting factor in device performance. For example: the HOMO level of poly(9,9-dioctylfluorene)(PFO) has been estimated to be ≈5.9 eV, so that , the hole injection energy barrier between PEDOT:PSS and PFO can be as large as 0.7-0.9 eV. Therefore, it is necessary to develop hole injecting materials whose the work function is close to or even below the HOMO level of semiconducting polymers which are located next to the hole injection layer. [189]

Also, the acidity of the PEDOT:PSS with a PH ranging from 1.5 to 2.5, is suspected to dissolve ITO layer in regular structure devices, which can migrate from the anode into the buffer layer, and even into the active layer; and recently, it has be demonstrated that the hydrophilic character of the PEDOT:PSS, can affect the performance of the devices. [190, 191] Actually, the research around the hole injection materials, point to the use of p-semiconductor oxides like NiO, V_2O_5 or MoO_3, that have approached good results. [192-195]

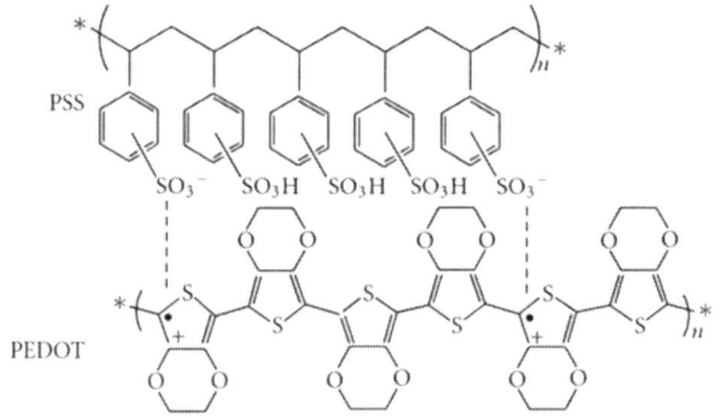

Figure 12. Molecular structure of poly(3,4-ethylenedioxithiophene):poly(styrene sulfonate): (PEDOT:PSS).

3.3. Improving Performance of OSC

The morphology control of the active layer in solar cells is of great importance in improving the device efficiencies. Since exciton dissociation occurs at the interface of the donor-acceptor materials, a large interfacial area should allow maximum exciton dissociation.

- Bulk-heterojunction instead bilayer devices. In the double-layer structure, the photoexcitations in the photoactive material have to reach the p-n interface where charge transfer can occur, before the excitation energy of the molecule is lost via intrinsic radiative an non-radiative decay processes to the ground state. Because the exciton diffusion length of the organic material is in general limited to 10-20nm, only absorption of light within a very thin layer around the interface contributes to the photovoltaic effect.

- Ratio between polymer and fullerene. In bulk heterojunction devices is of great importance; the optimum ratio depends of the polymer mixed with PCBM, for MEH-PPV, the ratio which gave higher efficiency was 1:3 and 1:4, similar for the MDMO-PPV. [196] For the system P3HT-PCBM the better ratios are around 1:1.
- Annealing temperature. Post-production annealing and pretreatment annealing have been carried out. This showed that the pretreatment resulted in a rougher surface which causes an increase in the efficiency of the device due to larger contact area between the polymer and the electrode. [197] The post annealing also results in improvement of the efficiency caused by a reordering of the lamellar structure; annealing not only causes recrystallization but also reduces the free volume and the density of defects at the interfaces. P3HT recristallization has a positive effect on the mobility of holes. [198, 199]
- The choice of solvent. Is important for the mixing of polymer and fullerene, for the domain size and for charge transport. [200] A higher degree of P3HT side chain ordering can be achieved with chlorobenzene [201] instead chloroform.
- Molecular weight. Has a direct influence on the morphology, and consequently, over the efficiency of the device. Results obtained from the polythiophene show that high molecular weight and low polydispersity results in higher efficiencies due to a better morphology of the narrow molecular weight distribution polymer fractions [202], although the amorphous part of the P3HT is needed to obtain better contact with PCBM. [203]

Also these strategies, we can find a lot of work to optimize the OSC's, taking very different alternatives, like incorporating nanostructures of semiconductors in the active area, carbon nano-tubes, polymer nano-wires, metal nano-particles, block copolymers, and more recently, covalent donor-acceptor systems. [94, 204-208] The efficiency of polymer-based solar cells, have been growth significantly in the last years, the last achievement, was developed by Polyera, with a polymer/fullerene organic solar cell in an inverted bulk heterojunction architecture, with a 9.1% certified. [209]

CONCLUSION

The application of conducting organic polymers constitute a fundamental part of excitonic solar cells (Dye sensitized solar cells, hybrid solar cells and organic solar cells). The exceptional electronic and optical properties of these polymers permits their application as part of the main active components of the final photovoltaic device (acceptor, donors), but are also part of extracting layers (buffer layers) and confer the final device enhanced properties like better electron or hole conductivity. The synthesis and application of new polymers (e.g. low band gap), have emerged as a main source of materials for highly efficient solar cells, with values ranging nowadays 7% efficiency for solid state DSC and 10% efficiency for organic solar cells. The study of the interaction between conducting polymers and inorganic semiconductors or other organic semiconductors, can give the clue for higher solar cell performance, stability and device degradation. Nowadays, excitonic solar cells have already reached the goal of >10% efficiency at laboratory scale. It is expected that in a short-term basis, the technology will arrive to its scale-up as photovoltaic modules with competitive power conversion efficiencies. Nevertheless, the main issue to overcome is the enhancement of device lifetime. The application of combined characterization techniques for long-term degradation studies is an urgent need. It is also imperative to follow characterization standards developed at international level in order to characterize photovoltaic performance. Therefore, as conducting organic polymers are part of the active main component of excitonic solar cells, it can be predicted that the application of new materials continue to grow, in parallel with the enhancement of the power conversion efficiency and lifetime of excitonic solar cells.

ACKNOWLEDGMENTS

We thank CSIC for the JAE Doc contract awarded to Y.Y. To the MICINN for the scholarship awarded to I.G.-V and to CONACYT (México) for the Ph.D. scholarship awarded to G. T.-E. To the Consolider NANOSELECT project CSD2007-00041, to the Xarxa de Referència en Materials Avançats per a l'Energia, XaRMAE of the Catalonia Government (Spain).

REFERENCES

[1] C. K. Chiang, *Physical Review Letters*, 39, 1098-1101 (1977).
[2] H. Spanggaard, F. C. Krebs, *Solar Energy Materials and Solar Cells,*83,125-146(2004).
[3] T. A. Skotheim, *Handbook of conducting polymers,* Dekker, New York (1986).
[4] B.R. Weinberger, M. Akhtar, S. C. Gau, *Synthetic Materials*, 4, 187-197 (1982).
[5] S. Glenis, G. Tourillon, F. Garnier, *Thin Solid Films*, 139, 221-231 (1986).
[6] M. Shwoerer, *Synthetic Materials*, 54, 427-433 (1993).
[7] H. Antoniadis, B.R.H., M. A. Abkowtz, S. A. Jenekhe, M. Stolka, *Synthetic Materials,* 62, 265 (1994).
[8] A. B. Holmes, *Journal of Physics: Condensed Matter*, 6, 1379 (1994).
[9] G. Zu, C. Z., A. J. Heeger. *Applied Physics Letters*, 64, 1540 (1994).
[10] N.S. Sariciftci. *Applied Physics Letters,* 62, 585-587 (1993).
[11] G. Yu, *Science*, 270, 1789-1791 (1995).
[12] S. Alem, *Applied Physics Letters,* 84, 2178-2180 (2004).
[13] S. Gunes, H. Neugebauer, N. S. Sariciftci, *Chemical Reviews*, 107, 1324-1338 (2007).
[14] G. Dennler, R. Gaudiana, C. J. Brabec, *Abstracts of Papers of the American Chemical Society*, 238 (2009).
[15] E. Perzon, *Synthetic Metals*, 154, 53-56 (2005).
[16] M. Al-Ibrahim, *Applied Physics Letters*, 86 (2005).
[17] H. Tributsc M. Calvin, *Photochemistry and Photobiology*, 14, 95 (1971).
[18] B. Oregan, M. Gratzel, *Nature*, 353, 737-740 (1991).
[19] Q. Yu, *Acs Nano*, 4, 6032-6038 (2010).
[20] Yella, *Science*, 334, 629-634 (20110.
[21] T. N. Murakami, M. Grätzel, *Inorganica Chimica Acta*, 361, 572-580 (2008).

[22] T. N. Murakami, *Journal of the Electrochemical Society*, 153, A2255-A2261 (2006).
[23] Y. Saito, *Chemistry Letters,* 10, 1060-1061 (2002).
[24] P. M. Sirimanne, *Thin Solid Films*, 518, 2871-2875 (2010).
[25] J. M. Pringle, V. Armel, D. R. MacFarlane, *Chemical Communications*, 46, 5367-5369 (2010).
[26] J. Wu, *Journal of Power Sources*, 181, 172-176 (2008).
[27] K. S. Lee, *Chemsuschem*, 5, 379-382 (2012).
[28] Q. H. Li, *Electrochemistry Communications*, 10, 1299-1302 (2008).
[29] J. Zhang, *Electrochimica Acta*, 55, 3664-3668 (2010).
[30] Z. P. Li, *Electrochemistry Communications*, 11, 1768-1771 (2009).
[31] Q. Qin, *Polymer Engineering and Science*, 51, 663-669 (2011).
[32] S. J. Peng, *Rsc Advances*, 2, 652-657 (2012).
[33] N. Ikeda, K. Teshima, T. Miyasaka, *Chemical Communications*, 16, 1733-1735 (2006).
[34] L. C. P. Almeida, *Journal of Materials Science*, 45, 5054-5060 (2010).
[35] H. C. Sun, *Journal of Physical Chemistry C*, 114, 11673-11679 (2010).
[36] K. C. Huang, *Journal of Materials Chemistry*, 21, 10384-10389 (2011).
[37] J. Chen, *Electrochimica Acta*, 56, 4624-4630 (2011).
[38] B. Fan, *Applied Physics Letters*, 93 (2008).
[39] K. M. Lee, *Thin Solid Films*,.518, 1716-1721 (2010).
[40] Y. Saito, *Chemical Communications*, 15, 1704-1705 (2004).
[41] U. Bach, *Nature,* 395, 583-585 (1998).
[42] K. Tennakone, *Journal of Physics D-Applied Physics*, 31, 1492-1496 (1998).
[43] L. Sicot, *Synthetic Metals*, 102, 991-992 (1999).
[44] D. Gebeyehu, *Synthetic Metals*, 125, 279-287 (2001).
[45] S. Spiekermann, *Synthetic Metals*, 121, 1603-1604 (2001).
[46] C. D. Grant, *Journal of Electroanalytical Chemistry*, 522,40-48 (2002).
[47] Y. X. Li, J. Hagen, D. Haarer, *Synthetic Metals*, 94, 273-277 (1998).
[48] J. Chen, *Journal of Physical Chemistry* C, 115, 23198-23203 (2011).
[49] T. Kitamura, *Chemistry Letters*, 10, 1054-1055 (2001).
[50] P. R. Somani, S. Radhakrishnan, *Journal of Materials Science-Materials in Electronics*, 15, 75-79 (2004).
[51] C. -P. Lee, *Journal of Materials Chemistry*, 20, 2356-2361 (2010).
[52] H. Al-Dmour, D. M. Taylor, J. A. Cambridge, *Journal of Physics D-Applied Physics,* 40, 5034-5038 (2007).
[53] N. Kudo, *Applied Physics Letters*, 90 (2007).
[54] Z. J. Wang, *Applied Surface Science*, 255, 1916-1920 (2008).
[55] K. -J. Jiang, *Advanced Functional Materials*, 19, 2481-2485 (2009).
[56] R. Zhu, *Advanced Materials*, 21, 994 (2009).
[57] K. Murakoshi, *Solar Energy Materials and Solar Cells*, 55, 113-125 (1998).
[58] J. Xia, *Journal of the American Chemical Society*, 130, 1258-1263 (2008).
[59] J. B. Xia, *Journal of Physical Chemistry* C, 112, 11569-11574 (2008).
[60] M. Lira-Cantu, C.L.R.M.S.C.S.U. Barbe, Editor. 2008. p. 249-257.
[61] S. Yanagida, Y. Yu, K. Manseki, *Accounts of Chemical Research*, 42,1827-1838(2009).
[62] Z. Mousavi, *Journal of Electroanalytical Chemistry*, 633, 246-252 (2009).
[63] Y. Yu, M. Lira-Cantu, "Solid state dye sensitized solar cells applying conducting organic polymers as hole conductors" VI Encuentro Franco-Espanol De Quimica Y

Fisica Del Estado Solido - VI Rencontre Franco-Espagnole Sur La Chimie Et La Physique De L Etat Solide, J.J.A.M.D.F. Carvajal, 22-27 (2010).
[64] H. Weller, *Angewandte Chemie-International Edition in English,* 32,. 41-53 (1993).
[65] W. U. Huynh, J. J. Dittmer, A. P. Alivisatos, *Science*, 295, 2425-2427 (2002).
[66] E. Arici, *International Journal of Photoenergy*, 5,199-208 (2003).
[67] N. C. Greenham, X. G. Peng, A. P. Alivisatos, *Physical Review B*, 54,17628-17637 (1996).
[68] W. U. Huynh, X. G. Peng, A. P. Alivisatos, *Advanced Materials*, 11, 92 (1999).
[69] P. Alivisatos, *Science*, 271, 933-937 (1996).
[70] S. A. McDonald, *Nature Materials*, 4, 138-U14 (2005).
[71] S. Gunes, *Solar Energy Materials and Solar Cells*, 91, 420-423 (2007).
[72] M. L. Steigerwald, L. E. Brus, *Accounts of Chemical Research*, 23, 183-188 (1990).
[73] S. Empedocles, M. Bawendi, *Accounts of Chemical Research*, 32, 389-396 (1999).
[74] S. Guenes, H. Neugebauer, N. S. Sariciftci, *Chemical Reviews*, 107, 1324-1338 (2007).
[75] P. Ravirajan, *Applied Physics Letters*, 86 (2005).
[76] D. C. Olson, *Thin Solid Films,* 496, 26-29 (2006).
[77] D. C. Olson, *Advanced Functional Materials*, 17, 264-269 (2007).
[78] J. Breeze, *Physical Review B,* 64 (2001).
[79] K. M. Coakley, M. D. McGehee, *Applied Physics Letters*, 83, 3380-3382 (2003).
[80] Y. Kwong, *Nanotechnology*, 15, 1156-1161 (2004).
[81] T. W. Zeng, *Nanotechnology*, 17, 5387-5392 (2006).
[82] C. Arango, *Advanced Materials*, 12, 1689 (2000).
[83] L. H. Slooff, M. M. Wienk, J. M. Kroon, *Thin Solid Films*, 451, 634-638 (2004).
[84] M. Y. Song, K. J. Kim, D. Y. Kim, *Solar Energy Materials and Solar Cells,* 85, 31-39 (2005).
[85] D. Yang, K. H. Yoon, *Synthetic Metals*, 142, 21-24 (2004).
[86] M. D. McGehee, *Mrs Bulletin*, 34, 95-100 (2009).
[87] M. Lira-Cantu, *Chemistry of Materials*, 18, 5684-5690 (2006).
[88] B. R. Saunders, M. L. Turner, A*dvances in Colloid and Interface Science*, 138, 1-23 (2008).
[89] P. L. Ong, I. A. Levitsky, *Energies,* 3, 313-334 (2010).
[90] Y. F. Zhou, M. Eck, M. Kruger, *Energy and Environmental Science*, 3, 1851-1864 (2010).
[91] L. B. Roberson, *Coordination Chemistry Reviews*, 248, 1491-1499 (2004).
[92] W. J. E. Beek, M. M. Wienk, R. A. J. Janssen, *Journal of Materials Chemistry*, 15, 2985-2988 (2005).
[93] J. Boucle, P. Ravirajan, J. Nelson, *Journal of Materials Chemistry*,17,3141-3153(2007).
[94] Gonzalez-Valls, M. Lira-Cantu, *Energy and Environmental Science*, 2, 19-34 (2009).
[95] S. D. Oosterhout, *Nature Materials*, 8, 818-824 (2009).
[96] W. J. E. Beek, M. M. Wienk, R. A. J. Janssen, *Advanced Materials*, 16, 1009 (2004).
[97] D. C. Olson, *Journal of Physical Chemistry C*, 111, 16640-16645 (2007).
[98] P. Atienzar, *Journal of Physical Chemistry Letters,* 1, 708-713 (2010).
[99] P. Ravirajan, *Journal of Applied Physics*, 95, 1473-1480 (2004).
[100] J. Boucle, *Advanced Functional Materials*, 18, 622-633 (2008).
[101] W. Osikowicz, *Applied Physics Letters*, 88, (2006).
[102] E. M. J. Johansson, *Journal of Physical Chemistry C,*111, 8580-8586 (2007).

[103] H. Ishii, *Advanced Materials*, 11, 605 (1999).
[104] I. G. Hill, *Applied Surface Science,* 166, 354-362 (2000).
[105] G. Koller, *Science*, 317, 351-355 (2007).
[106] E. M. J. Johansson, *Chemical Physics Letters*, 515, 146-150 (2011).
[107] J. Boucl, J. Ackermann, *Polymer International*, 61, 355-373 (2012).
[108] M. P. T. Christiaans, *Synthetic Metals*, 1999. 101(1-3): p. 265-266.
[109] A. C. Arango,, S. A. Carter, P. J. Brock, *Applied Physics Letters*, 74, 1698-1700 (1999).
[110] W.R. Duncan, O. V. Prezhdo, *Annual Review of Physical Chemistry*, 58,143-184(2007).
[111] N. M. Dimitrijevic, *Journal of Physical Chemistry B*, 110, 25392-25398 (2006).
[112] L. Goris, *Journal of Materials Science*, 40, 1413-1418 (2005).
[113] J. J. Benson-Smith, *Advanced Functional Materials*, 17, 451-457 (2007).
[114] D. Veldman, *Journal of the American Chemical Society*, 130, 7721-7735 (2008).
[115] L. Goris, *Applied Physics Letters*, 88, (2006).
[116] K. Vandewal, *Advanced Functional Materials*, 18, 2064-2070 (2008).
[117] B. C. Thompson, J. M. J. Frechet, *Angewandte Chemie-International Edition*, 47, 58-77 (2008).
[118] M. Hallermann, S. Haneder, E. Da Como, *Applied Physics Letters*, 93, (2008).
[119] Haeldermans, *Applied Physics Letters*, 93, (2008).
[120] T. J. Savenije,, J. M. Warman, A. Goossens, *Chemical Physics Letters*, 287, 148-153 (1998).
[121] C. Goh, S. R. Scully,M. D. McGehee, *Journal of Applied Physics*, 101 (2007).
[122] H. Wang, *Applied Physics Letters*, 87,(2005).
[123] J. Boucle, H. J. Snaith, N. C. Greenham, *Journal of Physical Chemistry C*, 114, 3664-3674 (2010).
[124] C. H. Chang, *Journal of Materials Chemistry*, 18, 2201-2207 (2008).
[125] Y. Y. Lin, *Journal of the American Chemical Society*, 131, 3644-3649 (2009).
[126] Y. C. Huang, *Journal of Materials Chemistry*, 21, 4450-4456 (2011).
[127] D. J. D. Moet, *Chemistry of Materials*, 19, 5856-5861 (2007).
[128] Dhanabalan, *Advanced Functional Materials*, 11, 255-262 (2001).
[129] Dhanabalan, *Synthetic Metals*, 119, 169-170 (2001).
[130] H. Neugebauer, *Journal of Chemical Physics*, 110, 12108-12115 (1999).
[131] Winder, N. S. Sariciftci, *Journal of Materials Chemistry*, 14, 1077-1086 (2004).
[132] M. X. Chen, *Applied Physics Letters*, 84, 3570-3572 (2004).
[133] A. Ajayaghosh, *Chemical Society Reviews*, 32, 181-191 (2003).
[134] M. Jorgensen, F. C. Krebs, *Polymer Bulletin*, 51, 23-30 (2003).
[135] F. C. Krebs, *Polymer Bulletin*, 52, 49-56 (2004).
[136] F. C. Krebs, *Macromolecules*, 35, 10233-10237 (2002).
[137] J. L. Bredas, A. J. Heeger, *Chemical Physics Letters*, 217, 507-512 (1994).
[138] K. G. Jespersen, *Journal of Chemical Physics,* 121, 12613-12617 (2004).
[139] K. G. Jespersen, *Synthetic Metals*, 155, 262-265 (2005).
[140] N. K. Persson, *Journal of Chemical Physics*, 123 (2005).
[141] L. Yang, *Polymer*, 46, 9955-9964 (2005).
[142] S. C. Rasmussen, B. D. S., J. E. Hutchison, ACS Symposium series 735, 1999. (Eds B.R. Hsieh Y. Wei) American Chemical Society Washington D.C. .
[143] K. Sivula, *Journal of the American Chemical Society*, 128, 13988-13989 (2006).
[144] A. J. Ferguson, *Journal of Physical Chemistry C*, 112, 9865-9871 (2008).

[145] A. P. Zoombelt, *Organic Letters*, 11, 903-906 (2009).
[146] X. A. Cao, *Applied Physics Letters,* 89, (2006).
[147] X. Gong, *Science*, 325, 1665-1667 (2009).
[148] S. H. Park, *Nature Photonics*, 3, 297-U5 (2009).
[149] Y. Y. Liang, L. P. Yu, *Accounts of Chemical Research*, 43, 1227-1236 (2010).
[150] Y. Lu, *Advanced Materials*, 22, 1407 (2010).
[151] L. Shen, *Applied Physics Letters*, 92, (2008).
[152] D. Baek, *Applied Physics Letters*, 86 (2005).
[153] T. Yamanari, *Solar Energy Materials and Solar Cells*, 93, 759-761 (2009).
[154] J. H. Huang, *Organic Electronics*, 10, 1109-1115 (2009).
[155] P. Wang, *Nano Letters*, 6, 1789-1793 (2006).
[156] C. Soci, Advanced Functional Materials, 17, 632-636 (2007).
[157] P. Coppo, *Macromolecules*, 36, 2705-2711 (2003).
[158] Y. J. Xia, *Applied Physics Letters,* 89 (2006).
[159] Y. Y. Liang, *Advanced Materials*, 22, E135 (2010).
[160] S. Admassie, *Synthetic Metals*, 156, 614-623 (2006).
[161] D. Muhlbacher, *Advanced Materials*, 18, 2931-2931 (2006).
[162] S. S. Li, *Journal of the American Chemical Society*, 133, 11614-11620.
[163] L. A. A. Pettersson,, L. S. Roman, O. Inganas, *Journal of Applied Physics*, 86 487-496 (1999).
[164] J. A. Rodriguez, *Physical Review B,* 65 (2002).
[165] J. A. Rodriguez, *Theoretical Chemistry Accounts*, 107 117-129 (2002).
[166] H. J. Freund, *Faraday Discussions*, 114, 1-31 (1999).
[167] L. Giordano, J. Goniakowski, G. Pacchioni, *Physical Review B*, 64, 075417 (2001).
[168] S. Abbet, *Journal of the American Chemical Society*, 122, 3453-3457 (2000).
[169] E. Z. Kurmaev, *Physical Review B,* 60, 15100-15106 (1999).
[170] P. J. D. Lindan, *Physical Review B*, 55, 15919-15927 (1997).
[171] J. Rodriguez, *Journal of Physical Chemistry B*, 104, 3630-3638.
[172] E. L. D. Hebenstreit, *Surface Science*, 486, L467-L474 (2001).
[173] E. L. D. Hebenstreit, W. Hebenstreit, U. Diebold, *Surface Science*, 470, 347-360(2001).
[174] K. M. Coakley, *Advanced Functional Materials*, 13, 301-306 (2003).
[175] C. Melis, A. Mattoni, L. Colombo, *Journal of Physical Chemistry C*, 114, 3401-3406 (2010).
[176] D. G. Castner, K. Hinds, D. W. Grainger, *Langmuir*, 12, 5083-5086 (1996).
[177] T. Umebayashi, *Applied Physics Letters*, 81, 454-456 (2002).
[178] E. L. D. Hebenstreit, W. Hebenstreit, U. Diebold, *Surface Science*, 461, 87-97 (2000).
[179] J. T. Mayer, *Journal of Electron Spectroscopy and Related Phenomena*, 73, 1-11(1995).
[180] S. E. Shaheen, M. S. White, D. C. Olsen, N. Kopidakis, D. S. Ginley, "Inverted bulk-heterojunction plastic solar cells." SPIE-the international society for optical engineering newsroom, 10.1117/2.1200705.0756 (2007).
[181] S. K. Hau, *Applied Physics Letters*, 93, (2008).
[182] B. Ratier, *Polymer International*, 61, 342-354 (2012).
[183] M. C. Scharber, *Advanced Materials*, 18, 789 (2006).
[184] N. Blouin, *Journal of the American Chemical Society*, 130, 732-742 (2008).
[185] S. Leach, *Chemical Physics*, 160, 451-466 (1992).
[186] Y. P. Zou, *Journal of the American Chemical Society*, 132, 5330 (2010).

[187] C. J. Brabec, *Advanced Functional Materials*, 11, 374-380 (2001).
[188] Y. J. He, *Journal of the American Chemical Society*, 132, 5532-5532 (2010).
[189] T. W. Lee, *Applied Physics Letters*, 87, (2005).
[190] D. M. Tanenbaum, *Rsc Advances*, 2, 882-893 (2012).
[191] R. Rosch et al. *Energy and Environmental Science*, 5, 6521-6540 (2012).
[192] V. Shrotriya, *Applied Physics Letters*, 88, (2006).
[193] T. Hori, *Thin Solid Films*, 518, 522-525 (2009).
[194] J. Meyer, *Journal of Applied Physics,* 110, (2011).
[195] J. S. Huang, *Organic Electronics*, 10, 1060-1065 (2009).
[196] H. Hoppe, N. S. Sariciftci, *Journal of Materials Research*, 19, 1924-1945 (2004).
[197] G. Li, *Journal of Applied Physics*, 98, (2005).
[198] T. Ahn, H. Lee, S. H. Han, *Applied Physics Letters*, 80, 392-394 (2002).
[199] P. J. Brown, *Physical Review B*, 67, (2003).
[200] S. E. Shaheen, *Applied Physics Letters,* 78, 841-843 (2001).
[201] M. Al-Ibrahim, *Solar Energy Materials and Solar Cells,* 85, 13-20 (2005).
[202] P. Schilinsky, *Chemistry of Materials*, 17, 2175-2180 (2005).
[203] S. Guillerez, *European Conference on hybrid and organic solar cells.* ECHOS 06 June 28-30 2006. Paris, France(28-O1-3).
[204] M. Skompska, *Synthetic Metals*, 160, 1-15 (2010).
[205] T. T. Xu, Q. Q. Qiao, *Energy and Environmental Science*, 4, 2700-2720 (2011).
[206] B. Ratier, *Polymer International*, 61, 342-354 (2012).
[207] D. Wrobel, A. Graja, *Coordination Chemistry Reviews*, 255, 2555-2577 (2011).
[208] J. T. Chen, C. S. Hsu, *Polymer Chemistry*, 2, 2707-2722 (2011).
[209] Polyera, *Polyera achieves world-record organic solar cell performance.* Certify by Newport Corporation´s PV Cell Lab, 2012.

INDEX

A

D

E

F

G

H

I

J

K

L

M

N

O

P

Q

R

S

T

U

V

W

X

Y

Z